Lecture Notes in Physics

Editorial Board

R. Beig, Wien, Austria
W. Beiglböck, Heidelberg, Germany
W. Domcke, Garching, Germany
B.-G. Englert, Singapore
U. Frisch, Nice, France
P. Hänggi, Augsburg, Germany
G. Hasinger, Garching, Germany
K. Hepp, Zürich, Switzerland
W. Hillebrandt, Garching, Germany
D. Imboden, Zürich, Switzerland
R. L. Jaffe, Cambridge, MA, USA
R. Lipowsky, Golm, Germany
H. v. Löhneysen, Karlsruhe, Germany
I. Ojima, Kyoto, Japan
D. Sornette, Nice, France, and Zürich, Switzerland
S. Theisen, Golm, Germany
W. Weise, Garching, Germany
J. Wess, München, Germany
J. Zittartz, Köln, Germany

The Lecture Notes in Physics

The series Lecture Notes in Physics (LNP), founded in 1969, reports new developments in physics research and teaching – quickly and informally, but with a high quality and the explicit aim to summarize and communicate current knowledge in an accessible way. Books published in this series are conceived as bridging material between advanced graduate textbooks and the forefront of research to serve the following purposes:

• to be a compact and modern up-to-date source of reference on a well-defined topic;
• to serve as an accessible introduction to the field to postgraduate students and nonspecialist researchers from related areas;
• to be a source of advanced teaching material for specialized seminars, courses and schools.

Both monographs and multi-author volumes will be considered for publication. Edited volumes should, however, consist of a very limited number of contributions only. Proceedings will not be considered for LNP.

Volumes published in LNP are disseminated both in print and in electronic formats, the electronic archive is available at springerlink.com. The series content is indexed, abstracted and referenced by many abstracting and information services, bibliographic networks, subscription agencies, library networks, and consortia.

Proposals should be sent to a member of the Editorial Board, or directly to the managing editor at Springer:

Dr. Christian Caron
Springer Heidelberg
Physics Editorial Department I
Tiergartenstrasse 17
69121 Heidelberg/Germany
christian.caron@springer.com

Pratip Bhattacharyya   Bikas K. Chakrabarti

Modelling Critical and Catastrophic Phenomena in Geoscience

A Statistical Physics Approach

Editors

Pratip Bhattacharyya
Physics Department
Gurudas College
Narkeldanga
Kolkata 700054, India
and
Centre for Applied Mathematics
and Computational Science
Saha Institute of Nuclear Physics
1/AF, Bidhannagar
Kolkata 700064, India
E-mail: Pratip.Bhattacharyya@saha.ac.in

Bikas K. Chakrabarti
Theoretical
Condensed Matter Physics Division
and
Centre for Applied Mathematics
and Computational Science
Saha Institute of Nuclear Physics
1/AF, Bidhannagar
Kolkata 700064, India
Email: bikask.chakrabarti@saha.ac.in

P. Bhattacharyya and B.K. Chakrabarti et al., *Modelling Critical and Catastrophic Phenomena in Geoscience*, Lect. Notes Phys. 705 (Springer, Berlin Heidelberg 2006),
DOI 10.1007/b11766995

Library of Congress Control Number: 2006928276

ISSN 0075-8450
ISBN-10 3-540-35373-9 Springer Berlin Heidelberg New York
ISBN-13 978-3-540-35373-7 Springer Berlin Heidelberg New York

This work is subject to copyright. All rights are reserved, whether the whole or part of the material is concerned, specifically the rights of translation, reprinting, reuse of illustrations, recitation, broadcasting, reproduction on microfilm or in any other way, and storage in data banks. Duplication of this publication or parts thereof is permitted only under the provisions of the German Copyright Law of September 9, 1965, in its current version, and permission for use must always be obtained from Springer. Violations are liable for prosecution under the German Copyright Law.

Springer is a part of Springer Science+Business Media
springer.com
© Springer-Verlag Berlin Heidelberg 2006

The use of general descriptive names, registered names, trademarks, etc. in this publication does not imply, even in the absence of a specific statement, that such names are exempt from the relevant protective laws and regulations and therefore free for general use.

Typesetting: by the authors and techbooks using a Springer LATEX macro package
Cover design: *design & production* GmbH, Heidelberg

Printed on acid-free paper   SPIN: 11766995   54/techbooks   5 4 3 2 1 0

Preface

Geophysics, or physics modelling of geological phenomena, is as old and as established as geoscience itself. The statistical physics modelling of various geophysical phenomena, earthquake in particular, is comparatively recent. This book intends to cover these recent developments in modelling various geophysical phenomena, including the imposing classic phenomenon of earthquakes, employing various statistical physical ideas and techniques. This first book on statistical physics modelling of geophysical phenomena contains extensive reviews by almost all the leading experts in the field and should be widely useful to the graduate students and researchers in geoscience and statistical physics. It grew out of the lecture notes from a workshop on "Models of Earthquakes: Physics Approaches", held in Saha Institute of Nuclear Physics, Kolkata, under the auspices of its Centre for Applied Mathematics and Computational Science in December 2005.

The book is divided in four parts. In the first part, tutorial materials are introduced. Chakrabarti introduces the fracture nucleation processes, their (extreme) statistics in disordered solids, in fibre bundle models and in the two fractal overlap models of earthquakes. In the next two chapters, Hemmer et al. and Kun et al. review the avalanche or quake statistics and the breaking dynamics in simple (mean-field like) fibre bundle models and in their extended versions, respectively. Hansen and Mathiesen discuss the scale invariance properties of the random and fractured surfaces.

In part II, physics models of earthquake and their statistical analysis are discussed in detail. Burridge recounts some of the early and very successful attempts like the spring-block models. Bhattacharyya discusses the recently introduced geometric models of earthquakes and their successes in capturing the statistics. The solid–solid friction and stick-slip models of earthquakes are discussed next by Matsukawa and Saito. Corral puts forward an intriguing analysis of the statistical correlations in various observed catalogue data for earthquakes. Similar spatio-temporal correlations in data and their analysis in the context of spring-block models are discussed by Kawamura. Spatio-temporal correlations between earthquakes and aftershocks are examined in detail by de Rubeis et al. In view of such correlations, the possibilities of short-term predictions for relatively stronger earthquakes are then examined by Tabar et al. Finally, following a detailed survey of the inadequacies of our knowledge of faults, fracture, etc., and of their dynamics and statistics, Kagan argues why physics may still fail in making precise long-term predictions of earthquakes.

In the third part, some related modelling efforts are reviewed. Herrmann discusses the sand-dune formations and their physics models. Mehta reviews the dynamics of sand-piles and of ripple formation in the same, in the next chapter. Dynamics of plastic flow, the Portevin-Le Châtelier effect in particular, of stick-slips as in the peeling of adhesive tapes, etc., are discussed by Ananthakrishna and De. Next, Pradhan and Chakrabarti reviewed the statistical nature of various possible precursors in some established models of catastrophic failures in sand-piles or of fractures in composites.

In the final part, we include some short notes on some interesting and occasionally speculative analysis of phenomena or models in all these related fields.

As mentioned already, these up-to-date, detailed and penetrating reviews by the leading experts are expected to make this volume a profound guide book for the graduate students and researchers in the related fields. We are extremely thankful to these contributors for their intensive work and pleasant cooperations. We are also very much indebted to Arnab Das for his help in compiling and editing this book. Finally, we express our gratitude to Johannes Zittartz, Series Editor, LNP, and Christian Caron of Physics Editorial Department of Springer for their encouragement and support.

Kolkata *Pratip Bhattacharyya*
April 2006 *Bikas K. Chakrabarti*

Contents

Part I Tutorial: Introductory Material

Statistical Physics of Fracture and Earthquake
B.K. Chakrabarti 3
1 Introduction 3
   1.1 Models of Fracture in Disordered Solids and Statistics 3
   1.2 Earthquake Models and Statistics 5
2 Fracture Statistics of Disordered Solids 7
   2.1 Griffith Energy Balance
       and Brittle Fracture Strength of Solids 7
   2.2 Extreme Statistics of the Fracture Stress 11
   2.3 Failure Statistics in Fibre Bundles 12
3 Two-Fractal Overlap Model of Earthquake and Its Statistics 19
4 Summary and Discussions 23
References 25

Rupture Processes in Fibre Bundle Models
P.C. Hemmer, A. Hansen, S. Pradhan 27
1 Introduction 27
2 Equal-Load-Sharing Fibre Bundles 30
   2.1 Burst Distribution: The Generic Case 31
   2.2 Burst Distribution: Nongeneric Cases 34
   2.3 Mapping onto a Random Walk Problem 36
   2.4 Crossover Behavior Near Criticality 38
   2.5 Avalanche Distribution at Criticality 41
   2.6 Recursive Dynamics 42
3 Fibre Bundles with Local Load Redistribution 46
   3.1 Stress Alleviation by Nearest Neighbors 46
   3.2 Intermediate Load-Sharing Models 48
   3.3 Elastic Medium Anchoring 49
References 54

Extensions of Fibre Bundle Models
F. Kun, F. Raischel, R.C. Hidalgo, H.J. Herrmann 57
1 Introduction 57
2 Fibre Bundle Model of Materials Failure 58
   2.1 Why Extensions are Necessary? 63
3 Gradual Degradation of Fibre Strength........................... 65
   3.1 Continuous Damage Fibre Bundle Model 66
   3.2 Macroscopic Constitutive Behavior 67
   3.3 Simulation Techniques..................................... 69
   3.4 Bursts of Fibres Breakings 70
4 Variable Range of Load Sharing 72
   4.1 Load Transfer Function 72
   4.2 Macroscopic Strength of Bundles 73
   4.3 Microstructure of Damage 75
5 Damage Enhanced Creep in Fibre Bundles 77
   5.1 Viscoelastic Fibre Bundle 78
   5.2 Macroscopic Response 79
   5.3 Microscopic Damage Process 80
   5.4 Universality Classes of Creep Rupture 81
6 Failure of Glued Interfaces of Solid Blocks 82
   6.1 The Beam Model of Interface Failure Under Shear 82
   6.2 Constitutive Behavior 83
   6.3 Simulation Techniques..................................... 85
   6.4 Microscopic Damage Process 87
7 Discussion and Outlook 88
References 89

Survey of Scaling Surfaces
A. Hansen, J. Mathiesen... 93
1 Introduction 93
2 Self-Affine Surfaces and Brownian Walks 93
3 Fractional Noise – White Noise.................................. 95
4 Lévy Flights 98
5 Fractal Surfaces 100
6 Multifractals.. 103
7 Multiaffine Surfaces... 105
8 Anomalous Scaling of Self-Affine Surfaces 106
9 Conclusion 109
References 110

Part II Physics Models of Earthquake

Some Early Earthquake Source Models
R. Burridge 113
1 Introduction 113
2 The Double-Couple Point Source 115
    2.1 The Calculation 116
    2.2 Action, Reaction, etc. 116
    2.3 The Double Couple 117
    2.4 The Double-Couple Radiation Pattern 117
3 The Block-and-Spring Model 118
    3.1 Computational Model 122
    3.2 Equations of Motion (Newton) 124
    3.3 Energy Balance 124
    3.4 Numerical Experiment 124
    3.5 Shock and Aftershock 125
    3.6 Potential Energy During an Aftershock Sequence
         and at Other Times 126
    3.7 Omori's Formula for Aftershock Rate 126
    3.8 Peak Kinetic Energy and Energy Radiated Versus Drop
         in Potential Energy 127
4 Continuum Model: Numerical 128
    4.1 Mathematical Formulation 129
    4.2 Setting Up the Integral Equations 130
    4.3 Integral Equation for Anti-Plane Strain 130
    4.4 Discretizing the Kernel 131
    4.5 The Discretization of K 131
    4.6 The Integral Equation for v 132
    4.7 The Solution for $\dot{v} = \dot{u}_y$ 132
    4.8 Numerical Solution for Nucleation at a Point 134
    4.9 Plane Strain 134
    4.10 The Numerical Scheme 136
    4.11 The Numerical Solution 136
    4.12 The Exact Static Solution for Comparison 137
    4.13 Another Numerical Solution 137
5 A Dynamic Shear Crack with Friction 138
    5.1 The Setup 138
    5.2 Initial Stress 139
    5.3 Static Friction 140
    5.4 Dynamic Friction 140
    5.5 The Mathematical Formulation 140
    5.6 Symmetry 141
    5.7 Analysis Confined to the Plane $x = +0$ 141
    5.8 The Basic Integral Relationships 142

	5.9	The Stress Ahead of the Crack 142
	5.10	The Crack Edge 143
	5.11	The Stopping Locus 143
	5.12	A Simple Example: $T = 1 - y^2$ 144
	5.13	The Residual Static Stress Drop 144
	5.14	Radiated Far-Field Pulses 146
6	A Model for Repeating Events 147
	6.1	Equations of Motion 147
	6.2	On the Fault Plane 149
	6.3	Nondimensionalization 149
	6.4	Rate of Strain Between Events 150
	6.5	The Parameters for Repeating Events 151

References 152

Geometric Models of Earthquakes
P. Bhattacharyya 155

1 Introduction 155
2 The Cantor Set 157
3 The Simplest Fractal-Overlap Model of a Fault 157
4 Time Series of Overlap Magnitudes 159
5 Analysis of the Time-Series 162
6 Emergence of a Power Law 166
References 168

Friction, Stick-Slip Motion and Earthquake
H. Matsukawa, T. Saito 169

1 Introduction 169
2 Friction 170
   2.1 Velocity and Waiting Time Dependence of Frictional Force and Earthquake 173
   2.2 Mechanism of Velocity and Waiting Time Dependence of Frictional Force 175
3 Stick-Slip Motion 178
4 Numerical Study of the Burridge–Knopoff Model 181
   4.1 Model 182
   4.2 Numerical Results 183
5 Summary and Discussion 186
References 188

Statistical Features of Earthquake Temporal Occurrence
Á. Corral 191

1 The Gutenberg–Richter Law and the Omori Law 192
2 Recurrence-Time Distributions and Scaling Laws 193
   2.1 Scaling Laws for Recurrence-Time Distributions 195
   2.2 Relation with the Omori Law 196
   2.3 Gamma Fit of the Scaling Function 197

	2.4	Universal Scaling Law for Stationary Seismicity............198
	2.5	Universal Scaling Law for Omori Sequences 201
3	\multicolumn{2}{l}{The Paradox of the Decreasing Hazard Rate and the Increasing}
	\multicolumn{2}{l}{Time Until the Next Earthquake ....203}
	3.1	Decreasing of the Hazard Rate 203
	3.2	Increasing of the Residual Time Until the Next Earthquake ...204
	3.3	Direct Empirical Evidence 205
4	\multicolumn{2}{l}{Scaling Law Fulfillment as Invariance}
	\multicolumn{2}{l}{Under a Renormalization-Group Transformation ....207}
	4.1	Simple Model to Renormalize 208
	4.2	Renormalization-Group Invariance of the Poisson Process 209
5	\multicolumn{2}{l}{Correlations in Seismicity ....211}
	5.1	Correlations Between Recurrence Times 211
	5.2	Correlations Between Recurrence Time and Magnitude 212
	5.3	Correlations Between Magnitudes 214
	5.4	Correlations Between Recurrence Times and Distances 216
6	\multicolumn{2}{l}{Bak et al's Unified Scaling Law ....216}
7	\multicolumn{2}{l}{Conclusions...219}
\multicolumn{3}{l}{References ....219}

Spatiotemporal Correlations of Earthquakes

H. Kawamura 223

1 Introduction 223
2 The Model, the Simulation and the Catalog......................226
  2.1 The Spring-Block Model of Earthquakes 226
  2.2 The Numerical Simulation 228
  2.3 The Seismic Catalog of Japan 229
3 Statistical Properties of Earthquakes............................229
  3.1 The Magnitude Distribution 229
  3.2 The Local Recurrence-Time Distribution....................233
  3.3 The Global Recurrence-Time Distribution 237
  3.4 Spatiotemporal Seismic Correlations before the Mainshock 239
  3.5 Spatiotemporal Seismic Correlations after the Mainshock 246
  3.6 The Time-Dependent Magnitude Distribution 250
4 Summary and Discussion 252
References 256

Space-time Combined Correlation Between Earthquakes and a New, Self-Consistent Definition of Aftershocks

V. De Rubeis, V. Loreto, L. Pietronero, P. Tosi 259

1 Introduction 259
2 Scale Invariance in Space 261
3 Scale Invariance in Time 264
4 Space-Time Combined Correlation Integral 264
5 Time Combined Correlation Dimension 268

6	Space Combined Correlation Dimension	273
7	Discussion and Conclusion	276
References		279

Short-Term Prediction of Medium- and Large-Size Earthquakes Based on Markov and Extended Self-Similarity Analysis of Seismic Data
M.R.R. Tabar, M. Sahimi, F. Ghasemi, K. Kaviani, M. Allamehzadeh, J. Peinke, M. Mokhtari, M. Vesaghi, M.D. Niry, A. Bahraminasab, S. Tabatabai, S. Fayazbakhsh, M. Akbari 281

1	Introduction	282
2	Analysis of Seismic Time Series as Markov Process	287
3	The Extended Self-Similarity of Seismic Data	289
4	Test of the Method	292
5	Summary	298
References		300

Why Does Theoretical Physics Fail to Explain and Predict Earthquake Occurrence?
Y.Y. Kagan 303

1	Introduction	303
2	Seismological Background	305
	2.1 Earthquakes	305
	2.2 Description of Earthquake Catalogs	306
	2.3 Earthquake Temporal Occurrence: Quasi-Periodic, Poisson, or Clustered?	311
	2.4 Earthquake Faults: One Fault, Several Faults, or an Infinite Number of Faults?	312
	2.5 Statistical and Physical Models of Seismicity	314
	2.6 Laboratory and Theoretical Studies of Fracture	316
3	Modern Earthquake Catalogs	317
4	Earthquake Size Distribution	317
	4.1 Magnitude Versus Seismic Moment	318
	4.2 Seismic Moment Distribution	319
	4.3 Seismic Moment Sum Distribution	322
5	Temporal Earthquake Distribution	326
6	Earthquake Location Distribution	328
	6.1 Multipoint Spatial Moments	328
	6.2 Correlation Dimension	329
	6.3 Spatial Scaling	332
7	Focal Mechanism Orientation and Stress Distribution	333
	7.1 Focal Mechanism Distribution	333
	7.2 Random Stress Tensor	334
8	Stochastic Processes and Earthquake Occurrence	338
	8.1 Earthquake Clustering	338

	8.2	Several Problems and Challenges 341
	8.3	Critical Continuum-state Branching Model
		of Earthquake Rupture 342
	8.4	Earthquake Forecasting Attempts 347
9	Discussion 349
References 352

Part III Modelling Related Phenomena

Aeolian Transport and Dune Formation
H.J. Herrmann 363
1 Introduction 363
2 The Aeolian Field 364
3 Aeolian Transport of Sand 370
4 Inclusion of the Slip Face 379
5 Dunes 380
6 Conclusion 384
References 384

Avalanches and Ripples in Sandpiles
A. Mehta 387
1 Introduction 387
2 Avalanches in a Rotating Cylinder 388
    2.1 The Model 388
    2.2 Results 391
    2.3 Discussion 400
3 Coupled Continuum Equations: The Dynamics of Sandpile Surfaces . 404
    3.1 Case A: The Edwards-Wilkinson Equation with Flow 407
    3.2 Case B: When Moving Grains Abound 409
    3.3 Case C: Tilt Combined with Flowing Grains 411
    3.4 Discussion of Coupled Equations 412
    3.5 Application to Ripple Formation 413
References 419

Dynamics of Stick-Slip: Some Universal and Not So Universal Features
G. Ananthakrishna, R. De 423
1 Introduction 424
2 Solid Friction 425
3 Stick-Slip Instability During Plastic Flow: The Portevin–
    Le Chatelier (PLC) Effect 428
    3.1 Dynamical Interpretation of Negative Strain Rate Sensitivity .. 431
    3.2 The Ananthakrishna's Model 431
    3.3 Slow Manifold Analysis 432
4 Peeling of Adhesive Tapes: Stick-Slip Instabilities 435

	4.1 Equation of Motion	436
5	Algorithm	439
	5.1 Results	440
6	Predictability of Slip Events in Power Law States	444
7	Conclusions	451
	References	455

Search for Precursors in Some Models of Catastrophic Failures
S. Pradhan, B.K. Chakrabarti 459

1	Introduction	459
2	Precursors in Failure Models	459
	2.1 Composite Material Under Stress: Fibre Bundle Model	459
	2.2 Electrical Networks within a Voltage Difference: Random Fuse Model	467
	2.3 SOC Models of Sandpile	468
	2.4 Fractal Overlap Model of Earthquake	471
3	Conclusions	475
	References	476

Part IV Miscellaneous Short Notes

Nonlinear Analysis of Radon Time Series Related to Earthquake
N.K. Das, H. Chauduri, R.K. Bhandari, D. Ghose, P. Sen, B. Sinha ... 481

1	Introduction	481
2	Methods of Analysis	482
	2.1 Phase Space Plot	482
	2.2 Fractal Dimension	483
	2.3 Correlation Dimension	484
	2.4 Largest Lyapunov Exponent (λ)	485
3	Results and Discussions	486
4	Conclusion and Outlook	489
	References	490

A Thermomechanical Model of Earthquakes
B. Bal, K. Ghosh 491

1	Introduction	491
2	Earthquakes and Relaxation Oscillation	492
	2.1 Relaxation Oscillation in Pressure Cooker	493
	2.2 Relaxation Oscillation in Pop-pop Boat	495
3	Earthquake Prediction	496
	References	498

Fractal Dimension of the 2001 El Salvador Earthquake Time Series
Md. Nurujjaman, R. Narayanan, A.N.S. Iyengar 499
1 Introduction 499
2 Multifractal Analysis 499
    2.1 Wavelet Analysis 500
    2.2 Singularity Detection 501
    2.3 Wavelet Transform Modulus Maxima 501
3 Results and Discussion 501
References 504

Approach in Time to Breakdown in the RRTN Model
A.K. Sen, S. Mozumdar 507
1 Introduction 507
    1.1 The Model 508
2 Some Features of the Model 508
    2.1 Percolation Threshold 508
    2.2 Non-linear Response 509
    2.3 Dielectric Breakdown as a Paradigm 509
3 Approach to Breakdown in Time 509
4 Analysis of Data and Conclusion 512
References 513

Critical Behaviour of Mixed Fibres with Uniform Distribution
U. Divakaran, A. Dutta 515
References 520

Index 521

List of Contributors

M. Allamehzadeh
Department of Seismology
The International Institute
of Earthquake Engineering
and Seismology
IIEES, P.O. Box 19531
Tehran, Iran
Mallam@iiees.ac.ir

G. Ananthakrishna
Materials Research Centre and
Centre for Condensed Matter Theory
Indian Institute of Science
Bangalore-560012, India
garani@mrc.iisc.ernet.in

M. Akbari
Department of Seismology
The International Institute
of Earthquake Engineering
and Seismology
IIEES, P.O. Box 19531
Tehran, Iran
mary.Akbari@gmail.com

A. Bahraminasab
ICTP
Strada Costiera 11
I-34100 Trieste, Italy
abahrami@ictp.it

Bijay Bal
Saha Institute of Nuclear Physics
1/AF Bidhannagar
Kolkata-700064
India
bijaybhushan.bal@saha.ac.in

Rakesh K. Bhandari
Variable Energy Cyclotron Center
1/AF, Bidhannagar
Kolkata-700064

Pratip Bhattacharyya
Physics Department
Gurudas College
Narkeldanga
Kolkata-700054
and
Centre for Applied Mathematics
and Computational Science
Saha Institute of Nuclear Physics
Kolkata-700064
pratip.bhattacharyya@saha.ac.in

Robert Burridge
Institute for Mathematics
and its Applications
400 Lind Hall
207 Church Street S.E.
Minneapolis, MN 55455-0436
and

Earth Resources Laboratory
Massachusetts Institute of
Technology
42 Carleton Street E34 Cambridge
MA 02142-1324
burridge@erl.mit.edu

Bikas K. Chakrabarti
Theoretical Condensed Matter
Physics Division and Centre
for Applied Mathematics
and Computational Science
Saha Institute of Nuclear Physics
1/AF Bidhannagar, Kolkata-700064
India
bikask.chakrabarti@saha.ac.in

Hirok Chauduri
Saha Institute of Nuclear Physics
1/AF, Bidhannagar
Kolkata-700064

Álvaro Corral
Departament de Física
Facultat de Ciències
Universitat Autònoma de Barcelona
E-08193 Bellaterra
Spain
Alvaro.Corral@uab.es

Nisith K. Das
Variable Energy Cyclotron Center
1/AF, Bidhannagar
Kolkata-700064
nkdas@veccal.ernet.in.

Rumi De
Materials Research Centre
Indian Institute of Science
Bangalore-560012, India
and
Department of Materials
and Interfaces
Weizmann Institute of Science
Rehovot 76100, Israel
rumi.de@weizmann.ac.il

Uma Divakaran
Department of Physics
Indian Institute of Technology
Kanpur 208016, India
udiva@iitk.ac.in

Amit Dutta
Department of Physics
Indian Institute of Technology
Kanpur 208016, India
dutta@iitk.ac.in

S. Fayazbakhsh
Department of Physics
Sharif University of Technology
P.O. Box 11365-9161, Tehran, Iran
S_Fayazbakhsh@mehr.sharif.edu

F. Ghasemi
Department of Physics
Sharif University of Technology
P.O. Box 11365-9161, Tehran, Iran
f_ghasemi@mehr.sharif.edu

Debasis Ghose
Saha Institute of Nuclear Physics
1/AF, Bidhannagar
Kolkata-700064, India

Kuntal Ghosh
Saha Institute of Nuclear Physics
1/AF Bidhannagar, Kolkata-700064
India
kuntal.ghosh@saha.ac.in

Alex Hansen
Department of Physics
Norwegian University of Science
and Technology
N-7491 Trondheim
Norway
alex.hansen@ntnu.no

Per C. Hemmer
Department of Physics
Norwegian University of Science
and Technology
N-7491 Trondheim, Norway
per.hemmer@ntnu.no

Hans J. Herrmann
Departamento de Fisica
Univesridade Federal do Ceara
Campus do Pici
60451-970 CE, Brazil
hans@ica1.uni-stuttgart.de

J. Herrmann
Department of Theoretical Physics
University of Debrecen
P.O. Box: 5, H-4010 Debrecen
Hungary

R.C. Hidalgo
Department of Fundamental Physics
University of Barcelona
Franques 1, 08028 Barcelona
Spain

A.N. Sekar Iyengar
Saha Institute of Nuclear Physics
1/AF, Bidhannagar
Kolkata-700064, India

Yan Y. Kagan
Department of Earth
and Space Sciences
University of California
Los Angeles, California
USA
kagan@equake.ess.ucla.edu

K. Kaviani
Department of Physics
Az-zahra University
P.O.Box 19834, Tehran
Iran
kaviani@scintist.com

Hikaru Kawamura
Department of Earth
and Space Science
Faculty of Science
Osaka University
Toyonaka 560-0043, Japan
kawamura@ess.sci.osaka-u.ac.jp

F. Kun
Department of Theoretical Physics
University of Debrecen
P. O. Box: 5, H-4010 Debrecen
Hungary
feri@dtp.atomki.hu

Vittorio Loreto
"La Sapienza" University
Physics Department
and INFM
Center for Statistical Mechanics
and Complexity
Roma, Italy
pietronero@roma1.infn.it
and
loreto@roma1.infn.it

Joachim Mathiesen
Department of Physics
Norwegian University
of Science and Technology
N-7491 Trondheim, Norway
Joachim.Mathiesen@ntnu.no

Hiroshi Matsukawa
Department of Physics and
Mathematics
Aoyamagakuin University
5-10-1 Fuchinobe
Sagamihara 229-8558
Japan
hm@phys.aoyama.ac.jp

Anita Mehta
SN Bose National Centre for
Basic Sciences
Block JD, Sector III Salt Lake
Calcutta 700 098, India
anita@bose.res.in

M. Mokhtari
Department of Seismology
The International Institute
of Earthquake Engineering

and Seismology, (IIEES)
P.O. Box 19531
Tehran, Iran
Mokhtari@iiees.ac.ir

Shubhankar Mozumdar
South Point High School
82/7A Ballygunge Place
Kolkata-700019, India

Ramesh Narayanan
Saha Institute of Nuclear Physics
1/AF, Bidhannagar
Kolkata-700064, India

M.D. Niry
Department of Physics
Sharif University of Technology
P.O. Box 11365-9161
Tehran, Iran
mdniry@mehr.sharif.edu

Md. Nurujjaman
Saha Institute of Nuclear Physics
1/AF, Bidhannagar
Kolkata-700064, India

J. Peinke
Carl von Ossietzky University
Institute of Physics
D-26111 Oldendurg, Germany
peinke@uni-olden-burg.de

Luciano Pietronero
"La Sapienza" University
Physics Department
and INFM
Center for Statistical Mechanics
and Complexity
Roma, Italy
pietronero@roma1.infn.it

Srutarshi Pradhan
Department of Physics
Norwegian University
of Science and Technology
N-7491 Trondheim, Norway
pradhan.srutarshi@ntnu.no

F. Raischel
Institute for Computational Physics
University of Stuttgart
Pfaffenwaldring 27
D-70569 Stuttgart, Germany

Valerio De Rubeis
Istituto Nazionale di Geofisica e
Vulcanologia
Roma, Italy
derubeis@ingv.it
tosi@ingv.it

Muhammad Sahimi
Department of Chemical
Engineering and Material Science
University of Southern California
Los Angeles
CA 90089, USA
moe@iran.usc.edu

Tatsuro Saito
Department of Physics
and Mathematics
Aoyamagakuin University
5-10-1 Fuchinobe
Sagamihara 229-8558
Japan

Asok K. Sen
TCMP Division
Saha Institute of Nuclear Physics
1/AF Bidhannagar
Kolkata-700064
India
asokk.sen@saha.ac.in

Prasanta Sen
Saha Institute of Nuclear Physics
1/AF, Bidhannagar
Kolkata-700064
India
prasanta.sen@saha.ac.in

Bikash Sinha
Variable Energy Cyclotron Center
1/AF, Bidhannagar, Kolkata-700064
India
and
Saha Institute of Nuclear Physics
1/AF, Bidhannagar
Kolkata-700064
India

S. Tabatabai
Institute of Geophysics
University of Tehran
Iran
S.Tabatai@tabagroup.com

M. Reza Rahimi Tabar
Department of Physics
Sharif University of Technology
P.O. Box 11365-9161
Tehran, Iran

and
CNRS UMR 6202
Observatoire de la
Côte d'Azur
BP 4229
06304 Nice Cedex 4, France
rahimitabar@sharif.edu

Patrizia Tosi
Istituto Nazionale di
Geofisica e Vulcanologia
Roma, Italy
derubeis@ingv.it and
tosi@ingv.it

M. Vesaghi
Department of Physics
Sharif University of Technology
P.O. Box 11365-9161
Tehran, Iran
vesaghi@sharif.edu

Part I

Tutorial: Introductory Material

Statistical Physics of Fracture and Earthquake

B.K. Chakrabarti

Theoretical Condensed Matter Physics Division and Centre for Applied
Mathematics and Computational Science, Saha Institute of Nuclear Physics,
1/AF Bidhannagar, Kolkata 700064, India
bikask.chakrabarti@saha.ac.in

1 Introduction

1.1 Models of Fracture in Disordered Solids and Statistics

If one applies tensile stress on a solid, the solid elongates and gets strained. The stress (σ) – strain (ϵ) relation is linear for small stresses (Hooke's law) after which nonlinearity appears in most cases. Finally at a critical stress σ_f, depending on the material, amount of disorder, the specimen size, etc., the solid breaks into pieces; fracture occurs. In the case of brittle solids, the fracture occurs immediately after the Hookean linear region, and consequently the linear elastic theory can be applied to study the essentially nonlinear and irreversible static fracture properties of brittle solids [1]. With extreme perturbation, therefore, the mechanical or electrical properties of solids tend to get destabilised and failure or breakdown occurs. In fact, these instabilities in the solids often nucleate around disorder, which then plays a major role in the breakdown properties of the solids. The growth of these nucleating centres, in turn, depends on various statistical properties of the disorder, namely the scaling properties of percolating structures, its fractal dimensions, etc. These statistical properties of disorder induce some scaling behaviour for the breakdown of the disordered solids [2, 3].

Obviously with more and more random voids, the linear response of, for example, the modulus of elasticity Y (say, the Young's modulus) of the solid decreases. So also does the breaking strength of the material: the fracture strength σ_f of the specimen. For studying most of these mechanical (elastic) breakdown problems of randomly disordered solids, one can take the lattice model of disordered solids. In these lattice models, a fraction p of the bonds (or sites) is intact, with the rest $(1 − p)$ being randomly broken or cut. Fluctuations in the random distribution give rise to random clusters of springs inside the bulk, for which the statistics is well developed [4], and one can investigate the effect of the voids or impurity clusters on the ultimate strength of the (percolating) elastic network of the bulk solid [2, 3]. One can also consider and compare the results for failure strength of the solids with random bond strength distribution, as for example, in random fibre bundle models (see e.g. [6, 7, 8]).

As is well known, the initial variations (decreases) of the linear responses like the elastic constant Y so the breakdown strengths σ_f are analytic with the impurity (dilution) concentration. Near the percolation threshold [4] p_c, up to (and at) which the solid network is marginally connected through the nearest neighbour occupied bonds or sites and below which the macroscopic connection ceases, the variations in these quantities with p are expected to become singular, the leading singularities being expressed by the respective critical exponents. The exponents for the modulus of elasticity $Y \sim \Delta p^{T_e}$ $\Delta p = (p - p_c)/p_c$ are well-known and depend essentially on the dimension d of the system (see, e.g., [4]). One kind of investigation searches for the corresponding singularities for the essentially nonlinear and irreversible properties of such mechanical breakdown strengths for p near the percolation threshold p_c: to find the exponent T_f for the average fracture stress $\sigma_f \sim \Delta p^{T_f}$ for p near p_c. Very often, one maps [5] the problem of breakdown to the corresponding linear problem (assuming brittleness up to the breaking point) and then derives [2] the scaling relations giving the breakdown exponents T_f in terms of the linear response exponent (T_e) and other lattice statistical exponents (see next section).

Unlike that for the "classical" linear responses of such solids, the extreme nature of the breakdown statistics, nucleating from the weakest point of the sample, gives rise to a non-self-averaging property. We will discuss (in the next section) these distribution functions $F(\sigma)$, giving the cumulative probability of failure of a disordered sample of linear size L. We show that the generic form of the function $F(\sigma)$ can be either the Weibull [2] form

$$F(\sigma) \sim 1 - \exp\left[-L^d \left(\frac{\sigma}{\Lambda(p)}\right)^{1/\phi}\right] \qquad (1)$$

or the Gumbel [2] form

$$F(\sigma) \sim 1 - \exp\left[-L^d \exp\left(\frac{\Lambda(p)}{\sigma}\right)^{1/\phi}\right] \qquad (2)$$

where $\Lambda(p)$ is determined by the linear response like the elasticity of the disordered solid by some other lattice statistical quantity etc. and ϕ is an exponent discussed in the next section.

In another kind of model for disordered systems, a loaded bundle of fibres represents the various aspects of fracture process through its self-organised dynamics. The fibre bundle model study was initiated by Pierce [6] in the context of testing the strength of cotton yarns. Since then, this model has been studied from various points of view. Fibre bundles are of two classes with respect to the time dependence of fibre strength: The "static" bundles contain fibres whose strengths are independent of time, whereas the "dynamic" bundles are assumed to have time-dependent elements to capture the creep rupture and fatigue behaviours. For simplicity, we will discuss here the "static" fibre models only. According to the load sharing rule, fibre bundles are being classified into two groups: Equal-load-sharing (ELS) bundles or democratic bundles and local load-sharing (LLS) bundles. In democratic or ELS bundles, intact fibres bear the applied load equally and in local load-sharing bundles the terminal load of the failed fibre is given equally to all the intact neighbours. The classic work of Daniels [7] on the strength of the "static" fibre bundles under ELS assumption initiated the probabilistic analysis of the model (see, e.g., [8]). The distribution of burst avalanches during fracture process is a marked feature of the fracture dynamics and can be observed in ultrasonic emissions during the fracture process. It helps characterizing different physical systems along with the possibility to predict the large avalanches. From a nontrivial probabilistic analysis, one gets [9] power law distribution of avalanches for static ELS bundles, whereas the power law exponent observed numerically for static LLS bundles differs significantly. This observation induces the possibility of presenting loaded fibre bundles as earthquake models (see Sect. 3). The phase transition [8] and dynamic critical behaviour of the fracture process in such bundles has been established through recursive formulation [8, 10] of the failure dynamics. The exact solutions [10] of the recursion relations suggest universal values of the exponents involved. Attempt has also been made [11] to study the ELS and LLS bundles from a single framework introducing a "range of interaction" parameter that determines the load transfer rule.

1.2 Earthquake Models and Statistics

The earth's solid outer crust, of about 20 km in average thickness, rests on the tectonic shells. Due to the high temperature–pressure phase changes and the consequent powerful convective flow in the earth's mantle, at several hundreds of kilometers of depth, the tectonic shell, divided into a small number (about 10) of mobile plates, has relative velocities of the order of a few centimetres per year [12, 13]. Over several tens of years, enormous elastic strains develop sometimes on the earth's crust when sticking (due to the solid–solid friction) to the moving tectonic plate. When slips occur between the crust and the tectonic plate, these stored elastic energies are released in "bursts", causing the damages during the earthquakes. Because of the uniform motion of the tectonic plates, the elastic strain energy stored in a portion of the crust (block), moving with the plate relative to a "stationary" neighbouring portion of the crust, can vary only due to the random strength of the solid–solid friction between the crust and the plate. The slip occurs when the accumulated stress exceeds the frictional force.

As in fracture (in fibre bundle model in particular), the observed distribution of the elastic energy release in various earthquakes seems to follow a power law. The number of earthquakes $N(m)$, having magnitude in the

Richter scale greater than or equal to m, is phenomenologically observed to decrease with m exponentially. This gives the Gutenberg–Richter law [12]

$$\ln N(m) = \text{constant} - a\, m\,,$$

where a is a constant. It appears [12, 13] that the amount of energy ϵ released in an earthquake of magnitude m is related to it exponentially:

$$\ln \epsilon = \text{constant} + b\, m\,,$$

where b is another constant. Combining therefore we get the power law giving the number of earthquakes $N(\epsilon)$ releasing energy equal to ϵ as

$$N(\epsilon) \sim \epsilon^{-\alpha}\,, \tag{3}$$

with $\alpha = a/b$. The observed value of the exponent (power) α in (3) is around unity (see, e.g., [13]).

Several laboratory and computer simulation models have recently been proposed [2] to capture essentially the above power law in the earthquake energy release statistics. In a very successful table-top laboratory simulation model of earthquakes, Burridge and Knopoff [14] took a chain of wooden blocks connected by identical springs to the neighbouring blocks. The entire chain was placed on a rigid horizontal table with a rough surface, and one of the end blocks was pulled very slowly and uniformly using a driving motor. The strains of the springs increase due to the creep motions of the blocks until one or a few of the blocks slip. The drops in the elastic energy of the chain during slips could be measured from the extensions or compressions of all the springs, and could be taken as the released energies in the earthquake. For some typical roughness of the surfaces (of the blocks and of the table), the distribution of these drops in the elastic energy due to slips indeed shows a power law behaviour with $\alpha \simeq 1$ in (3).

A computer simulation version of this model by Carlson and Langer [15] considers harmonic springs connecting equal mass blocks that are also individually connected to a rigid frame (to simulate other neighbouring portions of the earth's crust not on the same tectonic plate) by harmonic springs. The entire system moves on a rough surface with nonlinear velocity-dependent force (decreasing to zero for large relative velocities) in the direction opposite to the relative motion between the block and the surface. In the computer simulation of this model it is seen that the distribution of the elastic energy release in such a system can indeed be given by a power law like (3), provided the non-linearity of the friction force, responsible for the self-organisation, is carefully chosen [15].

The lattice automata model of Bak et al. [16] represents the stress on each block by a height variable at each lattice site. The site topples (the block slips) if the height (or stress) at this site exceeds a pre-assigned threshold value, and the height becomes zero there and the neighbours share the stress by increasing their heights by 1 unit. With this dynamics for the system, if any of the neighbouring sites of the toppled one was already at the threshold height, the avalanche continues. The boundary sites are considered to be all absorbing. With random addition of heights at a constant rate (increasing stress at a constant rate due to tectonic motion), such a system reaches its self-organised critical point where the avalanche size distributions follow a natural power law corresponding to this self-tuned critical state. Bak [16] identifies this self-organised critical state to be responsible for the Gutenberg–Richter-type power law. All these models are successful in capturing the Gutenberg–Richter power law, and the real reason for the self-similarity inducing the power law is essentially the same in all these different models: emergence of the self-oganised critical state for wide yet suitably chosen variety of non-linear many-body coupled dynamics. In this sense all these models incorporate the well-established fact of the stick-slip frictional instabilities between the earth's crust and the tectonic plate. It is quite difficult to check at this stage any further details and predictions of these models.

While the motion of the tectonic plate is surely an observed fact, and this stick-slip process should be a major ingredient of any bonafide model of earthquake, another established fact regarding the fault geometries of the earth's crust is the fractal nature of the roughness of the surfaces of the earth's crust and the tectonic plate. This latter feature is missing in any of these models discussed above. In fact, the surfaces involved in the process are results of large-scale fracture seperating the crust from the moving tectonic plate. Any such crack surface is observed to be a self-similar fractal, having the self-affine scaling property $z(\lambda x, \lambda y) \sim \lambda^\zeta z(x, y)$ for the surface coordinate z in the direction perpendicular to the crack surface in the (x, y) plane [2]. Various fractographic investigations indicate a fairly robust universal behaviour for such surfaces and the roughness exponent ζ is observed to have a value around $0.80-0.85$ (with a possible crossover to $\zeta \simeq 0.4$ for slow propagation of the crack-tip) [2, 3]. This widely observed scaling property of the fracture surfaces also suggests that the fault surfaces of the earth's crust or the tectonic plate should have similar fractal properties. In fact, some investigators of the earthquake dynamics have already pointed out that the fracture mechanics of the stressed crust of the earth forms self-similar fault patterns, with well-defined fractal dimensionalities near the contact areas with the major plates [2]. On the basis of these observations regarding the earthquake faults, we have developed a "two-fractal overlap" model [17, 18] discussed later.

2 Fracture Statistics of Disordered Solids

2.1 Griffith Energy Balance and Brittle Fracture Strength of Solids

One can easily show (see, e.g., [1, 2]) that in a stressed solid, the local stress at sharp notches or corners of the microcrack can rise to a level several times that of the applied stress. This indicates how the microscopic cracks or flaws within a solid might become potential sources of weakness of the solid. Although this stress concentration indicates clearly where the instabilities should occur, it is not sufficient to tell us when the instability does occur and the fracture propagation starts. This requires a detailed energy balance consideration.

Griffith in 1920, equating the released elastic energy (in an elastic continuum) with the energy of the surface newly created (as the crack grows), arrived at a quantitative criterion for the equilibrium extension of the microcrack already present within the stressed material [19]. We give below an analysis that is valid effectively for two-dimensional stressed solids with a single preexisting crack, as for example the case of a large plate with a small thickness. Extension to three-dimensional solids is straightforward.

Let us assume a thin linear crack of length $2l$ in an infinite elastic continuum subjected to uniform tensile stress (l perpendicular to the length of the crack (see Fig. 1). Stress parallel to the crack does not affect the stability of the crack and has not, therefore, been considered. Because of the crack (which cannot support any stress field, at least on its surfaces), the strain energy density of the stress field ($\sigma^2/2Y$) is perturbed in a region around the crack, having dimension of the length of the crack. We assume here this

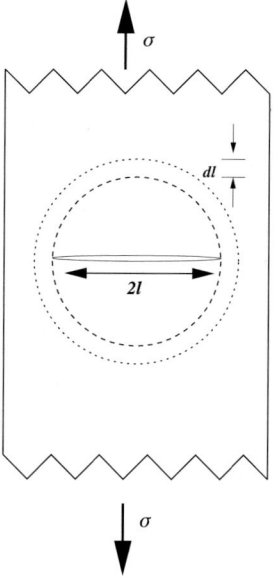

Fig. 1. A portion of a plate (of thickness w) under tensile stress σ (Model I loading) containing a linear crack of length $2l$. For a further growth of the crack length by $2dl$, the elastic energy released from the annular region must be sufficient to provide the surface energy $4\Gamma w dl$ (extra elastic energy must be released for finite velocity of crack propagation)

perturbed or stress-released region to have a circular cross section with the crack length as the diameter. The exact geometry of this perturbed region is not important here, and it determines only a (unimportant) numerical factor in the Griffith formula (see, e.g., [1]). Assuming therefore half of the stress energy of the annular or cylindrical volume, having the internal radius l and outer radius $l+dl$ and length w (perpendicular to the plane of the stress; here the width w of the plate is very small compared to the other dimensions), to be released as the crack propagates by a length dl, one requires this released strain energy to be sufficient for providing the surface energy of the four new surfaces produced. This suggests

$$\frac{1}{2}(\sigma^2/2Y)(2\pi w l dl) \geq \Gamma(4w dl) .$$

Here Y represents the Young's modulus of the solid and Γ represents the surface energy density of the solid, measured by the extra energy required to create unit surface area within the bulk of the solid.

We have assumed here, on average, half of the strain energy of the cylindrical region having a circular cross section with diameter $2l$ to be released. If this fraction is different or the cross section is different, it will change only some of the numerical factors, in which we are not very much interested here. Also, we assume here linear elasticity up to the breaking point, as in the case of brittle materials. The equality holds when energy dissipation, as in the case of plastic deformation or for the propagation dynamics of the crack, does not occur. One then gets

$$\sigma_{\mathrm{f}} = \frac{\Lambda}{\sqrt{2l}}; \qquad \Lambda = \left(\frac{4}{\sqrt{\pi}}\right)\sqrt{Y\Gamma} \qquad (4)$$

for the critical stress at and above which the crack of length $2l$ starts propagating and a macroscopic fracture occurs. Here Γ is called the stress-intensity factor or the fracture toughness. In fact, one can alternatively view the fracture occurring when the stress at the crack-tip (given by the stress intensity factor in (4)) exceeds the elastic stress limit for the medium.

In a three-dimensional solid containing a single elliptic disk-shaped planar crack perpendicular to the applied tensile stress direction, a straightforward extension of the above analysis suggests that the maximum stress concentration would occur at the two tips (at the two ends of the major axis) of the ellipse. The Griffith stress for the brittle fracture of the solid would therefore be determined by the same formula (4), with the crack length $2l$ replaced by the length of the major axis of the elliptic planar crack. Generally, for any dimension therefore, if a crack of length l already exists in an infinite elastic continuum, subject to uniform tensile stress σ perpendicular to the length of the crack, then for the onset of brittle fracture, Griffith equates (the differentials of) the elastic energy E_l with the surface energy E_s:

$$E_l \simeq \left(\frac{\sigma^2}{2Y}\right) l^d = E_\mathrm{s} \simeq \Gamma l^{d-1} , \qquad (5)$$

where Y represents the elastic modulus appropriate for the strain, Γ the surface energy density and d the dimension. Equality holds when no energy dissipation (due to plasticity or crack propagation) occurs and one gets

$$\sigma_f \sim \frac{\Lambda}{\sqrt{l}} \; ; \; \Lambda \sim \sqrt{Y\Gamma} \qquad (6)$$

for the breakdown stress at (and above) which the existing crack of length l starts propagating and a macroscopic fracture occurs. It may also be noted that the above formula is valid in all dimensions ($d \geq 2$).

This quasi-static picture can be extended [20] to fatigue behavior of crack propagation for $\sigma < \sigma_f$. At any stress σ less than σ_f, the cracks can still nucleate for a further extension at any finite temperature $k_B T$ with a probability $\sim \exp[-E/k_B T]$ and consequently the sample fails within a failure time τ given by

$$\tau^{-1} \sim \exp[-E(l_0)/k_B T],$$

where

$$E(l_0) = E_s + E_l \sim \Gamma l_0^2 - \frac{\sigma^2}{Y} l_0^3$$

is the crack (of length l_0) nucleation energy. One can therefore express τ as

$$\tau \sim \exp[A(1 - \frac{\sigma^2}{\sigma_f^2})], \qquad (7)$$

where (the dimensionless parameter) $A \sim l_0^3 \sigma_f^2/(Y k_B T)$ and σ_f is given by 6. This immediately suggests that the failure time τ grows exponentially for $\sigma < \sigma_f$ and approaches infinity if the stress σ is much less than σ_f when the temperature $k_B T$ is small, whereas τ becomes vanishingly small as the stress σ exceeds σ_f; see, e.g., [21] and [22].

For disordered solids, let us model the solid by a percolating system. As mentioned earlier, for the occupied bond/site concentration $p > p_c$, the percolation threshold, the typical pre-existing cracks in the solid will have the dimension (l) of correlation length $\xi \sim \Delta p^{-\nu}$ and the elastic strength $Y \sim \Delta p^{T_e}$ [4]. Assuming that the surface energy density Γ scales as ξ^{d_B}, with the backbone (fractal) dimension d_B [4], equating E_l and E_s as in (5), one gets $\left(\frac{\sigma_f^2}{2Y}\right) \xi^d \sim \xi^{d_B}$. This gives

$$\sigma_f \sim (\Delta p)^{T_f}$$

with

$$T_f = \frac{1}{2}[T_e + (d - d_B)\nu] \qquad (8)$$

for the "average" fracture strength of a disordered solid (of fixed value) as one approaches the percolation threshold. Careful extensions of such scaling relations (8) and rigorous bounds for T_f has been obtained and compared extensively in [2, 3].

2.2 Extreme Statistics of the Fracture Stress

The fracture strength σ_f of a disordered solid does not have self-averaging statistics; most probable and the average σ_f may not match because of the extreme nature of the statistics. This is because the "weakest point" of a solid determines the strength of the entire solid, not the average weak points! As we have modelled here, the statistics of clusters of defects are governed by the random percolation processes. We have also discussed how the linear responses, like the elastic moduli of such random networks, can be obtained from the averages over the statistics of such clusters. This was possible because of the self-averaging property of such linear responses. This is because the elasticity of a random network is determined by all the "parallel" connected material portions or paths, contributing their share in the net elasticity of the sample. However, the fracture or breakdown property of a disordered solid is determined by only the weakest (often the longest) defect cluster or crack in the entire solid. Except for some indirect effects, most of the weaker or smaller defects or cracks in the solid do not determine the breakdown strength of the sample. The fracture or breakdown statistics of a solid sample are therefore determined essentially by the extreme statistics of the most dangerous or weakest (largest) defect cluster or crack within the sample volume.

We discuss now more formally the origin of this extreme statistics. Let us consider a solid of linear size L, containing n cracks within its volume. We assume that each of these cracks has a failure probability $f_i(\sigma), i = 1, 2, \ldots, n$ to fail or break (independently) under an applied stress σ on the solid, and that the perturbed or stress-released regions of each of these cracks are seperate and do not overlap. If we denote the cumulative failure probability of the entire sample, under stress σ, by $F(\sigma)$, then [2]

$$1 - F(\sigma) = \prod_{i=1}^{n}(1 - f_i(\sigma)) \simeq \exp\left[-\sum_i f_i(\sigma)\right] = \exp\left[-L^d \tilde{g}(\sigma)\right] \quad (9)$$

where $\tilde{g}(\sigma)$ denotes the density of cracks within the sample volume L^d (coming from the sum $\sum_i$ over the entire volume), which starts propagating at and above the stress level σ. The above equation comes from the fact that the sample survives if each of the cracks within the volume survives. This is the essential origin of the above extreme statistical nature of the failure probability $F(\sigma)$ of the sample.

Noting that the pair correlation $g(l)$ of two occupied sites at distance l on a percolation cluster decays as $\exp(-l/\xi(p))$, and connecting the stress σ with the length l by using Griffith's law (4) that $\sigma \sim \frac{A}{l^\phi}$, one gets $\tilde{g}(\sigma) \sim \exp\left(-\frac{A^{1/\phi}}{\xi \sigma^{1/\phi}}\right)$ for $p \to p_c$. This, put in (9), gives the Gumbel distribution (2) given earlier [2]. If, on the other hand, one assumes a power law decay of $g(l)$: $g(l) \sim l^{-w}$, then using Griffith's law (4), one gets $\tilde{g}(\sigma) \sim \left(\frac{\sigma}{A}\right)^{-n}$, giving the Weibull distribution (1), from (9), where $m = w/\phi$ gives the Weibull modulus [2]. The variation of $F(\sigma)$ with σ in both the cases has the generic form shown

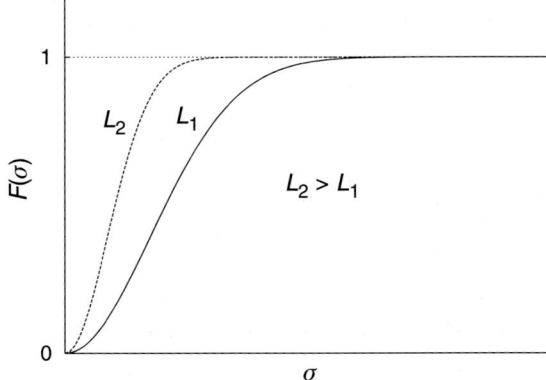

Fig. 2. Schematic variation of failure probability $F(\sigma)$ with stress σ for a disordered solid with volumes L_1^d or L_2^d ($L_2 > L_1$)

in Fig. 2. $F(\sigma)$ is non-zero for any stress $\sigma > 0$ and its value (at any σ) is higher for larger volume (L^d). This is because the possibility of a larger defect (due to fluctuation) is higher in a larger volume and, consequently, its failure probability is higher. Assuming $F(\sigma_f)$ is finite for failure, the most probable failure stress σ_f becomes a decreasing function of volume if extreme statistics are at work.

The precise ranges of the validity of the Weibull or Gumbel distributions for the breakdown strength of disordered solids are not well established yet. However, analysis of the results of detailed experimental and numerical studies of breakdown in disordered solids seems to suggest that the fluctuations of the extreme statistics dominate for small disorder [3]. Very near to the percolation point, the percolation statistics take over and the statistics become self-averaging. One can argue [19] that arbitrarily close to the percolation threshold, the fluctuations of the extreme statistics will probably get suppressed and the percolation statistics should take over and the most probable breaking stress becomes independent of the sample volume (its variation with disorder being determined, as in (8), by an appropriate breakdown exponent). This is because the appropriate competing length scales for the two kinds of statistics are the Lifshitz scale $\ln L$ (coming from the finiteness of the volume integral of the defect probability: $L^d(1-p)^l$ finite, giving the typical defect size $l \sim \ln L$) and the percolation correlation length ξ. When $\xi < \ln L$, the above scenario of extreme statistics should be observed. For $\xi > \ln L$, the percolation statistics are expected to dominate.

2.3 Failure Statistics in Fibre Bundles

The fibre bundle (see Fig. 3) consists of N fibres or Hook springs, each having identical spring constant κ. The bundle supports a load $W = N\sigma$ and the

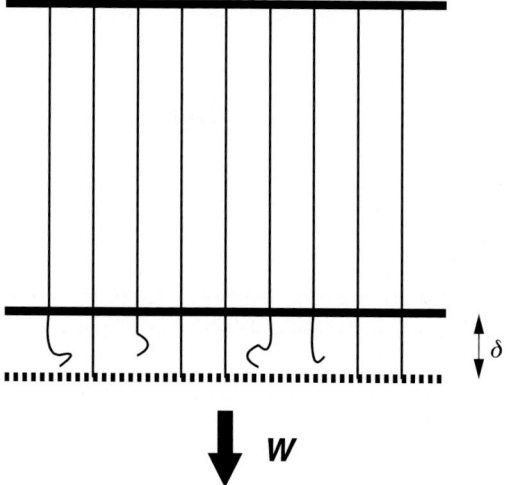

Fig. 3. The fibre bundle consists initially of N fibres attached in parallel to a fixed and rigid plate at the *top* and a downwardly movable platform from which a load W is suspended at the *bottom*. In the equal-load-sharing model considered here, the platform is absolutely rigid and the load W is consequently shared equally by all the intact fibres

breaking threshold $(\sigma_{\text{th}})_i$ of the fibres are assumed to be different for different fibre (i). For the ELS model we consider here, the lower platform is absolutely rigid, and therefore no local deformation and hence no stress concentration occurs anywhere around the failed fibres. This ensures ELS, i.e., the intact fibres share the applied load W equally and the load per fibre increases as more and more fibres fail. The strength of each of the fibre $(\sigma_{\text{th}})_i$ in the bundle is given by the stress value it can bear, and beyond which it fails. The strength of the fibres are taken from a randomly distributed normalised density $\rho(\sigma_{\text{th}})$ within the interval 0 and 1 such that

$$\int_0^1 \rho(\sigma_{\text{th}}) d\sigma_{\text{th}} = 1 \;.$$

The ELS assumption neglects "local" fluctuations in stress (and its redistribution) and renders the model as a mean-field one.

The breaking dynamics starts when an initial stress σ (load per fibre) is applied on the bundle. The fibres having strength less than σ fail instantly. Due to this rupture, total number of intact fibres decreases and rest of the (intact) fibres have to bear the applied load on the bundle. Hence effective stress on the fibres increases and this compels some more fibres to break. These two sequential operations, namely the stress redistribution and further breaking of fibres, continue till an equilibrium is reached, where either the surviving fibres are strong enough to bear the applied load on the bundle or all fibres fail.

This breaking dynamics can be represented by recursion relations in discrete time steps. For this, let us consider a very simple model of fibre bundles where the fibres (having the same spring constant κ) have a white or uniform strength distribution $\rho(\sigma_{\text{th}})$ upto a cut-off strength normalized to unity, as shown in Fig. 4: $\rho(\sigma_{\text{th}}) = 1$ for $0 \leq \sigma_{\text{th}} \leq 1$ and $= \rho(\sigma_{\text{th}}) = 0$ for $\sigma > \sigma_{\text{th}}$. Let us also define $U_t(\sigma)$ to be the fraction of fibres in the bundle that survive after (discrete) time step t, counted from the time $t = 0$ when the load is put (time step indicates the number of stress redistributions). As such, $U_t(\sigma = 0) = 1$ for all t and $U_t(\sigma) = 1$ for $t = 0$ for any σ; $U_t(\sigma) = U^*(\sigma) \neq 0$ for $t \to \infty$ and $\sigma < \sigma_f$, the critical or failure strength of the bundle, and $U_t(\sigma) = 0$ for $t \to \infty$ if $\sigma > \sigma_f$. Therefore $U_t(\sigma)$ follows a simple recursion relation (see Fig. 4)

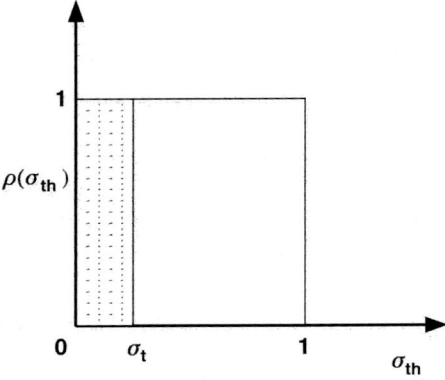

Fig. 4. The simple model considered here assumes uniform density $\rho(\sigma_{\text{th}})$ of the fibre strength distribution up to a cut-off strength (normalized to unity). At any load per fibre level σ_t at time t, the fraction σ_t fails and $1 - \sigma_t$ survives

$$U_{t+1} = 1 - \sigma_t; \quad \sigma_t = \frac{W}{U_t N}$$

$$\text{or} \quad U_{t+1} = 1 - \frac{\sigma}{U_t}. \tag{10}$$

At the equilibrium state ($U_{t+1} = U_t = U^*$), the above relation takes a quadratic form of U^*:

$$U^{*^2} - U^* + \sigma = 0 .$$

The solution is

$$U^*(\sigma) = \frac{1}{2} \pm (\sigma_f - \sigma)^{1/2}; \quad \sigma_f = \frac{1}{4} .$$

Here σ_f is the critical value of initial applied stress beyond which the bundle fails completely. The solution with $(+)$ sign is the stable one, whereas the one with $(-)$ sign gives unstable solution [10]. The quantity $U^*(\sigma)$ must be real valued as it has a physical meaning: it is the fraction of the original bundle that remains intact under a fixed applied stress σ when the applied stress lies in the range $0 \leq \sigma \leq \sigma_f$. Clearly, $U^*(0) = 1$. Therefore the stable solution can be written as

$$U^*(\sigma) = U^*(\sigma_f) + (\sigma_f - \sigma)^{1/2}; \quad U^*(\sigma_f) = \frac{1}{2} \text{ and } \sigma_f = \frac{1}{4}. \quad (11)$$

For $\sigma > \sigma_f$ we cannot get a real-valued fixed point as the dynamics never stops until $U_t = 0$ when the bundle breaks completely.

(a) At $\sigma < \sigma_f$

It may be noted that the quantity $U^*(\sigma) - U^*(\sigma_f)$ behaves like an order parameter that determines a transition from a state of partial failure ($\sigma \leq \sigma_f$) to a state of total failure ($\sigma > \sigma_f$) [10]:

$$O \equiv U^*(\sigma) - U^*(\sigma_f) = (\sigma_f - \sigma)^\beta; \quad \beta = \frac{1}{2}. \quad (12)$$

To study the dynamics away from criticality ($\sigma \to \sigma_f$ from below), we replace the recursion relation (10) by a differential equation

$$-\frac{dU}{dt} = \frac{U^2 - U + \sigma}{U}.$$

Close to the fixed point we write $U_t(\sigma) = U^*(\sigma) + \epsilon$ (where $\epsilon \to 0$). This, following (10), gives [10]

$$\epsilon = U_t(\sigma) - U^*(\sigma) \approx \exp(-t/\tau), \quad (13)$$

where $\tau = \frac{1}{2}\left[\frac{1}{2}(\sigma_f - \sigma)^{-1/2} + 1\right]$. Near the critical point we can write

$$\tau \propto (\sigma_f - \sigma)^{-\alpha}; \quad \alpha = \frac{1}{2}. \quad (14)$$

Therefore the relaxation time diverges following a power law as $\sigma \to \sigma_f$ from below [10].

One can also consider the breakdown susceptibility χ, defined as the change of $U^*(\sigma)$ due to an infinitesimal increment of the applied stress σ [10]

$$\chi = \left|\frac{dU^*(\sigma)}{d\sigma}\right| = \frac{1}{2}(\sigma_f - \sigma)^{-\gamma}; \quad \gamma = \frac{1}{2} \quad (15)$$

from (10). Hence the susceptibility diverges as the applied stress σ approaches the critical value $\sigma_f = \frac{1}{4}$. Such a divergence in χ had already been observed in the numerical studies.

(b) At $\sigma = \sigma_f$

At the critical point ($\sigma = \sigma_f$), we observe a different dynamic critical behaviour in the relaxation of the failure process. From the recursion relation (10), it can be shown that decay of the fraction $U_t(\sigma_f)$ of unbroken fibres that remain intact at time t follows a simple power law decay [10]:

$$U_t = \frac{1}{2}\left(1 + \frac{1}{t+1}\right), \tag{16}$$

starting from $U_0 = 1$. For large t ($t \to \infty$), this reduces to $U_t - 1/2 \propto t^{-\delta}$, $\delta = 1$, a strict power law that is a robust characterisation of the critical state.

Universality Class of the Model

The universality class of the model has been checked [10] taking two other types of fibre strength distributions: (I) linearly increasing density distribution and (II) linearly decreasing density distribution within the (σ_{th}) limit 0 and 1. One can show that while σ_f changes with different strength distributions ($\sigma_f = \sqrt{4/27}$ for case I and $\sigma_f = 4/27$ for case II), the critical behaviour remains unchanged: $\alpha = 1/2 = \beta = \gamma$, $\delta = 1$, for all these ELS models.

Nonlinear Stress–Strain Relation for the Bundle

One can now consider a slightly modified strength distribution of the ELS fibre bundle, showing typical non-linear deformation characteristics [7, 10]. For this, we consider a uniform density distribution of fibre strength, having a lower cut-off. Until failure of any of the fibres (due to this lower cut-off), the bundle shows linear elastic behaviour. As soon as the fibres start failing, the stress–strain relationship becomes non-linear. The dynamic critical behaviour remains essentially the same and the static (fixed point) behaviour shows elastic–plastic like deformation before rupture of the bundle.

Here the fibres are elastic in nature having identical force constant κ and the random fibre strengths distributed uniformly in the interval $[\sigma_L, 1]$ with $\sigma_L > 0$; the normalised distribution of the threshold stress of the fibres thus has the form (see Fig. 5):

$$\rho(\sigma_{\text{th}}) = \begin{cases} 0, & 0 \leq \sigma_{\text{th}} \leq \sigma_L \\ \dfrac{1}{1-\sigma_L}, & \sigma_L < \sigma_{\text{th}} \leq 1 \end{cases}. \tag{17}$$

For an applied stress $\sigma \leq \sigma_L$ none of the fibres break, although they are elongated by an amount $\delta = \sigma/\kappa$. The dynamics of breaking starts when applied stress σ becomes greater than σ_L. Now, for $\sigma > \sigma_L$ the fraction of unbroken fibres follows a recursion relation (for $\rho(\sigma_{\text{th}})$ as in Fig. 5):

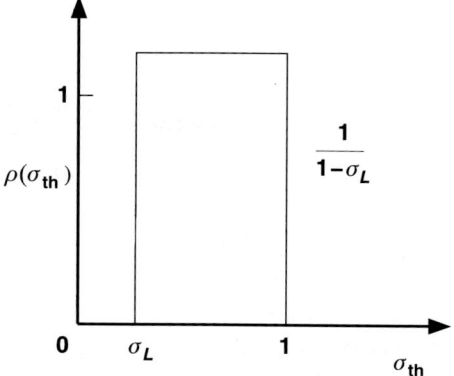

Fig. 5. The fibre breaking strength distribution $\rho(\sigma_{\text{th}})$ considered for studying elastic–plastic-type non-linear deformation behaviour of the equal-load-sharing model

$$U_{t+1} = 1 - \left[\frac{F}{NU_t} - \sigma_L\right]\frac{1}{1-\sigma_L} = \frac{1}{1-\sigma_L}\left[1 - \frac{\sigma}{U_t}\right], \qquad (18)$$

which has stable fixed points:

$$U^*(\sigma) = \frac{1}{2(1-\sigma_L)}\left[1 + \left(1 - \frac{\sigma}{\sigma_f}\right)^{1/2}\right]; \quad \sigma_f = \frac{1}{4(1-\sigma_L)}. \qquad (19)$$

The model now has a critical point $\sigma_f = 1/[4(1-\sigma_L)]$ beyond which total failure of the bundle takes place. The above equation also requires that $\sigma_L \leq 1/2$ (to keep the fraction $U^* \leq 1$). As one can easily see, the dynamics of U_t for $\sigma < \sigma_f$ and also at $\sigma = \sigma_f$ remains the same as discussed in the earlier section. At each fixed point there will be an equilibrium elongation $\delta(\sigma)$ and a corresponding stress $S = U^*\kappa\delta(\sigma)$ develops in the system (bundle). This $\delta(\sigma)$ can be easily expressed in terms of $U^*(\sigma)$. This requires the evaluation of σ^*, the internal stress per fibre developed at the fixed point, corresponding to the initial (external) stress σ ($= F/N$) per fibre applied on the bundle when all the fibres were intact. Expressing the effective stress σ^* per fibre in terms of $U^*(\sigma)$, one can write from (18)

$$U^*(\sigma) = 1 - \frac{\sigma^* - \sigma_L}{(1-\sigma_L)} = \frac{1-\sigma^*}{1-\sigma_L},$$

for $\sigma > \sigma_L$. Consequently,

$$\kappa\delta(\sigma) = \sigma^* = 1 - (1-\sigma_L)U^*(\sigma).$$

It may be noted that the internal stress σ_f^* ($= \sigma_f/U^*(\sigma_f)$) is universally equal to $1/2$ (independent of σ_L; from (19)) at the failure point $\sigma = \sigma_f$ of the bundle. This finally gives the stress–strain relation for the model:

$$S = \begin{cases} \kappa\delta, & 0 \leq \sigma \leq \sigma_L \\ \kappa\delta(1-\kappa\delta)/(1-\sigma_L), & \sigma_L \leq \sigma \leq \sigma_f \\ 0, & \sigma > \sigma_f \end{cases} . \qquad (20)$$

This stress–strain relation is schematically shown in Fig. 6, where the initial linear region has slope κ (the force constant of each fibre). This Hooke's region for stress S continues up to the strain value $\delta = \sigma_L/\kappa$, until which no fibres break ($U^*(\sigma) = 1$). After this, non-linearity appears due to the failure of a few of the fibres and the consequent decrease of $U^*(\sigma)$ (from unity). It finally drops to zero discontinuously by an amount $\sigma_f^* U^*(\sigma_f) = 1/[4(1-\sigma_L)] = \sigma_f$ at the breaking point $\sigma = \sigma_f$ or $\delta = \sigma_f^*/\kappa = 1/2\kappa$ for the bundle. This indicates that the stress drop at the final failure point of the bundle is related to the extent (σ_L) of the linear region of the stress–strain curve of the same bundle.

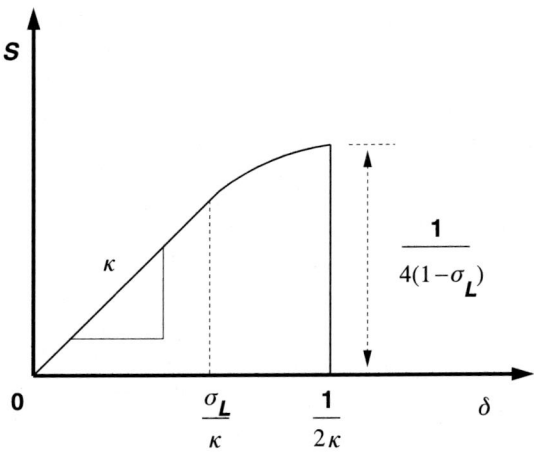

Fig. 6. Schematic stress (S)–strain (δ) curve of the bundle (shown by the *solid line*), following (20), with the fibre strength distribution (17) (as shown in Fig. 5). Note that the model gives analytic solution for the full non-linear stress–strain relationship of the bundle (*disordered solid*), including its failure (*fracture*) stress or strain

Strength of the LLS Fibre Bundles

So far, we studied models with fibres sharing the external load equally. This type of model shows (both analytically and numerically) existence of a critical strength (non-zero σ_f) of the macroscopic bundle [10] beyond which it collapses. The other extreme model, i.e., the LLS model has been proved to be difficult to tackle analytically.

It is clear, however, that the extreme statistics come into play for such LLS models, for which the strength $\sigma_f \to 0$ as the bundle size (N) approaches infinity. Essentially, for any finite load (σ), depending on the fibre strength distribution, the size of the defect cluster can be estimated using Lifshitz argument (see Sect. 2.2) as $\ln N$, giving the failure strength $\sigma_f \sim 1/(\ln N)^a$, where the exponent a assumes a value appropriate for the model (see, e.g., [8]). If a fraction f of the load of the failed fibre goes for global redistribution and the rest (fraction $1-f$) goes to the fibres neighbouring to the failed one, then we see (see Pradhan et al. [10]) that there is a crossover from extreme to self-averaging statistics at a finite value of f.

3 Two-Fractal Overlap Model of Earthquake and Its Statistics

Overlapping fractals form a whole class of models to simulate earthquake dynamics. These models are motivated by the observation that a fault surface, like a fractured surface, is a fractal object [2, 3]. Consequently, a fault, may be viewed as a pair of overlapping fractals. Fractional Brownian profiles have been commonly used as models of fault surfaces [3]. In that case the dynamics of a fault is represented by one Brownian profile drifting on another and each intersection of the two profiles corresponds to an earthquake [23]. However the simplest possible model of a fault – from the fractal point of view – was proposed by Chakrabarti and Stinchcombe [17]. This model is a schematic representation of a fault by a pair of dynamically overlapping Cantor sets. It is not realistic but, as a system of overlapping fractals, it has the essential feature. Since the Cantor set is a fractal with a simple construction procedure, it allows us to study in detail the statistics of the overlap of one fractal object on another. The two fractal overlap magnitude changes in time as one fractal moves over the other. The overlap (magnitude) time series can, therefore, be studied as a model time series of earthquake avalanche dynamics [15].

The statistics of overlaps between two fractals are not studied much yet, although their knowledge is often required in various physical contexts. It has been established recently that since the fractured surfaces have got well-characterized self-affine properties, the distribution of the elastic energies released during the slips between two fractal surfaces (earthquake events) may follow the overlap distribution of two self-similar fractal surfaces [8, 17]. Chakrabarti and Stinchcombe [17] had shown analytically by renormalization group calculations that for regular fractal overlap (Cantor sets and carpets) the contact area distribution $\rho(s)$ follows a simple power law decay:

$$\rho(s) \sim s^{-\gamma}; \quad \gamma = 1 . \tag{21}$$

In this so-called Chakrabarti–Stinchcombe model [18], the solid–solid contact surfaces of both the earth's crust and the tectonic plate are considered as average self-similar fractal surfaces. We then consider the distribution of contact areas, as one fractal surface slides over the other. We relate the total contact area between the two surfaces to be proportional to the elastic strain energy that can be grown during the sticking period, as the solid–solid friction force arises from the elastic strains at the contacts between the asperities. We then consider this energy to be released as one surface slips over the other and sticks again to the next contact or overlap between the rough surfaces. Since the two fractured surfaces are well-known fractals, with established (self-affine) scaling properties (see Fig. 7) Considering that such slips occur at intervals proportional to the length corresponding to that area, we obtain a power law for the frequency distribution of the energy releases. This compares quite well with the Gutenberg–Richter law.

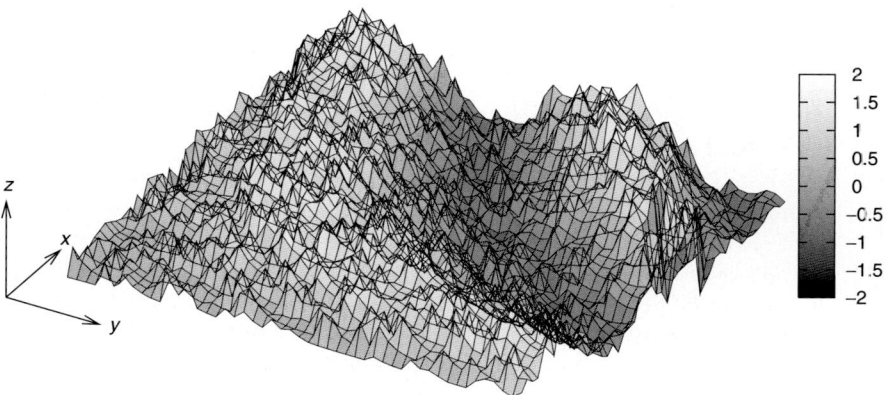

Fig. 7. A typical fracture surface: it has the self-affine scaling property $z(\lambda x, \lambda y) = \lambda^\zeta z(x,y)$ where the roughness exponent ζ has some universal value (e.g., $\zeta \sim 0.8$ for $(2+1)$-dimensional fractured surface)

In order to proceed with the estimate of the number density $n(\epsilon)$ of earthquakes releasing energy ϵ in our model, we first find out the distribution $\rho(s)$ of the overlap or contact area s between two self-similar fractal surfaces. We then relate s with ϵ and the frequency of slips as a function of s, giving finally $n(\epsilon)$. To start with a simple problem of contact area distribution between two fractals, we first take two Cantor sets [17] to model the contact area variations of two (nonrandom and self-similar) surfaces as one surface slides over the other. Figure 8a depicts structure in such surfaces at a scale that corresponds to only the second generation of iterative construction of two displaced Cantor sets, shown in Fig. 8b. It is obvious that with successive iterations, these surfaces will acquire self-similarity at every length scale, when the generation number goes to infinity. We intend to study the distribution of the total overlap s (shown by the shaded regions in Fig. 8b) between the two Cantor sets, in the infinite generation limit.

Let the sequence of generators G_l define our Cantor sets within the interval $[0,1]$: $G_0 = [0,1], G_1 \equiv RG_0 = [0,a] \bigcup [b,1]$ (i.e., the union of the intervals

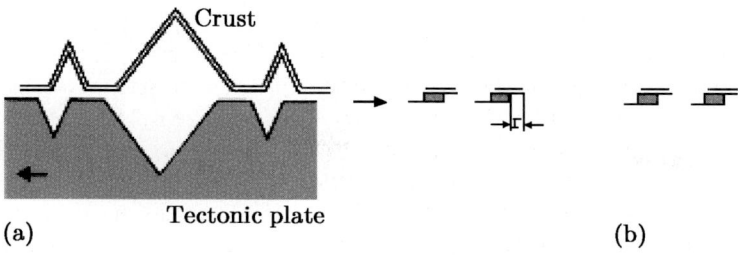

Fig. 8. (a) Schematic representations of a portion of the rough surfaces of the earth's crust and the supporting (moving) tectonic plate. (b) The one-dimensional projection of the surfaces form Cantor sets of varying contacts or overlaps as one surface slides over the other

$[0, a]$ and $[b, 1])$, ..., $G_{l+1} = RG_l$, ... ,. If we represent the mass density of the set G_l by $D_l(r)$, then $D_l(r) = 1$ if r is in any of the occupied intervals of G_l, and $D_l(r) = 0$ elsewhere. The required overlap magnitude between the sets at any generation l is then given by the convolution form $s_l(r) = \int dr' D_l(r') D_l(r - r')$. This form applies to symmetric fractals (with $D_l(r) = D_l(-r)$); in general, the argument of the second D_l should be $D_l(r + r')$.

One can express the overlap integral s_1 in the first generation by the projection of the shaded regions along the vertical diagonal in Fig. 9a. That gives the form shown in Fig. 9b. For $a = b \leq \frac{1}{3}$, the non-vanishing $s_1(r)$ regions do not overlap, and are symmetric on both sides with the slope of the middle curve being exactly double those on the sides. One can then easily check that the distribution $\rho_1(s)$ of overlap s at this generation is given by Fig. 9c, with both c and d greater than unity, maintaining the normalisation of the probability ρ_1 with $cd = 5/3$. The successive generations of the density $\rho_l(s)$ may therefore be represented by Fig. 10, where

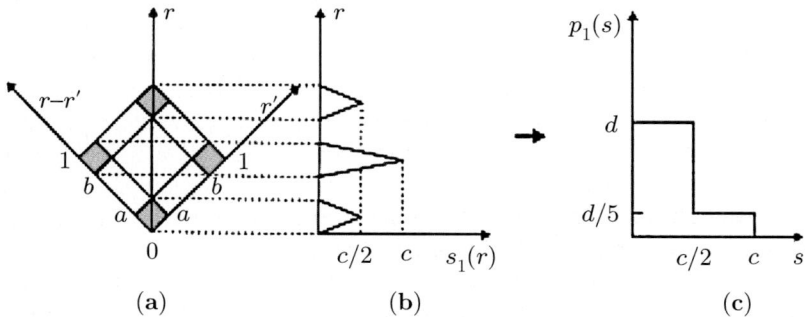

Fig. 9. (a) Two cantor sets (in their first generation) along the axes r and $r - r'$. (b) This gives the overlap $s_1(r)$ along the diagonal. (c) The corresponding density $\rho_1(s)$ of the overlap s at this generation

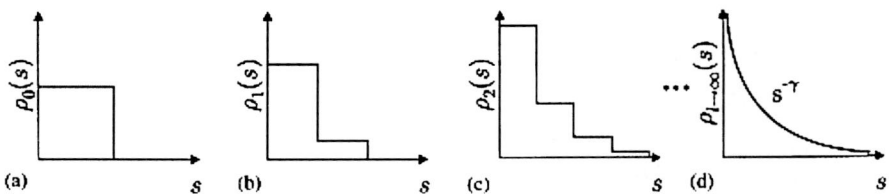

Fig. 10. The overlap densities $\rho(s)$ at various generations of the Cantor sets: at the zeroth (a), first (b), second (c) and at the infinite (or fixed point) (d) generations

$$\rho_{l+1}(s) = \tilde{R}\rho_l(s) \equiv \frac{d}{5}\rho_l\left(\frac{s}{c}\right) + \frac{4d}{5}\rho_l\left(\frac{2s}{c}\right). \tag{22}$$

In the infinite generation limit of the renormalisation group (RG) equation, if $\rho^*(s)$ denotes the fixed point distribution such that $\rho^*(s) = \tilde{R}\rho^*(s)$, then assuming $\rho^*(s) \sim s^{-\gamma}\tilde{\rho}(s)$, one gets $(d/5)c^\gamma + (4d/5)(c/2)^\gamma = 1$. Here $\tilde{\rho}(s)$ represents an arbitrary modular function, which also includes a logarithmic correction for large s. This agrees with the above-mentioned normalisation condition $cd = 5/3$ for the choice $\gamma = 1$. This result for the overlap distribution (21)

$$\rho^*(s) \equiv \rho(s) \sim s^{-\gamma}; \quad \gamma = 1,$$

is the general result for all cases that we have investigated and solved by the functional rescaling technique (with the $\log s$ correction for large s, renormalising the total integrated distribution).

The above study is for continuous relative motion of one Cantor set over the other. Study of the time (t) variation of contact area (overlap) $s(t)$ between two well-characterised fractals having the same fractal dimension as one fractal moves over the other with constant velocity, with discrete (minimum element in that generation) steps has been studied, for finite generations [18]. Bhattacharyya [18] studied this overlap distribution for two Cantor sets with periodic boundary conditions and each having dimension $\log 2 / \log 3$ (see Fig. 11). It was shown, using exact counting, that if $s \equiv 2^{n-k}$ (n is the generation number) then the probability $\tilde{\rho}(s)$ to get an overlap s is given by a binomial distribution [18]

$$\tilde{\rho}(2^{n-k}) = \binom{n}{n-k}\left(\frac{1}{3}\right)^{n-k}\left(\frac{2}{3}\right)^k \sim \exp(-r^2/n); \; r \to 0, \tag{23}$$

where $r^2 = \left[\frac{3}{2}\left(\frac{2}{3}n - k\right)\right]^2$. Expressing therefore r by $\log s$ near the maxima of $\tilde{\rho}(s)$, one can again rewrite (23) as

$$\tilde{\rho}(s) \sim \exp\left(-\frac{(\log s)^2}{n}\right); \; n \to \infty. \tag{24}$$

Noting that $\tilde{\rho}(s)d(\log s) \sim \rho(s)ds$, we find $\rho(s) \sim s^{-\gamma}$, $\gamma = 1$, as in (21), as the binomial or Gaussian part becomes a very weak function of s as $n \to \infty$

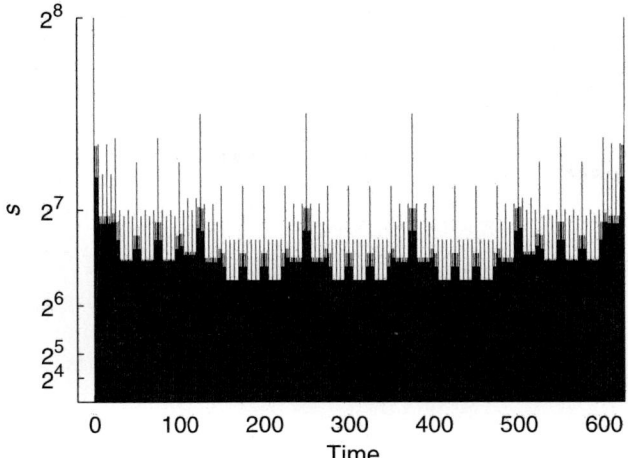

Fig. 11. For two (*regular*) Cantor sets, one moving uniformly over the other (with periodic boundary condition), the total measure of the (*shaded region* in Fig. 8(b) contributing to the) overlap s has the time variation as shown here (for $n = 4$)

[24]. It may be noted that this exponent value $\gamma = 1$ is independent of the dimension of the Cantor sets considered (here $\log 2/\log 3$) or, for that matter, independent of the fractals employed. It also denotes the general validity of (21) even for disordered fractals, as observed numerically [8].

Identifying the contact area or overlap s between the self-similar (fractal) crust and tectonic plate surfaces as the stored elastic energy E released during the slip, the distribution (21), of which a derivation is partly indicated here, reduces to the Gutenberg–Richter law (3) observed.

4 Summary and Discussions

Unlike the elastic constants (e.g., Y in (4)) of a solid, which can be estimated from the interatomic interactions and lattice structures of a solid and while the effect of disorder on them can be accommodated using simple analytic formulas, the fracture strength (σ_f) of a solid cannot be estimated easily from there and such estimates are orders of magnitude higher than those observed. The reason is that cracks nucleate around a defect or disorder in the solid and the variation of σ_f with the defect size l can be precisely estimated in a brittle solid using the Griffith's formula (4) in Sect. 2.1.

For disordered solids, the failure strength distribution $F(\sigma)$ depends on the sample volume and is given by the extreme statistics: Gumbel [2] or Weibull [2] type (see Fig. 2 for their generic behaviour). The average strength of a (finite volume) sample can be estimated from a modified Griffith-like formula

(8). When percolation correlation length exceeds the Lifshitz length, for large disorder, self-averaging statistics takes over (see Sect. 2.2) and the average strength, given by (8), becomes precisely defined (even for infinite system size) and σ_f becomes volume independent (as in the ELS fibre bundle model).

The inherent mean-field nature of the ELS models (discussion in Sect. 2.3) enables us to construct recursion relations ((10) for example) which captures essentially all the intriguing features of the failure dynamics. Although we have identified $O \equiv U^*(\sigma) - U^*(\sigma_f) \propto (\sigma_f - \sigma)^\beta$ as the order parameter (with exponent $\beta = 1/2$) for the continuous transition in such models, unlike in the conventional phase transitions it does not have a real-valued existence for $\sigma > \sigma_f$. The "type" of phase transition in such models has been a controversial issue. Earlier it was suggested to be a first-order phase transition, because the surviving fraction of fibres has a discontinuity at the breakdown point of the bundles. However, as the susceptibility shows divergence ($\chi \propto (\sigma_f - \sigma)^{-\gamma}; \gamma = 1/2$) at the breakdown point, the transition has been later identified to be of second order [10]. The dynamic critical behaviour of these models and the universality of the exponent values are straightforward. Here, divergence of relaxation time (τ) at the critical point ($\tau \propto (\sigma_f - \sigma)^{-\alpha}; \alpha = 1/2$) indicates "critical slowing" of the dynamics that is characteristic of conventional critical phenomena. At the critical point, one observes power law decay of the surviving fraction in time ($U_t(\sigma_f) \propto t^{-\delta}; \delta = 1$). We demonstrated the universality of the failure behaviour near $\sigma = \sigma_f$, for three different distributions: uniform, linearly increasing and linearly decreasing distributions of fibre strength. The critical strengths of the bundles differ in each case: $\sigma_f = 1/4, \sqrt{4/27}$ and $4/27$, respectively, for these three distributions. However, the critical behaviour of the order parameter O, susceptibility χ, relaxation time τ and of the time decay at σ_f, as given by the exponents β, γ, α and δ remain unchanged: $\alpha = 1/2 = \beta = \gamma$ and $\delta = 1$ for all three distributions.

The model also shows realistic nonlinear deformation behaviour with a shifted (by σ_L, away from the origin) uniform distribution of fibre strengths (see Sect. 2.3.2). The stress–strain curve for the model clearly shows three different regions: elastic or linear part (Hooke's region) when none of the fibres break ($U^*(\sigma) = 1$), plastic or nonlinear part due to the successive failure of the fibres ($U^*(\sigma) < 1$) and then finally the stress drops suddenly (due to the discontinuous drop in the fraction of surviving fibres from $U^*(\sigma_f)$ to zero) at the failure point $\sigma_f = 1/[4(1-\sigma_L)]$. This non-linearity in the response (stress–strain curve in Fig. 1.6) results from the linear response of the surviving fibres who share the extra load uniformly. The LLS bundles (see Sect. 2.3.3), on the other hand, show "zero" critical strength as the bundle size goes to infinity in one dimension (extreme statistic stakes over). It is not clear at this stage if, in higher dimensions, LLS bundles are going to have non-zero critical strength. In any case, the associated dynamics of failure of these higher dimensional bundles with variable range load transfer should be interesting.

We believe, the elegance and simplicity of the model, its common-sense appeal, the exact solubility of its critical behaviour in the mean field (ELS) limit, its demonstrated universality, etc., would promote the model eventually to a level competing with the Ising model of magnetic critical behaviour.

As emphasized already, we consider the the Gutenberg-Richter law (3) and similar power-laws in geophysics (see, e.g., [21]) to be very significant. Like the previous attempts [14, 15, 16], the model developed here [17] captures this important feature in its resulting statistics. Here, the established self-similarity of the fault planes are captured using fractals, Cantor sets in particular. Hence we consider in Sect. 3 this "Chakrabarti–Stinchcombe" model [18], where one Cantor set moves uniformly over another similar set (with periodic boundary conditions). The resulting overlap s (meaning the set of real numbers common in both the Cantor sets) changes with time: see, for example, Fig. 11 for a typical time variation overlap s for $n = 4$. The number density of such overlaps seems to follow a Gutenberg–Richter-type law (21). Judging from the comparisons of the exponent values α in (3) and γ in (21), the model succeeds at least as well as the earlier ones. More importantly, our model incorporates both the geologically observed facts: fractal nature of the contact surfaces of the crust and of the tectonic plate, and the stick-slip motion between them. However, the origin of the power law in the quake statistics here is the self-similarity of the fractal surfaces, and not any self-organisation directly in their dynamics. In fact, the extreme non-linearity in the nature of the crack propagation is responsible for the fractal nature of the rough crack surfaces of the crust and the tectonic plate. This, in turn, leads here to the Gutenberg–Richter-like power law in the earthquake statistics.

Acknowledgements

The author thanks M. Acharyya, K.K. Bardhan, L.G. Benguigui, P. Bhattacharyya, A. Chatterjee, D. Chowdhury, M.K. Dey, A. Hansen, S.S. Manna, S. Pradhan, P. Ray, D. Stauffer and R.B. Stinchcombe for collaborations at different stages.

References

1. B. Lawn, *Fracture of Brittle Solids*, Cambridge University Press, Cambridge (1993).
2. B.K. Chakrabarti and L.G. Benguigui, *Statistical Physics of Fracture and Breakdown in Disorder Systems*, Oxford University Press, Oxford (1997).
3. H.J. Herrmann and S. Roux (Eds.), *Statistical Models for the Fracture of Disordered Media*, Elsevier, Amsterdam (1990); M. Sahimi, *Heterogeneous Materials*, Vol. II, Springer, New York (2003).
4. D. Stauffer and A. Aharony, *Introduction to Percolation Theory*, Taylor and Francis, London (1992).

5. A.A. Griffith, Phil. Trans. Roy. Soc. London A **221** 163 (1920).
6. F.T. Pierce, J. Textile Inst. **17**, T355–368 (1926).
7. H.E. Daniels, Proc. R. Soc. London A **183** 405 (1945); S.L. Phoenix, SIAM J. Appl. Math. **34** 227 (1978); Adv. Appl. Prob. **11** 153 (1979).
8. S. Pradhan and B.K. Chakrabarti, Int. J. Mod. Phys. B **17** 5565 (2003).
9. P.C. Hemmer and A. Hansen, J. Appl. Mech. **59** 909 (1992); M. Kloster, A. Hansen and P.C. Hemmer, Phys. Rev. E **56** 2615 (1997); S. Pradhan, A. Hansen and P.C. Hemmer, Phys. Rev. Lett. **95** 125501 (2005); F. Raischel, F. Kun and H.J. Herrmann, cond-mat/0601290 (2006).
10. S. Pradhan and B.K. Chakrabarti, Phys. Rev. E **65** 016113 (2001); S. Pradhan, P. Bhattacharyya and B.K. Chakrabarti, Phys. Rev. E **66** 016116 (2002); P. Bhattacharyya, S. Pradhan and B.K. Chakrabarti, Phys. Rev. E **67** 046122 (2003).
11. R.C. Hidalgo, F. Kun and H.J. Herrmann, Phys. Rev. E **64** 066122 (2001); S. Pradhan, B.K. Chakrabarti and A. Hansen, Phys. Rev. E **71** 036149 (2005).
12. B. Gutenberg and C.F. Richter, *Seismicity of the Earth and Associated Phenomena*, Princeton University Press, Princeton, NJ (1954).
13. L. Knopoff, Proc. Natl. Acad. Sci. USA **97** 11880 (2000); Y.Y. Kagan, Physica D **77** 160 (1994); C.H. Scholz, *The Mechanics of Earthquake and Faulting*, Cambridge University Press, Cambridge (1990); B.V. Kostrov and S. Das, *Principles of Earthquake Source Mechanics*, Cambridge University Press, Cambridge (1988).
14. R. Burridge and L. Knopoff, Bull. Seis. Soc. Am. **57** 341–371 (1967).
15. J.M. Carlson and J.S. Langer, Phys. Rev. Lett. **62** 2632–2635 (1989); J.M. Carlson, J.S. Langer and B.E. Shaw, Rev. Mod. Phys. **66** 657–670 (1994); G. Ananthakrishna and H. Ramachandran in *Nonlinearity and Breakdown in Soft Condensed Matter*, Eds. K.K. Bardhan, B.K. Chakrabarti and A. Hansen, LNP **437**, Springer Verlag, Heidelberg (1994); T. Mori and H. Kawamura, Phys. Rev. Letts. **94** 058501 (2005).
16. P. Bak, *How Nature Works*, Oxford University Press, Oxford (1997).
17. B.K. Chakrabarti and R.B. Stinchcombe, Physica A **270** 27 (1999); S. Pradhan, B.K. Chakrabarti, P. Ray and M.K. Dey, Phys. Scripta T **106** 77 (2003).
18. P. Bhattacharyya, Physica A **348** 199 (2005).
19. D.J. Bergman and D. Stroud, in *Solid State Physics*, **46** Eds. H. Ehrenreich and D. Turnbull, Academic Press, New York, p. 147 (1992).
20. S. Pradhan and B. K. Chakrabarti, Phys. Rev. E **67** 046124 (2003)
21. D. Sornette, *Critical Phenomena in Natural Sciences, Chaos, Fractals, Self-Organization and Disorder: Concepts and Tools*, 2nd Ed., Springer, Heidelberg (2004)
22. A. Politi, S. Ciliberto and R. Scorretti, Phys. Rev. E **66** 026107 (2002)
23. V. de Rubeis, R. Hallgass, V. Loreto, G. Paladin, L. Pietronero and P. Tosi, Phys. Rev. Lett. **76** 2599 (1996).
24. B.K. Chakrabarti and A. Chatterjee, in Proc. The Seventh International Conference on Vibration Problems ICOVP-2005, Istanbul, Ed. E. Inan (Springer, 2006), arXiv:cond-mat/0512136; P. Bhattacharyya, A. Chatterjee and B.K. Chakrabarti, arXiv:physics/0510038.

Rupture Processes in Fibre Bundle Models

P.C. Hemmer, A. Hansen, and S. Pradhan

Department of Physics, Norwegian University of Science and Technology,
7491 Trondheim, Norway.
per.hemmer@ntnu.no
alex.hansen@ntnu.no
pradhan.srutarshi@ntnu.no

1 Introduction

Fibre bundles with statistically distributed thresholds for breakdown of individual fibres are interesting models of the statics and dynamics of failures in materials under stress. They can be analyzed to an extent that is not possible for more complex materials. During the rupture process in a fibre bundle avalanches, in which several fibres fail simultaneously, occur. We study by analytic and numerical methods the statistics of such avalanches, and the breakdown process for several models of fibre bundles. The models differ primarily in the way the extra stress caused by a fibre failure is redistributed among the surviving fibres.

When a rupture occurs somewhere in an elastic medium, the stress elsewhere is increased. This may in turn trigger further ruptures, which can cascade to a final complete breakdown of the material. To describe or model such breakdown processes in detail for a real material is difficult, due to the complex interplay of failures and stress redistributions. Few analytic results are available, so computer simulations is the main tool (see [1, 2, 3] for reviews). Fibre bundle models, on the other hand, are characterized by simple geometry and clear-cut rules for how the stress caused by a failed element is redistributed on the intact fibres. The attraction and interest of these models lies in the possibility of obtaining exact results, thereby providing inspiration and reference systems for studies of more complicated materials.

In this review we survey theoretical and numerical results for several models of bundles of N elastic and parallel fibres, clamped at both ends, with statistically distributed thresholds for breakdown of individual fibres (Fig. 1). The individual thresholds x_i are assumed to be independent random variables with the same cumulative distribution function $P(x)$ and a corresponding density function $p(x)$:

$$\text{Prob}(x_i < x) = P(x) = \int_0^x p(u)\, du \ . \tag{1}$$

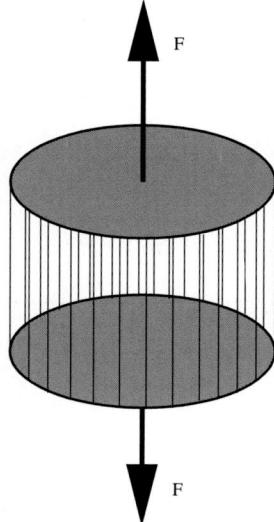

Fig. 1. A fibre bundle of N parallel fibres clamped at both ends. The externally applied force is F

Whenever a fibre experiences a force equal to or greater than its strength threshold x_i, it breaks immediately and does not contribute to the strength of the bundle thereafter. The maximal load the bundle can resist before complete breakdown of the whole bundle is called the *critical* load. The models differ in the probability distribution of the thresholds. Two popular examples of threshold distributions are the uniform distribution

$$P(x) = \begin{cases} x/x_r & \text{for} \quad 0 \leq x \leq x_r \\ 1 & \text{for} \quad x > x_r \,, \end{cases} \qquad (2)$$

and the Weibull distribution

$$P(x) = 1 - \exp(-(x/x_r)^k) \,. \qquad (3)$$

Here $x \geq 0$, x_r is a reference threshold and the dimensionless number k is the Weibull index (Fig. 2).

Much more fundamental, however, is the way the models differ in the mechanism for how the extra stress caused by a fibre failure is redistributed among the unbroken fibres. The simplest models are the equal-load-sharing models, in which the load previously carried by a failed fibre is shared equally by all the remaining intact bonds in the system. That some exact results could be extracted for this model was demonstrated by Daniels [4] in a classic work some sixty years ago. Local-load-sharing models, on the other hand, are relevant for materials in which the load originally carried by a failed fibre is shared by the surviving fibres in the immediate vicinity of the ruptured fibre.

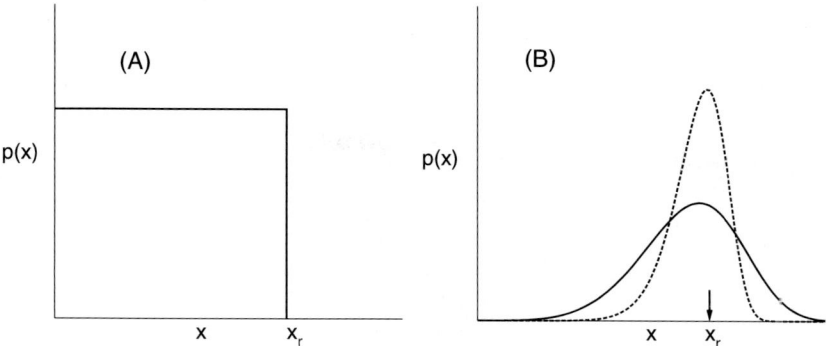

Fig. 2. The uniform distribution (**A**) and Weibull distribution (**B**) with $k = 5$ (*solid line*) and $k = 10$ (*dotted line*)

The main property of the fibre bundle breakdown process to be studied in the present review is the distribution of the sizes of the burst avalanches. The *burst distribution* $D(\Delta)$ is defined as the expected number of bursts in which Δ fibres break simultaneously when the bundle is stretched until complete breakdown. For the equal-load-distribution models that we consider in Sect. 2 Hemmer and Hansen [5] showed that the generic result is a power law,

$$\lim_{N \to \infty} \frac{D(\Delta)}{N} \propto \Delta^{-\xi}, \qquad (4)$$

for large Δ, with $\xi = 5/2$. In Sect. 2.2 we will show, however, that for some rather unusual threshold distributions the power law (4) is not obeyed. More importantly, we show in Sect. 2.4 that when the whole bundle at the outset is close to being critical, the exponent ξ crosses over to the value $\xi = 3/2$. In Sect. 2.5 we pay particular attention to the rupture process at criticality, i.e., just before the whole bundle breaks down.

The average strength of the bundle for a given load can be viewed as the result of a sequential process. In the first step all fibres that cannot withstand the applied load break. Then the stress is redistributed on the surviving fibres, which compels further fibres to burst. This starts an iterative process that continues until equilibrium is reached, or all fibres fail. When equilibrium exists, it characterizes the *average* strength of the bundle for the given load. This recursive dynamics can be viewed as a fixed-point problem, with interesting properties when the critical load is approached. We review such recursive dynamics in Sect. 2.6.

For other stress redistribution principles than equal-load-sharing, the avalanche distributions are different from the power law (4). In Sect. 3 we study examples of such systems. Special cases are local-stress-distribution models in which the surviving nearest neighbors to a failed fibre share all the extra stress, and a model in which the fibres are anchored to an elastic clamp.

2 Equal-Load-Sharing Fibre Bundles

This is the fibre-bundle model with the longest history. It was used by Pierce, in the context of testing the strength of cotton yarn [6]. The basic assumptions are that the fibres obey Hookean elasticity right up to the breaking point, and that the load distributes itself *evenly* among the surviving fibres. The model with this democratic load redistribution is similar to mean-field models in statistical physics.

At a force x per surviving fibre, the total force on the bundle is

$$F(x) = Nx[1 - \phi(x)], \qquad (5)$$

where $\phi(x)$ is the fraction of failed fibres. In Fig. 3 we show an example of a F vs. x. We have in mind an experiment in which the force F, our control parameter (Fig. 1), is steadily increasing. This implies that not all parts of the $F(x)$ curve are physically realized. The experimentally relevant function is

$$F_{ph}(x) = \text{LMF } F(x), \qquad (6)$$

the least monotonic function not less than $F(x)$. A horizontal part of $F_{ph}(x)$ corresponds to an avalanche, the size of which is characterized by the number of maxima of $F(x)$ within the corresponding range of x (Fig. 3).

It is the fluctuations in $F(x)$ that create avalanches. For a large sample the fluctuations will be small deviations from the average macroscopic characteristics $\langle F \rangle$. This *average* total force is given by

$$\langle F \rangle(x) = Nx[1 - P(x)]. \qquad (7)$$

Let us for the moment assume that $\langle F \rangle(x)$ has a single maximum. The maximum corresponds to the value $x = x_c$ for which $d\langle F \rangle/dx$ vanishes. This gives

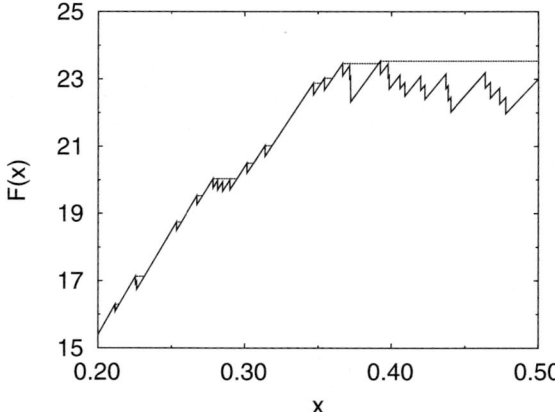

Fig. 3. $F(x)$ vs. x curve. Avalanches are shown as *horizontal lines*

$$1 - P(x_c) - x_c\, p(x_c) = 0 \,. \tag{8}$$

The threshold x_c corresponding to the maximum of F is denoted the *critical threshold*. Because of fluctuations, however, the maximum value of the force may actually occur at a slightly different value of x.

2.1 Burst Distribution: The Generic Case

That the rupture process produces a power-law decay of the burst distribution $D(\Delta)$ is seen at once by simulation experiments. Figure 4 shows results for the uniform threshold distribution (2) and the Weibull distribution (3) with index $k = 5$.

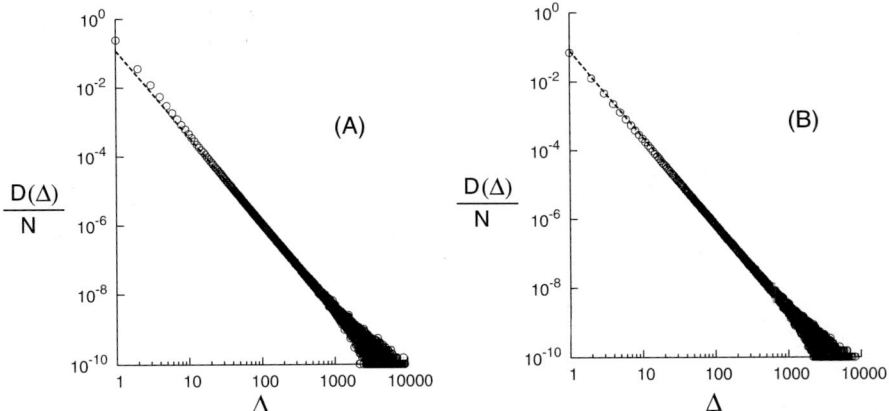

Fig. 4. The burst distribution $D(\Delta)/N$ for the uniform distribution (**A**) and the Weibull distribution with index 5 (**B**). The *dotted lines* represent the power law with exponent $-5/2$. Both figures are based on 20000 samples of bundles each with $N = 10^6$ fibres

In order to derive analytically the burst distribution, let us start by considering a small threshold interval $(x, x + dx)$ in a range where the average force $\langle F \rangle(x)$ increases with x. For a large number N of fibres the expected number of surviving fibres is $N[1 - P(x)]$. And the threshold values in the interval, of which there are $Np(x)dx$, will be Poisson distributed. When N is arbitrarily large, the burst sizes can be arbitrarily large in any finite interval of x.

Assume that an infinitesimal increase in the external force results in a break of a fibre with threshold x. Then the load that this fibre suffered, will be redistributed on the $N[1 - P(x)]$ remaining fibres; thus they experience a load increase

$$\delta x = \frac{x}{N[1 - P(x)]} \,. \tag{9}$$

The *average* number of fibres that break as a result of this load increase is

$$a = a(x) = Np(x) \cdot \delta x = \frac{xp(x)}{1 - P(x)} . \qquad (10)$$

For a burst of size Δ the increase in load per fibre will be a factor Δ larger than the quantity (9), and an average number $a(x)\Delta$ will break. The probability that precisely $\Delta - 1$ fibres break as a consequence of the first failure is given by a Poisson distribution with this average, i.e., it equals

$$\frac{(a\Delta)^{\Delta-1}}{(\Delta - 1)!} e^{-a\Delta} . \qquad (11)$$

This is not sufficient, however. We must ensure that the thresholds for these $\Delta - 1$ fibres are not so high that the avalanche stops before reaching size Δ. This requires that at least n of the thresholds are in the interval $(x, x + n\delta x)$, for $1 \leq n \leq \Delta - 1$. In other words, if we consider the Δ intervals $(x, x + \delta x)$, $(x + \delta x, x + 2\delta x)$, ..., $(x + (\Delta - 1)\delta, x + \Delta\delta x)$, we must find at most $n - 1$ thresholds in the n last intervals. There is the same a priori probability to find a threshold in any interval. The solution to this combinatorial problem is given in [7]. The resulting probability to find all intermediate thresholds weak enough equals $1/\Delta$. Combining this with (11), we have for the probability $\phi(\Delta, x)$ that the breaking of the first fibre results in a burst of size Δ:

$$\phi(\Delta, x) = \frac{\Delta^{\Delta-1}}{\Delta!} a(x)^{\Delta-1} e^{-a(x)\Delta} . \qquad (12)$$

This gives the probability of a burst of size Δ, as a consequence of a fibre burst due to an infinitesimal increase in the external load. However, we still have to ensure that the burst actually *starts* with the fibre in question and is not part of a larger avalanche starting with another, weaker, fibre. Let us determine the probability $P_b(x)$ that this initial condition is fulfilled.

For that purpose consider the $d-1$ fibres with the largest thresholds below x. If there is no strength threshold in the interval $(x - \delta x, x)$, at most one threshold value in the interval $(x - 2\delta x, x)$, ..., at most $d - 1$ values in the interval $(x - d\delta x, x)$, then the fibre bundle can not at any of these previous x-values withstand the external load that forces the fibre with threshold x to break. The probability that there are precisely h fibre thresholds in the interval $(x - \delta x\, d, x)$ equals

$$\frac{(ad)^h}{h!} e^{-ad} .$$

Dividing the interval into d subintervals each of length δx, the probability $p_{h,d}$ that these conditions are fulfilled is exactly given by $p_{h,d} = 1 - h/d$ (see [7]). Summing over the possible values of h, we obtain the probability that the avalanche can not have started with the failure of a fibre with any of the d nearest-neighbor threshold values below x:

$$P_b(x|d) = \sum_{h=0}^{d-1} \frac{(ad)^h}{h!} e^{-ad} \left(1 - \frac{h}{d}\right) = (1-a)e^{-ad}\sum_{h=0}^{d-1}\frac{(ad)^h}{h!} + \frac{(ad)^d}{d!} e^{-ad} .\tag{13}$$

Finally we take the limit $d \to \infty$, for which the last term vanishes. For $a > 1$ the sum must vanish since the left-hand side of (13) is non-negative, while the factor $(1-a)$ is negative. For $a < 1$, on the other hand, we find

$$P_b(x) = \lim_{d\to\infty} P_b(x|d) = 1 - a ,\tag{14}$$

where $a = a(x)$. The physical explanation of the different behavior for $a > 1$ and $a \leq 1$ is straightforward: The maximum of the total force on the bundle occurs at x_c for which $a(x_c) = 1$, see (8) and (10), so that $a(x) > 1$ corresponds to x values almost certainly involved in the final catastrophic burst. The region of interest for us is therefore when $a(x) \leq 1$, where avalanches on a microscopic scale occur. This is accordance with what we found in the beginning of this section, viz. that thin burst of a fibre with threshold x leads immediately to a average number $a(x)$ of additional failures.

Summing up, we obtain the probability that the fibre with threshold x is the first fibre in an avalanche of size Δ as the product

$$\Phi(x) = \phi(\Delta, x) P_b(x) = \frac{\Delta^{\Delta-1}}{\Delta!} a(x)^{\Delta-1} e^{-a(x)\Delta}[1 - a(x)] ,\tag{15}$$

where $a(x)$ is given by (10),

$$a(x) = \frac{x\,p(x)}{1 - P(x)} .$$

Since the number of fibres with threshold values in $(x, x+\delta x)$ is $N\,p(x)\,dx$, the burst distribution is given by

$$\frac{D(\Delta)}{N} = \frac{1}{N}\int_0^{x_c} \Phi(x) p(x)\,dx = \frac{\Delta^{\Delta-1}}{\Delta!}\int_0^{x_c} a(x)^{\Delta-1} e^{-a(x)\Delta}[1 - a(x)]\,p(x)\,dx .\tag{16}$$

For large Δ the maximum contribution to the integral comes from the neighborhood of the upper integration limit, since $a(x)\,e^{-a(x)}$ is maximal for $a(x) = 1$, i.e., for $x = x_c$. Expansion around the saddle point, using

$$a^\Delta e^{-a\Delta} = \exp\left[\Delta\left(-1 - \frac{1}{2}(1-a)^2 + \mathcal{O}(1-a)^3\right)\right] ,\tag{17}$$

as well as $a(x) \simeq 1 + a'(x_c)(x - x_c)$, produces

$$\frac{D(\Delta)}{N} = \frac{\Delta^{\Delta-1} e^{-\Delta}}{\Delta!} a'(x_c) \int_0^{x_c} p(x_c)\, e^{-a'(x_c)^2 (x_c-x)^2 \Delta/2} (x - x_c)\,dx .\tag{18}$$

The integration yields the asymptotic behavior

$$D(\Delta)/N \propto \Delta^{-5/2}, \tag{19}$$

universal for those threshold distributions for which the assumption of a single maximum of $\langle F \rangle(x)$ is valid.

Note that if the experiment had been stopped before complete breakdown, at a force per fibre x less than x_c, the asymptotic behavior would have been dominated by an *exponential* fall-off rather than a power law:

$$D(\Delta)/N \propto \Delta^{-5/2} e^{-[a(x) - 1 - \ln a(x)]\Delta}. \tag{20}$$

When x is close to x_c the exponent is proportional to $(x_c - x)^2 \Delta$. The burst distribution then takes the scaling form

$$D(\Delta) \propto \Delta^{-\eta} G(\Delta^\nu (x_c - x)), \tag{21}$$

with a Gaussian function G, a power law index $\eta = \frac{5}{2}$ and $\nu = \frac{1}{2}$. Thus the breakdown process is similar to critical phenomena with a critical point at total breakdown [5, 8, 26].

2.2 Burst Distribution: Nongeneric Cases

What happens when the average strength curve, $\langle F \rangle(x)$, does *not* have a unique maximum? There are two possibilities: (i) it has *several* parabolic maxima, or (ii) it has *no* parabolic maxima.

When there are several parabolic maxima, and the absolute maximum does not come first (i.e. at the lowest x value), then there will be several avalanche series each terminating at a local critical point with an accompanying burst of macroscopic size, while the breakdown of the bundle occurs when the absolute maximum is reached [9]. The power law asymptotics (19) of the avalanche distribution is thereby unaffected, however.

The second possibility, that the average strength curve has no parabolic maxima, is more interesting. We present here two model examples of such threshold distributions.

The threshold distribution for model I is, in dimensionless units,

$$P(x) = \begin{cases} 0 & \text{for } x \leq 2 \\ 1 - (x-1)^{-1/2} & \text{for } x > 2, \end{cases} \tag{22}$$

while model II corresponds to

$$P(x) = \begin{cases} 0 & \text{for } x \leq 1 \\ 1 - x^{-\alpha} & \text{for } x > 1, \end{cases} \tag{23}$$

with a positive parameter α.

For the two models the corresponding macroscopic bundle strength per fibre is, according to (16),

$$\frac{\langle F \rangle}{N} = \begin{cases} x & \text{for } x \le 2 \\ \dfrac{x}{\sqrt{x-1}} & \text{for } x > 2 \end{cases} \qquad (24)$$

for model I, and

$$\frac{\langle F \rangle}{N} = \begin{cases} x & \text{for } x \le 1 \\ x^{1-\alpha} & \text{for } x > 1 \end{cases} \qquad (25)$$

for model II (see Fig. 5).

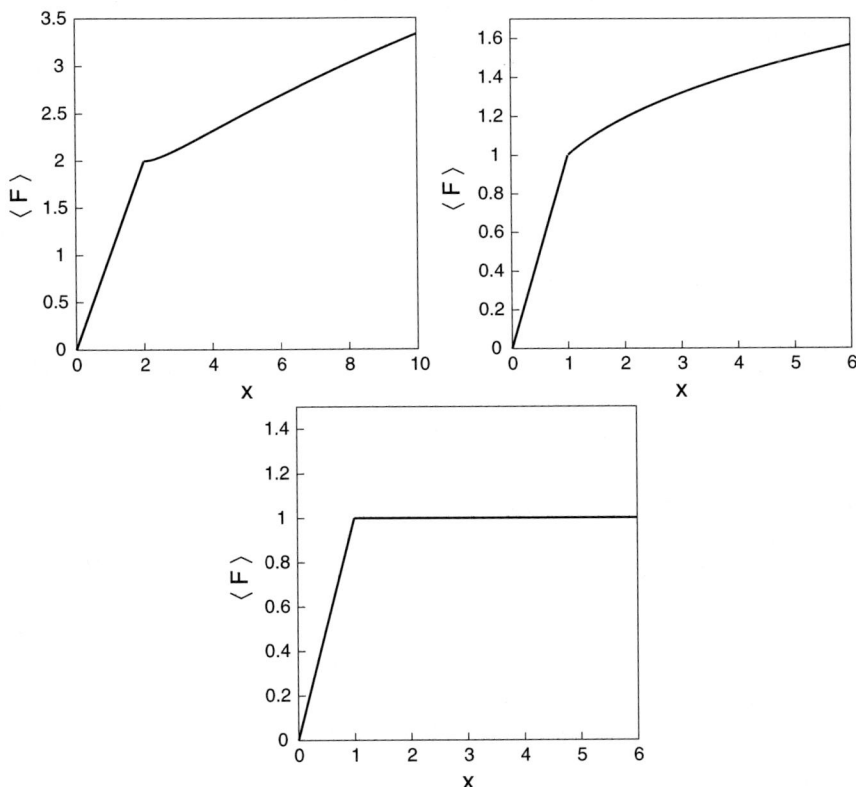

Fig. 5. Average force on the fibre bundle for model I (*upper left*), model II for $\alpha = 3/4$ (*upper right*), and model II in the limit $\alpha \to 1$ (*bottom figure*)

To calculate the avalanche distribution we use (16), in both cases with $x_c = \infty$ as the upper limit in the integration. For Model I the $\langle F \rangle$ graph has at $x = 2+$ an extremum, viz. a minimum. At the minimum we have $a(x) = \dfrac{x}{2(x-1)} = 1$. Since the Δ-dependent factor $a^\Delta e^{-a\Delta}$ in the integrand of (16) has a maximum for $a = 1$, we obtain

$$D(\Delta)/N \propto \Delta^{-5/2} \tag{26}$$

for large Δ. Even if the macroscopic load curve does not have a maximum at any finite x in this case, the generic power law (19) holds.

For model II (16) gives

$$\frac{D(\Delta)}{N} = \frac{1-\alpha}{\alpha} \frac{\Delta^{\Delta-1}}{\Delta!} \left[\alpha e^{-\alpha}\right]^\Delta \propto \Delta^{-3/2} \left[\alpha e^{1-\alpha}\right]^\Delta . \tag{27}$$

For $\alpha = 3/4$ as in Fig. 5, or more generally $\alpha < 1$, the avalanche distribution does *not* follow a power law, but has an exponential cut-off in addition to a $\Delta^{-3/2}$ dependence.

When $\alpha \to 1$, the average force (25) approaches a constant for $x > 1$, and the burst distribution (27) approaches a power law

$$\frac{D(\Delta)}{N} \propto \Delta^{-3/2} , \tag{28}$$

a result easily verified by simulation on the system with $P(x) = 1 - 1/x$ for $x \geq 1$. That a power law with exponent $3/2$, different from the generic burst distribution (19), appears when

$$\frac{d}{dx}\langle F \rangle \to 0 , \tag{29}$$

will be apparent when we in Sect. 2.5 study what happens at criticality.

2.3 Mapping onto a Random Walk Problem

Let F_k be the force on the bundle when the kth fibre fails. It is the non-monotonicities in the sequence $F_1, F_2, \ldots$ that produce avalanches of size $\Delta > 1$. Let us consider the probability distribution of the force increase $\Delta F = F_{k+1} - F_k$ between two consecutive bursts, the first taking place at a force $x = x_k$ per fibre, so that $F_k = (N - k + 1)x$.

Since $\Delta F = (N - k)(x_{k+1} - x) - x$, it follows that

$$\Delta F \geq -x . \tag{30}$$

The probability to find the $k + 1$th threshold in the interval $(x_{k+1}, x_{k+1} + dx_{k+1})$, for given $x = x_k$, equals

$$(N - k - 1)\frac{[1 - P(x_{k+1})]^{N-k-2}}{[1 - P(x)]^{N-k-1}} p(x_{k+1}) \, dx_{k+1} . \tag{31}$$

By use of the connection $x_{k+1} = x + (\Delta F + x)/(N - k)$ this probability density for x_{k+1} is turned into the probability density $\rho(\Delta F) \, d\Delta F$ for ΔF:

$$\rho(\Delta F) = \frac{N-k-1}{N-k} \frac{[1-P(x+(\Delta F+x)/(N-k))]^{N-k-2}}{[1-P(x)]^{N-k-1}} p\left(x + \frac{\Delta F + x}{N-k}\right), \quad (32)$$

which is properly normalized to unity. For large $N - k$ this simplifies to

$$\rho(\Delta F) = \begin{cases} 0 & \text{for } \Delta F < -x \\ \frac{p(x)}{1-P(x)} \exp\left[-\frac{(\Delta F + x)}{1-P(x)} p(x)\right] & \text{for } \Delta F \geq -x \end{cases}. \quad (33)$$

The values $F_1, F_2, F_3, \ldots$ of the force on the bundle can be considered as the positions of a random walker with the probability $\rho(\Delta F)$ for the length of the next step [8]. It is a random walk of an unusual type: The step length is variable, with the steps in the negative direction are limited in size (Fig. 6). In general the walk is *biased* since

$$\langle \Delta F \rangle = \int \Delta F\, \rho(\Delta F)\, d\Delta F = \frac{1 - P(x) - xp(x)}{p(x)} \quad (34)$$

is zero, e.g. unbiased, *only* at the critical threshold x_c, given by (8). That the random walk is unbiased at criticality is to be expected, of course, since the average bundle strength $\langle F \rangle$ as function of x is stationary here.

The probability that ΔF is positive equals

$$\text{Prob}(\Delta F > 0) = \int_0^\infty \rho(\Delta F)\, d\Delta F = \exp\left[-\frac{xp(x)}{1-P(x)}\right]. \quad (35)$$

That ΔF is positive implies that the burst has the length $\Delta = 1$. The result (35) is identical to the previously determined probability $\phi(1, x)$, (12), for a burst of length 1, when we have not ensured that the burst actually *starts* with the fibre in question and is not part of a larger avalanche.

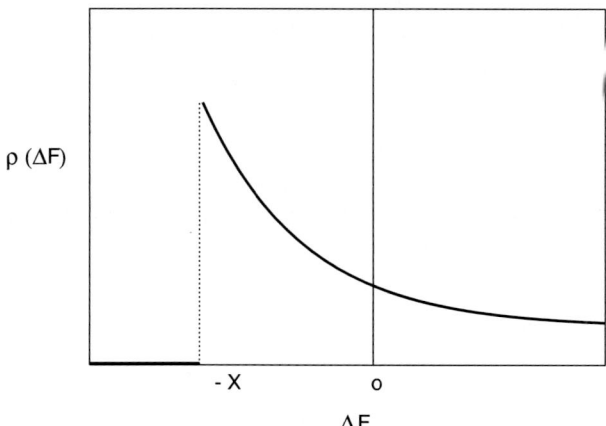

Fig. 6. The probability distribution $\rho(\Delta F)$ of the step length in the random walk

In Sect. 2.5 we will see that the random-walk analogy can be used in a quantitative way to predict the avalanche distribution power-law exponent at criticality.

2.4 Crossover Behavior Near Criticality

When *all* fibre failures are recorded we have seen that the burst distribution $D(\Delta)$ follows the asymptotic power law $D \propto \Delta^{-5/2}$. If we just sample bursts that occur near criticality, a different behavior is seen [10, 11]. As an illustration we consider the uniform threshold distribution, and compare the complete burst distribution with what one gets when one samples merely bursts from breaking fibres in the threshold interval $(0.9x_c, x_c)$. Figure 7 shows clearly that in the latter case a different power law is seen.

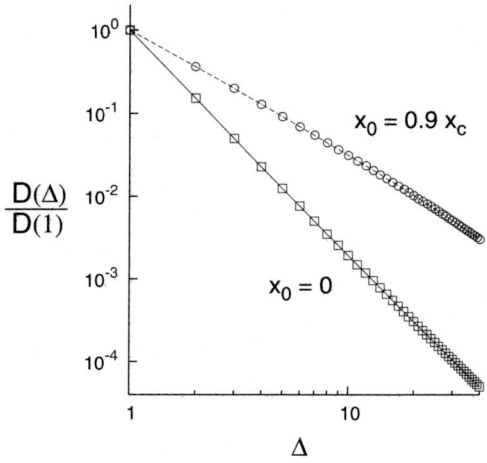

Fig. 7. The distribution of bursts for threshold's uniformly distributed in an interval (x_0, x_c), with $x_0 = 0$ and with $x_0 = 0.9x_c$. The figure is based on 50 000 samples, each with $N = 10^6$ fibres

If we want to study specifically the contribution from failures occurring when the bundle is nearly critical, we evaluate the expression (18) for the burst distribution over a small interval (x_0, x_c), rather than integrating from 0 to x_c. The argument in Sect. 2.1 that the major contribution to the integral comes from the critical neighborhood is still valid. We obtain

$$\frac{D(\Delta)}{N} = \frac{\Delta^{\Delta-2} e^{-\Delta}}{\Delta!} \frac{p(x_c)}{a'(x_c)} \left[1 - e^{-\Delta/\Delta_c}\right], \tag{36}$$

with

$$\Delta_c = \frac{2}{a'(x_c)^2 (x_c - x_0)^2}. \tag{37}$$

By use of Stirling approximation $\Delta! \simeq \Delta^\Delta e^{-\Delta}\sqrt{2\pi\Delta}$, – a reasonable approximation even for small Δ – the burst distribution (36) may be written

$$\frac{D(\Delta)}{N} = C\Delta^{-5/2}\left(1 - e^{-\Delta/\Delta_c}\right), \tag{38}$$

with a nonzero constant

$$C = (2\pi)^{-1/2}p(x_c)/a'(x_c). \tag{39}$$

We see from (38) that there is a crossover at a burst length around Δ_c, so that

$$\frac{D(\Delta)}{N} \propto \begin{cases} \Delta^{-3/2} & \text{for } \Delta \ll \Delta_c \\ \Delta^{-5/2} & \text{for } \Delta \gg \Delta_c \end{cases} \tag{40}$$

The difference between the two power-law exponents is unity, as suggested by Sornette's "sweeping of an instability" mechanism [12]. Such a difference in avalanche power law exponents has been observed numerically by Zapperi et al. in a fuse model [13].

We have thus shown the existence of a crossover from the generic asymptotic behavior $D \propto \Delta^{-5/2}$ to the power law $D \propto \Delta^{-3/2}$ near criticality, i.e., near global breakdown. The crossover is a universal phenomenon, independent of the threshold distribution $p(x)$. In addition we have located where the crossover takes place.

For the uniform distribution $\Delta_c = (1 - x_0/x_c)^{-2}/2$, so for $x_0 = 0.8 x_c$, we have $\Delta_c = 12.5$. For the Weibull distribution $P(x) = 1 - \exp(-(x-1)^{10})$, where $1 \leq x \leq \infty$, we get $x_c = 1.72858$ and for $x_0 = 1.7$, the crossover point will be at $\Delta_c \simeq 14.6$. Such crossover is clearly observed (Fig. 8) near the expected values $\Delta = \Delta_c = 12.5$ and $\Delta = \Delta_c = 14.6$, respectively, for the above distributions.

The simulation results shown in the figures are based on *averaging* over a large number of fibre bundles with moderate N. For applications it is important that crossover signals are seen also in a single sample. We show in Fig. 9 that equally clear power laws are seen in a *single* fibre bundle when N is large.

An important question in strength considerations of materials is how to obtain signatures that can warn of imminent system failure. This is of uttermost importance in, e.g., the dimond mining industry where sudden failure of the mine can be very costly in terms of lives. These mines are under continuous acoustic surveillance, but at present there are no tell-tale acoustic signature of imminent catastrophic failure. The same type of question is of course also central to earthquake prediction. The crossover seen here in our fibre bundle models is such a signature, it signals that catastrophic failure is imminent. The same type of crossover phenomenon is also seen in the burst distribution of a two-dimensional model of fuses with stochastically distributed current tresholds [10]. This signal does not hinge on observing rare events, and is seen also in a single system (Fig. 9). It has, therefore, a strong potential as a

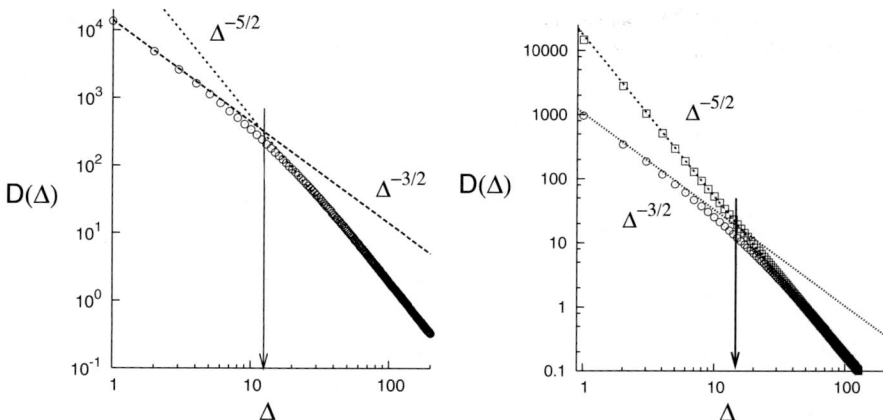

Fig. 8. The distribution of bursts for the uniform threshold distribution (*left*) with $x_0 = 0.80 x_c$ and for a Weibull distribution (*right*) with $x_0 = 1$ (*square*) and $x_0 = 1.7$ (*circle*). Both the figures are based on 50000 samples with $N = 10^6$ fibres each. The *straight lines* represent two different power laws, and the arrows locate the crossover points $\Delta_c \simeq 12.5$ and $\Delta_c \simeq 14.6$, respectively

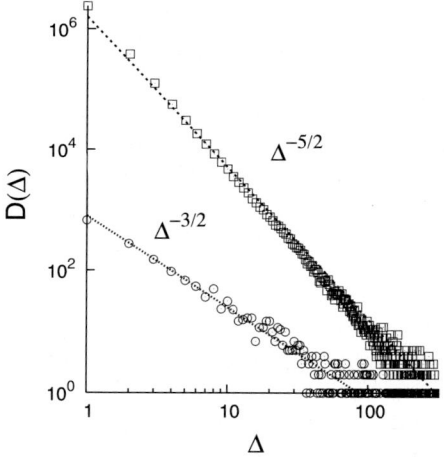

Fig. 9. The distribution of bursts for the uniform threshold distribution for a single fibre bundle with 10^7 fibres. Results with $x_0 = 0$ (recording all avalanches), are shown as squares, the circles stand for avalanches near the critical point ($x_0 = 0.9 x_c$)

useful detection tool. It is interesting that most recently, Kawamura [14] has observed a decrease in exponent value of the local magnitude distribution of earthquakes as the mainshock is approached (see Fig. 20 of [14]), analysing earthquakes in Japan (from JUNEC catalog).

Obviously, one cannot count bursts all the way to complete breakdown to have a useful detection tool. It suffices to sample bursts in finite intervals (x_0, x_f), with $x_f < x_c$. In this case we obtain the avalanche distribution by restricting the integration in (18) to the appropriate intervals. When such an interval is in the neighborhood of x_c we obtain

$$\frac{D(\Delta)}{N} \simeq \frac{\Delta^{\Delta-1} e^{-\Delta} p(x_c) a'(x_c)}{\Delta!} \int_{x_0}^{x_f} e^{-a'(x_c)^2 (x_c-x)^2 \Delta/2} (x - x_c) \, dx \quad (41)$$

$$\propto \Delta^{-5/2} \left(e^{-\Delta(x_c - x_f)^2/a} - e^{-\Delta(x_c - x_0)^2/a} \right), \quad (42)$$

with $a = 2/a'(x_c)^2$. This shows a crossover:

$$\frac{D(\Delta)}{N} \propto \begin{cases} \Delta^{-3/2} & \text{for } \Delta \ll a/(x_c - x_0)^2 \\ \Delta^{-5/2} & \text{for } a/(x_c - x_0)^2 \ll \Delta \ll a/(x_c - x_f)^2 \end{cases}, \quad (43)$$

with a final exponential behavior when $\Delta \gg a/(x_c - x_f)^2$.

The 3/2 power law will be seen only when the beginning of the interval, x_0, is close enough to the critical value x_c. Observing the 3/2 power law is therefore a signal of imminent system failure.

2.5 Avalanche Distribution at Criticality

Precisely *at* criticality ($x_0 = x_c$) the crossover takes place at $\Delta_c = \infty$, and consequently the $\xi = 5/2$ power law is no longer present. We will now argue, using the random walk representation in Sect. 2.3, that precisely at criticality the avalanche distribution follows a power law with exponent 3/2.

At criticality the distribution (33) of the step lengths in the random walk simplifies to

$$\rho_c(\Delta F) = \begin{cases} 0 & \text{for } \Delta F < -x_c \\ x_c^{-1} e^{-1} e^{-\Delta F/x_c} & \text{for } \Delta F \geq -x_c \end{cases} \quad (44)$$

A first burst of size Δ corresponds to a random walk in which the position after each of the first $\Delta-1$ steps is *lower* than the starting point, but after step no. Δ the position of the walker exceeds the starting point. The probability of this equals

$$\text{Prob}(\Delta) = \int_{-x_c}^{0} \rho_c(\delta_1) d\delta_1 \int_{-x_c}^{-\delta_1} \rho_c(\delta_2) d\delta_2 \int_{-x_c}^{-\delta_1-\delta_2} \rho_c(\delta_3) d\delta_3 \ldots$$

$$\int_{-x_c}^{-\delta_1-\delta_2\ldots-\delta_{\Delta-2}} \rho_c(\delta_{\Delta-1}) d\delta_{\Delta-1} \int_{-\delta_1-\delta_2\ldots-\delta_{\Delta-1}}^{\infty} \rho_c(\xi_\Delta) d\delta_\Delta. \quad (45)$$

To simplify the notation we have introduced $\delta_n \equiv \Delta F_n$. In [11] we have evaluated the multiple integral (45), with the result

$$\text{Prob}(\Delta) = \frac{\Delta^{\Delta-1} e^{-\Delta}}{\Delta!} \simeq \frac{1}{\sqrt{2\pi}} \Delta^{-3/2}. \qquad (46)$$

We note in passing that for the standard unbiased random walk with constant step length we obtain a *different* expression for the burst probability, but again with a 3/2 power law for large Δ:

$$\text{Prob}(\Delta) = \frac{1}{2^{\Delta-1}\Delta} \binom{\Delta-2}{\frac{1}{2}\Delta-1} \simeq \frac{1}{\sqrt{2\pi}} \Delta^{-3/2}. \qquad (47)$$

At completion of the first burst, the force, i.e., the excursion of the random walk, is larger than all previous values. Therefore one may use this point as a new starting point to find, by the same calculation, the distribution of the next burst, etc. Consequently the complete burst distribution is essentially proportional to $\Delta^{-3/2}$, as expected. The simulation results exhibited in Fig. 10 are in excellent agreement with these predictions.

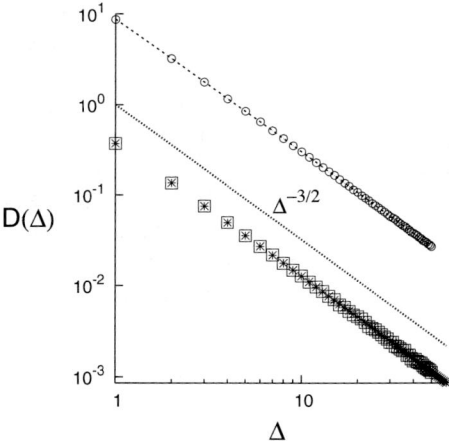

Fig. 10. Distributions of the first bursts (*squares*) and of all bursts (*circles*) for the uniform threshold distribution with $x_0 = x_c$. The simulation results are based on 10^6 samples with 80000 fibres each. The crosses stand for the analytic result (46)

One of the unusual threshold distributions we studied in Sect. 2.2 corresponded to an constant average force, $\langle F \rangle /N$ independent of x. Such a bundle is not critical at a single point x_c, but in a whole interval of x. That the burst exponent for this model takes the critical value 3/2 is therefore no surprise.

2.6 Recursive Dynamics

The relation between the number of ruptured fibres and a given external load per fibre, $\sigma = F/N$, can be viewed as the result of a sequential process. In the first step all fibres with thresholds less than $x_1 = \sigma$ must fail. Then the load is redistributed on the surviving fibres, which forces more fibres to burst, etc. This starts an iterative process that goes on until equilibrium is reached, or all fibres rupture [16, 17, 18].

Assume all fibres with thresholds less than x_t break in step number t. The expected number of intact fibres is then

$$U_t = 1 - P(x_t), \qquad (48)$$

so that the load per fibre is increased to σ/U_t. In step number $t+1$, therefore, all fibres with threshold less than

$$x_{t+1} = \frac{\sigma}{1 - P(x_t)} \qquad (49)$$

must fail. This iteration defines the recursive dynamics [16, 17, 18]. Alternatively an iteration for the U_t can be set up:

$$U_{t+1} = 1 - P(\sigma/U_t); \qquad U_0 = 1. \qquad (50)$$

If the iteration (49) converges to a finite fixed-point x^*,

$$\lim_{t \to \infty} x_t = x^*,$$

the fixed-point value must satisfy

$$x^* = \frac{\sigma}{1 - P(x^*)}. \qquad (51)$$

This fixed-point relation,

$$\sigma = x^* [1 - P(x^*)], \qquad (52)$$

is identical to the relation (7) between the average force per fibre, $\langle F \rangle / N$, and the threshold value x.

Equation (52) shows that a necessary condition to have a finite positive fixed-point value x^* is

$$\sigma \leq \sigma_c \equiv \max_x \{x [1 - P(x)]\} \qquad (53)$$

Thus σ_c is the critical value of the external load per fibre, beyond which the bundle fails completely.

As a simple example take the uniform threshold distribution (2) with $x_r = 1$, for which the iterations take the form

$$x_{t+1} = \frac{\sigma}{1 - x_t} \quad \text{and} \quad U_{t+1} = 1 - \frac{\sigma}{U_t}. \qquad (54)$$

Moreover, $\sigma_c = 1/4$, and the quadratic fixed-point equation for U^* has the solution

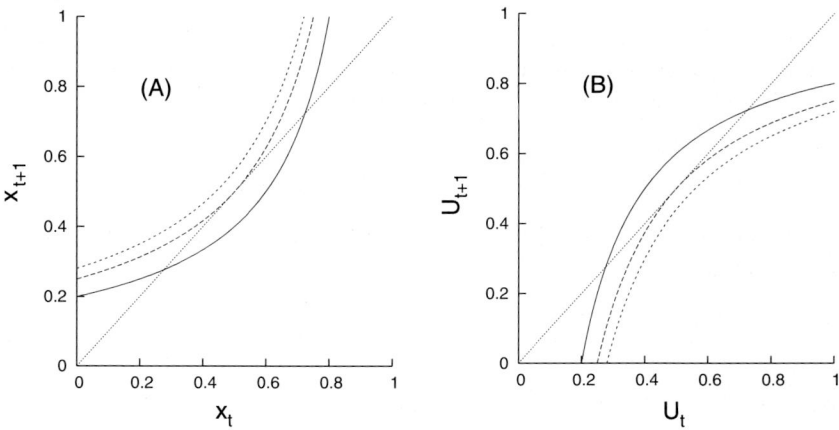

Fig. 11. Graphical representation of the iterations for x (**A**) and for U (**B**). The graphs are shown for different values of the external stress: $\sigma = 0.2$ (*solid*), 0.25 (*dashed*), and 0.28 (*dotted*), respectively. The intersections between the graph and the straight 45° line define possible fixed points

$$U^*(\sigma) = \frac{1}{2} \pm (\sigma_c - \sigma)^{1/2} = U^*(\sigma_c) \pm (\sigma_c - \sigma)^{1/2} , \tag{55}$$

since $U^*(\sigma_c) = 1/2$. The iterations are sketched in Fig. 11.

A fixed point is attractive if $|dU_{t+1}/dU_t|$ is less than 1 at the fixed point and repulsive if the derivative exceeds unity. We see in Fig. 11 that the stable fixed point corresponds to the smallest value of x^*, and to the largest value of U^* (the plus sign in (55)):

$$U^*(\sigma) - U^*(\sigma_c) = (\sigma_c - \sigma)^\beta, \quad \text{with} \quad \beta = \frac{1}{2} . \tag{56}$$

Thus $U^*(\sigma) - U^*(\sigma_c)$ behaves like an order parameter, signalling total bundle failure when it is negative, partial failure when it is positive.

Close to a stable fixed point the iterated quantity changes with tiny amounts, so one may expand in the difference $\epsilon_t = U_t - U^*$. To first order (54) yields

$$\epsilon_{t+1} = \epsilon_t \cdot \frac{\sigma}{U^{*2}} . \tag{57}$$

Thus the fixed point is approached monotonously, with exponentially decreasing steps:

$$\epsilon_t \propto e^{-t/\tau} , \tag{58}$$

with

$$\tau = \frac{1}{\ln(U^{*2}/\sigma)} . \tag{59}$$

Precisely *at* the critical point, where $U^* = 1/2$ and $\sigma_c = 1/4$, the relaxation parameter τ is infinite, signalling a non-exponential approach to the fixed point. Close to the critical point one easily shows that

$$\tau \propto (\sigma_c - \sigma)^{-\alpha}, \quad \text{with} \quad \alpha = \frac{1}{2}. \tag{60}$$

One may define a *breakdown susceptibility* χ by the change of $U^*(\sigma)$ due to an infinitesimal increment of the applied stress σ,

$$\chi = -\frac{dU^*(\sigma)}{d\sigma} = \frac{1}{2}(\sigma_c - \sigma)^{-\gamma}, \quad \text{with} \quad \gamma = \frac{1}{2}. \tag{61}$$

The susceptibility diverges as the applied stress σ approaches its critical value. Such a divergence was noted in previous numerical studies [15, 19].

When at criticality the approach to the fixed point is not exponential, what is it? Putting $U_t = U_c + \epsilon_t$ in the iteration (54) for $\sigma = 1/4$, it may be rewritten as follows

$$\epsilon_{t+1}^{-1} = \epsilon_t^{-1} + 2, \quad \text{with} \quad \epsilon_0 = \frac{1}{2}, \tag{62}$$

with solution $\epsilon_t^{-1} = 2t + 2$. Thus we have, *exactly*,

$$U_t = \frac{1}{2} + \frac{1}{2t+2}. \tag{63}$$

For large t this follows a power-law approach to the fixed point, $U_t - U_c = \frac{1}{2}t^{-\delta}$, with $\delta = 1$.

These critical properties are valid for the uniform distribution, and the natural question is how general the results are. In [18] two other threshold distributions were investigated, and all critical properties, quantified by the indicies α, β, γ and δ were found to be the same as for the uniform threshold distribution. This suggests strongly that the critical behavior is universal, which we now prove.

When an iteration is close to the fixed point, we have for the deviation

$$\epsilon_{t+1} = U_{t+1} - U^* = P\left(\frac{\sigma}{U^*}\right) - P\left(\frac{\sigma}{U^* + \epsilon_t}\right) = \epsilon_t \cdot \frac{\sigma}{U^{*2}} p(\sigma/U^*), \tag{64}$$

to lowest order in ϵ_t.

This guarantees an exponential relaxation to the fixed point, $\epsilon_t \propto e^{-t/\tau}$, with parameter

$$\tau = 1 \Big/ \ln\left(\frac{U^{*2}}{\sigma p(\sigma/U^*)}\right). \tag{65}$$

Criticality is determined by the extremum condition (8), which by the relation (48) takes the form

$$U_c^2 = \sigma p(\sigma/U_c)$$

Thus $\tau = \infty$ at criticality. To study the relaxation at criticality we must expand (64) to second order in ϵ_t since to first order we simply get the useless equation $\epsilon_{t+1} = \epsilon_t$. To second order we obtain

$$\epsilon_{t+1} = \epsilon_t - C\epsilon_t^2,$$

with a positive constant C. This is satisfied by

$$\epsilon_t = \frac{1}{Ct} + \mathcal{O}(t^{-2}) .$$

Hence in general the dominating critical behavior for the approach to the fixed point is a power law with $\delta = 1$. The values $\alpha = \beta = \gamma = \frac{1}{2}$ can be shown to be consequences of the parabolic maximum of the load curve, (7) or (49), at criticality: $|x_c - x^*| \propto (\sigma_c - \sigma)^{\frac{1}{2}}$.

Thus all threshold distributions for which the macroscopic strength function has a single parabolic maximum, are in this universality class.

3 Fibre Bundles with Local Load Redistribution

The assumption that the extra stress caused by a fibre failure is shared equally among all surviving fibres is often unrealistic, since fibres in the neighborhood of the failed fibre are expected to take most of the load increase. One can envisage many systems for such local stress redistributions. A special case is the model with a one-dimensional geometry where the two nearest-neighbor fibres take up all extra stress caused by a fibre failure (Sect. 3.1). It is special for two reasons: It is an extreme case because the range of the stress redistribution is minimal, and, secondly, it is amenable to theoretical analysis. In other models, treated in Sect. 3.2 and 3.3, the stress redistribution occurs over a larger region. In Sect. 3.3 this comes about by considering a clamp to be an elastic medium.

3.1 Stress Alleviation by Nearest Neighbors

The simplest model with nearest-neighbor stress redistribution is one-dimensional, with the N fibres ordered linearly, with or without periodic boundary conditions. Thus two fibres, one of each side, take up, and divide equally, the extra stress caused by a failure. The force on a fibre surrounded by n_l broken fibres on the left-hand side and n_r broken fibres on the right-hand side is then

$$\frac{F_{tot}}{N}\left(1 + \frac{1}{2}n_l + \frac{1}{2}n_r\right) \equiv f(2 + n_l + n_r) , \qquad (66)$$

where F_{tot} is the total force on the bundle, and $f = F_{tot}/2N$, one-half the force-per-fibre, is a convenient forcing parameter. The model has been discussed previously in a different context [20, 21, 22, 23, 24, 25]. Preliminary simulation studies [26, 27] showed convincingly that this local model is *not* in the same universality class as the equal-load-sharing fibre bundles. For the uniform threshold distribution and for $1 \leq \Delta \leq 10$ an effective exponent between 4 and 5 was seen, much larger than $5/2$.

Avalanches in this model, and in similar local stress-redistribution models, have a character different from bursts in the equal-load-sharing models. In the present model the failure of one fibre can by a domino effect set in motion a fatal avalanche: If the failing fibre has many previously failed fibres as neighbors, the load on the fibres on each side is high, and if they burst, the load on the new neighbors is even higher, etc., which may produce an unstoppable avalanche.

In [7] the burst distribution was determined analytically for the uniform threshold distribution,

$$P(x) = \begin{cases} x & \text{for } 0 \leq x \leq 1 \\ 1 & \text{for } x > 1 \end{cases} \tag{67}$$

In this model there is an upper limit to the size Δ of an avalanche that the bundle can survive. Since the threshold values are at most unity, it follows from (66) that if

$$\Delta > f^{-1} - 2, \tag{68}$$

then the bundle breaks down completely. Consequently an asymptotic power law distribution of the avalanche sizes is not possible.

In the analytic derivation periodic boundary conditions were used. The fairly elaborate procedure was based on a set of recursion relations between configurations and events at fixed external force. In Fig. 12 we show the

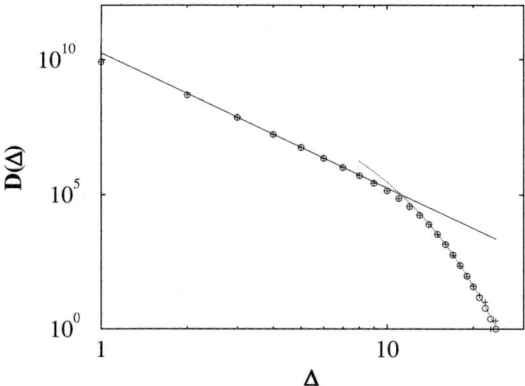

Fig. 12. Burst distribution in a local-load-sharing model for a bundle of $N = 20000$ fibres. Theoretical results are shown as circles, and simulation results (for 4000000 samples) are shown as crosses. The *straight line* shows the power law Δ^{-5}

resulting burst distribution for a bundle of $N = 20000$ fibres, compared with simulation results. The agreement is extremely satisfactory.

We see from Fig. 12 that, as expected, the burst distribution does *not* follow a power law for large Δ, but falls off faster. For Δ less than 10 the burst distribution follows approximately a power law, $D(\Delta) \propto \Delta^{-\xi}$ with ξ of the order of 5.

In [7] the maximum load F_{max} the fibre bundle could tolerate was estimated to have the following size dependence,

$$F_{max} \propto \frac{N}{\ln N} . \tag{69}$$

This is different from the equal-load-sharing model, for which $F_{max} \propto N$. A similar logarithmic size-dependence of the bundle strength for local-stress-redistribution models has been proposed by other authors [28, 29, 30].

The qualitative explanation of the non-extensive result is that for large N the probability of finding a weak region somewhere is high. Since, as discussed above, a weak region is the seed to complete bundle failure, it may be reasonable that the maximum load the bundle can carry does not increase proportional to N, but slower than linear.

3.2 Intermediate Load-Sharing Models

It might be interesting to study models that interpolates between global and nearest-neighbor load sharing. The main question is whether the burst distribution changes from one behavior to the other in a continuous manner, or whether a discontinuous change occurs.

In [8] was introduced such an intermediate model, with the same one dimensional geometry as the nearest-neighbor model of the preceding section. When a fibre i fails in this model the elastic constants of the two nearest surviving neighbors l and r on both sides are updated as follows

$$\kappa_l \to \frac{1}{2}\lambda(\kappa_l + \kappa_r + \kappa_i) \tag{70}$$

$$\kappa_i \to 0 \tag{71}$$

$$\kappa_r \to \frac{1}{2}\lambda(\kappa_l + \kappa_r + \kappa_i) . \tag{72}$$

For $\lambda = 1$ this corresponds to the local load-sharing by surviving nearest-neighbors (see the preceding section). And with $\lambda = 0$ the intact nearesneighbor fibres to a failing fibre does not take part in the load-sharing. But since all the other surviving fibres then share the load equally, this limiting case must have the same behavior as the equal-load-sharing model. The numerics seems to suggest that there is a cross-over value of λ separating the universality classes of the local-load-sharing and the equal-load-sharing regimes.

A stress redistribution scheme that in a straightforward way interpolates between the two extreme models was recently proposed by Pradhan et al. [31]: A fraction g of the extra load caused by a fibre failure is shared by the nearest neighbors, and the remaining load is shared equally among all intact fibres. They show that in a one-dimensional geometry a crossover value g_c exists, such that for $g < g_c$ the bundle belongs to the equal-load-sharing regime, while for $g > g_c$ the system is like the local-load-sharing model of Sect. 3.1. The crossover value was determined to be $g_c = 0.79 \pm 0.01$.

It would be more realistic to have a stress redistribution whose magnitude falls off monotonically with the geometric distance r from the failed fibre. Hidalgo et al. [32] introduced such a model, for which the extra stress on a fibre followed a power law decay, proportional to $r^{-\gamma}$. In the limit $\gamma \to 0$ the equal-load-distribution model is recovered, while the limit $\gamma \to \infty$ corresponds to the nearest-neighbor model in Sect. 3.1. Again a crossover is observed, at a value $\gamma_c \simeq 2$ for the range parameter.

In the next section we consider a model with a different, but similar, interaction decaying with the distance from the failed fibre.

3.3 Elastic Medium Anchoring

In this section we generalize the fibre bundle problem to include more realistically the elastic response of the surfaces to which the fibres are attached. So far, these have been assumed to be infinitely stiff for the equal-load-sharing model, or their response has been modeled as very soft, but in a fairly unrealistic way in the local-load-sharing models, see Sect. 3.2. In [33], a realistic model for the elastic response of the clamps was studied. The model was presented as addressing the problem of failure of weldings. In this context, the two clamps were seen as elastic media glued together at a common interface. Without loss of generality, one of the media was assumed to be infinitely stiff whereas the other was soft.

The two clamps can be pulled apart by controlling (fixing) either the applied force or the *displacement*. The displacement is defined as the change in the distance between two points, one in each clamp positioned far from the interface. The line connecting these points is perpendicular to the average position of the interface. In our case, the pulling is accomplished by controlling the displacement. As the displacement is increased slowly, fibres – representing the glue – will fail, ripping the two surfaces apart.

The model consists of two two-dimensional square $L \times L$ lattices with periodic boundary conditions. The lower one represents the hard, stiff surface and the upper one the elastic surface. The nodes of the two lattices are matched (i.e. there is no relative lateral displacement). The thresholds of the fibres are taken from an uncorrelated uniform distribution. The spacing between the fibres is a in both the x and y directions. The force that each fibre is carrying is transferred over an area of size a^2 to the soft clamp: As the two clamps are separated by controlling the displacement of the hard clamp, D, the forces carried by the fibres increase. As for the fibre bundle models studied in the previous sections, when the force carried by a fibre reaches its breaking threshold, it breaks irreversibly and the forces redistribute. Hence, the fibres are broken one by one until the two clamps are no longer in mechanical contact. As this process is proceeding, the elastic clamp is of course deforming to accomodate the changes in the forces acting on it.

The equations governing the system are as follows. The force, f_i, carried by the ith fibre is given by

$$f_i = -k(u_i - D) \,, \tag{73}$$

where k is the spring constant and u_i is the deformation of the elastic clamp at site i. All unbroken fibres have $k = 1$ while a broken fibre has $k = 0$. The quantity $(u_i - D)$ is, therefore, the length fibre i is stretched. In addition, a force applied at a point on an elastic surface will deform this surface over a region whose extent depends on its elastic properties. This is described by the coupled system of equations,

$$u_i = \sum_j G_{i,j} f_j \,, \tag{74}$$

where the elastic Green function, $G_{i,j}$ is given by [34, 35]

$$G_{i,j} = \frac{1-s^2}{\pi e a^2} \int_{-a/2}^{+a/2} \int_{-a/2}^{+a/2} \frac{dx'\, dy'}{|(x-x', y-y')|} \,. \tag{75}$$

In this equation, s is the Poisson ratio, e the elastic constant, and the denominator $|\mathbf{i} - \mathbf{j}|$ is the distance between sites $i = (x, y)$ and $j = (x', y')$. The indices i and j run over all L^2 sites. The integration over the area a^2 is done to average the force from the fibres over this area. Strictly speaking, the Green function applies for a medium occupying the infinite half-space. However, with a judicious choice of elastic constants, we may use it for a finite medium if its range is small compared to L, the size of the system.

By combining (73) and (74), we obtain

$$(\mathbf{I} + \mathbf{KG})\mathbf{f} = \mathbf{KD} \,, \tag{76}$$

where we are using matrix-vector notation. $\mathbf{I}$ is the $L^2 \times L^2$ identity matrix, and $\mathbf{G}$ is the Green function represented as an $L^2 \times L^2$ dense matrix. The constant vector $\mathbf{D}$ is L^2 dimensional. The *diagonal* matrix $\mathbf{K}$ is also $L^2 \times L^2$. Its matrix elements are either 1, for unbroken fibres, or 0 for broken ones. Of course, $\mathbf{K}$ and $\mathbf{G}$ do not commute.

Once (76) is solved for the force $\mathbf{f}$, (74) easily yields the deformations of the elastic clamp.

Equation (76) is of the familiar form $\mathbf{Ax} = \mathbf{b}$. Since the Green function connects all nodes to all other nodes, the $L^2 \times L^2$ matrix $\mathbf{A}$ is dense which puts severe limits on the size of the system that may be studied.

The simulation proceeds as follows: We start with all springs present, each with its randomly drawn breakdown threshold. The two clamps are then pulled apart, the forces calculated using the Conjugate Gradient (CG) algorithm [36, 37], and the fibre which is the nearest to its threshold is broken, i.e. the corresponding matrix element it the matrix **K** is zeroed. Then the new forces are calculated, a new fibre broken and so on until all fibres have failed.

However, there are two problems that render the simulation of large systems extremely difficult. The first is that since **G** is a $L^2 \times L^2$ *dense* matrix, the number of operations per CG iteration scales like L^4. Even more serious is the fact that as the system evolves and springs are broken, the matrix $(\mathbf{I} + k\mathbf{G})$ becomes very ill-conditioned.

To overcome the problematic L^4 scaling of the algorithm, we note that the Green function is diagonal in Fourier space. Consequently, doing matrix-vector multiplications using FFTs the scaling is much improved and goes like $L^2 \ln(L)$. Symbolically, this can be expressed as follow:

$$(\mathbf{I} + \mathbf{K}\mathbf{F}^{-1}\mathbf{F}\mathbf{G})\mathbf{F}^{-1}\mathbf{F}\mathbf{f} = \mathbf{K}\mathbf{D} , \tag{77}$$

where **F** is the FFT operator and $\mathbf{F}^{-1}$ its inverse ($\mathbf{F}^{-1}\mathbf{F} = 1$). Since **I** and **K** are diagonal, operations involving them are performed in real space. With this formulation, the number of operations/iteration in the CG algorithm now scales like $L^2 \ln(L)$.

To overcome the runaway behavior due to the ill-conditioning we need to precondition the matrix [36, 38]. This means that instead of solving (77), we solve the equivalent problem

$$\mathbf{Q}(\mathbf{I} + \mathbf{K}\mathbf{F}^{-1}\mathbf{F}\mathbf{G})\mathbf{F}^{-1}\mathbf{F}\mathbf{f} = \mathbf{Q}\mathbf{K}\mathbf{D} , \tag{78}$$

where we simply have multiplied both sides by the arbitrary, positive definite preconditioning matrix **Q**. Clearly, the ideal choice is $\mathbf{Q_0} = (\mathbf{I} + \mathbf{K}\mathbf{G})^{-1}$ which would always solve the problem in one iteration. Since this is not possible in general, we look for a form for **Q** which satisfies the following two conditions: (1) It should be as close as possible to $\mathbf{Q_0}$, and (2) be fast to calculate. The choice of a good **Q** is further complicated by the fact that as the system evolves and fibres are broken, corresponding matrix elements of **K** are set to zero. So, the matrix $(\mathbf{I} + \mathbf{K}\mathbf{G})$ evolves from the initial form $(\mathbf{I} + \mathbf{G})$ to the final one **I**.

Batrouni et al. [33] chose the form

$$\mathbf{Q} = \mathbf{I} - (\mathbf{K}\mathbf{G}) + (\mathbf{K}\mathbf{G})(\mathbf{K}\mathbf{G}) - (\mathbf{K}\mathbf{G})(\mathbf{K}\mathbf{G})(\mathbf{K}\mathbf{G}) + \cdots \tag{79}$$

which is the Taylor series expansion of $\mathbf{Q_0} = (\mathbf{I} + \mathbf{K}\mathbf{G})^{-1}$. For best performance, the number of terms kept in the expansion is left as a parameter since it depends on the physical parameters of the system. It is important to emphasize the following points: (a) As fibres are broken, the preconditioning matrix evolves with the ill-conditioned matrix and, therefore, remains a good

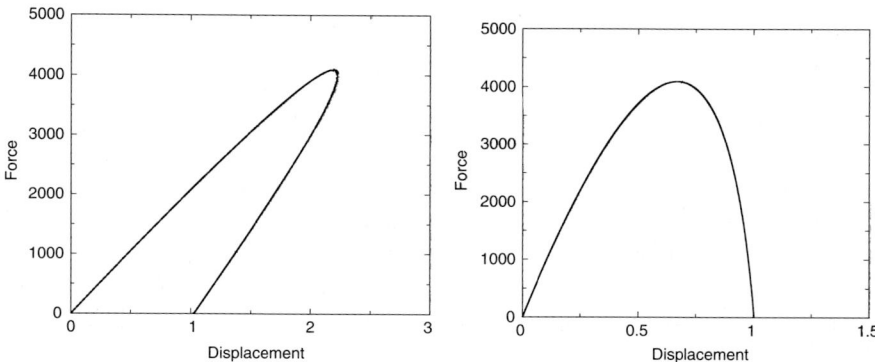

Fig. 13. Force-displacement curve, 128×128 systems with $e = 10$ (*left*) and $e = 100$ (*right*)

approximation to its inverse throughout the breaking process. (b) All matrix multiplications involving **G** are done using FFTs. (c) The calculation of **Q** can be easily organized so that it scales like $nL^2 \ln(L)$ where n is the number of terms kept in the Taylor expansion, (79). The result is a stable accelerated algorithm which scales essentially as the volume of the system.

Figure 13 (left) shows the force-displacement curve for a system of size 128×128 with elastic constant $e = 10$. Whether we control the applied force, F, or the displacement, D, the system will eventually suffer catastrophic collapse. However, this is not so when $e = 100$ as shown in Fig. 13 (right). In this case, only controlling the force will lead to catastrophic failure. In the limit when $e \to \infty$, the model becomes the equal-load-sharing fibre bundle model, where $F = (1 - D)D$. In this limit there are no spatial correlations and the force instability is due to the the decreasing total elastic constant of the system making the force on each surviving bond increase faster than the typical spread of threshold values. No such effect exists when controlling displacement D. However, when the elastic constant, e, is small, spatial correlations in the form of localization, where fibres that are close in space have a tendency to fail consequtively, do develop, and these are responsible for the displacement instability which is seen in Fig. 13.

We now turn to the study of the burst distribution. We show in Fig. 14 the burst distributions for $e = 10$ and $e = 100$. In both cases we find that the burst distribution follows a power law with an exponent $\xi = 2.6 \pm 0.1$. It was argued in [33] that the value of ξ in this case is indeed $5/2$ as in the global-load-sharing model.

As the failure process proceeds, there is an increasing competition between global failure due to stress enhancement and local failure due to local weakness of material. When the displacement, D, is the control parameter and e is sufficiently small (for example $e = 10$), catastrophic failure eventually occurs due to localization. The onset of this localization, i.e. the catastrophic

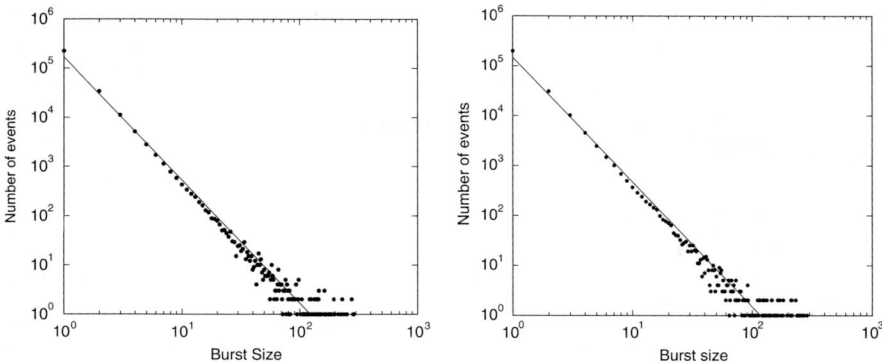

Fig. 14. Burst distribution for 128×128, for $e = 10$ (*left*) and $e = 100$ (*right*). The slope of the *straight lines* is -2.5

regime, occurs when the two mechanisms are equally important. This may be due to self organization [39] occuring at this point. In order to test whether this is the case, Batrouni et al. [33] measured the size distribution of broken bond clusters at the point when D reaches its maximum point on the $F - D$ characteristics, i.e. the onset of localization and catastrophic failure. The analysis was performed using a Hoshen-Kopelman algorithm [40]. The result is shown in Fig. 15, for 56 disorder realizations, $L = 128$ and $e = 10$.

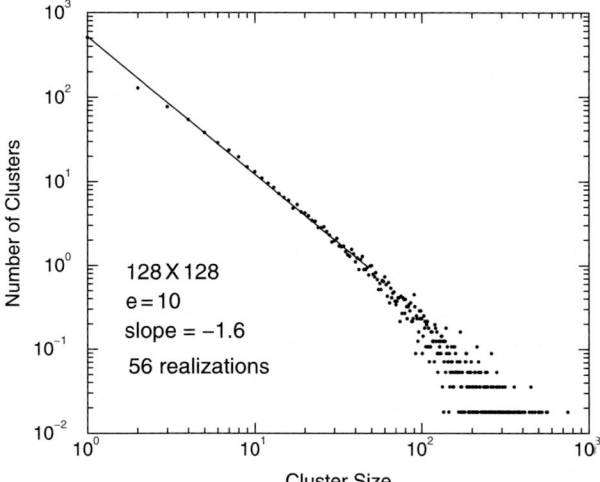

Fig. 15. Area distribution of zones where glue has failed for systems of size 128×128 and elastic constant $e = 10$. The *straight line* is a least square fit and indicates a power law with exponent -1.6

The result is consistent with a power law distribution with exponent -1.6, and consequently with self organization. If this process were in the universality class of percolation, the exponent would have been -2.05. Hence, we are dealing with a new universality class in this system.

Acknowledgement

S. P. thanks NFR (Research Council of Norway) for financial support through Grant No. 166720/V30.

References

1. H.J. Herrmann and S. Roux, eds. *Statistical Models for the Fracture of Disordered Media* (North-Holland, Amsterdam, 1990).
2. B.K. Chakrabarti and L.G. Benguigui *Statistical Physics and Breakdown in Disordered Systems* (Oxford University Press, Oxford, 1997).
3. D. Sornette *Critical Phenomena in Natural Sciences* (Springer Verlag, Berlin, 2000).
4. H.E. Daniels, Proc. Roy. Soc. London **A 183**, 405 (1945).
5. P.C. Hemmer and A. Hansen, ASME J. Appl. Mech. **59**, 909 (1992).
6. F.T. Peirce, J. Text. Ind. **17**, 355 (1926).
7. M. Kloster, A. Hansen, and P.C. Hemmer, Phys. Rev. **E 56**, 2615 (1997).
8. A. Hansen and P.C. Hemmer, Trends in Statistical Physics **1**, 213 (1994).
9. W. Lee, Phys. Rev. **E 50**, 3797 (1994).
10. S. Pradhan, A. Hansen, and P.C. Hemmer, Phys. Rev. Lett. **95**, 125501 (2005).
11. S. Pradhan, A. Hansen, and P.C. Hemmer, submitted to Phys. Rev. **E**, cond-mat/0512015 (2005).
12. D. Sornette, J. Phys. **I** France **4**, 209 (1994).
13. S. Zapperi, P.K.V.V. Nukula, and S. Simunovic, Phys. Rev. **E 71**, 026106 (2005).
14. H. Kawamura, arXiv:cond-mat/0603335, (2006).
15. R. de Silveira, Am. J. Phys. **67**, 1177 (1999).
16. S. Pradhan and B.K. Chakrabarti, Phys. Rev. **E 65**, 016113 (2001).
17. S. Pradhan, P. Bhattacharyya, and B.K. Chakrabarti, Phys. Rev. **E 66** 016116 (2002).
18. P. Bhattacharyya, S. Pradhan, and B.K. Chakrabarti, Phys. Rev. **E 67**, 046122 (2003).
19. S. Zapperi, P. Ray, H.E. Stanley, and A. Vespignani, Phys. Rev. Lett. **85**, 2865 (2000).
20. D.G. Harlow, Proc. Roy. Soc. Lond. Ser. **A 397**, 211 (1985).
21. D.G. Harlow and S.L. Phoenix, J. Mech. Phys. Solids **39**, 173 (1991).
22. P.M. Duxbury and P.M. Leath, Phys. Rev. **B 49**, 12676 (1994).
23. D.G. Harlow and S.L. Phoenix, Int. J. Fracture **17**, 601 (1981).
24. S.L. Phoenix and R.L. Smith, Int. J. Sol. Struct. **19**, 479 (1983).
25. C.C. Kuo and S.L. Phoenix, J. Appl. Prob. **24**, 137 (1987).
26. A. Hansen and P.C. Hemmer, Phys. Lett. **A 184**, 394 (1994).
27. S.D. Zhang and E.J. Ding, Phys. Lett. **A 193** 425 (1994).

28. R.L. Smith, Ann. Prob. **10**, 137 (1982).
29. S.D. Zhang and E.J. Ding, Phys. Rev. **B 53**, 646 (1996).
30. S.D. Zhang and E.J. Ding, J. Phys. **A 28**, 4323 (1995).
31. S. Pradhan, B.K. Chakrabarti, and A. Hansen, Phys. Rev. **E 71**, 036149 (2005).
32. R.C. Hidalgo, Y. Moreno, F. Kun, and H.J. Herrmann, Phys. Rev. **E 65**, 046148 (2002).
33. G.G. Batrouni, A. Hansen, and J. Schmittbuhl, Phys. Rev. **E 65**, 036126 (2002).
34. L. Landau and E.M. Lifshitz, *Theory of Elasticity* (Clarendon Press, Oxford, 1958).
35. K.L. Johnson, *Contact Mechanics* (Cambridge University Press, Cambridge, 1985).
36. G.G. Batrouni and A. Hansen, J. Stat. Phys. **52**, 747 (1988).
37. W.H. Press, S.A. Teukolsky, W.T. Vetterling, and B.P. Flannery, *Numerical Recipes in Fortran 77: The Art of Scientific Computing* (Cambridge University Press, Cambridge, 1992).
38. G.G. Batrouni, A. Hansen, and M. Nelkin, Phys. Rev. Lett. **57**, 1336 (1986).
39. P. Bak, C. Tang and K. Wiesenfeld, Phys. Rev. Lett. **59**, 381 (1987).
40. D. Stauffer and A. Aharony, *Introduction to Percolation Theory* (Taylor and Francis, London, 1992).

Extensions of Fibre Bundle Models

F. Kun[1], F. Raischel[2], R.C. Hidalgo[3], and H.J. Herrmann[2]

[1] Department of Theoretical Physics, University of Debrecen, P. O. Box: 5, 4010 Debrecen, Hungary
feri@dtp.atomki.hu
[2] Institute for Computational Physics, University of Stuttgart, Pfaffenwaldring 27, 70569 Stuttgart, Germany
[3] Department of Fundamental Physics, University of Barcelona, Franques 1, 08028 Barcelona, Spain

The fibre bundle model is one of the most important theoretical approaches to investigate the fracture and breakdown of disordered media extensively used both by the engineering and physics community. We present the basic construction of the model and provide a brief overview of recent results focusing mainly on the physics literature. We discuss the limitations of the model to describe the failure of composite materials and present recent extensions of the model which overcome these problems making the model more realistic: we gradually enhance the fibre bundle model by generalizing the failure law, constitutive behavior, deformation state and way of interaction of fibres. We show that beyond the understanding of the fracture of fibre reinforced composites, these extensions of the fibre bundle model also address interesting problems for the statistical physics of fracture.

1 Introduction

The damage and fracture of disordered materials is a very important scientific and technological problem which has attracted an intensive research over the past decades. One of the first theoretical approaches to the problem was the fibre bundle model introduced by Peires in 1927 to understand the strength of cotton yarns [46]. In his pioneering work, Daniels provided the sound probabilistic formulation of the model and carried out a comprehensive study of bundles of threads assuming equal load sharing after subsequent failures [12]. In order to capture fatigue and creep effects, Coleman proposed a time dependent formulation of the model [7], assuming that the strength of loaded fibres is a decreasing function of time. Later on these early works initiated an intense research in both the engineering [47] and physics [6, 23] communities making fibre bundle models one of the most important theoretical approaches to the damage and fracture of disordered materials.

The development of fibre bundle models encountered two kinds of challenges: on the one hand it is important to work out realistic models of materials failure, which have a detailed representation of the microstructure of the material, the local stress fields and their complicated introduction. Such models make possible to clarify the effect of microscopic material parameters on the macroscopic response of solids. In this context, fibre bundle models served as a starting point to develop more realistic micromechanical models of the failure of fibre reinforced composites widely used by the modern aerospace and automobile industry. Analytical methods and numerical techniques have been developed making possible realistic treatment of even large scale fibrous structures. On the other hand, the damage and fracture of disordered materials addresses several interesting problems also for statistical physics. Embedding the failure and breakdown of materials into the general framework of statistical physics and clarifying its analogy to phase transitions and critical phenomena still keep scientists fascinated. Here fibre bundle models provide an excellent testing ground of ideas offering also the possibility of analytic solutions.

Fibre bundle models have been introduced to describe materials' degradation and failure, however, due to their simplicity, during the last two decades they also served as a general model of the breakdown of a broad class of disordered systems of many interacting elements subject to various types of external loads. Examples can be mentioned from magnets driven by an applied field [11, 71] through scale-free networks [10, 38] to earthquakes [63, 67] and social phenomena [11].

In this paper first we present the basic formulation of the classical fibre bundle model and briefly summarize the most important recent results obtained on the macroscopic response and microscopic damage process of disordered materials focusing on the physics literature. We discuss limitations of the model to describe the fracture of fibre reinforced composites and propose extensions which make the model more realistic. We gradually improve components of the model construction by generalizing the damage law, constitutive behavior, deformation state and interaction law of fibres. Finally, we give an overview of the applications of the extended fibre bundle model.

2 Fibre Bundle Model of Materials Failure

The construction of fibre bundle models (FBMs) is based on the following simplifying assumptions [2, 12, 28, 49, 53, 62]:

Discretization The disordered solid is represented as a discrete set of parallel fibres of number N organized on a regular lattice (square, triangular, ...), see Fig. 1. The fibres can solely support longitudinal deformation which allows to study only loading of the bundle parallel to fibres.

Failure Law When the bundle is subjected to an increasing external load, the fibres are assumed to have perfectly brittle response, i.e. they have linearly

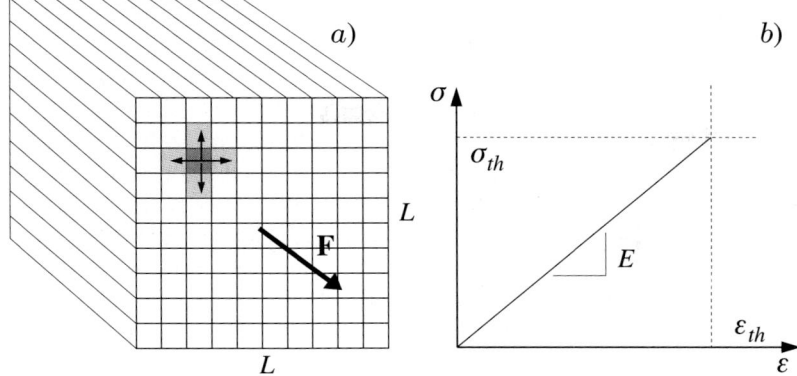

Fig. 1. (a) Set up of the classical fibre bundle model. Parallel fibres are assembled on a square lattice of size $L \times L$ loaded by an external force **F** parallel to the fibre direction. In the limiting case of very localized load sharing the load of a failed fibre (*dark grey*) is equally redistributed over its nearest intact neighbors (*light grey*). (b) Single fibres have linearly elastic behavior up to the failure load σ_{th}

elastic behavior until they break at a failure load σ_{th}^i, $i = 1, \ldots, N$ as it is illustrated in Fig. 2a. The elastic behavior of fibres is characterized by the Young modulus E, which is identical for all fibres. The failure of fibres is instantaneous and irreversible such that the load on broken fibres drops down to zero immediately at the instant of failure (see Fig. 2a), furthermore, broken fibres are never restored (no healing).

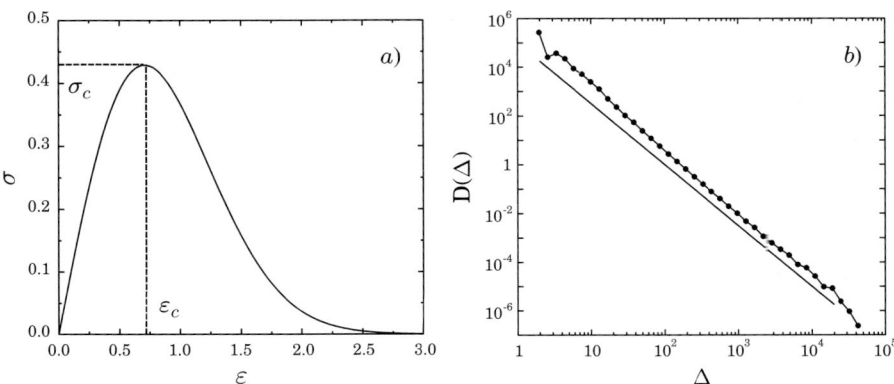

Fig. 2. (a) Macroscopic constitutive behavior of a fibre bundle with global load sharing (3) using Weibull distributed strength values σ_{th} ($m = 2$ and $\lambda = 1$). (b) Distribution D of burst sizes Δ obtained by computer simulations of a bundle of 10^7 fibres. A *straight line* of slope 2.5 is drawn to guide the eye

Load Sharing Rule After a fibre fails its load has to be shared by the remaining intact fibres. The range and form of interaction of fibres, also called load sharing rule, is a crucial component of the model which has a substantial effect on the micro and macro behavior of the bundle. Most of the studies in the literature are restricted to two extreme forms of the load sharing rule: (*i*) in the case of global load sharing (GLS), also called equal load sharing (ELS), the load is equally redistributed over all intact fibres in the bundle irrespective of their distance from the failed one. The GLS rule corresponds to the mean field approximation of FBM where the topology of the fibre bundle (like the square lattice structure in Fig. 2a) becomes irrelevant. Such a loading condition naturally arises when parallel fibres are loaded between perfectly rigid platens, like for the wire cable of an elevator. FBM with global load sharing is a usual starting point for more complex investigations since it makes possible to obtain the most important characteristic quantities of the bundle in closed analytic forms [22, 28, 53, 54, 62]. (*ii*) In the other extreme of the local load sharing (LLS), the entire load of the failed fibre is redistributed equally over its local neighborhood (usually nearest neighbors) in the lattice considered, leading to stress concentrations along failed regions (see Fig. 1a). Due to the non-trivial spatial correlations, the analytic treatment of LLS bundles has serious limitations [17, 20, 49], most of the studies here rely on large scale computer simulations [9, 18, 21, 26]. Such localized load sharing occurs when a bundle of fibres is loaded between plates of finite compliance [4, 13, 19].

Distribution of Failure Thresholds The strength of fibres σ_{th}^i, i.e. the value of the local load at which they break, is an independent identically distributed random variable with the probability density $p(\sigma_{th})$ and distribution function $P(\sigma_{th}) = \int_{\sigma_{th}^{min}}^{\sigma_{th}^{max}} p(x) dx$. The randomness of breaking thresholds is assumed to represent the disorder of heterogeneous materials, and hence, it is practically the only component of the classical FBM where material dependent features (e.g. amount of disorder) can be taken into account. For the disordered breaking thresholds the uniform distribution between 0 and 1 with the density and distribution functions

$$p(\sigma_{th}) = 1, \qquad P(\sigma_{th}) = \sigma_{th}, \tag{1}$$

is a usual starting point in most of the investigations. Another widely used distribution in FBMs is the Weibull distribution

$$P(\sigma_{th}) = 1 - \exp\left[-\left(\frac{\sigma_{th}}{\lambda}\right)^m\right], \tag{2}$$

where m and λ denote the Weibull index and scale parameter, respectively. Besides its physical ground [6, 23], the Weibull distribution has the advantage from the viewpoint of modelling that the amount of disorder can be controlled by varying the Weibull index m.

Time Dependence According to the time dependence of the fibre strength, two classes of FBMs can be distinguished: in static fibre bundles the breaking thresholds are constant during the entire loading history of the bundle. In order to model a certain type of creep rupture and fatigue behavior of materials, time dependent strength of fibres has been introduced [7, 34, 35, 42, 65].

In spite of their simplicity, FBMs provided a deeper understanding of the breakdown of heterogeneous materials revealing important basic features of the failure process. In the following we briefly summarize the most important results of FBMs on the microscopic failure process and macroscopic response of disordered materials under quasistatic loading conditions. Loading of a parallel bundle of fibres can be performed in two substantially different ways: when the deformation ε of the bundle is controlled externally, the load on single fibres σ_i is always determined by the externally imposed deformation ε as $\sigma_i = E\varepsilon$, i.e. no load sharing occurs and consequently the fibres break one-by-one in the increasing order of their breaking thresholds. At a given deformation ε the fibres with breaking thresholds $\sigma_{th}^i < E\varepsilon$ are broken, furthermore, all intact fibres keep the equal load $E\varepsilon$. Hence, the macroscopic constitutive behavior $\sigma(\varepsilon)$ of the FBMs can be cast in the form

$$\sigma(\varepsilon) = E\varepsilon \left[1 - P(E\varepsilon)\right] , \qquad (3)$$

where $[1 - P(E\varepsilon)]$ is the fraction of intact fibres at the deformation ε [11, 62]. A representative example of $\sigma(\varepsilon)$ is presented in Fig. 2a for the case of Weibull distributed strength values with $m = 2$ and $\lambda = 1$. Under stress controlled conditions when the load is gradually increased externally, a more complex damage process occurs starting with the breaking of the weakest fibre. After each fibre breaking the load dropped by the broken fibre has to be redistributed over the surviving intact ones. Assuming global load sharing (equal load sharing), all intact fibresndexfibre bundle model receive the same load increment which might cause secondary fibre breakings. The subsequent load redistribution after consecutive fibre failures can lead to an entire avalanche of breakings which can stop after a certain number of fibres keeping the integrity of the bundle, or can be catastrophic resulting in the macroscopic failure of the entire system. Hence, under stress controlled loading, the constitutive curve (3) can only be realized up to the maximum, where a catastrophic avalanche occurs breaking all the remaining fibres [22, 24, 28, 54]. It has been shown analytically that for a broad class of disorder distributions P under GLS conditions the macroscopic constitutive curve $\sigma(\varepsilon)$ of FBMs has a quadratic maximum whose position and value define the critical strain ε_c and stress σ_c of the bundle [11, 62], respectively (see Fig. 2a). Increasing the size N of finite bundles, the global strength $\sigma_c(N)$ rapidly converges to the finite non-zero strength of the infinite bundle [12, 49, 62].

When the load sharing is localized the macroscopic response of the bundle becomes more brittle, i.e. the constitutive curve $\sigma(\varepsilon)$ of the LLS bundle follows its GLS counterpart but the macroscopic failure occurs at a lower critical stress $\sigma_c^{LLS} < \sigma_c^{GLS}$, which is preceded by only a weak non-linearity [26]. Large scale computer simulations revealed that in the thermodynamic limit $N \to \infty$, the strength of LLS bundles tends to zero as $1/(\ln N)$.

On the micro-level the spatial and temporal evolution of damage shows also a strong dependence on the range of load redistribution: when the excess load of failed fibres is redistributed globally (GLS) no spatial correlations arise, i.e. fibre breaking proceeds in a completely stochastic manner in space which results in randomly nucleated clusters of broken fibres analogous to percolation. Under stress controlled loading conditions, the subsequent load redistribution results in a bursting activity preceding the macroscopic failure of the bundle. For equal load sharing it has been proven analytically that the distribution $D(\Delta)$ of burst sizes Δ recorded during the entire course of loading has a universal power law behavior

$$D(\Delta) \sim \Delta^{-\alpha}, \qquad (4)$$

with an exponent $\alpha = 5/2$ independent of the disorder distribution P [22, 28, 51], see Fig. 2b. The power law distribution of bursts prevails also for more complicated long range interactions (like in the fuse model [3, 45]) but localized load sharing results in a rapid decrease of $D(\Delta)$ and in a dependence on the specific form of disorder P [18].

In recent years several novel aspects of breakdown phenomena have been revealed in the framework of FBMs: It has been shown that thermal noise leads to the reduction of the strength of materials, furthermore, thermally activated cracking gives rise to sub-critical crack growth and a finite lifetime of materials [59, 61]. When avalanches of fibre failures are solely recorded in the vicinity of the point of macroscopic failure, i.e. the strength distribution of the remaining intact fibres is close to critical, the avalanche size distribution $D(\Delta)$ proved to show a crossover from the power law of exponent 5/2 to another power law regime with a lower exponent 3/2 [55, 56]. The connectivity properties of the bundle turned out to play an important role in breakdown processes, i.e. determining the local interacting partners of fibres in a bundle by a Barabasi-Albert network instead of a regular lattice, the failure process is substantially different [10].

FBMs played also a crucial role to clarify the fundamental problem of the relation of breakdown of disordered materials to critical phenomena [2, 5, 37, 68, 69, 71] and self organized criticality [36]. When fibre bundles with global load sharing approach the point of macroscopic failure under a quasistatically increasing load, it was found that microscopic quantities of the bundle such as the distribution of bursts of fibre breakings show a scaling behavior typical for continuous phase transitions [5, 19, 22, 28, 37, 51, 52, 53]. Since at the same time macroscopic quantities like the Young modulus of the bundle proved to have a finite jump, it was suggested that the macroscopic failure of GLS systems occurs analogous to first order phase transitions close to a spinodal [33, 69, 70, 71]. An interesting mapping of the fracture process of fibre bundles to Ising-like models widely studied in statistical physics has been suggested

[53, 57, 66], which provides further hints to embed fracture phenomena into the general framework of statistical physics.

2.1 Why Extensions are Necessary?

A very important type of materials for which fibre bundle models are of outstanding importance are the so-called fibre reinforced composites (FRC) which have two ingredients, i.e. a matrix or carrier material in which thin fibres of another material are embedded in a certain geometrical arrangement. The modern automotive and aerospace industries use a large amount of fibre reinforced composites due to their very good mass specific properties, i.e. FRCs provide a very high strength at a relatively low mass and they sustain these properties also under extreme conditions (high temperature, pressure). The mechanical performance of FRC can be well controlled by the properties of the fibre and matrix materials, and by the fabrication of the fibre-matrix interface and the geometrical structure of the composite. This flexibility makes possible the design and tailoring of special purpose structural materials that best fit to the demands of applications. Fig. 3a) presents the structure of the carbon fibre reinforced silicon carbide composite (C/C-SiC) extensively used by the aerospace industry. It can be observed that parallel bundles of long carbon fibres (diameter $\sim 10\,\mu m$) are embedded in the silicon-carbide ceramic matrix [30]. Reinforcement by fibrous structures can also be found in nature in various types of plants. The structure of soft-wood of the species spruce is shown in Fig. 3b) as a representative example [15]. Theoretical studies of the fracture of composites encounter various challenges: on the one hand, applications of materials in construction components require the development of

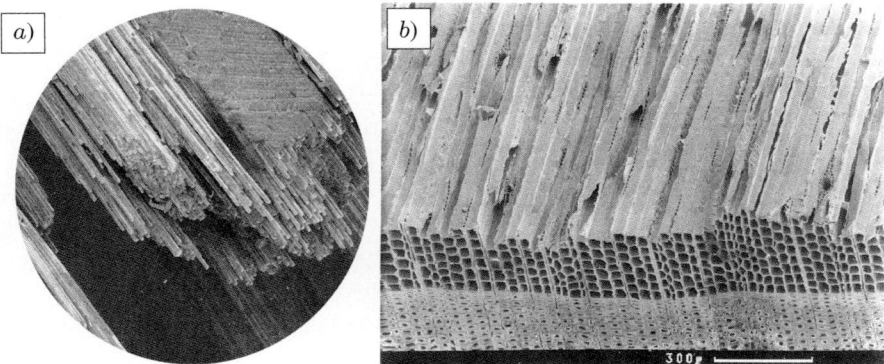

Fig. 3. (a) Structure of a fibre reinforced composite (Carbon fibre reinforced silicon-carbide). Long parallel fibres are embedded in a matrix material [30]. (b) The soft-wood is an example of natural fibrous materials. The fibres are tubes whose wall gets thicker as the tree grows [15]

analytical and numerical models which are able to predict the damage histories of loaded specimens in terms of the characteristic microscopic parameters of the constituents. On the other hand, it is important to reveal universal aspects of the fracture of composites, which are independent of specific material properties relevant on the microlevel. Such universal quantities can help to extract the relevant information from measured data and make possible to design monitoring techniques of the gradual degradation of composites' strength and construct methods of forecasting catastrophic failure events.

When fibre composites are subject to an external load parallel to the fibre direction, most of the load is carried by the fibres, the matrix material and properties of the fibre-matrix interface mainly determine the interaction (load transfer) among fibres, which make FBMs the most adequate approach to describe the fracture process of FRCs [8, 30, 48, 50]. As it has been presented above, in their basic setup FBMs provide a rather idealized representation of the behavior of disordered brittle materials. In order to obtain a more realistic description especially for the failure of composites, FBMs have to be enhanced by capturing a more detailed picture of the mechanical behavior of the constituents, the evolution of the local stress field and its interaction with the disordered material properties. In the following we outline important aspects of the mechanical performance of FRCs which cannot be described in the framework of the classical FBM but call for extensions of the model.

In applications, long fibre composites loaded parallel to the direction of fibres are often found to undergo a gradual degradation process such that the macroscopic constitutive curve $\sigma(\varepsilon)$ of the composite develops a plateau regime and the global failure is preceded by strain hardening. This effect becomes especially important when the fibre bundle has a hierarchical organization and the failure mechanisms relevant at the lower length scales (at the scale of fibres) gradually activate the breaking of higher order substructures (sub-bundles, bundles, and plies) of the system. When fibres are embedded in a matrix material, after a fibre breaks, the fibre-matrix interface debonds in the vicinity of the crack [49]. Due to the frictional contact at the interface, the load of failed fibres builds up again over a certain stress recovery length so that the broken fibre can still contribute to the overall load bearing capacity of the system.

Under high steady stresses, materials may undergo time dependent deformation resulting in failure called creep rupture, which determines the long time performance and limits the lifetime of construction elements. This complex time dependent macroscopic behavior emerges as the result of the interplay of the time dependent deformation of the constituents and the gradual accumulation of damage.

In applications solid blocks are often joined together by welding or glueing of the interfaces which are expected to sustain various types of external loads. Interfacial failure also occurs in fibre reinforced composites, where debonding of the fibre-matrix interface can even be the dominating damage mechanism when the composite is sheared. When interfaces of solids are subject to shear the interface elements suffer not only longitudinal deformation (compression and elongation) but also bending deformation. Such complex deformation states cannot be captured by discretizing the interface in terms of fibres which can only support longitudinal deformation.

As it has been discussed in Sect. 2, the range and functional form of load redistribution of fibres play a crucial role in the failure process of the model. A realistic description of load transfer following fibre failure should obviously fall somewhere between the limiting cases of the widely studied global and local load redistributions. The importance of the accurate description of the load transfer from broken to intact fibres has already been recognized by Daniels for cotton yarns [12].

In the following we present extensions of the classical fibre bundle model gradually improving components of the model construction outlined in Sect. 2: (*I*) In order to understand the damage mechanism that lead to plateau formation and strain hardening of the macroscopic response of quasi-brittle materials, we modify the failure law of fibres in FBMs assuming that individual fibres undergo a gradual degradation process reducing their Young modulus in a multiplicative way in consecutive failure events. (*II*) To describe the damage enhanced creep of fibrous materials we introduce time dependent deformation behavior for the fibres and combine it with strain controlled breaking in a global load sharing framework. (*III*) We show that more complex deformation states (besides the longitudinal deformation) and failure processes can be described by FBMs when fibres are substituted by beams, which can be stretched, compressed and bent at the two ends. We apply the bundle of beams to investigate the fracture of glued interfaces of solid blocks. (*IV*) Motivated by the results of fracture mechanics on the stress distribution around cracks, we introduce a one-parameter load transfer function for fibres, which can interpolate between the limiting cases of global and local load sharing. The value of the parameter of the load transfer function controls the effective range of interaction of broken and intact fibres. We apply the variable range of load sharing approach also to clarify universal aspects of the creep rupture of composites. Figure 4 gives an overview of the structure of the remaining part of the paper. We note that since in the framework of the classical fibre bundle model no effect of the matrix material between fibres is taken into account, the model is also called dry fibre bundle model (DFBM) in the literature in order to distinguish it from the improved variants.

3 Gradual Degradation of Fibre Strength

In the following we introduce a so-called continuous damage fibre bundle model as an extension of the classical FBM by generalizing the damage law of fibres. We assume that the stiffness of fibres gradually decreases in consecutive failure events. We show that varying its parameters, the model provides a broad spectrum of macroscopic constitutive behaviors from plasticity to

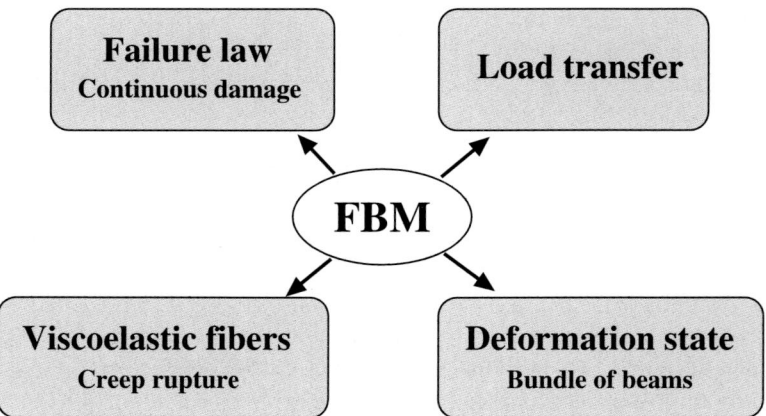

Fig. 4. Extensions of fibre bundle models: we gradually enhance FBMs by generalizing the damage law, the load transfer, the constitutive behavior and deformation state of the fibres

strain hardening in qualitative agreement with experiments. To reveal the microscopic process of damage we construct a simulation techniques and obtain the statistics of bursts of fibre breakings. We find that burst distributions show a power law behavior but surprisingly the exponent can be different from the well know mean field exponent of FBMs [24, 33].

3.1 Continuous Damage Fibre Bundle Model

Based on the classical fibre bundle model, the continuous damage model is composed of N parallel fibres with identical Young-modulus E and random failure thresholds σ_{th}^i, $i = 1, \ldots, N$ of probability density p and distribution function P (see also Fig. 1 for illustration). The fibres are assumed to have linear elastic behavior up to breaking (brittle failure). Under uniaxial loading of the specimen a fibre fails if it experiences a load larger than its breaking threshold σ_{th}^i. We generalize the failure law of DFBM by assuming that at the failure point the stiffness of the fibre gets reduced by a factor a, where $0 \leq a < 1$, i.e. the stiffness of the fibre after failure is aE. In principle, a fibre can now fail more than once and the maximum number $k_{\max}$ of failures allowed for fibres is a parameter of the model. Once a fibre has failed, its damage threshold σ_{th}^i can either be kept constant for the further breakings (quenched disorder) or new failure thresholds of the same distribution can be chosen (annealed disorder), which can model some microscopic rearrangement of the material after failure. The damage law of the model is illustrated in Fig. 5 for both types of disorder, which should be compared to Fig. 1. The characterization of damage by a continuous parameter corresponds to describe the system on length scales larger than the typical crack size. This can be interpreted such

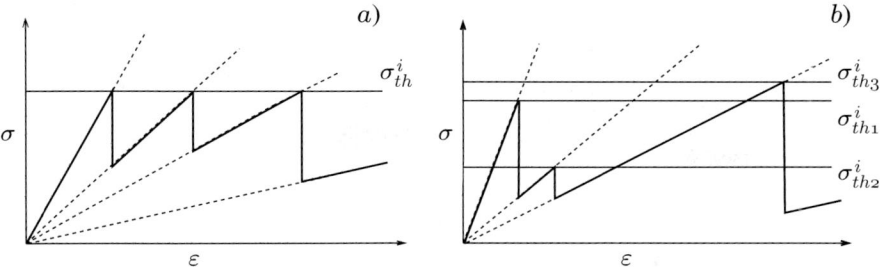

Fig. 5. The damage law of a single fibre of the continuous damage model when multiple failure is allowed (**a**) for quenched, and (**b**) for annealed disorder. The *horizontal lines* indicate the damage threshold σ_{th}^i. See also [24, 33]

that the smallest elements of the model are fibres and the continuous damage is due to cracking inside fibres. However, the model can also be considered as the discretization of the system on length scales larger than the size of single fibres, so that one element of the model consists of a collection of fibres with matrix material in between. In this case the microscopic damage mechanism resulting in multiple failure of the elements is the gradual cracking of matrix and the breaking of fibres. In the following we refer to the elements of the continuous damage FBM as fibres, but we have the above two possible interpretations in mind. After failure the fibre skips a certain amount of load which has to be taken by the other fibres. For the load redistribution we assume infinite range of interaction among fibres (mean field approach), furthermore, equal strain condition is imposed which implies that stiffer fibres of the system carry more load. At a strain ε, the load of fibre i that has failed $k(i)$ times reads as

$$f_i(\varepsilon) = E a^{k(i)} \varepsilon, \qquad (5)$$

where $Ea^{k(i)}$ is the actual stiffness of fibre i. It is important to note that, in spite of the infinite interaction range, (5) is different from the usual global load sharing (GLS) where all the intact fibres carry always the same amount of load (see Sect. 2). In the following the initial fibre stiffness E will be set to unity.

3.2 Macroscopic Constitutive Behavior

The key quantity to determine the macroscopic constitutive behavior of the continuous damage fibre bundle model (CDFBM) and to characterize its microscopic damage process, is the probability $P_k(\varepsilon)$ that during the loading of a specimen an arbitrarily chosen fibre failed precisely k-times at a strain ε. Here $k = 0, \ldots, k_{\max}$ denotes the failure index, and $k = 0$ is assigned to the intact fibres. $P_k(\varepsilon)$ can be cast in the following form for *annealed disorder*

$$P_k(\varepsilon) = [1 - P(a^k\varepsilon)] \prod_{j=0}^{k-1} P(a^j\varepsilon), \quad (6)$$

for $\quad 0 \leq k \leq k_{\max} - 1$,

and $\quad P_{k_{\max}}(\varepsilon) = \prod_{j=0}^{k_{\max}-1} P(a^j\varepsilon)$,

and for *quenched disorder*

$$P_0(\varepsilon) = 1 - P(\varepsilon),$$
$$P_k(\varepsilon) = P(a^{k-1}\varepsilon) - P(a^k\varepsilon), \text{ for } 1 \leq k \leq k_{\max} - 1, \quad (7)$$
$$\text{and} \quad P_{k_{\max}}(\varepsilon) = P(a^{k_{\max}-1}\varepsilon).$$

It can easily be seen that the probabilities (6,7) fulfill the normalization condition $\sum_{k=0}^{k_{\max}} P_k(\varepsilon) = 1$. Average quantities of the fibre ensemble during a loading process can be calculated using the probabilities (6,7). For instance, the average load on a fibre F/N at a given strain ε reads as

$$\frac{F}{N} = \varepsilon \left[\sum_{k=0}^{k_{\max}} a^k P_k(\varepsilon) \right], \quad (8)$$

which provides the macroscopic constitutive behavior of the model. The single terms in the sum give the load carried by the subset of fibres of failure index k. The variants of fibre bundle models used widespread in the literature can be recovered by special choices of the parameters $k_{\max}$ and a of the model. A micromechanical model of composites [8, 30, 48, 50] can be obtained with the parameter values $k_{\max} = 1$, $a \neq 0$

$$\frac{F}{N} = \varepsilon [1 - P(\varepsilon)] + a\varepsilon P(\varepsilon), \quad (9)$$

while setting $k_{\max} = 1$, $a = 0$, i.e. skipping the second term in (9) results in the classical dry bundle model of Daniels [12] with the constitutive behavior (3). In Fig. 6 we show the explicit form of the constitutive law with annealed disorder for different values of $k_{\max}$ in the case of the Weibull distribution (2). The parameter values are set to $m = 2$, $\lambda = 1$ in all the calculations.

Note that in the constitutive equation (8) the term of the highest failure index $k_{\max}$ can be conceived such that the fibres have a residual stiffness of $a^{k_{\max}}$ after having failed $k_{\max}$ times. This residual stiffness results in a hardening of the material, hence, the F/N curves in Fig. 6a asymptotically tend to straight lines with a slope $a^{k_{\max}}$. Increasing $k_{\max}$ the hardening part of the constitutive behavior is preceded by a longer and longer plastic plateau, and in the limiting case of $k_{\max} \to \infty$ the materials behavior becomes completely plasticity. A similar plateau and asymptotic linear hardening has been observed in brittle matrix composites, where the multiple cracking of matrix

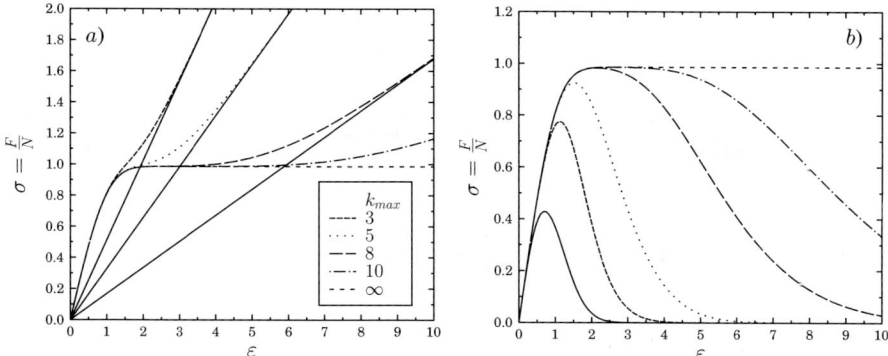

Fig. 6. Constitutive behavior of the model of annealed disorder (**a**) with (**b**) without residual stiffness at $a = 0.8$ for different values of $k_{\max}$. In (**b**) the lowest curve presents the constitutive behavior of the classical dry bundle model for comparison

turned out to be responsible for the relatively broad plateau of the constitutive behavior, and the asymptotic linear part is due to the linear elastic behavior of fibres remaining intact after matrix cracking [16].

In order to describe macroscopic cracking and global failure of a specimen instead of hardening, the residual stiffness of the fibres has to be set to zero after a maximum number k^* of allowed failures [33, 72]. In this case the constitutive law can be obtained from the general form (8) by replacing $k_{\max}$ in the upper limit of the sum by $k^* - 1$. A comparison of the constitutive laws of the dry and continuous damage FBM with global failure is presented in Fig. 6b. One can observe that the dry FBM constitutive law has a relatively sharp maximum, while the continuous damage FBM curves exhibit a plateau whose length increases with increasing k^*. Note that the maximum value of F/N corresponds to the macroscopic strength of the material, furthermore, in stress controlled experiments the plateau and the decreasing part of the curves cannot be reached. However, in strain controlled experiments the plateau and the decreasing regime can also be realized. The value of the driving stress $\sigma \equiv F/N$ corresponding to the plastic plateau, and the length of the plateau are determined by the damage parameter a, and by $k_{\max}, k^*$: Decreasing a at a fixed $k_{\max}, k^*$, or increasing $k_{\max}, k^*$ at a fixed a increase the plateau's length.

3.3 Simulation Techniques

Due to the difficulties of the analytic treatment, we develop a simulation technique and explore numerically the properties of bursts of fibre breakings in our continuous damage fibre bundle model. The interaction of fibres, i.e. the way of load redistribution is crucial for the avalanche activity. A very important property of CDFBM is that in spite of the infinite range of interaction the load on intact fibres is not equal, but stiffer fibres carry more load, furthermore, for quenched disorder damage localization occurs, which can affect also the avalanche activity [24, 33].

To implement the quasi-static loading of a specimen of N fibres in the framework of CDFBM, the local load on the fibres f_i has to be expressed in terms of the external driving F. Making use of (5) it follows that

$$F = \sum_{i=1}^{N} f_i = \varepsilon \sum_{i=1}^{N} a^{k(i)}, \tag{10}$$

and hence, the strain and the local load on fibres can be obtained as

$$\varepsilon = \frac{F}{\sum_{i=1}^{N} a^{k(i)}}, \quad f_i = F \frac{a^{k(i)}}{\sum_{i=1}^{N} a^{k(i)}}, \tag{11}$$

when the external load F is controlled. The simulation of the quasi-static loading proceeds as follows: in a given stable state of the system we determine the load on the fibres f_i from the external load F using (11). The next fibre to break can be found as

$$r = \frac{\sigma_{th}^{i^*}}{f_{i^*}} = \min_i \frac{\sigma_{th}^{i}}{f_i}, \quad r > 1, \tag{12}$$

i.e. that fibre breaks for which the ratio σ_{th}^{i}/f_i is the smallest. Here i^* denotes the index of the fibre to break, σ_{th}^{i} is the damage threshold of fibre i, and f_i is the local load on it. To ensure that the local load of a fibre is proportional to its stiffness, the external load has to be increased in a multiplicative way, so that $F \to rF$ is imposed, and the failure index of fibre i^* is increased by one $k(i^*) \to k(i^*) + 1$. After the breaking of fibre i^*, the load f_i carried by the fibres has to be recalculated making use of (11), which provides also the correct load redistribution of the model. If there are fibres in the state obtained, whose load exceeds the local breaking threshold, they fail, i.e. their failure index is increased by 1 and the local load is again recalculated until a stable state is obtained. A fibre cannot break any longer if its failure index k has reached k^* or $k_{\max}$ during the course of the simulations. This dynamics gives rise to a complex avalanche activity of fibre breaks, which is also affected by the type of disorder. The size of an avalanche Δ is defined as the number of breakings initiated by a single failure due to an external load increment.

3.4 Bursts of Fibres Breakings

Computer simulations revealed that varying the two parameters of the model $k_{\max}, a$, or k^*, a and the type of disorder, the CDFBM shows an interesting variety of avalanche activities, characterized by different shapes of the avalanche

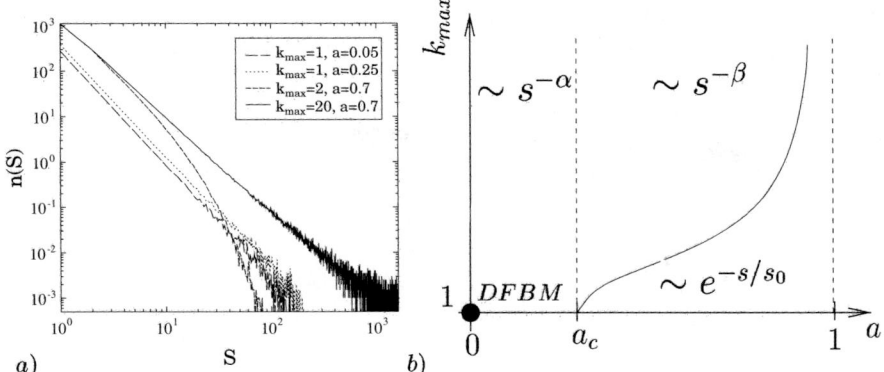

Fig. 7. (a) Avalanche size distributions for different values of $k_{\max}$ and a when fibres have remaining stiffness and the disorder is annealed. The number of fibres was $N = 1600$ and averages were made over 2000 samples. The number of avalanches D of size Δ are shown to demonstrate also how the total number of avalanches changes. (b) Phase diagram for the continuous damage model with remaining stiffness for both types of disorder. The functional form of the avalanche statistics is given in the parameter regimes. The location of DFBM in the parameter space is also indicated

size distributions. In Fig. 7a the histograms $D(\Delta)$ of the avalanche sizes Δ are shown which were obtained for a system of remaining stiffness and annealed disorder with Weibull parameters $m = 2$, $\lambda = 1$. Since in the limiting case of $a \to 0$ the CDFBM recovers the global load sharing dry fibre bundle model, in Fig. 7a the curves with small a and $k_{\max} = 1$ are power laws with an exponent $\alpha = 5/2$ in agreement with the analytic results [22, 28, 54]. Increasing the value of a at a fixed $k_{\max}$ only gives rise to a larger number of avalanches, i.e. parallel straight lines are obtained on a double logarithmic plot, but the functional form of $D(\Delta)$ does not change. However, when a exceeds a critical value a_c ($a_c \approx 0.3$ was obtained with the Weibull parameters specified above), the avalanche statistics drastically changes. At a fixed $a > a_c$ when $k_{\max}$ is smaller than a specific value $k_c(a)$, the avalanche sizes show exponential distribution, while above $k_c(a)$ the distribution takes a power law form with an exponent $\beta = 2.12 \pm 0.05$. Based on the above results of simulations a phase diagram is constructed which summarizes the properties of avalanches with respect to the parameters of the model. Figure 7 demonstrates the existence of three different regimes. If the damage parameter a is smaller than a_c, the dynamics of avalanches is close to the simple DFBM characterized by a power law of the mean field exponent $\alpha = 5/2$. However, for $a > a_c$ the avalanche size distribution depends on the number of failures $k_{\max}$ allowed. The curve of $k_c(a)$ in the phase diagram separates two different regimes. For the parameter regime below the curve, avalanche distributions with an exponential shape were obtained. However, the parameter regime above $k_c(a)$

is characterized by a power law distribution of avalanches with a constant exponent $\beta = 2.12 \pm 0.05$ significantly different from the mean field exponent $\alpha = 5/2$ [22, 28, 54]. It is important to emphasize that the overall shape of the phase diagram is independent of the type of the disorder (annealed or quenched), moreover, the specific values $a_c \approx 0.3$ and $k_c(a)$ depend on the details of the disorder distribution $p(\sigma_{th})$.

4 Variable Range of Load Sharing

In this section we introduce a one-parameter load transfer function to obtain a more realistic description of the interaction of fibres. Varying its parameter, the load transfer function interpolates between the two limiting cases of load redistribution, i.e. the global and the local load sharing schemes widely studied in the literature. We show that varying the effective range of interaction a crossover occurs from mean field to short range behavior. To explore the properties of the two regimes and the emergence of the crossover in between, a comprehensive numerical study of the model is performed. We study the dependence of the ultimate strength of the material on the system size and found that the system has only one nonzero critical point in the thermodynamic limit. When no critical point exits, the ultimate strength of the material goes to zero exactly as in local load sharing models as $\sim 1\ln(N)$, with increasing system size N. We also study the distribution of avalanches and cluster sizes for the two distinct regimes and perform a moments analysis to accurately obtain the crossover value [26].

4.1 Load Transfer Function

The fracture of heterogeneous systems is characterized by the highly localized concentration of stresses at the crack tips that makes possible the nucleation of new micro-cracks at these regions leading to the growth of the crack and final failure of the system. In elastic materials, the stress distribution around cracks follows a power law

$$\sigma_{add} \sim r^{-\gamma}, \qquad (13)$$

where σ_{add} is the stress increase on a material element at a distance r from the crack tip. Motivated by the above result of fracture mechanics we extend the fibre bundle model introducing a load sharing rule of the form of (13).

In the model we suppose that, in general, all intact fibres have a nonzero probability of being affected by the ongoing failure event, and that the additional load received by an intact fibre i depends on its distance r_{ij} from fibre j which has just been broken. Furthermore, elastic interaction is assumed between fibres such that the load received by a fibre follows the power law form.

Hence, in our discrete model the stress-transfer function $F(r_{ij}, \gamma)$ takes the form

$$F(r_{ij}, \gamma) = Z r_{ij}^{-\gamma}, \qquad (14)$$

where γ is an adjustable parameter and r_{ij} is the distance of fibre i to the rupture point (x_j, y_j), i.e., $r_{ij} = \sqrt{(x_i - x_j)^2 + (y_i - y_j)^2}$ in 2D. Z is given by the normalization condition $Z = (\sum_{i \in I} r_{ij}^{-\gamma})^{-1}$, where the sum runs over the set I of all intact elements. For simplicity, periodic boundary conditions with the minimum image convention are used (no Ewald summation is performed [1]) so that the largest r value is $R_{\max} = \frac{\sqrt{2}(L-1)}{2}$, where L is the linear size of the system. It is easy to see that in the limits $\gamma \to 0$ and $\gamma \to \infty$ the load transfer function (14) recovers the two extreme cases of load redistribution of fibre bundle models, i.e. the global and the local load sharing, respectively. We should note here that, strictly speaking, for all γ different from the two limits above, the range of interaction covers the whole lattice. However, when changing the exponent, one moves from a very localized *effective* interaction to a truly global one as γ approaches zero. So, we will refer henceforth to a change in the *effective* range of interaction.

4.2 Macroscopic Strength of Bundles

We have carried out large scale computer simulations of the model described above in two dimensions. The fibres with Weibull distributed strength values (2) are organized on a square lattice of linear size L using periodic boundary conditions (see also Fig. 1). Computer simulations were performed varying the effective range of interaction γ over a broad interval recording the avalanche size distribution, the cluster size distribution and the ultimate strength of the bundle for several system sizes L. Each numerical simulation was repeated over at least 50 different realizations of the disorder distribution with the parameter values $m = 2$ and $\lambda = 1$ of the Weibull distribution.

Figure 8a shows the ultimate strength σ_c of the fibre bundle for different values of the parameter γ and several system sizes from $L = 33$ to $L = 257$. Clearly, two distinct regimes can be distinguished: For small γ, the strength σ_c is independent, within statistical errors, of both the effective range of interaction γ and the system size L. At a given point $\gamma = \gamma_c$ a crossover is observed, where γ_c falls in the vicinity of $\gamma = 2$. The region $\gamma > \gamma_c$ might eventually be further divided into two parts, the first region characterized by the dependence of the ultimate strength of the bundle on both the system size and the effective range of interaction; and a second region where σ_c only depends on the system size. This would mean that there might be two transition points in the model, for which the system displays qualitatively and quantitatively different behaviors. For $\gamma \leq \gamma_c$ the ultimate strength of the bundle behaves as in the limiting case of global load sharing, whereas for $\gamma \geq \gamma_c$ the local load sharing behavior seems to prevail. Nevertheless, the most important feature is that when decreasing the effective range of interaction in the thermodynamic

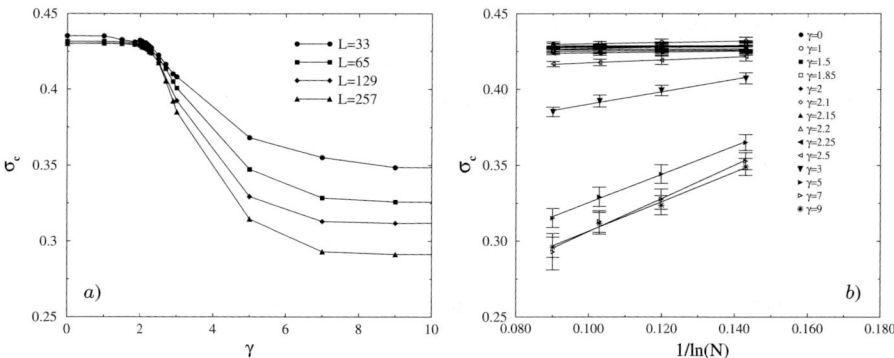

Fig. 8. (a) Macroscopic strength σ_c of fibre bundles of different size L as a function of the exponent γ of the load transfer function (14). (b) Variation of σ_c with the number of fibres of the bundle N at different γ values

limit, for $\gamma > \gamma_c$, the critical load is zero. This observation is further supported by Fig. 8(b), where we have plotted the evolution of σ_c as a function of $1/\ln N$ for different values of the exponent γ. Here, the two limiting cases are again clearly differentiated. For large γ all curves decreases when $N \to \infty$ as

$$\sigma_c(N) \sim \frac{\alpha}{\ln N} \;. \qquad (15)$$

This qualifies for a genuine short range behavior as found in LLS models where the same relation was obtained for the asymptotic strength of the bundle [49]. It is worth noting that in the model we are analyzing, the limiting case of local load sharing corresponds to models in which short range interactions are considered to affect the nearest and the next-nearest neighbors only. In the transition region, the maximum load the system can support also decreases as we approach the thermodynamic limit, but in this case much slower than for $\gamma \gg \gamma_c$. It has been pointed out that for some modalities of stress transfer, which can be considered as intermediate between GLS and LLS, σ_c decreases for large system sizes following the relation $\sigma_c \sim 1/\ln(\ln N)$ as in the case of hierarchical load transfer models [41]. In our case, we have fitted our results with this relation but we have not obtained a single collapsed curve because the slopes continuously vary until the LLS limit is reached. Finally, the region where the ultimate stress does not depend on the system size shows the behavior expected for the standard GLS model, where the critical load can be exactly computed as $\sigma_c = (me)^{-1/m}$ for the Weibull distribution. The numerical values obtained for $m = 2$ are in good agreement with this later expression.

4.3 Microstructure of Damage

In order to characterize the fracture process under the quasistatically increasing external load we analyzed the cluster structure of broken fibres. The clusters formed during the evolution of the fracture process are sets of spatially connected broken sites on the square lattice [33, 70, 71]. We consider the clusters just before the global failure and they are defined taking into account solely nearest neighbor connections. As it has been discussed in Sect. 2, the case of global load sharing does not assume any spatial structure of fibres since it corresponds to the mean field approach. However, in our case it is obtained as a limiting case of a local load sharing model on a square lattice, which justifies the cluster analysis also for GLS. Figure 9 illustrates how the cluster structure just before complete breakdown changes for various values

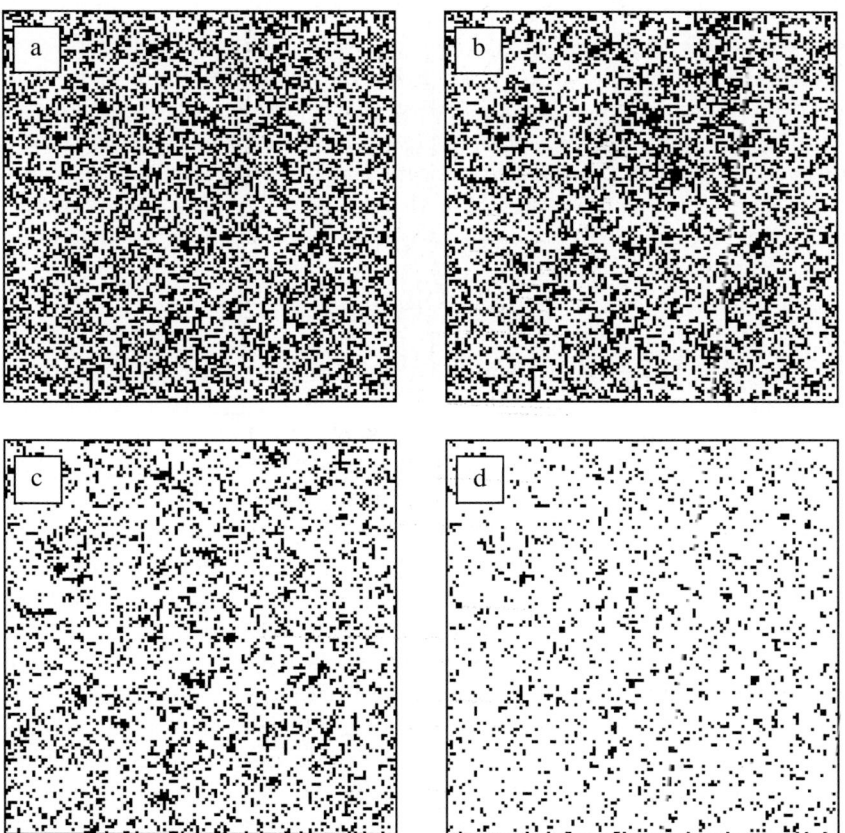

Fig. 9. Snapshots of the clusters of broken fibres (*black dots*) just before complete breakdown occurs. A clear change of the cluster structure can be observed. The values of γ are: (**a**) $\gamma = 0$, (**b**) $\gamma_c = 2.2$, (**c**) $\gamma = 3.0$, and (**d**) $\gamma = 9.0$

of γ. In the limit where the long range interaction dominates, the clusters are randomly distributed on the lattice indicating that there is no correlated crack growth in the system and no stress concentration occurs. The cluster structure of the limiting case of $\gamma = 0$ can be mapped to percolation clusters on a square lattice generated with the probability $0 < P(\sigma_c) < 1$, where σ_c is the fracture strength of the fibre bundle. However, the value of $P(\sigma_c)$ depends on the Weibull exponent m (2) and is normally different from the critical percolation probability $p_c = 0.592746$ of the square lattice. The equality $P(\sigma_c) = p_c$ is obtained for $m = 1.1132$, hence, for physically relevant values $m \geq 2$ used in simulations, the system is below p_c at complete breakdown. This picture radically changes when the short range interaction prevails. In this case, the stress transfer is limited to a neighborhood of the failed elements and there appear regions where a few isolated cracks drive the rupture of the material by growth and coalescence. The differences in the structure of clusters also explain the lack of a critical strength when N goes to infinity in models with local stress redistribution. Since in the GLS model the clusters are randomly dispersed across the entire lattice, the system can tolerate a larger amount of damage and a higher stress, whereas for LLS models a small increment of the external field may provoke a run away event ending with the macroscopic breakdown of the material. All the above numerical results suggest that the crossover between the mean field and the short range regimes occurs in the vicinity of $\gamma = 2$. Further support for the precise value of γ_c can be obtained by studying the change in the cluster structure of broken fibres. The moments of $n(s_c)$ defined as

$$m_k \equiv \int s_c^k n(s_c) ds , \qquad (16)$$

where m_k is the kth moment, describe much of the physics associated with the breakdown process. We study these moments to quantitatively characterize the point where the crossover from mean field to short range behavior takes place. The zero moment $n_c = m_o$ is the total number of clusters in the system and is plotted in Fig. 10a as a function of the parameter γ. Figure 10b presents the variation of the first moment of the distribution m_1, which is equal to the total number of broken fibres $N_c = m_1$. It turns out that up to a certain value of the effective range of interaction γ, N_c remains constant and then it decreases fast until a second plateau seems to arise. Note that the constant value of N_c for small γ is in agreement with the value of the fraction of broken fibres just before the breakdown of the material in mean field models [33, 62]. This property clearly indicates a change in the evolution of the failure process and may serve as a criterion to determine the crossover point. However, a more abrupt change is observed in the average cluster size $\langle s_c \rangle$ as a function of γ. According to the moments description, the average cluster size is equal to the second moment of the cluster distribution divided by the total number of broken sites, i.e. $\langle s_c \rangle = m_2/m_1$. It can be seen in Fig. 10d that $\langle s_c \rangle$ has a sharp maximum at $\gamma = 2.2 \pm 0.1$, and thus the average cluster size drastically

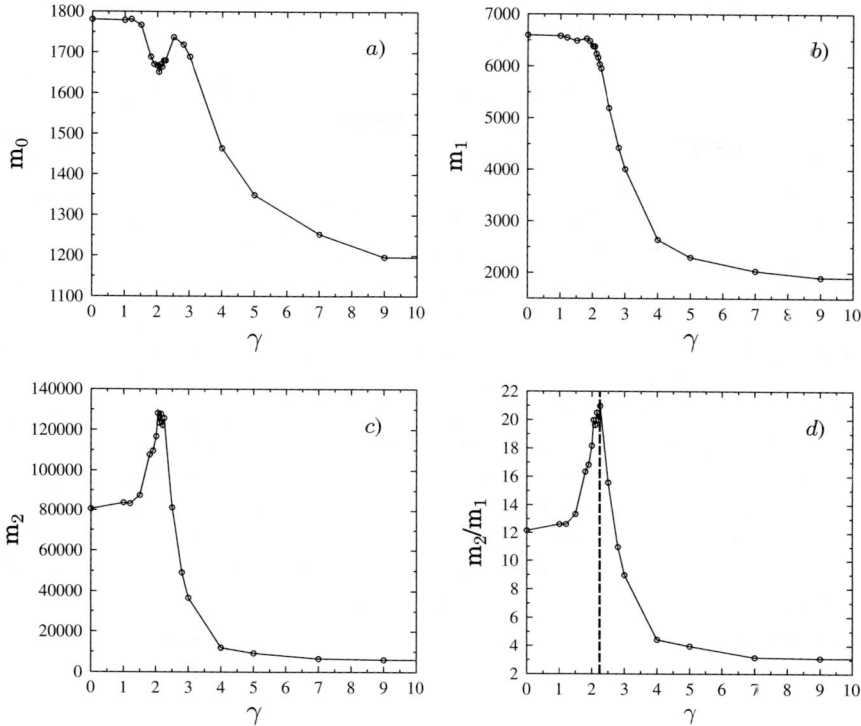

Fig. 10. Moments m_k of the clusters of broken fibres as a function of γ. The 0th and first moments are equal to the total number of clusters $n_c = m_0$ (**a**) and the total number of broken fibres $N_c = m_1$ (**b**), respectively. The average cluster size $\langle s_c \rangle$ can be determined as the ratio of the second (**c**) and first moments $\langle s_c \rangle = m_2/m_1$ (**d**)

changes at this point, which again suggests the crossover point to be in the vicinity of $\gamma_c = 2$.

5 Damage Enhanced Creep in Fibre Bundles

Under high steady stresses, materials may undergo time dependent deformation resulting in failure called creep rupture which limits their lifetime, and hence, has a high impact on their applicability in construction elements. Creep failure tests are usually performed under uniaxial tensile loading when the specimen is subjected to a constant load σ_0 and the time evolution of the damage process is followed by recording the strain ε of the specimen and the acoustic signals emitted by microscopic failure events. In order to describe the time dependent macroscopic response and finite lifetime of materials in the framework of FBMs, we extend the classical fibre bundle model presented in Sect. 2 by assuming that the fibres are viscoelastic, i.e. they exhibit time dependent deformation under a constant external load and fail in a strain controlled manner. In the framework of the viscoelastic fibre bundle model we show by analytical and numerical calculations that there exists a critical load above which the deformation of the system monotonically increases in time resulting in global failure at a finite time t_f, while below the critical load the deformation tends to a constant value giving rise to an infinite lifetime. Our studies revealed that for global load sharing the transition between the two regimes occurs analogously to continuous phase transitions, while for local load sharing it becomes abrupt, defining two universality classes of creep rupture [25, 31, 32].

5.1 Viscoelastic Fibre Bundle

In order to capture time dependent macroscopic response in the framework of fibre bundle models, we assume that the fibres themselves are viscoelastic. For simplicity, the pure viscoelastic behavior of fibres is modelled by a Kelvin-Voigt element which consists of a spring and a dashpot in parallel and results in the constitutive equation $\sigma_0 = \beta \dot{\varepsilon} + E\varepsilon$, where σ_0 is the external load, β denotes the damping coefficient, and E is the Young modulus of fibres, respectively (see Fig. 11a).

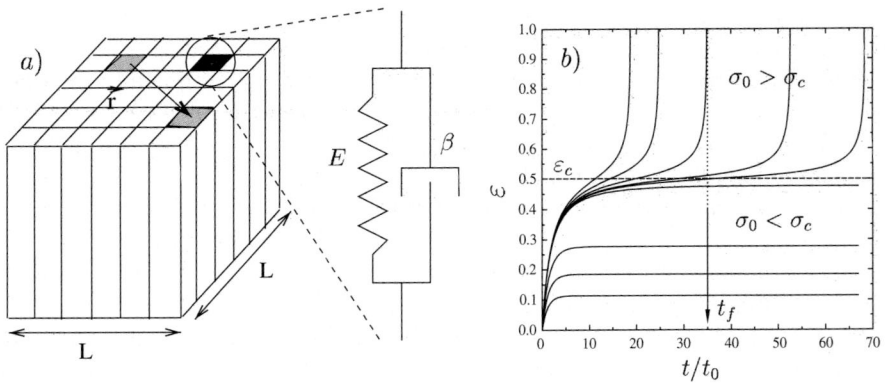

Fig. 11. (a) The viscoelastic fibre bundle model. Intact fibres are modelled by Kelvin-Voigt elements. (b) $\varepsilon(t)$ for several different values of the external load σ_0 below and above σ_c. t_0 denotes the characteristic timescale of the system $t_0 = \beta/E$.

In order to capture failure in the model a strain controlled breaking criterion is imposed, i.e. a fibre fails during the time evolution of the system when its strain exceeds a breaking threshold $\varepsilon_i, i = 1, \ldots, N$ drawn from a probability distribution. For the stress transfer between fibres following fibre failure we assume that the excess load is equally shared by all the remaining intact fibres (global load sharing), which provides a satisfactory description of load redistribution in continuous fibre reinforced composites. For the breaking thresholds of fibres a uniform distribution (1) between 0 and 1, and a Weibull distribution (2) were considered. The construction of the model is illustrated in Fig. 11a. In the framework of global load sharing most of the quantities describing the behavior of the fibre bundle can be obtained analytically. In this case the time evolution of the system under a steady external load σ_0 is described by the differential equation

$$\frac{\sigma_0}{1 - P(\varepsilon)} = \beta \dot{\varepsilon} + E\varepsilon , \tag{17}$$

where the viscoelastic behavior is coupled to the failure of fibres [25, 31, 32].

5.2 Macroscopic Response

The viscoelastic fibre bundle model with the equation of motion (17) can provide an adequate description of natural fibre composites like wood subjected to a constant load [15]. For the behavior of the solutions $\varepsilon(t)$ of (17), two distinct regimes can be distinguished depending on the value of the external load σ_0: when σ_0 falls below a critical value σ_c, (17) has a stationary solution ε_s, which can be obtained by setting $\dot{\varepsilon} = 0$, i.e. $\sigma_0 = E\varepsilon_s[1 - P(\varepsilon_s)]$. It means that as long as this equation can be solved for ε_s at a given external load σ_0, the solution $\varepsilon(t)$ of (17) converges to ε_s when $t \to \infty$, and the system suffers only a partial failure. However, when σ_0 exceeds the critical value σ_c no stationary solution exists, furthermore, $\dot{\varepsilon}$ remains always positive, which implies that for $\sigma_0 > \sigma_c$ the strain of the system $\varepsilon(t)$ monotonically increases until the system fails globally at a finite time t_f [25, 31, 32].

The behavior of is illustrated in Fig. 11b for several values of σ_0 below and above σ_c with uniformly distributed breaking thresholds between 0 and 1. It follows from the above argument that the critical value of the load σ_c is the static fracture strength of the bundle. The creep rupture of the viscoelastic bundle can be interpreted so that for $\sigma_0 \leq \sigma_c$ the bundle is partially damaged implying an infinite lifetime $t_f = \infty$ and the emergence of a stationary macroscopic state, while above the critical load $\sigma_0 \geq \sigma_c$ global failure occurs at a finite time t_f, but in the vicinity of σ_c the global failure is preceded by a long lived stationary state. The nature of the transition occurring at σ_c can be characterized by analyzing how the creeping system behaves when approaching the critical load both from below and above. For $\sigma_0 \leq \sigma_c$, the fibre bundle relaxes to the stationary deformation ε_s through a gradually decreasing breaking activity. It can be shown analytically that $\varepsilon(t)$ has an exponential relaxation to ε_s with a characteristic time scale τ that depends on the external load σ_0 as

$$\tau \sim (\sigma_c - \sigma_0)^{-1/2} , \tag{18}$$

for $\sigma_0 \leq \sigma_c$, i.e. when approaching the critical point from below the characteristic time of the relaxation to the stationary state diverges according to

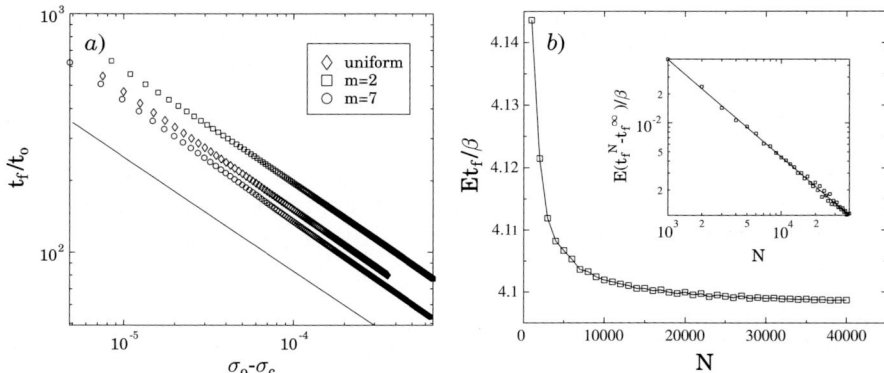

Fig. 12. (a) Lifetime t_f of the bundle as a function of the distance from the critical point $\sigma_0 - \sigma_c$ for three different disorder distributions, i.e. uniform distribution and Weibull distribution with $m = 2$ and $m = 7$ were considered. (b) t_f as a function of the number of fibres N at a fixed value of the external load σ_0. Results of computer simulations (*symbols*) are in good agreement with the analytic predictions (*solid lines*)

a universal power law with an exponent $1/2$ independent on the form of the disorder distribution P.

Above the critical point the lifetime defines the characteristic time scale of the system which can be cast in the form

$$t_f \sim (\sigma_0 - \sigma_c)^{-1/2}, \qquad (19)$$

for $\sigma_0 > \sigma_c$, so that t_f also has a power law divergence at σ_c with a universal exponent $1/2$ like τ below the critical point, see Fig. 12a. Hence, for global load sharing the system exhibits scaling behavior on both sides of the critical point indicating a continuous transition at the critical load σ_c. It can also be shown analytically that fixing the external load above the critical point, the lifetime t_f of the system exhibits a universal scaling with respect to the number N of fibres of the bundle (Fig. 12b). The lifetime of finite bundles t_f^N decreases to the lifetime of the infinite bundle as $t_f^N - t_f^\infty \sim 1/N$ [25, 31].

5.3 Microscopic Damage Process

The process of fibre breaking on the micro level can easily be monitored experimentally by means of the acoustic emission techniques. Except for the primary creep regime, where a large amount of fibres break in a relatively short time, the time of individual fibre failures can be recorded with a high precision. In order to characterize the process of fibre breaking in our viscoelastic fibre bundle model, we calculated numerically the distribution f of waiting times

Δt between consecutive breaks [32]. A detailed analysis revealed that $f(\Delta t)$ shows a power law behavior on both sides of the critical point

$$f(\Delta t) \sim \Delta t^{-b} \qquad (20)$$

[32]. The value of the exponent b is different on the two sides of the critical point, i.e. $b \approx 1.5$ and $b \approx 2.0$ were measured below and above the critical load; however, b proved to be independent of the disorder distribution of fibres, see Fig. 13.

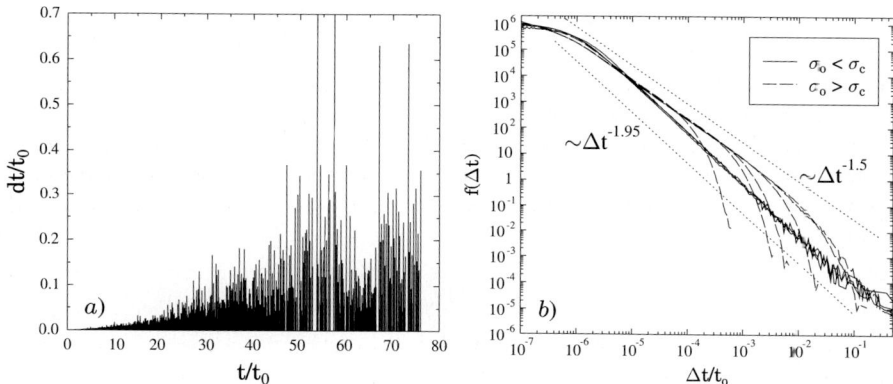

Fig. 13. (a) Representative example of waiting times Δt between consecutive fibre breakings for a load $\sigma_0 < \sigma_c$ plotted at the time t of breakings. (b) Distribution f of Δt below and above the critical load. Simulation results are presented for 10^7 fibres with uniformly distributed threshold values between 0 and 1

5.4 Universality Classes of Creep Rupture

In order to clarify the effect of the range of load sharing on damage enhanced creep, we carried out computer simulations applying the variable range of load sharing approach presented in Sect. 4 in the viscoelastic fibre bundle model (see also Fig. 11a for explanation). Simulation were carried out varying the exponent γ of the load transfer function (14) in a broad range. Computer simulations revealed that the behavior described above for global load sharing prevails in the range $0 \leq \gamma < \gamma_c$, where the value of γ_c turned to be the same as the one determined in Sect. 4 for dry fibre bundles. For $\gamma \geq \gamma_c$ the interaction of fibres becomes localized which results in an abrupt failure of the viscoelastic fibre bundle above the critical load $\sigma_0 > \sigma_c$, i.e. when σ_0 exceeds σ_c the lifetime of the bundle suddenly jumps to a finite value without any scaling behavior. It is interesting to note, however, that the power law distribution of waiting times (20) prevails for localized load sharing. The value of the exponent b below the critical load $\sigma_0 < \sigma_c$ is always $b \approx 2.0$ independent of the effective range of load redistribution γ, while for $\sigma_0 > \sigma_c$ the exponent b increases from ≈ 1.5 to ≈ 2.0 with increasing γ [32].

6 Failure of Glued Interfaces of Solid Blocks

Solid blocks are often joined together by welding or glueing of the interfaces which are expected to sustain various types of external loads. Besides welded joints, glued interfaces of solids play a crucial role in fibre reinforced composites where fibres are embedded in a matrix material. The properties of the fibre-matrix interface are controlled by the fabrication process of the composite and have an important effect on the mechanical performance of the system.

The classical setup of fibre bundle models have recently been applied to study the failure of glued interfaces when the external load is applied perpendicular to the interface of a rigid and a compliant block [4]. It was found that the compliance of one of the solid blocks introduces stress localization and results in a fractal structure of the failed glue just before macroscopic breakdown [4]. Glued interfaces can be teared apart by applying the external load locally at one of the ending point of the interface. Such loading conditions have also been studied in the framework of FBMs considering two stiff plates [29] and also the combination of a stiff plate and a compliant one [13, 14, 60].

In all the above studies, the external load resulted in a stretching deformation of material elements of the interface so that the classical fibre bundle model discussed in Sect. 2 provided an adequate framework for the theoretical description of the failure process. However, under shear loading more complex deformation states of material elements can arise leading to a more complex degradation process, which cannot be captured by FBMs. We propose a novel approach to the shear failure of glued interfaces by extending the classical fibre bundle model to model interfacial failure [58].

6.1 The Beam Model of Interface Failure Under Shear

In our model the interface is discretized in terms of elastic beams which can be elongated and bent when exposed to shear load, see Fig. 14. The beams are assumed to have identical geometrical extensions (length l and width d) and linearly elastic behavior characterized by the Young modulus E. In order to capture the failure of the interface, the beams are assumed to break when their deformation exceeds a certain threshold value. Under shear loading of the interface, beams suffer stretching and bending deformation resulting in two modes of breaking. The stretching and bending deformation of beams can be expressed in terms of a single variable, i.e. longitudinal strain, which enables us to map the interface model to the simpler fibre bundle models. The two breaking modes can be considered to be independent or combined in the form of a von Mises type breaking criterion. The strength of beams is characterized

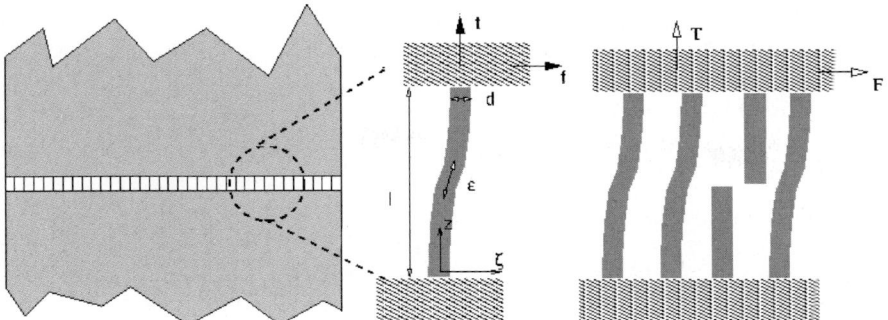

Fig. 14. Illustration of the model construction. The sheared interface is discretized in terms of beams (*left*), which suffer stretching and bending deformation (*middle*) and fail due to the two deformation modes (*right*)

by the two threshold values of stretching and bending a beam can withstand. The breaking thresholds are assumed to be randomly distributed variables of the joint probability distribution. The randomness of the breaking thresholds is supposed to represent the disorder of the interface material. After breaking of a beam the excess load has to be redistributed over the remaining intact elements. Coupling to the rigid blocks ensures that all the beams have the same deformation giving rise to global load sharing, i.e. the load is equally shared by all the elements, stress concentration in the vicinity of failed beams cannot occur.

6.2 Constitutive Behavior

Breaking of the beam is caused by two breaking modes, i.e. stretching and bending which can be either independent or coupled by an empirical breaking criterion. Assuming that the two breaking modes are independent, a beam breaks if either the longitudinal stress t or the bending moment m exceeds the corresponding breaking threshold, see Fig. 14. Since the longitudinal stress t and the bending moment m acting on a beam can easily be expressed as functions of the longitudinal deformation ε, the breaking conditions can be formulated in a transparent way in terms of ε. To describe the relative importance of the breaking modes, we assign to each beam two breaking thresholds $\varepsilon_1^i, \varepsilon_2^i, i = 1, \ldots, N$, where N denotes the number of beams. The threshold values ε_1 and ε_2 are randomly distributed according to a joint probability density function $p(\varepsilon_1, \varepsilon_2)$ between lower and upper bounds $\varepsilon_1^{\min}, \varepsilon_1^{\max}$ and $\varepsilon_2^{\min}, \varepsilon_2^{\max}$, respectively. The density function needs to obey the normalization condition

$$\int_{\varepsilon_2^{\min}}^{\varepsilon_2^{\max}} d\varepsilon_2 \int_{\varepsilon_1^{\min}}^{\varepsilon_1^{\max}} d\varepsilon_1 \, p(\varepsilon_1, \varepsilon_2) = 1 \; . \tag{21}$$

First, we provide a general formulation of the failure of a bundle of beams. We allow for two independent breaking modes of a beam that are functions f and g of the longitudinal deformation ε. Later on this case will be called the OR breaking rule. A single beam breaks if either its stretching or bending deformation exceed the respective breaking threshold ε_1 or ε_2, i.e. failure occurs if

$$\frac{f(\varepsilon)}{\varepsilon_1} \geq 1, \quad \text{or} \tag{22}$$

$$\frac{g(\varepsilon)}{\varepsilon_2} \geq 1, \tag{23}$$

where (22, 23) describe the stretching and bending breaking modes, respectively. The failure functions $f(\varepsilon)$ and $g(\varepsilon)$ can be determined from the elasticity equations of beams, but in general the only restriction for them is that they have to be monotonous. For our specific case of sheared beams they take the form

$$f(\varepsilon) = \varepsilon, \quad g(\varepsilon) = a\sqrt{\varepsilon}, \tag{24}$$

where a is a constant and the value of the Young modulus E is set to 1. In the plane of breaking thresholds each point $(\varepsilon_1, \varepsilon_2)$ represents a beam. For each value of ε those beams which survived the externally imposed deformation are situated in the area $f(\varepsilon) \leq \varepsilon_1 \leq \varepsilon_1^{\max}$ and $g(\varepsilon) \leq \varepsilon_2 \leq \varepsilon_2^{\max}$, as it is illustrated in Fig. 15a. The constitutive behavior of the interface can be obtained by integrating the load of single beams over the intact ones in the plane of breaking thresholds in Fig. 15a. For the OR criterion one gets

$$\sigma = \varepsilon \int_{g(\varepsilon)}^{\varepsilon_2^{\max}} d\varepsilon_2 \int_{f(\varepsilon)}^{\varepsilon_1^{\max}} d\varepsilon_1 \, p(\varepsilon_1, \varepsilon_2). \tag{25}$$

Assuming the thresholds of the two breaking modes to be independently distributed, the disorder distribution factorizes $p(\varepsilon_1, \varepsilon_2) = p_1(\varepsilon_1) p_2(\varepsilon_2)$ and $\sigma(\varepsilon)$ takes the simple form

$$\sigma = \varepsilon [1 - P_1(f(\varepsilon))][1 - P_2(g(\varepsilon))]. \tag{26}$$

The terms $1 - P_1(f(\varepsilon))$ and $1 - P_2(g(\varepsilon))$ provide the fraction of beams failed under the stretching and bending breaking modes, respectively.

When the two breaking modes are coupled by a von Mises type breaking criterion, a single beam breaks if its strain ε fulfils the condition

$$\left(\frac{f(\varepsilon)}{\varepsilon_1}\right)^2 + \frac{g(\varepsilon)}{\varepsilon_2} \geq 1. \tag{27}$$

This algebraic condition can be geometrically represented as it is illustrated in Fig. 15a. In this case the constitutive integral

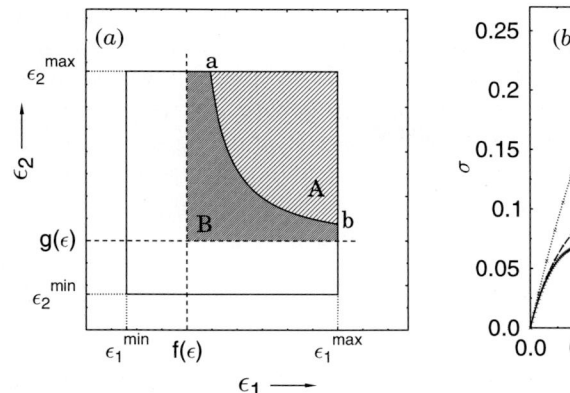
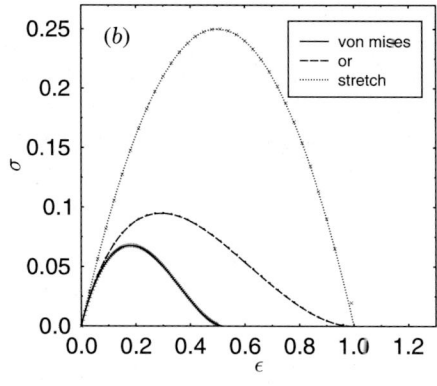

Fig. 15. (a) The plane of breaking thresholds. Fibres which are intact at a deformation ε fall in the rectangle bounded by $f(\varepsilon)$, $g(\varepsilon)$ and the maximum values of the two thresholds $\varepsilon_1^{\max}$, $\varepsilon_2^{\max}$ for the OR criterion (area $A+B$), and in the area bounded by the curve connecting a and b and the maximum thresholds for the von Mises type criterion (area A), respectively. The fibres which break due to the coupling of the two breaking modes in the von Mises criterion fall in area B. (b) Comparison of the constitutive curves of a simple dry fibre bundle and of the beam model with different breaking criteria using uniformly distributed breaking thresholds

$$\sigma = \epsilon \int_a^{\varepsilon_1^{\max}} d\varepsilon_1 \int_{\tilde{\varepsilon}_2(\varepsilon_1,\varepsilon)}^{\varepsilon_2^{\max}} d\varepsilon_2\, p(\varepsilon_1,\varepsilon_2) \tag{28}$$

cannot be performed explicitly with the integration limit $\tilde{\varepsilon}_2(\varepsilon_1,\varepsilon) = \varepsilon_1^2 g(\varepsilon)/(\varepsilon_1^2 - f^2(\varepsilon))$ in general so that $\sigma(\varepsilon)$ cannot be obtained in a closed analytic form. The constitutive curves obtained with the OR and von Mises breaking criteria are illustrated in Fig. 15b for the specific case of a uniform distribution between $\varepsilon_1^{\min} = \varepsilon_2^{\min} = 0$ and $\varepsilon_1^{\max} = \varepsilon_2^{\max} = 1$, i.e., $P_1(\varepsilon) = \varepsilon$ and $P_2(\varepsilon) = \varepsilon$.

It follows from the structure of (26) and can be seen in Fig. 15b that the existence of two breaking modes leads to a reduction of the strength of the material, both the critical stress σ_c and strain ε_c take smaller values compared to the case of a single breaking mode applied in simple fibre bundle models [5, 12, 24, 26, 28, 37, 52, 62]. The coupling of the two breaking modes by the von Mises form (27) gives rise to further reduction of the interface strength. Note that due to the non-linearity of (27) also the functional form of $\sigma(\varepsilon)$ changes.

6.3 Simulation Techniques

Based on the above equations analytic results can only be obtained for the simplest forms of disorder like the uniform distribution. In order to determine the behavior of the system for complicated disorder distributions and explore the microscopic failure process of the sheared interface, it is necessary to work out a computer simulation technique. For the simulations we consider an ensemble of N beams arranged on a square lattice (see Fig. 1). Two breaking thresholds $\varepsilon_1^i, \varepsilon_2^i$ are assigned to each beam i $(i = 1, \ldots, N)$ of the bundle from the joint probability distribution $p(\epsilon_1, \epsilon_2)$. For the OR breaking rule, the failure of a beam is caused either by stretching or bending depending on which one of the conditions (22, 23) is fulfilled at a lower value of the external load. In this way an effective breaking threshold ϵ_c^i can be defined for the beams as

$$\epsilon_c^i = \min(f^{-1}(\varepsilon_1^i), g^{-1}(\varepsilon_2^i)), \qquad i = 1, \ldots, N, \tag{29}$$

where f^{-1} and g^{-1} denote the inverse of f, g, respectively. A beam i breaks during the loading process of the interface when the load on it exceeds its effective breaking threshold ϵ_c^i. For the case of the von Mises type breaking criterion (27), the effective breaking threshold ϵ_c^i of beam i can be obtained as the solution of the algebraic equation

$$\left(\frac{f(\epsilon_c^i)}{\epsilon_1^i}\right)^2 + \frac{g(\epsilon_c^i)}{\epsilon_2^i} = 1, \qquad i = 1, \ldots, N. \tag{30}$$

In the case of global load sharing, the load and deformation of beams is everywhere the same along the interface, which implies that beams break in increasing order of their effective breaking thresholds. In the simulation, after determining ϵ_c^i for each beam, they are sorted in increasing order. Quasi-static loading of the beam bundle is performed by increasing the external load to break only a single element. Due to the subsequent load redistribution on the intact beams, the failure of a beam may trigger an avalanche of breaking beams. This process has to be iterated until the avalanche stops, or leads to catastrophic failure at the critical stress and strain.

In Fig. 15b the analytic results on the constitutive behavior obtained with uniform distribution of the breaking thresholds are compared to the corresponding results of computer simulations. As a reference, we also plotted the constitutive behavior of a bundle of fibres where the fibres fail solely due to simple stretching (i.e., DFBM) [5, 12, 24, 26, 28, 37, 52, 62]. It can be seen in the figure that the simulation results are in perfect agreement with the analytical predictions. It is important to note that the presence of two breaking modes substantially reduces the critical stress σ_c and strain ε_c (σ_c and ε_c are the value of the maximum of the constitutive curves) with respect to the case when failure of elements occurs solely under stretching [5, 12, 24, 26, 28, 37, 52, 62]. Since one of the failure functions is non-linear, the shape of the constitutive curve also changes, especially in the post-peak regime. The coupling of the two breaking modes in the form of the von Mises criterion gives rise to further reduction of the strength of the interface, see Fig. 15b.

6.4 Microscopic Damage Process

During the quasi-static loading process of an interface, avalanches of simultaneously failing beams occur. Inside an avalanche, however, the beams can break solely under one of the breaking modes when the OR criterion is considered, or the breaking can be dominated by one of the breaking modes in the coupled case of the von Mises type criterion. In order to study the effect of the disorder distribution of beams on the relative importance of the two breaking modes and on the progressive failure of the interface, we considered independently distributed breaking thresholds both with a Weibull distribution of exponents m_1 and m_2, and scale parameters λ_1 and λ_2 for the stretching and bending modes, respectively. It can be seen in Fig. 16a that increasing λ_2 of the bending mode, the beams become more resistant against bending so that the stretching mode starts to dominate the breaking of beams, which is indicated by the increasing fraction of stretching failure. In the limiting case of $\lambda_2 \gg \lambda_1$ the beams solely break under stretching. It is interesting to note that varying the relative importance of the two failure modes gives also rise to a change of the macroscopic constitutive behavior of the system. Shifting the strength distributions of beams, the functional form of the constitutive behavior remains the same, however, the value of the critical stress and strain vary in a relatively broad range. Varying the amount of disorder in the breaking thresholds, i.e. the Weibull exponents, has a similar strong effect on the macroscopic response of the system, see Fig. 16b. Applying the von Mises

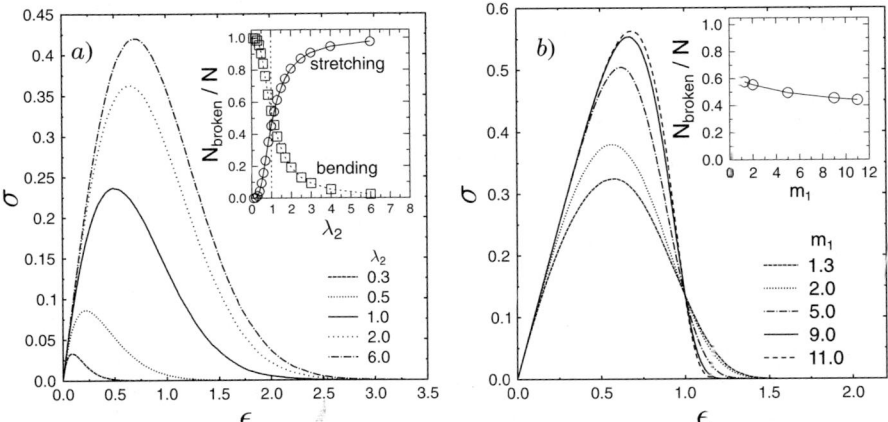

Fig. 16. (a) Constitutive behavior of a bundle of $N = 90000$ beams using the OR criterion for different values of the scale parameter λ_2 of the bending mode. The other parameters were fixed $m_1 = m_2 = 2.0$ and $\lambda_1 = 1.0$. Inset: fraction of beams broken under the two breaking modes. (b) Constitutive behavior of the same bundle changing the Weibull exponent m_1 of the stretching mode with $\lambda_1 = \lambda_2 = 1.0$ and $m_2 = 2.0$. Inset: the fraction of beams broken under the bending mode

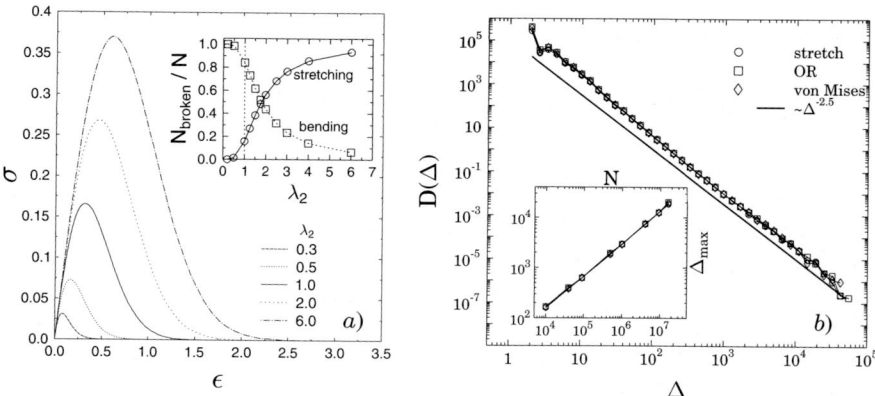

Fig. 17. (a) Constitutive behavior of the interface applying the von Mises breaking criterion at different values of the scale parameter λ_2 of the bending threshold. (b) Comparison of the avalanche size distribution for the classical DFBM, the OR and the von Mises type breaking criteria. The inset shows that the size of largest avalanche $\Delta_{\max}$ is proportional to the size of the bundle N

breaking criterion, the microscopic and macroscopic response of the interface show similar behavior (Fig. 17a). Our careful numerical analysis of the microscopic failure process of the interface revealed that the size of avalanches of simultaneously failing beams under stress controlled loading of the interface has a power law distribution (Fig. 17b). The exponent of the power law proved to be 5/2; it is equal to the mean field exponent characterizing the distribution of burst in simple fibre bundle models [28].

7 Discussion and Outlook

We presented an overview of recent extensions of the classical fibre bundle model in order to provide a more realistic description of the fracture and breakdown of disordered materials. We gradually improved the model by generalizing the damage law, constitutive behavior, deformation state and the way of interaction of the fibres. Analytical calculations and computer simulations have been carried out to explore the macroscopic response and the microscopic damage process of the extended model, which was then confronted with the classical FBM intensively studied in the literature.

Recently, several applications of the extended fibre bundle models have been proposed in a broad range of breakdown phenomena. The continuous damage fibre bundle model with the gradual degradation of fibre strength proved to adequately describe the relevant damage mechanism of various types of materials. CDFBM was successfully applied to understand the strength and fracture mechanism of nacre [43, 44], where even a quantitative comparison to experimental findings was possible. CDFBMs have also been considered to describe node breaking failures in scale free networks [38], and it has been further generalized considering time dependent fibre strength [34, 35].

An interesting study of the viscoelastic fibre bundle model was performed in [39, 40] where the model was further improved considering a more realistic time dependent response of single fibres instead of Kelvin-Voigt elements [40]. The authors carried out creep rupture experiments of fibre composites with randomly oriented short fibres and found a quantitative agreement of the model predictions and experimental results on the macro and micro behavior of the system [39]. The careful analyzes of the accumulation of damage in the viscoelastic fibre bundle model also helps to construct forecasting methods of the imminent macroscopic failure [64].

Based on the variable range of load sharing provided by the load transfer function (14), interesting results have recently been achieved on the analogy of damage and fracture to phase transitions and critical phenomena. Analyzing FBMs on scale free networks showed that besides the range of interaction also the structure of connectivity of the bundle is crucial to determine the universality class of the fracture transition [10]. Apart from fibre reinforced composites, short range interaction of fibres proved to be important also for the mechanical behavior and rupture of biological tissues [27].

Acknowledgements

This work was supported by the Collaborative Research Center SFB 381 and by the NATO Grant PST.CLG.977311. H.J. Herrmann is grateful for the Max Planck Prize. F. Kun acknowledges financial support of the Research Contracts NKFP 3A-043/2004, OTKA M041537, T049209 and of the György Békésy Foundation of the Hungarian Academy of Sciences. The authors are grateful to Y. Moreno for valuable discussions.

References

1. M.P. Allen and D.J. Tildesley, Editors, *Computer Simulation of Liquids*, Oxford University Press, Oxford (1984).
2. J.V. Andersen, D. Sornette, and K. Leung, Tricritical Behaviour in Rupture Induced by Disorder, Phys. Rev. Lett., **78** 2140–2143 (1997).
3. G.G. Batrouni and A. Hansen, Fracture in three-dimensional fuse networks, Phys. Rev. Lett., **80** 325 (1998).
4. G.G. Batrouni, A. Hansen, and J. Schmittbuhl, Heterogeneous interfacial failure between two elastic blocks, Phys. Rev. E, **65** 036126 (2002).
5. P. Bhattacharyya, S. Pradhan, and B.K. Chakrabarti, Phase transition in fibre bundle models with recursive dynamics, Phys. Rev. E, **67**(14) 046122 (2003).
6. B.K. Chakrabarti and L.G. Benguigui, *Statistical Physics of Fracture and Breakdown in Disordered Systems*, Oxford University Press, Oxford (1997).

7. B.D. Coleman, Time dependence of mechanical breakdown phenomena, J. Appl. Phys., **27** 862 (1956).
8. W.A. Curtin, The "tough" to brittle transition in brittle matrix composites, J. Mech. Phys. Solids, **41** 217 (1993).
9. W.A. Curtin, Size Scaling of Strength in Heterogeneous Materials, Phys. Rev. Lett., **80** 1445–1448 (1998).
10. D.-H. Kim, B.J. Kim, and H. Jeong, Universality class of the fibre bundle model on complex networks, Phys. Rev. Lett., **94** 025501 (2005).
11. Ravá da Silveira, An introduction to breakdown phenomena in disordered systems, Am. J. Phys., **67** 1177 (1999).
12. H.E. Daniels, The statistical theory of the strength of bundles of threads-I, Proc. R. Soc. London A, **183** 405–435 (1945).
13. A. Delaplace, S. Roux, and G. Pijaudier-Cabot, Damage cascade in a softening interface, Int. J. Solids Struct., **36** 1403–1426 (1999).
14. A. Delaplace, S. Roux, and G. Pijaudier-Cabot, Avalanche Statistics of Interface Crack Propagation in Fibre Bundle Model: Characterization of Cohesive Crack, J. Eng. Mech., **127** 646–652 (2001).
15. G. Dill-Langer, R.C. Hidalgo, F. Kun, Y. Moreno, S. Aicher, and H.J. Herrmann, Size dependency of tension strength in natural fibre composites, Physica A, **325** 547–560 (2003).
16. A.G. Evans and F.W. Zok, The physics and mechanics of fibre-reinforced brittle matrix composites, Journal of Materials Science, **29** 3857–3896 (1994).
17. J.B. Gómez, D. Iñiguez, and A.F. Pacheco, Solvable Fracture Model with Local Load Transfer, Phys. Rev. Lett., **71** 380 (1993).
18. A. Hansen and P.C. Hemmer, Burst Avalanches in Bundles of Fibres: Local Versus Global Load-Sharing, Phys. Lett. A, **184** 394–396 (1994).
19. A. Hansen and S. Roux, Statistical toolbox for damage and fracture. In D. Krajcinovic and J. van Mier, Editors, *Damage and Fracture of Disordered Materials*, number 410 in CISM Courses and Lectures, Chap. 2, pp. 17–101. Springer Verlag (2000).
20. D.G. Harlow and S.L. Phoenix, The Chain-of-Bundles Probability Model For the Strength of Fibrous Materials i: Analysis and Conjectures, J. Composite Materials, **12** 195 (1978).
21. D.G. Harlow and S.L. Phoenix, The Chain-of-Bundles Probability Model for the Strength of Fibrous Materials ii: A Numerical Study of Convergence, J. Composite Materials, **12** 314 (1978).
22. P.C. Hemmer and A. Hansen, The Distribution of Simultaneous Fibre Failures in Fibre Bundles, J. Appl. Mech., **59** 909–914 (1992).
23. H.J. Herrmann and S. Roux, Editors, *Statistical models for the fracture of disordered media*, Random materials and processes, Elsevier, Amsterdam (1990).
24. R.C. Hidalgo, F. Kun, and H.J. Herrmann, Bursts in a fibre bundle model with continuous damage, Phys. Rev. E, **64**(6) 066122 (2001).
25. R.C. Hidalgo, F. Kun, and H.J. Herrmann, Creep rupture of viscoelastic fibre bundles, Phys. Rev. E, **65** 032502 (2002).
26. R.C. Hidalgo, Y. Moreno, F. Kun, and H. J. Herrmann, Fracture model with variable range of interaction, Phys. Rev. E, **65** 046148 (2002).
27. B.J. Kim, Phase transitions in the modified fibre bundle model, Europhys. Letts., **66** 819 (2004).
28. M. Kloster, A. Hansen, and P.C. Hemmer, Burst avalanches in solvable models of fibrous materials, Phys. Rev. E, **56** 2615–2625 (1997).

29. J. Knudsen and A.R. Massih, Breakdown of disordered media by surface loads, Phys. Rev. E, **72** 036129 (2005).
30. F. Kun and H.J. Herrmann, Damage development under gradual loading of composites, Journal of Materials Science, **35** 4685 (2000).
31. F. Kun, R.C. Hidalgo, H.J. Herrmann, and K.F. Pal, Scaling laws of creep rupture of fibre bundles, Phys. Rev. E, **67**(6) 061802 (2003).
32. F. Kun, Y. Moreno, R.C. Hidalgo, and H.J. Herrmann, Creep rupture has two universality classes, Europhys. Lett., **63**(3) 347–353 (2003).
33. F. Kun, S. Zapperi, and H. J. Herrmann, Damage in fibre bundle models, Eur. Phys. J. B, **17** 269 (2000).
34. L. Moral, J.B. Gomez, and Y. Moreno, Exact numerical solution for a time-dependent fibre-bundle model with continuous damage, J. Phys. A-Math. Gen., **34** 9983 (2001).
35. L. Moral, Y. Moreno, J.B. Gomez, and A.F. Pacheco, Time dependence of breakdown in a global fibre-bundle model with continuous damage, Phys. Rev. E, **63** 066106 (2001).
36. Y. Moreno, J.B. Gómez, and A.F. Pacheco, Self-organized criticality in a fibre-bundle-type model, Physica A, **274** 400 (1999).
37. Y. Moreno, J.B. Gomez, and A.F. Pacheco, Fracture and second-order phase transitions, Phys. Rev. Lett., **85**(14) 2865–2868 (2000).
38. Y. Moreno, J.B. Gomez, and A.F. Pacheco, Instability of scale-free networks under node-breaking avalanches, Eur. Phys. Lett., **58** 630 (2002).
39. H. Nechad, A. Helmstetter, R. El Guerjouma, and D. Sornette, Andrade and critical time to failure laws in fibre-matrix composites: Experiments and model, J. Mech. Phys. Solids, **53** 1099 (2005).
40. H. Nechad, A. Helmstetter, R. El Guerjouma, and D. Sornette, Creep ruptures in heterogeneous materials, Phys. Rev. Lett., **94** 045501 (2005).
41. W.I. Newman and A.M. Gabrielov, Failure of hierarchical distributions of fibre bundles, International Journal of Fracture, **50** 1–14 (1991).
42. W.I. Newman, D.L. Turcotte, and A.M. Gabrielov, log-periodic behavior of a hierarchical failure model with applications to precursory seismic activation, Phys. Rev. E, **52** 4827 (1995).
43. P.K. Nukala and S. Simunovic, A continuous damage random thresholds model for simulating the fracture behavior of nacre, Biomaterials, **26** 6087 (2005).
44. P.K. Nukala and S.Simunovic, Statistical physics models for nacre fracture simulation, Phys. Rev. E, **72** 041919 (2005).
45. P.V.V. Nukala, S. Simunovic, and S. Zapperi, Percolation and localization in the random fuse model, J. Stat. Mech: Theor. Exp., p. P08001 (2004).
46. F.T. Peires, Tensile tests for cotton yarns. v.-'the weakest link', theorems on the strength of long composite specimens, J. Textile Inst., **17** T355–368 (1926).
47. S.L. Phoenix and I.J. Beyerlein, Statistical Strength Theory for Fibrous Composite Materials, volume 1 of *Comprehensive Composite Materials*, Sect. 1.19, pages 1–81. Pergamon, Elsevier Science, N. Y. (2000).
48. S.L. Phoenix, M. Ibnabdeljalil, and C.-Y. Hui, Int. J. Solids Structures, **34** 545 (1997).
49. S.L. Phoenix and I.J. Beyerlein, Statistical Strength Theory for Fibrous Composite Materials, In A. Kelly and C. Zweben, Editors, *Comprehensive Composite Materials*, Vol. 1, Sect. 1.19. Pergamon, Elsevier Science, New York (2000).
50. S.L. Phoenix and R. Raj, Scalings in fracture probabilities for a brittle matrix fibre composite, Acta metall. mater., **40** 2813 (1992).

51. S. Pradhan and B.K. Chakrabarti, Precursors of catastrophe in the Bak-Tang-Wiesenfeld, Manna, and random-fibre-bundle models of failure, Phys. Rev. E, **65** 016113 (2002).
52. S. Pradhan and B.K. Chakrabarti, Failure due to fatigue in fibre bundles and solids, Phys. Rev. E, **67**(14) 046124 (2003).
53. S. Pradhan and B. K. Chakrabarti, Failure properties of fibre bundle models, Int. J. Mod. Phys. B, **17** 5565–5581 (2003).
54. S. Pradhan, A. Hansen, and P.C. Hemmer, Crossover behavior in burst avalanches: Signature of imminent failure, Phys. Rev. Lett., **95** 125501 (2005).
55. S. Pradhan, A. Hansen, and P.C. Hemmer, Crossover behavior in burst avalanches: Signature of imminent failure, Phys. Rev. Lett., **95** 125501 (2005).
56. S. Pradhan and A. Hansen, Failure properties of loaded fibre bundles having a lower cutoff in fibre threshold distribution, Phys. Rev. E, **72** 026111 (2005).
57. S.R. Pride and R. Toussaint, Thermodynamics of fibre bundles, Physica A, **312** 159–171 (2002).
58. F. Raischel, F. Kun, and H.J. Herrmann, Simple beam model for the shear failure of interfaces, Phys. Rev. E, **72** 046126 (2005).
59. S. Roux, Thermally activated breakdown in the fibre-bundle model, Phys. Rev. E, **62** 6164 (2000).
60. S. Roux, A. Delaplace, and G. Pijaudier-Cabot, Damage at heterogeneous interfaces, Physica A, **270** 35–41 (1999).
61. R. Scorretti, S. Cilibreto, and A Guarino, Disorder enhances the effect of thermal of thermal noise in the fibre bundle model, Europhys. Lett., **55** 626–632 (2001).
62. D. Sornette, Elasticity and failure of a set of elements loaded in parallel, J. Phys. A, **22** L243–L250 (1989).
63. D. Sornette, Mean-field solution of a block-spring model of earthquakes, J. Phys. I. France, **2** 2089 (1992).
64. D. Sornette. Statistical physics of rupture in heterogeneous media. In S. Yip, editor, *Handbook of Materials Modeling*, Sect. 4.4. Springer Science and Business Media (2005).
65. D. Sornette and J.V. Andersen, Scaling with respect to disorder in time-to-failure, Eur. Phys. J. B, **1** 353 (1998).
66. R. Toussaint and S.R. Pride, Interacting damage models mapped onto Ising and percolation models, Phys. Rev. E, **71** 046127 (2005).
67. D.L. Turcotte and M.T. Glasscoe, A damage model for the continuum rheology of the upper continental crust, Tectonophysics, **383** 71–80 (2004).
68. E. Vives and A. Planes, Avalanches in a fluctuationless first-order phase transition in a random-bond ising model, Phys. Rev. B, **50** 3839 (1994).
69. S. Zapperi, P. Ray, H.E. Stanley, and A. Vespignani, First-Order Transition in the Breakdown of Disordered Media, Phys. Rev. Lett., **78** 1408 (1997).
70. S. Zapperi, P. Ray, H.E. Stanley, and A. Vespignani, Analysis of damage clusters in fracture processes, Physica A, **270** 57 (1999).
71. S. Zapperi, P. Ray, H.E. Stanley, and A. Vespignani, Avalanches in breakdown and fracture processes, Phys. Rev. E, **59** 5049 (1999).
72. S. Zapperi, A. Vespignani, and H.E. Stanley, Plasticity and avalanche behaviour in microfracturing phenomena, Nature, **388** 658 (1997).

Survey of Scaling Surfaces

A. Hansen and J. Mathiesen

Department of Physics, Norwegian University of Science and Technology
7491 Trondheim, Norway
Alex.Hansen@ntnu.no
Joachim.Mathiesen@ntnu.no

1 Introduction

Surfaces that show scale invariance are abundant in Nature and in the literature. There are many ways that such scale invariance may manifest itself and unfortunately this has led to the introduction of many different concepts in the literature. It is the aim of this short review to systematize and clarify the relations between these different concepts. Hence, we will describe and relate the following concepts:

- Self affinity,
- Fractional noise,
- Lévy flights,
- Fractality,
- Multifractality,
- Multiaffinity, and
- Anomalous scaling.

Foremost, our goal is to provide the reader with a toolbox and enough intuition to use it, and therefore we will not aim at mathematical rigor.

In the following, we will mostly describe a function $z = z(t)$ of one parameter t. For concreteness, we consider z to be some signal that varies with time t. Occasionally, we will address the properties of surfaces $z = z(x, y)$, where (x, y) is a point on a two-dimensional surface. To have a specific picture in mind, $z(x, y)$ might be the height of the surface above some chosen zero level, at (x, y).

2 Self-Affine Surfaces and Brownian Walks

The easiest way to approach the concept of self affinity is to start with the one-dimensional Brownian motion. Consider the position $z(t)$ of a Brownian particle at time t and with the initial condition $z(0) = 0$. If we now approach this motion from a statistical point of view, the fundamental quantity to consider, is the probability, $p(z;t)\,dz$, that the particle is in the interval $[z-dz/2, z+dz/2]$ at time t. The probability density $p(z;t)$ is easily found by solving the diffusion equation

$$\frac{\partial p}{\partial t} = \frac{\partial^2 p}{\partial z^2}\;,\tag{1}$$

where we have set the diffusion constant equal to unity for simplicity, and where the initial condition is the δ-function, $p(z;0)=\delta(z)$. The solution is the Gaussian distribution

$$p(z;t) = \frac{e^{-z^2/2t}}{\sqrt{2\pi t}}\;.\tag{2}$$

The Gaussian distribution shows the scaling property

$$\lambda^{1/2}\,p(\lambda^{1/2}z;\lambda t) = p(z;t)\;.\tag{3}$$

The meaning of this scale invariance is that the Brownian particle spreads out in space as the square root of time. In particular, the second moment of the position of the particle at time t is

$$\langle z^2 \rangle = \int_{-\infty}^{+\infty} dz\, z^2 p(z,t) = t\;.\tag{4}$$

A *self-affine* function is characterized statistically in the same way as the Brownian particle, i.e., by the probability density $p(z,t)$. The property that *defines* self affinity is the scale invariance

$$\lambda^H\,p(\lambda^H z;\lambda t) = p(z;t)\;.\tag{5}$$

The scaling exponent H is the *Hurst*, or *roughness* exponent. In the case of the Brownian particle, $H=1/2$. The function is called self affine when the Hurst exponent is in the range $0 \le H \le 1$. The upper limit ensures that the function grows slower than unity with time,

$$\frac{\sqrt{\langle z^2 \rangle}}{t} \propto \frac{t^H}{t} \to 0\;,\tag{6}$$

as $t \to \infty$. If $z(t)$ characterizes the height of a surface at the point t the property (6) implies that the surface is asymptotically *flat*. When $H<0$, $\langle z^2 \rangle$ approaches a constant which is larger than zero. The function z is then statistically stationary and does not fulfill the scaling invariance (5) and as a consequence, it is not self affine. Rather, it is then a *fractional noise*. We will discuss this in the next section.

There exist several methods to detect whether a function is self affine and to measure the Hurst exponent H. They may be grouped into two classes:

(1) *Intrinsic* methods and (2) *extrinsic methods*. The intrinsic methods measure the properties of the surface by investigating portions of various size, whereas the extrinsic methods investigate the entire surface. In the latter case, a change in scale corresponds to a change of the size of the entire surface. In general, the intrinsic methods are used when there are few samples, and it is not easy to change the sample size. This is the typical situation faced by experimentalists whereas theorists using stochastic methods often have no problems when changing the sample size. As a result, intrinsic methods are favored by experimentalists and extrinsic methods by theorists. As we will see in Sect. 8, it is not obvious that these two classes of methods give the same Hurst exponent – and when they do not, we are dealing with *anomalous* self affinity.

In our experience, the best intrinsic methods are (1) the Average Wavelet Coefficient Method (AWC) [1, 2], (2) Fourier Power Spectrum Method, (3) First Return Method [3], and (4) Detrended Fluctuation Analysis [4]. We strongly recommend that one uses more than one method as they react differently to the corrections to scaling that are almost always present. The differences between the Hurst exponents measured with the different methods give a good indication of the error bars that should be used. As for the extrinsic methods, there are two quite similar methods in use: (1) Plotting $\max z(t) - \min z(t)$ vs. system size (i.e., length of record), and (2) Plotting rms value of $z(t)$ vs. system size. Again, we advice to use both methods and compare the results.

3 Fractional Noise – White Noise

As was pointed out in Sect. 2, Hurst exponents may be negative. When this is the case, $\langle z^2 \rangle$, does not grow with the size of the interval over which it has been measured. One should therefore not refer to such surfaces as self affine. Usually, they are referred to as *fractional noises*. The concept of *fractional noise* is a direct generalization of the motion of a Brownian particle. Let us return to the Fokker-Planck description of the motion, (1). This concentrates on describing the statistical properties of an ensemble of trajectories $z(t)$ of a particle. An equivalent description is the Langevin approach, where one solves the equation

$$\frac{dz(t)}{dt} = \eta(t) , \qquad (7)$$

where $\eta(t)$ is an uncorrelated noise such that $\langle \eta(t) \rangle = 0$ and $\langle \eta(t)\eta(t') \rangle = \delta(t - t')$. From this equation we observe that the noise corresponds to the derivative of the Brownian particle trajectory. Intuitively, one would expect that as $\sqrt{\langle z^2 \rangle} \sim t^{1/2}$, its derivative should be characterized by an exponent $1/2 - 1 = -1/2$. Before showing that this intuition is indeed correct, we generalize the concept of a noise. Starting with a self-affine function $z(t)$ characterized by a Hurst exponent H, the corresponding *fractional noise* is

$\eta(t)$, which in turn is characterized by a Hurst exponent $H' = H - 1$. In order to show that this is so, let us work in terms of the probability density $p(z;t)$. Our goal is to demonstrate that if $\eta(t) = dz(t)/dt$, then it obeys the scale invariance expressed in (5) after replacing H by $H' = H - 1$. Let us denote $q(\eta;t)$ as the probability density for finding a value η at time t given that it was zero at $t = 0$. We need to construct the probability density $q(\eta;t)$ that z is $z + \eta\delta$ at $t + \delta$ when it was z at t,

$$q(\eta;t) = q_0(\eta) + \lim_{\delta \to 0} \delta \int_{-\infty}^{+\infty} dz\, p(z;t)\, p(z + \eta\delta; t + \delta) = q_0(\eta) + q_1(\eta;t) \,. \quad (8)$$

The term $q_0(\eta)$ is the value of the noise at $t + \delta$ which comes in addition to the cumulated noise in the interval t to $t + \delta$. By combining the scaling properties of (5) with this equation, we find that

$$\lambda^{H-1} q_1(\lambda^{H-1}\eta; \lambda t) = q_1(\eta;t) \,. \quad (9)$$

It is important to note that whereas a self-affine function is statistically *non stationary*, a fractional noise will be stationary. Statistical stationarity means that the moments of the statistical distribution approaches constant values with increasing sample sizes. As an example, take the Brownian motion described by the Gaussian distribution (2). The moments of the position of such a Brownian motion at time t is

$$\langle |z|^k \rangle = \int_{-\infty}^{+\infty} dz\, |z|^k\, p(z;t) \propto t^{k/2} \,, \quad (10)$$

i.e., they all increase with t. However, all moments of the noise, η converge towards constants with increasing t. We may illustrate this with white noise. For simplicity, let us assume that it is drawn from a flat distribution on the unit interval. The average value is then $\int_0^1 dx\, x = 1/2$ and the variance is $w^2 = \int_0^1 dx\, (x - 1/2)^2 = 1/12$. This average and variance is based on averaging over an ensemble containing an infinite number of samples. If we instead draw N random numbers and calculate the variance over this finite sample, we find using order statistics [6]

$$w^2(N) = \sum_{k=1}^{N} \frac{1}{N+1}\left(\frac{k}{N+1} - \mu\right)^2 = \frac{1}{12} - \frac{1}{4N} + \mathcal{O}\left(\frac{1}{N^2}\right) \,. \quad (11)$$

where $\mu = 1/N \sum_{i=1}^{N} i/(N+1)$. The leading term here is the constant $1/12$. However, the leading correction scales as N^{-1}. This we recognize as $N^{2(H-1)}$, where $H = 1/2$ is the Hurst exponent of the corresponding random walk. We should therefore interpret $w \sim t^{H'} = t^{H-1}$ not as a diverging term in the limit $t \to 0$, but as the scaling of the leading correction to a constant term in the limit $t \to \infty$.

In light of the remarks just made, we now return to (9). We calculate w^2 using this equation,

$$w^2(\delta) = \int_{-\infty}^{+\infty} d\eta \; \eta^2 \; q_0(\eta) + \int_{-\infty}^{+\infty} d\eta \; \eta^2 \; q_1(\eta;t) \;. \tag{12}$$

The first term is responsible for the leading term, which is a constant. However, the second term is a correction, which becomes zero in the limit $\delta \to 0$! In order to interpret it, we rewrite $q_1(\eta;t)$ as follows:

$$q_1(\eta;t) = \lim_{\delta \to 0} \delta^{1-H} \int_{-\infty}^{+\infty} dz \; p(z;t/\delta) p(z + \eta\delta; t/\delta + 1) \;. \tag{13}$$

By *not* taking the limit $\delta = 0$ directly, but rather letting $t/\delta \to \infty$, i.e. we measure t in units of δ in the limit of vanishing δ, we see that the second term in (12) is a correction term which scales as t^{H-1}.

Is fractional noise ever encountered in Nature? One very good example is the force distribution at the interface between two fracture surfaces that are squeezed together. Fracture surfaces are self affine – and in the case of brittle fractures, it is now quite well established that their Hurst exponent has a universal value $H = 0.8$. The force distribution is essentially the derivative of the fracture surface, and hence is a fractional noise described by a Hurst exponent $H' = -0.2$. Recently, this has been measured directly for the stress distribution prior to the 1996 Kobe earth quake [7].

There is a very subtle point in connection with measuring the Hurst exponent of fractional noises, which does have a bearing on how they are determined experimentally [5]. Note first that the Fourier power spectrum of a one-dimensional signal $z(t)$ is

$$\tilde{P}(f) \sim \frac{1}{f^{1+2H}} \;, \tag{14}$$

whereas that of a two-dimensional signal $z(x,y)$ is

$$\tilde{P}(\mathbf{k}) = \frac{1}{|\mathbf{k}|^{2+2H}} \;. \tag{15}$$

The power spectrum of white noise is constant. Hence, in one dimension, the corresponding Hurst exponent is $-1/2$, as already noted. However, in two dimensions, it is $H = -1$. Suppose that we draw a line in some direction on the two-dimensional surface (x,y), Now, let us assume that z is a fractional noise characterized by a Hurst exponent $H = -1/2$. We are now faced with the paradoxical situation that if we measure along the one-dimensional cut, we find white noise, whereas measuring the signal as a two-dimensional signal, it is *not* white noise, but rather a correlated fractional noise. The solution to the paradox [5] is that it is *impossible* to detect a Hurst exponent $H \leq -1/2$ in one-dimensional cuts through two-dimensional surfaces; one will find an effective $H_{\text{eff}} = \max(H, -1/2)$. Using a two-dimensional method, such as the two-dimensional Fourier power spectrum, there is no such limitation on H. On the other hand, if the trace is one-dimensional but not a cut through a two-dimensional surface, $H_{\text{eff}} = H$, also in the range $-1 \leq H \leq -1/2$. In numerical and in "real" experiments, care has to be taken in measuring fractional noise in two dimensions, as it is very tempting to analyze one-dimensional cuts.

4 Lévy Flights

In this section, we focus our attention on the step sizes of self-affine signals and fractional noises. We will study the histogram of $\Delta z = z(t+\delta) - z(t)$ in the limit of $\delta \to 0$. Think of a Brownian motion. Discretize this, and we have a random walker. The question we pose in this discretized limit is the shape of the histogram of step sizes of the random walker. When the step-size distribution does not have a power law tail towards large values, we have an ordinary random walk. When there *is* a power law tail, we have a Lévy flight [8]. The derivative of a Lévy flight is a power-law distributed, uncorrelated noise, $\eta(t)$. The probability density distribution of values of $|\eta|$ follows the power law

$$p(|\eta|) \sim |\eta|^{-\alpha-1} \text{, for } |\eta| \to \infty \,, \tag{16}$$

and where $\alpha > 0$. The Lévy walk itself is therefore also characterized by the exponent α.

Let us now concentrate on a power-law distributed noise. We assume that it is discretized, $\eta(t_n) = \eta_n$, where $t_n = n\Delta t$ and $0 \leq n \leq N$. Hence, time runs between zero and $T = N\Delta t$. Using extreme statistics [9], we may estimate the largest η-value to expect as

$$P(\eta_{\max}) = \int^{\eta_{\max}} d\eta \, p(|\eta|) = 1 - \frac{1}{N} \,. \tag{17}$$

Hence, $P(|\eta|)$ is the cumulative distribution. The kth moment of this distribution is therefore

$$\langle |\eta|^k \rangle = \int^{\eta_{\max}} d\eta |\eta|^k p(\eta) \sim \eta_{\max}^{k-\alpha} \,, \tag{18}$$

when $k > \alpha$. Solving (17) with respect to $\eta_{\max}$, we find

$$\eta_{\max} \propto T^{1/\alpha} \,. \tag{19}$$

Combining this equation with (18), gives

$$\langle |\eta|^k \rangle \sim T^{k/\alpha - 1} \,. \tag{20}$$

Therefore, based on the scaling of the second moment, $k = 2$, alone it is not possible to distinguish between a power-law distributed noise and a self-affine function. However, if we vary k, the exponent does not vary as k times a constant, but rather as k times a constant minus another constant. We will return to the significance of this in Sect. 7.

The Lévy flight is the integral of the power-law distributed noise. It is a random walk with power-law distributed step sizes.

The Gaussian distribution, (2), has its importance from being the only distribution which is invariant under summation and for which all moments exist: If z_1 and z_2 are distributed according to a Gaussian, then the sum $z_1 + z_2$ is also distributed according to a Gaussian. In terms of random walks, this translates into (2) being the asymptotic distribution. If the condition that all moments are to exist (i.e., not diverge), the Gaussian distribution becomes one of a family of distributions, the Lévy distributions. They are characterized by the exponent β where $0 < \beta < 2$ and may be written

$$p_\beta(z;t) = \int_{-\infty}^{+\infty} df\, \tilde{p}_\beta(f;t)\, e^{ifz}, \qquad (21)$$

where

$$\tilde{p}_\beta(f;t) = e^{-t|f|^\beta}. \qquad (22)$$

Setting $z = z_1 + z_2$ and $t = t_1 + t_2$, we find

$$\begin{aligned}
p_\beta(z;t) &= \int_{-\infty}^{+\infty} dz_1 p_\beta(z - z_1; t_2) p_\beta(z_1, t_1) \\
&= \int_{-\infty}^{+\infty} df\, e^{ifz} e^{-t_1|f|^\beta} e^{-t_2|f|^\beta} \\
&= \int_{-\infty}^{+\infty} df\, e^{ikz} e^{-(t_1+t_2)|f|^\beta} \\
&= \int_{-\infty}^{+\infty} dk\, e^{ifz} \tilde{p}_\beta(f,t).
\end{aligned} \qquad (23)$$

When $\beta = 2$, we recover the Gaussian distribution, (2). In the range $0 < \beta < 2$, the distribution $p_\beta(z;t)$ has a power law tail towards large $|z|$ values, $p_\beta(z;t) \sim 1/|z|^{\beta+1}$ as $|z| \to \infty$. Moments of the distribution, corresponding to (10) scale as

$$\langle |z|^k \rangle = \int_{-\infty}^{+\infty} dz\, |z|^k\, p_\beta(z;T) \sim T^{k/\beta}, \qquad (24)$$

when $k < \beta$ to ensure convergence. In terms of the moments, a Lévy flight is indistinguishable from a self affine-function. This is, however, not all: Combining (21) and (22) gives the scale invariance

$$\lambda^{1/\beta} p_\beta(\lambda^{1/\beta} z; \lambda t) = p_\beta(z;t). \qquad (25)$$

This is the same scale invariance as in (5) when we identify $1/\beta$ with the Hurst exponent H. Hence, there are two ways that self affinity may arise:

(1) It may arise through a power-law distribution in the values of $|\eta|$ without spatial correlations (i.e., the Lévy flights just discussed), or (2) it may arise through correlations in the sign of η. When there are no correlations in the sign and no power law in the distribution of $|\eta|$, we have a random walk with Hurst exponent $H = 1/2$. The two mechanisms will lead to functions that qualitatively *look* very differently.

We will in the next section relate the exponents β, defined in (22), and α, defined in (16).

5 Fractal Surfaces

In the older literature on rough surfaces, the concepts of self affinity and fractality were often confused. It is not uncommon still today to see this confusion linger on in published work.

Regarding fractals, several definitions exists of a *fractal dimension* and we refer to the literature for surveys of these. We will in the following use the box-counting dimension. If we cover the object by boxes of size b^d, where d is the embedding dimension, and rescale $b \to \lambda b$, the number of such boxes needed, N, scales as

$$N \sim \lambda^{-D}, \tag{26}$$

where D is the box-counting fractal dimension.

We now use this definition to study a self-affine function $z(t)$ [10]. We cover it by boxes of size $b \times b$. If z changes by an amount $|\Delta z|$ over the size of a box, b, we will need

$$N \sim \frac{\max(|\Delta z|, b)}{b} \times \frac{1}{b} = \frac{1}{b^2} \max(|\Delta z|, b) \tag{27}$$

boxes, where the use of the max function should be noted: When $|\Delta z|$ is less than b, we still need one box to cover it along the z direction.[1] As $\Delta z \sim b^H$, we find

$$N \sim \begin{cases} b^{-2+H} & \text{if } |\Delta z| \gg b, \\ b^{-1} & \text{if } |\Delta z| \ll b. \end{cases} \tag{28}$$

Hence, we find that the fractal dimension is $D = 2 - H$ on small scales, crossing over to $D = 1$ on large scales. The box-counting dimension is, in other words, not an appropriate tool to study the scaling properties of self-affine functions.

A more appropriate way to discuss the fractal properties of a function $z(t)$ is to measure its total length over an interval $0 < t < T$,

$$\ell = \int_0^T dt \sqrt{1 + \left(\frac{dz}{dt}\right)^2}. \tag{29}$$

---

[1] In the literature, the crossover due to the max function is typically overlooked, see e.g., [10].

If we find that $\ell \sim T^D$, the function is fractal with fractal dimension D.

We may rewrite (29) as

$$\ell = T \int_0^1 dt \sqrt{1 + \frac{1}{T^2}\left(\frac{dz(Tt)}{dt}\right)^2} . \tag{30}$$

If we now assume that the derivative $|dz/dt| = |\eta|$ is bounded by M, we see that

$$\ell \leq T \int_0^1 dt \sqrt{1 + \frac{M^2}{T^2}} \sim T \text{ as } T \to \infty . \tag{31}$$

Hence, we have that $D = 1$, the fractal dimension is one. If, on the other hand, the derivative is *not* bounded, what does it take to produce a fractal dimension larger than unity? We assume that the derivative $dz/dt = \eta$ is distributed following an uncorrelated power law, $p(|\eta|) \sim |\eta|^{-1-\alpha}$, i.e., the same situation as studied in Sect. 4. Dividing the interval T into N pieces Δt, we now use order statistics [6] to estimate the integral in (29). We start by discretizing (29),

$$\ell = \sum_{n=1}^{N} \Delta t \sqrt{1 + \eta_n^2} . \tag{32}$$

We now order the noise $|\eta_n| \to |\eta|_{(n)}$ where

$$|\eta|_{(1)} \leq |\eta|_{(2)} \leq \cdots \leq |\eta|_{(N)} . \tag{33}$$

We estimate the value of $|\eta|_{(n)}$ by generalizing (17),

$$P(|\eta|_{(n)}) = \int^{|\eta|_{(n)}} d\eta \, p(|\eta|) = 1 - \frac{n}{N} . \tag{34}$$

Hence, using that $p(|\eta|) \propto |\eta|^{-\alpha-1}$, (32) becomes

$$\ell = \sum_{n=1}^{N} \Delta t \sqrt{1 + \eta_n^2} \sum_{n=1}^{N} \Delta t \sqrt{1 + \eta_{(n)}^2} \sum_{n=1}^{N} \Delta t \sqrt{1 + \left(\frac{n}{N}\right)^{-2/\alpha}} . \tag{35}$$

This sum is dominated by small-n terms only when $\alpha < 1$. Hence, we find that

$$\ell \sim \begin{cases} T^{1/\alpha} & \text{if } \alpha < 1 , \\ T^1 & \text{if } \alpha \geq 1 . \end{cases} \tag{36}$$

The fractal dimension is therefore

$$D = \max\left(\frac{1}{\alpha}, 1\right) . \tag{37}$$

We see clearly how the fractal dimension is a measure of the strength α of the singularity connected with the distribution of η values.

It should be noted that the fractal dimension D in (37) may exceed 2, i.e., the embedding dimension of the function $z(t)$. This happens when $\alpha < 1/2$ and is caused by the fractal curve not being isotropic. If it were, its length could not grow faster than any linear length scale raised to the power of the embedding dimension. However, in the present case, the t and z directions scale very differently – if the t direction is scaled by λ, then the z direction must be scaled by λ^D. For this reason, it may be somewhat counterintuitive to use the term fractal dimension for (37).

What about the Hurst exponent when η is distributed according to a power law with low enough exponent to produce a fractal length? We then have the same mechanism as the one producing an effective Hurst exponent for Lévy flights, see Sect. 4. Let us suppose that the sign of η changes randomly. $z(t)$ is then a Lévy flight. We calculate

$$w^2(T) = \langle z(T)^2 \rangle = \int_0^T dt_1\, dt_2\, \langle \eta(t_1)\eta(t_2) \rangle \,. \tag{38}$$

We discretize and order as was done in (35) and get

$$w^2 = \sum_{n=1}^N \Delta t\, \eta_n^2 = \sum_{n=1}^N \Delta t\, \eta_{(n)}^2 = \sum_{n=1}^N \Delta t \left(\frac{n}{N}\right)^{-2/\alpha} . \tag{39}$$

Hence, in the same way as in (36), we find

$$w \sim \begin{cases} T^{1/\alpha} & \text{if } \alpha < 1, \\ T^1 & \text{if } \alpha \geq 1. \end{cases} \tag{40}$$

The effective Hurst exponent is therefore

$$H = \max\left(\frac{1}{\alpha}, \frac{1}{2}\right) . \tag{41}$$

Whereas the fractal dimension differs from the trivial value for 1 for $\alpha < 1$, the effective Hurst exponent differs from the random walk value already for $\alpha < 2$. Furthermore, for $\alpha < 1$, the effective Hurst exponent becomes larger than unity, signaling that the surface is no longer asymptotically flat.

The promised relation between β, defined in (22), and α, defined in (16) is therefore simply

$$\beta = \alpha \,, \tag{42}$$

as a result of (24) combined with (41).

We note that when $\alpha < 1/2$, the corresponding effective Hurst exponent becomes larger than unity – see (41). $z(t)$ is therefore not asymptotically flat. This is closely related to the fact that $D > 2$ when $\alpha < 1/2$ as already remarked.

When overhangs in $z(t)$ are present as shown in Fig. 1, the function is no longer single valued. In order to render the function single valued, one may

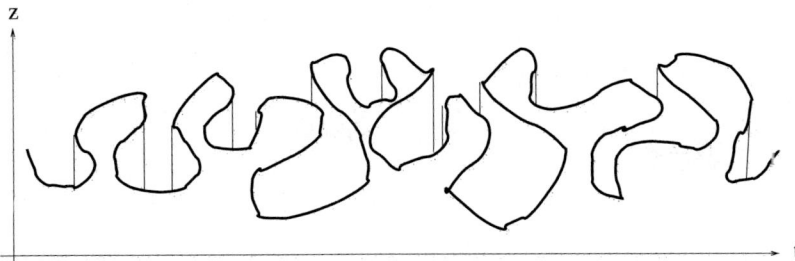

Fig. 1. A fractal curve in $t-z$-plane, causing overhangs when viewed from above (i.e., line of sight parallel with the negative z axis). The overhangs causes the function $z(t)$ to be multivalued. By choosing the branch of $z(t)$ with the largest z value for every t, jumps are generated in the derivative of $z(t)$. There is a connection between the fractal dimension of the curve and the jumps size distribution caused by the overhangs

use different tricks. One such trick is to use the solid-on-solid approximation. Imagine that we watch the function from "above" – that is, we look along the negative z axis. The $z(t)$ that we see for any t from this point of view, will always be the largest $z(t)$ value whenever there are multiple values of z for a given t. As shown in the figure, this produces overhangs and as a consequence jumps in the position of z. As was shown in [11] and repeated in [12], if the multivalued function $z(t)$ has a fractal dimension D_e (i.e., its length ℓ scales as the linear extent T as $\ell \sim T^{D_e}$), then the jump size distribution is given by the asymptotic power law $p(|\eta|) \sim |\eta|^{-D_e-1}$ for large $|\eta|$ values. Hence, in this case $\alpha = D_e$. Since $1 \leq D_e \leq 2$, we see that α is in the range where $D = 1$, from (37). As was shown in [12], there is a relation between the fractal dimension D_e and the Hurst exponent of the solid-on-solid approximated function. It is $H = 2 - D_e$. The reader should note that there is no crossover phenomenon here in contrast to the very similar formula relating Hurst exponent to an effective fractal dimension in (28). Overhangs in a wide range of growth models has been studied in [13].

6 Multifractals

In the theory of *multifractals* [14] one considers distributions lying upon fractal sets. In the previous section, our starting point was a function $z(t)$ whose derivative $\eta(t)$ has values that are distributed according to a power law $p(|\eta|) \propto |\eta|^{-\alpha-1}$. We then derived the fractal dimension D given by (37). Suppose now that there are more than one power law distribution. Each power law is characterized by an exponent $\alpha = \alpha(\tilde{D})$. The parameter $\tilde{D}$ refers to the fractal dimension of the support of the power law characterized by $\alpha(\tilde{D})$. That is, for each t the corresponding $\eta(t)$ has its value drawn from the power law distribution associated with that given t value. All the t values associated with a given power law form a fractal subset of all possible t values. The fractal subset is characterized by a fractal dimension $\tilde{D}$. In the previous section, there was only one power law α and the fractal dimension of its support was $\tilde{D} = 1$.

We now wish to calculate the total length of the function $z(t)$, (29). When there was only one distribution – singularity – at play, we could use order statistics, (35). We could attempt the same here. We could substitute the spatially ordered sequence of η values, η_n, by the ordered sequence $\eta_{(n)}$. The additional complication we are then facing is that the position n dictates which statistical distribution the corresponding η_n has been drawn from. There exists, however, a better way of approaching this problem. We divide the range of η values into pieces of size $\Delta\eta$ and sum over these. This necessitates the introduction of the number of η values in each interval $[\eta, \eta + \Delta\eta]$. We denote this number $\mathcal{N}(\eta, N)\Delta\eta$. Hence, (35) may now be written

$$\ell = \sum_{n=1}^{N} \Delta t \sqrt{1 + \eta_n^2} = \sum_{n=1}^{\infty} \Delta\eta\, \mathcal{N}(n\Delta\eta, N) \sqrt{1 + (n\Delta\eta)^2} \,. \tag{43}$$

In the continuum limit, this becomes

$$\ell = \int_0^\infty d\eta\, \mathcal{N}(\eta, T) \sqrt{1 + \eta^2} \,. \tag{44}$$

In the case when there is only one singularity, α and its support has fractal dimension $\tilde{D} = 1$, $\mathcal{N}(\eta, T) = T|\eta|^{-\alpha-1}$. When there are several singularities, this expression generalizes to

$$\mathcal{N}(\eta, T) = \sum_{\tilde{D}} T^{\tilde{D}} |\eta|^{-\alpha(\tilde{D})-1} \,. \tag{45}$$

Let us now approach the limit $T \to \infty$ and $|\eta| \to \infty$, keeping the ratio $\gamma = \log|\eta|/\log T$ constant. The sum (45) may then be written

$$\mathcal{N}(\eta, T) = \sum_{\tilde{D}} T^{\tilde{D} - (\alpha(\tilde{D})+1)\gamma} = T^{\tilde{D}(\gamma) - (\alpha(\gamma)+1)\gamma} = T^{f(\gamma)} \,. \tag{46}$$

The right-hand side of this equation is the result of a saddle point calculation. Hence, $\tilde{D}(\gamma)$, and thus $\alpha(\gamma)\alpha(\tilde{D}(\gamma))$, is the solution of the equation

$$\frac{d}{d\tilde{D}}\left(\tilde{D} - \left(\alpha(\tilde{D}) + 1\right)\gamma\right) = 0 \,. \tag{47}$$

The length of $z(t)$, (44), may also be determined through a similar saddle point approximation,

$$\ell = \int_0^\infty d\eta\, T^\gamma \mathcal{N}(T^\gamma, T) = T^{D_1} \,, \tag{48}$$

in the limit $T \to \infty$, while keeping the ratio $\gamma = \log|\eta|/\log T$ constant. In this limit, $\sqrt{1+\eta^2} \to |\eta|$. Hence, the fractal dimension of the length of $z(t)$ is

$$D_1 = f(\gamma(1)) + \gamma(1) , \qquad (49)$$

where $\gamma(1)$ is the solution of the equation

$$\frac{d}{d\gamma}(f(\gamma) + \gamma) = 0 . \qquad (50)$$

The function $f(\gamma)$ was defined in (46).

Equation (48) turns out to define one fractal dimension out of an infinite hierarchy. We calculate

$$\langle |\eta|^k \rangle = \int_0^\infty d\eta \, \eta^k \mathcal{N}(\eta, T) = T^{kD_k} . \qquad (51)$$

The right-hand side of this equation is the asymptotic behavior of the integral, and the saddle-point approximation gives that

$$kD_k = f(\gamma(k)) + k\gamma(k) , \qquad (52)$$

where $\gamma(k)$ is the solution of the equation

$$\frac{d}{d\gamma}(f(\gamma) + k\gamma) = 0 . \qquad (53)$$

Equations (52) and (53) form the core of the so-called $f - \alpha$ formalism.

But, what *is* a multifractal? The object that we have presented in this Section is very complicated. Is there any way of conveying in a compact way an intuitive feel for multifractals? The answer is yes. Scale invariant and spatially uncorrelated statistical distributions are simple power laws. Scale invariant and spatially *correlated* statistical distributions are multifractals.

7 Multiaffine Surfaces

Multiaffine surfaces are described by a hierarchy of scaling exponents [15]. The concept is usually tied to the kth order height-height correlation function

$$C_k(T) = \langle |z(t+T) - z(t)|^k \rangle \sim T^{kH_k} . \qquad (54)$$

If H_k is not a constant with respect to k, according to standard terminology, $z(t)$ is multiaffine. However, the situation here is reminiscent of the one encountered in Sect. 4, where we defined defined an *effective* Hurst exponent even though the function was not self affine. We are faced with the same situation here. Even though we find scaling behaviour as described in (54) with a non-trivial H_k, the multiaffinity may only be an *effective* one In particular, when $z(t)$ is multifractal, the hierarchy of fractal dimensions, defined in (51), will also be the hierarchy of exponents defined in (54). This is most easily seen by rewriting this equation slightly,

$$\left\langle \left| \int_0^T dt\, \eta(t) \right|^k \right\rangle \sim T^{kH_k}, \qquad (55)$$

and comparing it to (51) after it has been rewritten as

$$\left\langle \int_0^T dt\, |\eta(t)|^k \right\rangle \sim T^{kD_k}. \qquad (56)$$

If we now assume that the sign of $\eta(t)$ changes randomly, we will have that

$$\langle \eta(t_1) \cdots \eta(t_k) \rangle \propto \begin{cases} 0 & \text{if } k \text{ is odd}, \\ \sum_{\text{pairs}} \prod_{i<j} \delta(t_i - t_j) & \text{if } k \text{ is even}, \end{cases} \qquad (57)$$

where the sum over pairs run over all combinations of pairs of indices. We will therefore find terms that demand all k times t_i to be equal or $k-2$ times to be equal to each other and two other times to be different from the first $k-2$ but equal to each other, and so on. Hence, by combining the correlations in (57) with (55) and (56), we find

$$\left\langle \left| \int_0^T dt\, \eta(t) \right|^k \right\rangle = A_0 \left\langle \int_0^T dt\, |\eta(t)|^k \right\rangle$$
$$+ A_2 \left\langle \int_0^T dt\, |\eta(t)|^{k-2} \right\rangle \left\langle \int_0^T dt\, |\eta(t)|^2 \right\rangle + \cdots, \qquad (58)$$

where A_0, A_2 and so on is a sequence of constants and k is even. The sum runs over all combinations of even numbered exponents that add up to k. When k is odd, $\langle | \int_0^T dt\, \eta(t)|^k \rangle = 0$. As a result, $H_2 = D_2/2$. However, for $k > 2$, the correlation function (54) will *not* scale when there is multifractality in the length ℓ of $z(t)$.

Does "real" multiaffinity exist, multiaffinity which is not just another name for multifractality in some form? Self affinity has either to do with the distribution of the sign of $\eta(t)$ or with the distribution of the absolute value of η. Multifractality is caused by the second mechanism as described in the previous section. Is it possible to have multiaffinity caused solely by correlations in the sign of η alone?

8 Anomalous Scaling of Self-Affine Surfaces

In this last section, we will consider another form of additional scaling behavior that may occur in self-affine surfaces. This is *anomalous scaling* [16, 17]. The easiest way to build an intuition concerning this phenomenon, is to consider self affine-function $z(t)$ on the interval $[0,T]$. The length of the interval, T may be a decorrelation time, or it may be the extent of the entire signal. Let us now consider an interval ΔT such that $\Delta T < T$. Using the structure function, $C_2(\Delta T)$, we find the scaling

$$C_2(\Delta T) = A(T)(\Delta T)^{2H} , \tag{59}$$

where $A(T)$ is a prefactor. The crucial observation now is that the prefactor may have a power law dependence on the interval T,

$$A(T) \sim T^{2(H_g - H)} , \tag{60}$$

where H_g is a scaling exponent different from H – the Hurst exponent. If this is the case, we find that

$$C_2(\Delta T) \sim \begin{cases} (\Delta T)^{2H} & \text{if } \Delta T < T , \\ (\Delta T)^{2H_g} & \text{if } \Delta T = T . \end{cases} \tag{61}$$

It is actually of great importance to verify whether a given self-affine function shows anomalous scaling or not. The reason for this is that intrinsic methods for measuring the Hurst exponent (see Sect. 2) are based on a changing window size ΔT, as thus measure H, whereas the extrinsic methods are based on setting $\Delta T = T$ and changing T, resulting in H_g being measured.

It is very important to realize that anomalous scaling implies that a change of T, which is a *global* change, changes the prefactor A, which in turn changes the roughness *locally*. There are two ways this can occur [18]. The first way is that the prefactor A changes physically when T is changed. The second way, which is much more subtle, is that A changes due to statistical nonstationarity. That is, for a given ΔT, the roughness is an average over $T/\Delta T$ samples. If this average is not stationary, it will depend on this ratio, with the result that A changes.

Using wavelets [19], it is possible to determine whether anomalous scaling is present or not due to statistical nonstationarity.

Given the function $z(t)$, its wavelet transform is defined by

$$w_z(a,b) = \frac{1}{a^{1/2}} \int_0^T \psi^*_{a,b}(t) z(t) dt , \tag{62}$$

where the wavelet is constructed from the single compact function

$$\psi_{a,b}(t) = \psi\left(\frac{t-b}{a}\right) , \tag{63}$$

and where a is the scale and b is the position. The function may be reconstructed from the wavelet coefficients through

$$z(t) = \sum_{a,b} w_z(a,b) \psi_{a,b}(t) . \tag{64}$$

If the field z is self affine so that it statistically scales as

$$\lambda^{-H} z(\lambda t) = z(t) , \qquad (65)$$

then the wavelet coefficients (62) will obey statistically the invariance [2]

$$\lambda^{-(H+1/2)} w_z(\lambda a, \lambda b) = w_z(a, b) . \qquad (66)$$

This equation forms the basis for the Average Wavelet Coefficient (AWC) method. This consists in measuring

$$\langle |w_z(a,b)| \rangle = \frac{1}{N_a} \sum_b |w_z(a,b)| , \qquad (67)$$

where $N_a = T/a$. Equation (66) implies the scaling behavior

$$\langle |w_z(a,b)| \rangle \sim a^{H+1/2} . \qquad (68)$$

It is this scaling relation that is measured when determining H using the AWC method.

Equation (66) implies that the wavelet coefficients may be written in the form

$$w_z(a,b) = a^{(H+1/2)} \omega_z\left(\frac{b}{a}\right) , \qquad (69)$$

where $\omega_z(t)$ is a function of a single argument.

Suppose now that the function $\omega_z(t)$ is in itself self affine with respect to its argument, so that it statistically obeys the scaling relation

$$\lambda^{-H_p} \omega_z(\lambda t) = \omega_z(t) . \qquad (70)$$

Combining this equation with (64) gives the expected local scaling for $z(t)$,

$$\begin{aligned} z(\lambda t) &= \sum_{a,b} \frac{a^{H+1/2}}{a^{1/2}} \omega_z\left(\frac{b}{a}\right) \psi_{a,b}(\lambda t) \\ &= \sum_{a,b} (\lambda a)^H \omega_z\left(\frac{\lambda b}{\lambda a}\right) \psi_{\lambda a, \lambda b}(\lambda t) \\ &= \lambda^H \sum_{a,b} a^H \omega_z\left(\frac{b}{a}\right) \psi_{a,b}(t) = \lambda^H z(t) . \end{aligned} \qquad (71)$$

We investigate the global scaling behavior by noting that if we rescale the interval T on which $z(t)$ exists by λ, the number of wavelet coefficients at a given scale a, $N_a = T/a$, will remain the same if a is also rescaled by λ. The average of the wavelet coefficients at scale a is

$$\langle |w_z(a,b)| \rangle = \frac{1}{N_a} \sum_b |w_z(a,b)| \frac{1}{N_a} \sum_b a^{H+1/2} \left| \omega_z\left(\frac{b}{a}\right) \right| . \qquad (72)$$

We now rescale T and a by λ so that $N_a = T/a$ remains unchanged. The average wavelet coefficients are then rescaled as

$$\langle |w_z(a,b)| \rangle \to \langle |w_z(\lambda a, b)| \rangle = \frac{1}{N_a} \sum_b (\lambda a)^{H+1/2} \left| \omega_z \left(\frac{b}{\lambda a} \right) \right|$$

$$= \frac{1}{N_a} \sum_b \lambda^{H+1/2} a^{H+1/2} \lambda^{-H_p} \left| \omega_z \left(\frac{b}{a} \right) \right|$$

$$= \lambda^{(H-H_p)+1/2} \langle |w_z(a,b)| \rangle . \tag{73}$$

What is the meaning of this scaling relation? Keeping the ratio N_a fixed means that we are comparing the average value of the wavelet coefficients at a given *relative* level before and after the change of T. For example, using the Daubechie discrete wavelet basis, there are two wavelet coefficients for $a = T/2$, the first level, four wavelet coefficients with $a = T/4$, the second level, and so on. We then compare the two wavelet coefficients at the first level before the rescaling with the two wavelet coefficients at the first level after the rescaling. The way they change measures the global Hurst exponent H_g, which was first introduced in (60) and (61). We find

$$H_g = H - H_p . \tag{74}$$

It should be noted that the lack of statistical stationarity implied by (70) leads to $z(t)$ not being statistically translationally invariant. As a result, when we measure $\langle \omega_z \rangle$ over two different intervals A and B finding that $\langle \omega_z \rangle_A > \langle \omega_z \rangle_B$, $z(t)$ will be rougher in interval A than in interval B.

9 Conclusion

The picture we have presented here is a complex one. Many of the notions are quite involved and it is difficult to develop an intuition for them. It does not help that the literature is quite confusing, e.g., self affinity and fractality have quite often been mixed, and a relation like $D = 2 - H$ connecting a fractal dimension and a Hurst exponent is common. As we have seen in Sect. 5 the fractal dimension is only an effective one and the relation is only valid on small scales. It is interesting to note that both multiaffinity and anomalous scaling are concepts that have been recently claimed to have been observed for brittle fracture surfaces. Whether such surfaces actually *do* show such complex scaling behaviour is still, in the opinion of the present authors, an open question. Under any circumstances, the task of determining geometrical scaling exponents is a surprisingly difficult task. To measure the necessary exponents to pin down even more complex scaling such as multiaffinity and anomalous scaling is considerably more difficult.

A last piece of advice, seemingly trivial, but nevertheless important: Before you do any quantitative analysis of a surface, *look at it*. Are there many pronounced spikes or is it rather smooth? Are overhangs prominent? This pins down which tool should be used for your quantitative analysis.

References

1. A.R. Mehrabi, H. Rassamdana and M. Sahimi, Phys. Rev. E **56**, 712 (1997).
2. I. Simonsen, A. Hansen and O.M. Nes, Phys. Rev. E **58**, 2779 (1998).
3. A. Hansen, K.J. Måløy and T. Engøy, Fractals, **2**, 527 (1994).
4. C.-K. Peng, S.V. Buldyrev, S. Havlin, H.E. Stanley and A.L. Goldberger, Phys. Rev. E **49**, 1685 (1994).
5. A. Hansen, J. Schmittbuhl and G.G. Batrouni: Phys. Rev. E **63**, 062102 (2001).
6. H.A. David: *Order Statistics*, 2nd edn, John Wiley, New York (1981).
7. J. Schmittbuhl, G. Chambon, A. Hansen and M. Bouchon, Geophys. Res. Lett. submitted (2006).
8. B.B. Mandelbrot, *The Fractal Geometry of Nature*, W.H. Freeman, San Fransisco (1982).
9. E.J. Gumbel, *Statistical Theory of Extreme Values and Some Practical Applications*, National Bureau of Standards Appl. Math. Series, **33** (1954).
10. J. Feder, *Fractals*, Plenum Press, New York (1988).
11. L. Furuberg, A. Hansen, E.L. Hinrichsen, J. Feder and T. Jøssang, Phys. Script. T **38**, 91 (1991).
12. A. Hansen, G.G. Batrouni, T. Ramstad and J. Schmittbuhl, Condmat/0511545.
13. J. Asikainen, M. Dubé, J. Heinonen and T. Ala-Nissila, Eur. Phys. J. B **30**, 253 (2002).
14. T.C. Halsey, M.H. Jensen, L.P. Kadanoff, I. Procaccia, and B.I. Shraiman, Phys. Rev. A **33**, 1141 (1986).
15. A.L. Barabasi and H.E. Stanley, *Fractal Growth Phenomena*, Cambridge University Press, Cambridge (1995).
16. J.M. López, M.A. Rodríguez and R. Cuerno, Phys. Rev. E **56**, 3993 (1997).
17. J. López and J. Schmittbuhl, Phys. Rev. E **57**, 6405 (1998).
18. A. Hansen, J. Kalda and K.J. Måløy, preprint, 2006.
19. I. Daubechie, *Ten Lectures on Wavelets*, SIAM, Philadelphia (1992).

Part II

Physics Models of Earthquake

Some Early Earthquake Source Models

R. Burridge

Institute for Mathematics and its Applications, 400 Lind Hall, 207 Church Street S.E. Minneapolis, MN 55455-0436
and
Earth Resources Laboratory, Massachusetts Institute of Technology, 42 Carleton Street E34, Cambridge, MA 02142-1324
`burridge@erl.mit.edu`

We describe five early models for earthquake sources considered by the author in the 1960's and 70's. In order of increasing complexity they are:

1. The time-step, double-couple source, which is shown to represent a source consisting of instantaneous (tangential) slip on a small region a fault embedded in an isotropic elastic medium.
2. A spring-and-block model consisting of a chain of blocks connected by springs. The blocks rest on a rough table and the chain of blocks is drawn along slowly by an irresistible constant-velocity drive at the leading end of the spring at one end of the chain.
3. A numerical simulation of an earthquake model in which the region of slip, and the forces acting are specified in advance. The problem is to find the amount of slip and the displacement field.
4. An analytic solution of the antiplane-strain problem of strike slip on a vertical fault, imbedded in an isotropic elastic space, and driven by an initial shear stress. Here we specify a special fracture criterion and find the region of slip as part of the solution as well as the displacements on the fault.
5. An analytical treatment of a repetitive earthquake model in which the preceding model is embedded in a highly viscous viscoelastic halfspace driven by a constant shear stress at infinity.

1 Introduction

We describe some early models for earthquake sources described by the author and coworkers in earlier papers starting with the double-couple point source [1] and ending with the still idealized two-dimensional repetitive fracture mechanical model for frictional strike slip on a vertical fault embedded in a viscoelastic half space [8]. We will not consider here other more recent and more sophisticated models treated by Aki, Madariaga, Rice, and Dietrich among others. Except for the simple double-couple source, the models have in common the feature that a fault (or its analog) is held in equilibrium until increasing tangential traction breaks the static limiting friction. Once static friction is broken sliding occurs and is resisted by the smaller dynamic friction. It is the instability induced by this drop in friction that leads to the characteristic earthquake-like behaviour. The net driving force during the motion is the drop in traction from the initial traction to the lower traction dictated by dynamic friction. After sliding begins static friction plays no further role. Other phenomena of this type are the vibrations of strings of the violin and related instruments, the squealing of brakes, and the creaking of door hinges. In the musical examples the periodicity is dictated by resonance of the finite vibrating system, whereas in earthquakes the system is effectively infinite leading to aperiodic motion.

More specifically we consider the following models. In order of increasing complexity they are:

1. The time-step, double-couple source, which was shown in [1] to represent an instantaneous (tangential) slip on a small region of a fault embedded in an isotropic elastic medium. We give a brief derivation of this result in the special case of a plane fault and illustrate the associated far-field radiation pattern.
2. The spring and block model of [2]. This model in its simplest form consists of a chain of blocks connected by springs. The blocks rest on a rough table and the chain of blocks is drawn along slowly by an irresistible constant-velocity drive at the leading end of the spring at the (front) end of the chain. The blocks are observed to move in a jerky fashion reminiscent of earthquakes. In fact this system mimics many qualitative features of earthquakes. If we interpret each dynamic episode as an earthquake the number of blocks slipping in one event may represent the extent of rupture at the source. Then the magnitude may be represented by the drop in the potential energy of the springs (or its logarithm) and the frequency-magnitude relation is reminiscent of the celebrated Gutenberg–Richter relation for real earthquakes. While this model can be set up as a simple bench-top demonstation, computational models simulating it and its generalizations are richer and lend themselves to more precise analysis. For instance, by replacing the frictional resistance by a viscous resistance at one or more of the blocks aftershock sequences may be generated and analysed statistically. The chain of blocks may be replaced by a two dimensional array, and the system may be driven in more complicated ways so that it is more like a discretization of a sliding fault driven by remote relative motion of the elastic halfspaces on either side of it. Other forms may be imagined in which several faults interact. See [3] for a good account of some extensions and variations of the model and also other papers by the same authors for further developments. Also Chakrabarti and Benguigui give good review

of this model in Chap. 4 of [4] and related models viewed in the context of the physics of disordered systems.
3. A numerical simulation of a continuum earthquake model in which the region of slip is specified and the problem is to find the amount of slip assuming as known the region of slip, the initial stresses, and the frictional forces. We reduce this proplem to the solution of a system of singular integral equations and we discuss its numerical solution, and in particular the discretization of the non-integrable kernel. We show illustrations of the displacements obtained both for antiplane and for plane strain [5].
4. An analytic solution of the two-dimensional problem of strike slip on a vertical fault driven by an initial shear stress. The initial configuration is held in equilibrium by static friction until the traction at some point on the fault exceeds the static limit of friction. Thereafter a region of slip, which is found as part of the solution, expands about the point of nucleation in general running initially with the shear-wave speed both up and down. The downward-traveling "crack" tip suddenly decelerates to a subsonic speed owing to the increasing friction with depth, and then slows gradually to a stop. The complete sliding motion also stops around this time. The full sliding displacement is calculated by solving analytically certain Abel integral equations, which arise from restricting to the fault plane the Green function for the two-dimensional wave equation. The far field pulse shapes are calculated and a very strong breakout phase is identified which dominates in directions close to vertically down. The subsonic part of the downward travelling crack-edge trajectory is interrupted by this breakout phase and runs again briefly at the shearwave speed. This work [6] was influenced by work of Boris Kostrov from the 1960s especially [7].
5. A repetitive earthquake model consisting of the previous model imbedded in a viscoelastic halfspace and driven by a shear stress at infinity [8]. The time scale for relaxation of the viscoelastic medium is very much greater than the time scale of the slip process. The relative magnitudes of static friction, dynamic friction, and the stress at infinity determine the ratio of the repetition period to the time scale of the seismic events.

These models represent the main contributions of the present author to the theory of earthquake mechanisms. Although the main thrust of this conference is on recent developments in the statistical behavior of earthquake models, and closely related to our spring-and-block second model, it is desirable to see this branch of the theory in the wider context of, and close relation to, the more deterministic mechanical studies represented by our models 3, 4, and 5.

2 The Double-Couple Point Source

It was found early on that the radiation from earthquakes was nowhere near spherically symmetric and a four-lobed pattern was observed for P first motions alternating outward movement from the source with inward motion, and an intertwining pattern of S-wave polarizations. The pattern seemed to have the symmetry of a dipole source for those events small enough or far enough that they appeared to emanate from point sources, indicating the mechanism may be a shearing event across a fault plane. The observed seismograms, especially when S waves were taken into account, indicated a somewhat higher degree of symmetry than a single dipole in that there were two orthogonal planes of odd symmetry and the radiation patterns did not seem to distinguish one as the preferred plane of slip. The question then arose as to whether the body-force equivalent should be a single dipole with couple or a pair of crossed dipoles. We now know that it is the latter, called the double couple, that simulates slip on a small patch of a fault plane imbedded in an isotropic elastic medium. We next show why this is so in the restricted case of a planar fault.

2.1 The Calculation

One way to see this is as follows: Suppose the plane $z=0$ is the fault and the jump $[\boldsymbol{u}]$ in displacement $\boldsymbol{u}$ in passing across $z=0$ from $z<0$ to $z>0$ is

$$[\boldsymbol{u}](x,y,z,t) = \begin{pmatrix} U(x,y,t) \\ 0 \\ 0 \end{pmatrix}. \tag{1}$$

The body force $\boldsymbol{f}$ acting in an unfaulted medium to produce the displacement field $\boldsymbol{u}$ is given by

$$\boldsymbol{f} = \rho \ddot{\boldsymbol{u}} - \nabla \cdot \boldsymbol{\tau}, \text{ where } \boldsymbol{\tau} = \lambda \nabla \cdot \boldsymbol{u} \mathbf{1} + \mu [\nabla \boldsymbol{u} + (\nabla \boldsymbol{u})^T] \tag{2}$$

Here λ and μ are the Lamé constants and $\boldsymbol{\tau}$ is the stress field.

Write $\boldsymbol{u} = (u, v, w)$ and then

$$\begin{aligned} u &= U(x,y,t)\mathrm{H}(z) + V_1(x,y,t)\mathrm{R}(z) + C_1(x,y,z,t), \\ v &= V_2(x,y,t)\mathrm{R}(z) + C_2(x,y,z,t), \\ w &= V_3(x,y,t)\mathrm{R}(z) + C_3(x,y,z,t), \end{aligned} \tag{3}$$

where H and R are the unit step function and its integral, the ramp function, and C_k are functions continuous with their first z-derivatives at $z=0$ and such that $\rho \ddot{\boldsymbol{u}} - \nabla \cdot \boldsymbol{\tau} = \mathbf{0}$ for $z \neq 0$.

U is the only non-zero component of relative displacement across the fault, and the V_k allow for possible discontinuities in the first derivatives.

2.2 Action, Reaction, etc.

We invoke Newton's law that action and reaction are equal and opposite to demand that τ_{k3} are continuous across $z=0$. Then using (3) and (2) and setting the coefficients of $\mathrm{H}(z)$ to zero in τ_{k3} we get

$$V_1 = 0, \; V_2 = 0, \; \lambda U_x + (\lambda + 2\mu)V_3 = 0 \; . \tag{4}$$

On substituting for $\boldsymbol{u}$ in the expression for $\boldsymbol{f}$ in (2) and retaining only the terms having support on $z = 0$, because $\rho\ddot{\boldsymbol{u}} - \nabla\cdot\boldsymbol{\tau} = \boldsymbol{0}$ for $z \neq 0$, we find that

$$\boldsymbol{f} = \begin{pmatrix} -\mu\, U(x,y,t)\, \delta'(z) \\ 0 \\ -\mu\, U_x(x,y,t)\, \delta(z) \end{pmatrix} . \tag{5}$$

2.3 The Double Couple

If we specialize to a very small area of slip and write

$$U(x,y,t) = \delta(x)\delta(y)\mathrm{H}(t) \tag{6}$$

$\boldsymbol{f}$ becomes

$$\boldsymbol{f} = \begin{pmatrix} -\mu\, \delta(x)\, \delta(y)\, \delta'(z)\, \mathrm{H}(t) \\ 0 \\ -\mu\, \delta'(x)\, \delta(y)\, \delta(z)\, \mathrm{H}(t) \end{pmatrix} . \tag{7}$$

This is the classic double couple for instantaneous slip given in (6) in the x-direction on a fault normal to the z-direction. See [1] for a derivation of this result when the surface of discontinuity is an arbitrary curved surface.

2.4 The Double-Couple Radiation Pattern

In a uniform isotropic elastic solid the i-component of displacement radiated from an instantaneous point force acting at the origin in the j-direction

$$f_{ij}(x_1, x_2, x_3, t) = \delta_{ij}\delta(x_1)\delta(x_2)\delta(x_3)\delta(t) \tag{8}$$

is the Green's function

$$G_{ij} = \frac{\gamma_i\gamma_j}{4\pi\rho\alpha^2}\frac{\delta(t-r/\alpha)}{r} + \frac{\delta_{ij} - \gamma_i\gamma_j}{4\pi\rho\beta^2}\frac{\delta(t-r/\beta)}{r} \\ + \left(\frac{t}{4\pi\rho r}\right)_{,ij} \mathrm{H}\left(t, \frac{r}{\alpha}, \frac{r}{\beta}\right) . \tag{9}$$

where $\alpha = \sqrt{(\lambda + 2\mu)/\rho}$ and $\beta = \sqrt{\mu/\rho}$ are the P and S wave speeds, $\gamma_i = x_i/r$, and

$$\mathrm{H}(t, a, b) = \begin{cases} 1 & \text{for } a \leq t \leq b , \\ 0, & \text{otherwise} . \end{cases} \tag{10}$$

Here and in what follows subscripts after a comma denote partial differentiations with respect to the corresponding spatial coordinates, for instance $u_{,ij}$ stands for $\partial^2 u/\partial x_i \partial x_j$. To obtain the double-couple radiation we convolve $G_{i,j}$ with the double-couple body force equivalent, which amounts to forming

$$-\mu \int^t (G_{i1,3} + G_{i3,1}) \, dt \qquad (11)$$

and retaining the slowest decaying terms with distance:

$$u_i(t,x) = \frac{\mu \gamma_1 \gamma_3}{2\pi \rho \alpha^3} \times \frac{1}{r} \times \gamma_i \times \delta\left(t - \frac{r}{\alpha}\right)$$

$$+ \frac{\mu \sqrt{\gamma_1^2 + \gamma_3^2 - 4\gamma_1^2 \gamma_3^2}}{4\pi \rho \beta^3} \times \frac{1}{r} \qquad (12)$$

$$\times \frac{\delta_{i1} \gamma_3 + \delta_{i3} \gamma_1 - 2\gamma_1 \gamma_3 \gamma_i}{\sqrt{\gamma_1^2 + \gamma_3^2 - 4\gamma_1^2 \gamma_3^2}} \times \delta\left(t - \frac{r}{\beta}\right)$$

$$+ \{\text{terms decaying more rapidly with distance}\}.$$

The successive factors in each of the two terms are, respectively

- the radiation pattern,
- amplitude variation due to geometrical spreading,
- vector polarization (normalized),
- the pulse shape, i.e. the dependence on time.

The quantities r/α and r/β in the arguments of the δ-functions are the travel times of the P-wave and of the S-wave. It turns out that the S polarization multiplied by its amplitude is the gradient of the P wave amplitude on the unit sphere of directions (see Figs. 1-4).

3 The Block-and-Spring Model

The block-and-spring model in its simplest form consists of a chain of massive blocks connected by springs. The blocks rest on a rough horizontal plane. At one end of the chain, the leading end, another spring is attached with its distal end drawn along at a constant low velocity so as to extend the spring. After some time the net force on the leading block is sufficient to overcome limiting static friction at the first block, at which point the first block moves under the combined forces supplied by the first and second springs and dynamic friction, which is lower than static limiting friction. During the motion of the first block the tension in the second spring may or may not be sufficient to trigger slipping of the second, or further, blocks. In either case the system will

Some Early Earthquake Source Models 119

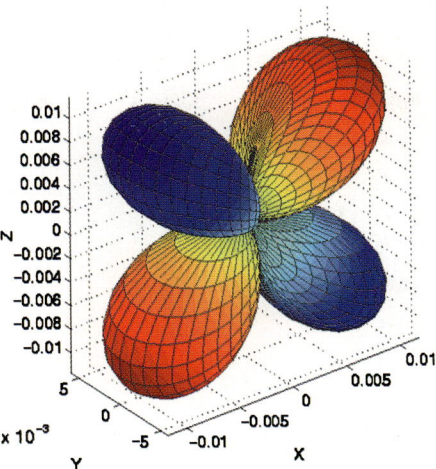

Fig. 1. This is the P radiation pattern, which is a (*spherical*) polar plot of the P amplitude $A_P(\boldsymbol{\omega})$ radiated in direction $\boldsymbol{\omega}$ as a function of $\boldsymbol{\omega}$, i.e. the surface $r = |A_P(\boldsymbol{\omega})|$. The shaded also represents the signed amplitude, on a scale of *dark red* = 1 to *dark blue* = −1

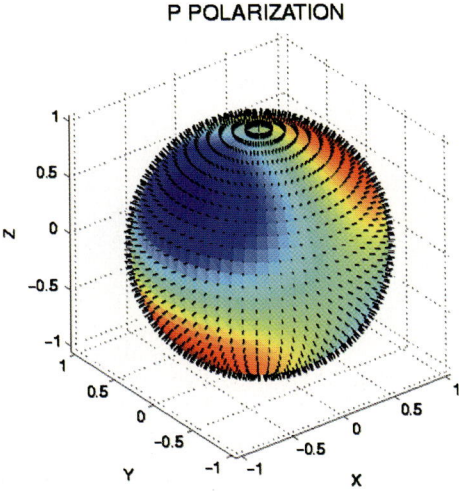

Fig. 2. The corresponding P polarization vector $\boldsymbol{u}_P$ is plotted as a field of vectors $\boldsymbol{u}_P(\boldsymbol{\omega})$ attached to the unit sphere of $\boldsymbol{\omega}$. The shaded represents the signed amplitude, on a scale of *dark red* = 1 to *dark blue* = −1

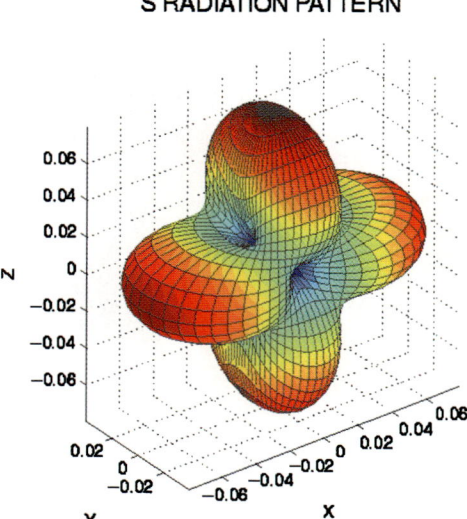

Fig. 3. This is the S radiation pattern, which is a (*spherical*) polar plot of the S amplitude $A_S(\boldsymbol{\omega})$ radiated in direction $\boldsymbol{\omega}$ as a function of $\boldsymbol{\omega}$, i.e. the surface $r = |A_P(\boldsymbol{\omega})|$. The shaded also represents the amplitude, on a scale of *dark red* = 1 to *dark blue* = 0. Compare the scales on the axes with those of Fig. 1 to see how much more energetic the S wave is than the P

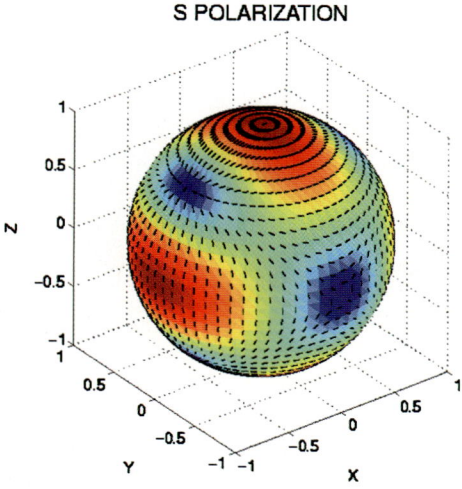

Fig. 4. The corresponding S polarization vector is also plotted as a field of vectors attached to the unit sphere of directions, on a scale of *dark red* = 1 to *dark blue* = 0

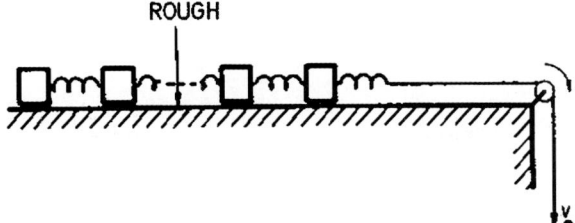

Fig. 5. The chain of blocks and springs drawn slowly from the right end

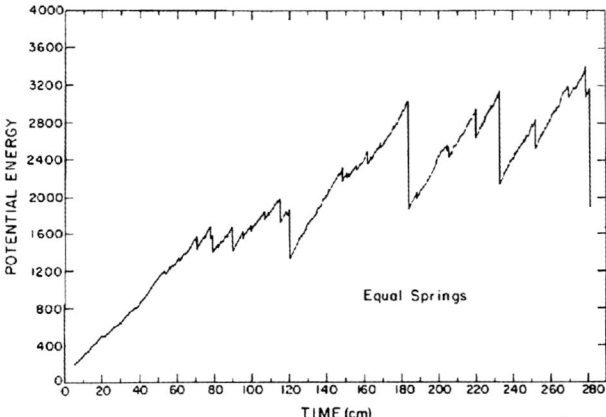

Fig. 6. Potential energy plotted against time for laboratory model. All springs are equal

come to rest and an analog earthquake will have occurred. After a long time, roughly inversely proportional to the low velocity of the drive, the system will undergo another quake and in the process the tension of springs farther and farther down the chain will be stressed and more blocks will slide. The number of blocks slipping may not be a monotone function of time, but there is an increasing trend until finally all the blocks will slide together in a single event, and the system will be discharged, subsequently to go through another similar "seismic" cycle. Potential energy in the springs increases over time between events with relatively sudden drops during them. Figure 5 shows the laboratory model. In Fig. 6 we show a plot of the potential energy in the springs calculated from observations of a sequence of events occurring for this model, and in Fig. 7 we show the energy-frequency relation.

If $N(M)$ is the number of (real) earthquakes with magnitudes greater than M, then the Gutenberg–Richter relation is observed:

$$\log_{10} N = a - b M \, . \tag{13}$$

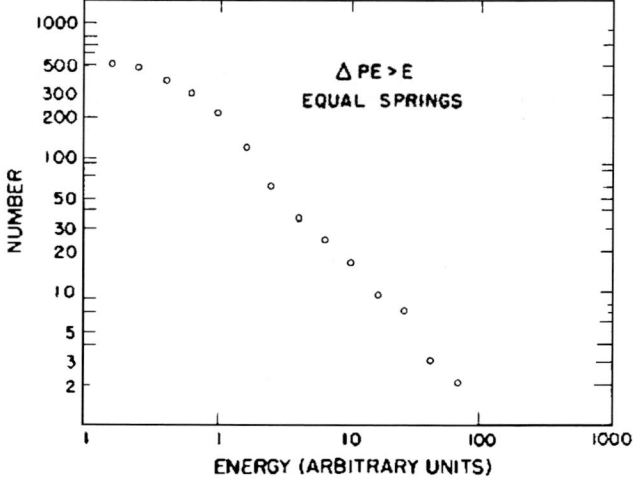

Fig. 7. Energy-frequency relation for the laboratory model

The energy-magnitude relation is

$$M = \alpha + \beta \log_{10} E. \tag{14}$$

These give

$$\log_{10} \frac{N}{N_0} = -b\beta \log_{10} E. \tag{15}$$

If we define the energy of an event in the model to be the drop in potential energy, then we observe a similar energy-frequency relation with a slope -1. See Fig. 7.

3.1 Computational Model

To gain greater control over the parameters of the model we turn now to numerical simulation, which at the same time gives greater flexibility in designing the model. We consider the simulation of the system indicated in Fig. 8.

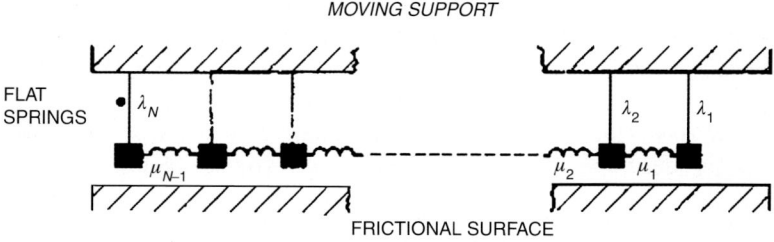

Fig. 8. Schematic diagram for the computational model

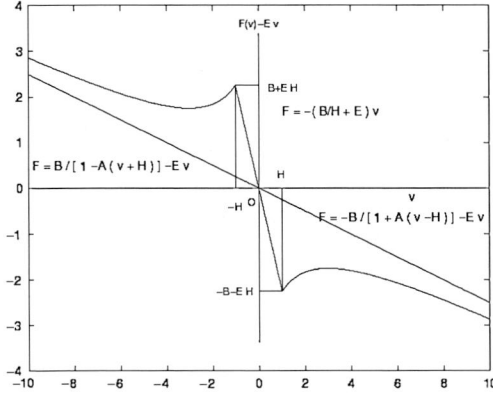

Fig. 9. Frictional force $F(v)$ as a function of relative velocity v

The drive now comes from the moving support. Friction is very important, especially the drop in resistance for significantly non-zero relative velocities. See Fig. 9.

The parameters defining the model are: N the number of blocks and of flat springs, $N - 1$ the number of connecting springs, m_j the mass of the j-th block, λ_j the constant of the j-th flat spring, μ_j the constant of the j-th connecting spring, and V the velocity of the moving support, which drives the system. If v_j is the velocity of the j-th block $F_j(v_j)$ is the velocity-dependent frictional force. The form of the function F_j is given in terms of the constants A_j, B_j, E_j, H_j, by the formula

$$F_j(v_j) = \begin{cases} \dfrac{B_j}{1 - A_j(v_j + H_j)} - E_j v_j & \text{for } v_j < -H_j \,, \\ -\left(\dfrac{B_j}{H_j} + E_j\right) v_j & \text{for } |v_j| \leq H_j \,, \\ \dfrac{B_j}{1 + A_j(v_j - H_j)} - E_j v_j & \text{for } v_j > H_j \,. \end{cases} \quad (16)$$

See Fig. 9 and compare (16) for the significance of the various constants. The term $-E_j v_j$ was inserted to represent radiation in the simplest possible way, namely the resistance due to radiation along stretched strings attached one to each block. The (small) interval $(-H_j, H_j)$ was introduced for generality and to make $f_j(v_j)$ a true function of v_j instead of having a range of values at $v_j = 0$. But for frictional blocks h_j is taken as very small, and when $-H_j < v_j < H_j$ the block is regarded as being at rest. However, certain blocks may be taken as viscous blocks. Then the interval $(-H_j, H_j)$ is expanded and the limiting static friction increased so that these blocks stay in the interval $(-H_j, H_j)$ sliding against a resistance proportional to their velocities.

3.2 Equations of Motion (Newton)

Let m_j be the mass of the j-th block, x_j its position relative to its equilibrium position with no tension in the springs, and y_j its velocity. Then, expressed as a first-order system,

$$\begin{aligned}\dot{x}_j &= y_j \, , \\ m_j \dot{y}_j &= \mu_j(x_{j+1} - x_j) + \mu_{j-1}(x_{j-1} - x_j) \\ &\quad + \lambda_j(Vt - x_j) + F_j(y_j) - E_j y_j \, .\end{aligned} \quad (17)$$

3.3 Energy Balance

The energy balance is as follows

$$\begin{aligned}\frac{\mathrm{d}}{\mathrm{d}t} \frac{1}{2} \sum \{m_j y_j^2 &+ \mu_j(x_{j+1} - x_j)^2 + \lambda_j(Vt - x_j)^2\} \\ &= \sum V\lambda_j(Vt - x_j) - \sum E_j y_j^2 + \sum y_j F_j(y_j) \, ,\end{aligned} \quad (18)$$

where we have introduced $\mu_0 = \mu_N = 0$, $x_0 = x_1$, $x_{N+1} = x_N$ for notational convenience, and the various terms have the following significance:

$$\begin{aligned}&\tfrac{1}{2} \sum m_j y_j^2 \quad \text{Kinetic Energy} \\ &\tfrac{1}{2} \sum \mu_j (x_{j+1} - x_j)^2 \quad \text{Potential Energy in connecting springs} \\ &\tfrac{1}{2} \sum \lambda_j (Vt - x_j)^2 \quad \text{Potential Energy in flat springs} \\ &\sum V\lambda_j(Vt - x_j) \quad \text{Power of Drive by flat springs} \\ &-\sum E_j y_j^2 \quad \text{Power Radiated along strings (not shown)} \\ &\sum y_j F_j(y_j) \quad \text{Power Dissipated by friction (negative)}\end{aligned} \quad (19)$$

3.4 Numerical Experiment

The numerical values of the parameters defining the model illustrated in Fig. 10 are listed in Table 1.

Let us note the peculiarity here that two regions of the model are separated by the two "viscous" blocks 4 & 5 whose parameters were so chosen that they never break static limiting friction $\pm(B_j + E_j H_j)$, but move slowly under a highly viscous retarding force, their velocities remaining in the interval $-H_j \leq v_j \leq H_j$ which is taken to be significantly larger than for the other frictional blocks and the viscous coefficient $B_j/H_j + E_j$ smaller but still large, 10^7 as opposed to 5×10^9. See Table 1.

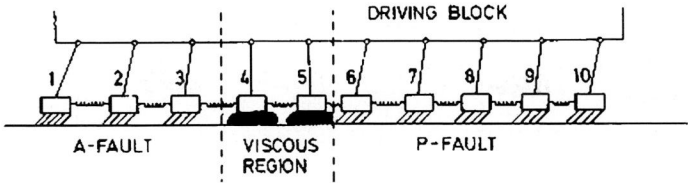

Fig. 10. Schematic diagram for the computational model with the parameters in Table 1

Table 1.

j	m_j	B_j	H_j	B_j/H_j	A_j	E_j	V	λ_j	μ_j
1	1	5	10^{-9}	5×10^9	10	1	10^{-8}	1	100
2	1	5	10^{-9}	5×10^9	10	1		1	100
3	1	5	10^{-9}	5×10^9	10	1		1	100
4	1	10^9	100	10^7	10	1		1	100
5	1	10^9	100	10^7	10	1		1	100
6	1	15	10^{-9}	1.5×10^{10}	10	1		1	100
7	1	10	10^{-9}	5×10^{10}	10	1		1	100
8	1	10	10^{-9}	5×10^{10}	10	1		1	100
9	1	10	10^{-9}	5×10^{10}	10	1		1	100
10	1	10	10^{-9}	5×10^{10}	10	1		1	

3.5 Shock and Aftershock

When this model was simulated computationally many "events" occurred, and we next show one sequence exhibiting aftershocks. The displacements are plotted against time in Figs. 11 & 12 first for the main event and then for

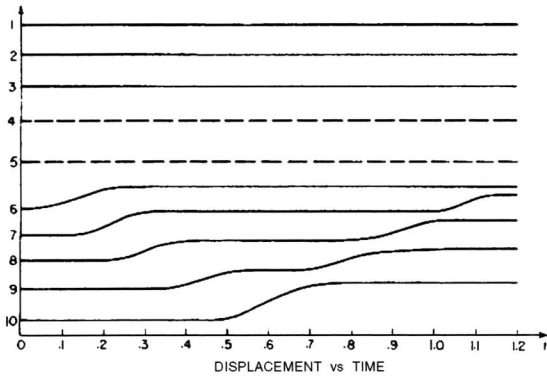

Fig. 11. Detailed motion of the blocks during a major shock. The motion persisted for about 1 time unit

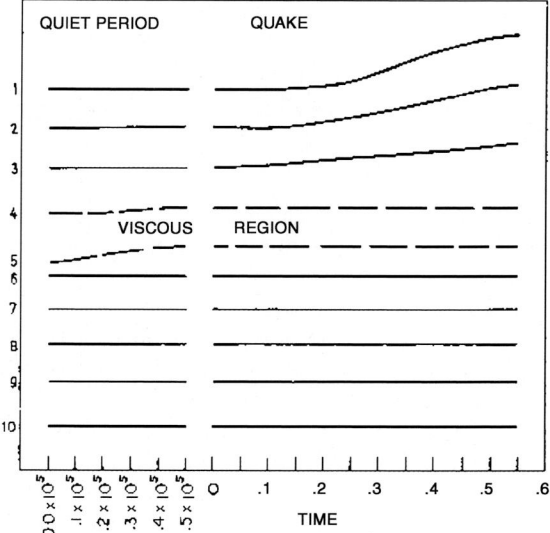

Fig. 12. Continuation of the motion of the blocks following the shock of Fig. 11. Only particles 4 and 5 move during the interval of about 0.4×10^5 time units. This motion is then followed by an aftershock among particles 1–3 enduring for about 1 time unit

the first ensuing aftershock. Notice particularly the slow triggering mechanism between the main event and the aftershock in which the viscous elements 4 & 5 are working.

3.6 Potential Energy During an Aftershock Sequence and at Other Times

It was found that during an aftershock sequence the potential energy decreased monotonically but it had an increasing trend at other times. This is explained by the greater rate of working of the viscous elements involved in the triggering mechanism during the aftershock sequence.

The statistics of aftershock times of occurrence agrees with Omori's form better than with an exponential function, which might seem more plausible at first sight.

3.7 Omori's Formula for Aftershock Rate

Best least-squares fit to Omori's form is

$$N = \frac{1}{0.030 + 0.16\,t} \, . \tag{20}$$

Best least-squares fit to an exponential is

$$N = 23.0\,e^{-.015\,t}\ . \tag{21}$$

The sum of squares of residuals for Omori's form is smaller and about two thirds that for the exponential.

3.8 Peak Kinetic Energy and Energy Radiated Versus Drop in Potential Energy

It is of interest to estimate energy released in an earthquake from the seismic traces. We find that for this model total (integrated) energy radiated is better correlated with drop in potential energy than peak kinetic energy. This may be true also for real earthquakes where drop in potential energy is not observable from the seismic records but kinetic energy may be estimated from them (see Figs. 13-16).

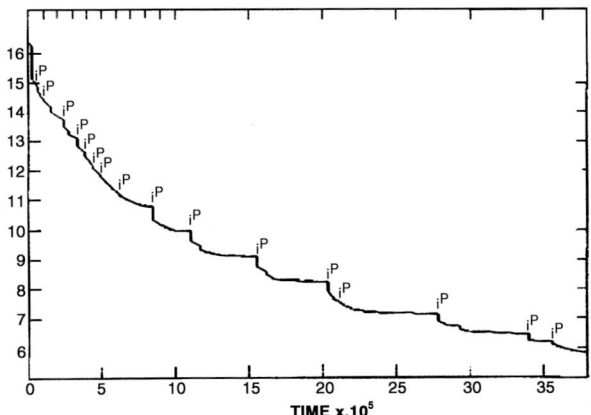

Fig. 13. Potential energy as a function of time for the aftershock sequence following a main shock. Shocks occurring on the principal segment of the fault are identified by the symbol P

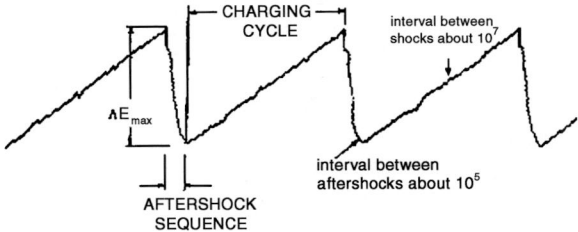

Fig. 14. Schematic plot of potential energy as a function of time for a longer period embracing several main events and their aftershock sequences

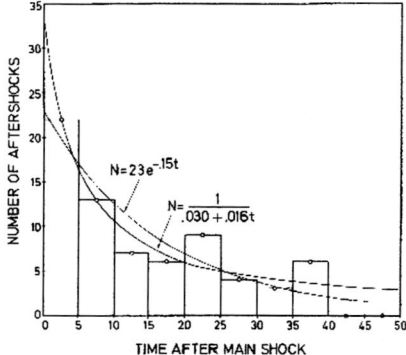

Fig. 15. Histogram of number of aftershocks in a time interval 5×10^5 time units long. Curves fitted by least squares are an exponential function and a function of the type suggested by Omori. The residual for Omori's form is the smaller of the two

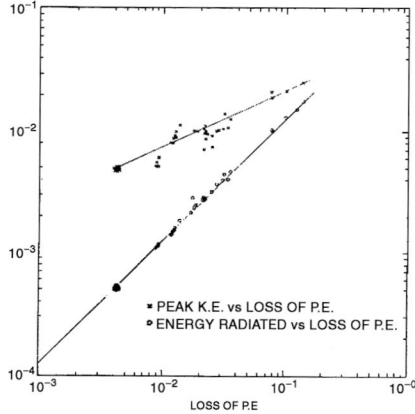

Fig. 16. Peak kinetic energy and energy radiated plotted against potential energy release in shocks of the charging cycle

4 Continuum Model: Numerical

We now go over to a continuum model discretized and solved numerically. We deal with the elastodynamic model of a spreading area of slip on a fault. Again the net driving force is the difference between static initial shear traction and the smaller dynamical friction.

Static friction must be sufficient to hold the system in its initial equilibrium. We prescribe the domain of slip as a function of time, but in the model of Sect. 5 we shall relax this and treat a model where the region of slip is found as part of the solution. We first set up the framework in three dimensions but specialize to two dimensions in the numerical examples.

Let $z = 0$ be the fault plane, $\boldsymbol{u} = (u_x, u_y, u_z)$ the displacement, and τ_{ij} be the stress change once motion has begun. Then u_x, u_y, τ_{zz} are odd functions of z, while $u_z, \tau_{zx}, \tau_{zy}$ are even. Wherever the odd functions are continuous across $z = 0$, they are zero there. Thus τ_{zz} is zero on the whole of $z = 0$ and u_x, u_y are zero on $z = 0$ but outside the region of slip. See Fig. 17.

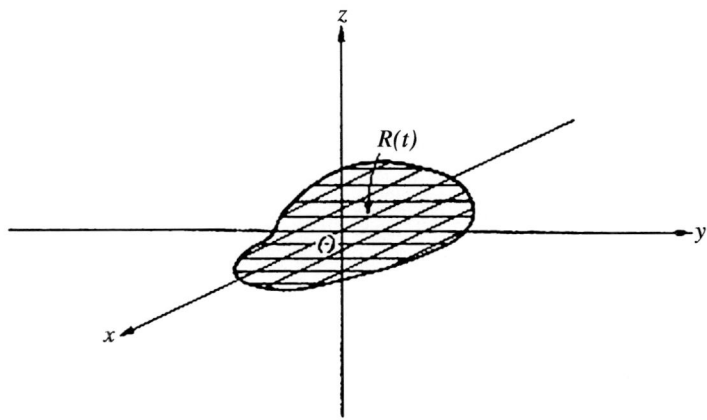

Fig. 17. This shows the crack region $R(t)$ lying in the plane $z = 0$. $R(t)$ is a growing region of slip which can support only reduced tangential tractions

4.1 Mathematical Formulation

For an isotropic medium with Lamé constants λ and μ, τ_{ij} is given by

$$\tau_{ij} = \lambda\, u_{k,k} \delta_{ij} + \mu\left(u_{i,j} + u_{j,i}\right). \tag{22}$$

Then the equation of motion and the initial condition are

$$\begin{aligned} \rho \ddot{\boldsymbol{u}} &= (\lambda + \mu)\nabla(\nabla \cdot \boldsymbol{u}) + \mu \nabla^2 \boldsymbol{u}\,, \\ \boldsymbol{u} &= 0, \quad \text{for} \quad t < 0\,. \end{aligned} \tag{23}$$

On the fault-plane $z = 0$ the z-component of the drop in traction is zero:

$$-\lambda \nabla \cdot \boldsymbol{u} - 2\mu u_{z,z} = 0 \tag{24}$$

On the region $R(t)$ of $z = 0$, i.e. the crack, the drops in tangential components of traction are prescribed functions T_x and T_y, and so on the region $R(t)$ of $z = 0$

$$\begin{aligned} \tau_{xz} &= \mu(u_{x,z} + u_{z,x}) = -T_x(t, x, y)\,, \\ \tau_{yz} &= \mu(u_{y,z} + u_{z,y}) = -T_y(t, x, y)\,, \\ \tau_z &= 0\,. \end{aligned} \tag{25}$$

On the plane $z = 0$ but outside $R(t)$ the tangential displacement is zero:
$$u_x = 0, \; u_y = 0 \, .$$

The problem is: given $R(t)$ and the stress drop, to find u_x and u_y on $R(t)$ and u_z on the whole plane $z = 0$.

4.2 Setting Up the Integral Equations

If (u_x, u_y) are given on $z = +0$ for $t > 0$ then the complete motion is determined and in particular (T_x, T_y) is determined as a translation independent linear function of (u_x, u_y). Hence we may write

$$\boldsymbol{\tau} = \boldsymbol{K} * \boldsymbol{u} \, , \qquad (26)$$

i.e.

$$\tau_{iz}(t, x, y) = \int K_{ij}(t - t', x - x', y - y') u_j(t', x', y') \, dt' \, dx' \, dy' \qquad (27)$$

for some kernel $K(t, x, y)$, which may be distribution valued. For given (t, x, y) the region in (t', x', y') of integration for u_x and u_y is the intersection of the region $R = \{(t', x', y') | (x', y') \in R(t)\}$ and the dependence domain $C(t, x, y)$ given by

$$\sqrt{(x - x')^2 + (y - y')^2} \le \alpha(t - t') \, , \qquad (28)$$

where $\alpha = \sqrt{([\lambda + 2\mu]/\rho}$ the P-wave speed. For u_z the integration is over the complete dependence domain.

$$\begin{aligned}
\tau_{xz} &= K_{xx} * u_x + K_{xy} * u_y + K_{xz} * u_z \, , \\
\tau_{yz} &= K_{yx} * u_x + K_{yy} * u_y + K_{yz} * u_z \, , \\
0 &= K_{zx} * u_x + K_{zy} * u_y + K_{zz} * u_z \, .
\end{aligned} \qquad (29)$$

But we can eliminate u_z to get

$$\begin{aligned}
\tau_{xz} &= (K_{xx} - K_{zz}^{-1} * K_{xz} * K_{zx}) * u_x \\
&\quad + (K_{xx} - K_{zz}^{-1} * K_{xz} * K_{zy}) * u_y \\
\tau_{yz} &= (K_{yx} - K_{zz}^{-1} * K_{yz} * K_{zx}) * u_x \\
&\quad + (K_{yy} - K_{zz}^{-1} * K_{yz} * K_{zy}) * u_y \, .
\end{aligned} \qquad (30)$$

4.3 Integral Equation for Anti-Plane Strain

We now specialize to the two-dimensional problem obtained by assuming that all quantities are independent of y and that the region $R(t)$ of slip is an infinite strip in the plane $z = 0$ parallel to the y-direction. Normalizing the shear velocity to 1 we then have

$$\tau_{yz} = \frac{1}{\pi} \int_{R \cap C} \frac{u_y(t', x')}{[(t - t')^2 - (x - x')^2]^{3/2}} \, dt' \, dx' \, . \qquad (31)$$

4.4 Discretizing the Kernel

Writing v for u_y and K for K_{yy} we have

$$-v_{,z}|_{z=0} = K * v|_{z=0} . \tag{32}$$

But v satisfies the scalar wave equation $v_{tt} = v_{xx} + v_{zz}$ so that

$$\begin{aligned} K * K * v|_{z=0} &= (\partial_t^2 - \partial_x^2) v|_{z=0} \\ &= [\delta''(t)\delta(x) - \delta(t)\delta''(x)] * v|_{z=0} . \end{aligned} \tag{33}$$

This equation for K may be discretized and solved numerically.

4.5 The Discretization of K

Setting $t = ih$ and $x = jh$

$$\delta''(t)\delta(x) - \delta(t)\delta''(x) = \frac{1}{h^2} \left\{ \begin{matrix} 0 & 1 & 0 \\ -1 & 0 & -1 \\ 0 & 1 & 0 \end{matrix} \right\} = H_{ij}, \quad \text{say}, \tag{34}$$

taking the top element as $i = 0, j = 0$ and i increasing downward. Then $K * K = (1/h^2)H$ and by considering the dependence domain we assume

$$K_{ij} = 0 \text{ for } |j| > i . \tag{35}$$

Then setting $K = (1/h)\bar{K}$, $H = (1/h^2)\bar{H}$

$$\bar{K}_{00}^2 = \bar{H}_{00} = 1 , \tag{36}$$

so that taking $i = 0, j = 0$

$$\bar{K}_{00} = \pm 1 . \tag{37}$$

Taking $\bar{K}_{00} = 1$ and writing down the discrete convolution equation for $i0 = 1, j0 = 1$ we obtain

$$\bar{K}_{00}\bar{K}_{11} + \bar{K}_{11}\bar{K}_{00} = -1 . \tag{38}$$

So $\bar{K}_{11} = -.5$, and in general for other values of $i0, j0$ we get a time-advancing explicit scheme

$$2\bar{K}_{00}\bar{K}_{i_0 j_0} = \bar{H}_{i_0 j_0} - \sum_{i=1}^{i_0-1} \sum_{j=\max(-i, -i_0+j_0+i)}^{\min(i, i_0+j_0-i)} \bar{K}_{ij}\bar{K}_{i_0-i, j_0-j} \tag{39}$$

for $|j_0| \leq i_0$. This leads to K_{ij} with non-zero values as listed in Table 2.

In this case K can also be found analytically as

$$\bar{K}_{ij} = (-1)^i \begin{pmatrix} \frac{1}{2} \\ \frac{i+j}{2} \end{pmatrix} \begin{pmatrix} \frac{1}{2} \\ \frac{i-j}{2} \end{pmatrix} . \tag{40}$$

but adaptations of the outlined numerical method can also be applied in more complicated situations such as plane strain and three dimensions where one cannot obtain analytic results.

Table 2

i\j	-5	-4	-3	-2	-1	0	1	2	3	4	5
0						1					
1					-.5		-.5				
2				-.125		.25		-.125			
3			-.0625		.0625		.0625		-.0625		
4		-.039063		.03125		.015625		.03125		-.039063	
5	-.027344		.019531		.007813		.007813		.019531		-.027344

4.6 The Integral Equation for v

Hence the discretized form of the integral equation for v can be written as a time-advancing explicit scheme

$$\bar{K}_{00} v_{i_0 j_0} = h T_{i_0 j_0} - \sum_{i=1}^{i_0} \sum_{j} \bar{K}_{ij} v_{i_0-i, j_0-j} . \qquad (41)$$

The region of integration (summation) is shown in Fig. 18 and can be reduced to region I minus region II of Fig. 19 when the point (x,t) lies in the region D of the figure.

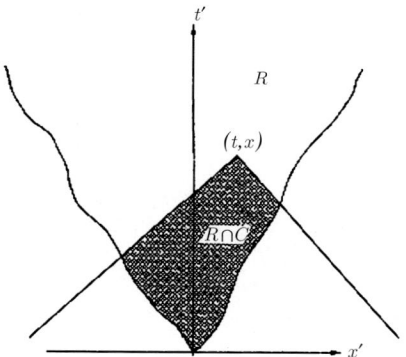

Fig. 18. This shows the crack region $R(t)$ lying in the plane $z = 0$. $R(t)$ is a growing region of slip which can support only reduced tangential tractions

4.7 The Solution for $\dot{v} = \dot{u}_y$

The particle velocity $\dot{v}$ satisfies the same integral equation as v, but the problem for $\dot{v}$ is slightly easier than that for v in that δ-functions are involved instead of unit step functions. We consider the case where the strip $-1 < x < 1$ becomes lubricated instantaneously at $t = 0$ over its whole width, and the crack does not extend further.

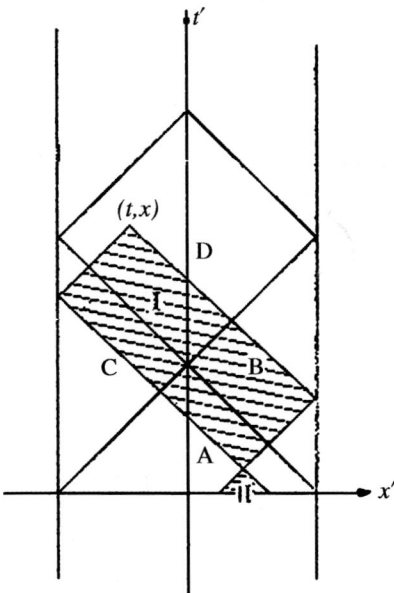

Fig. 19. This shows the reduced region of integration, hatched, in the (t', x')-plane

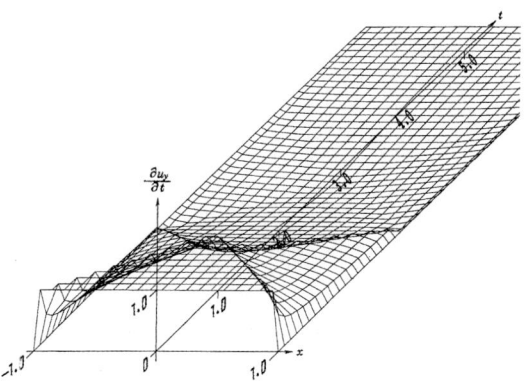

Fig. 20. A plot of $\dot{v} = \dot{u}_y$ calculated by solving the integral equation numerically

The numerical solution for $\dot{v}$ is shown in Fig. 20.

One may also obtain the corresponding analytic solution easily for $t < 3 - |x|$. This is shown in Fig. 21 as a check on the numerical method.

The displacement itself is plotted in Fig. 22.

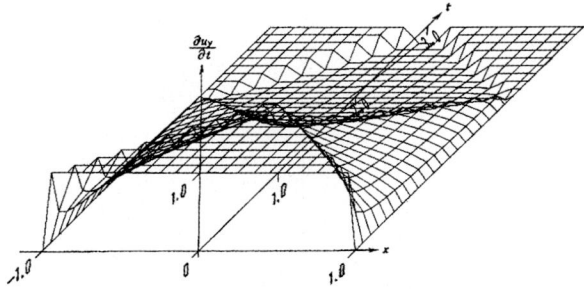

Fig. 21. A plot of $\dot{v} = \dot{u}_y$ calculated analytically in the region $t \leq 3 - |x|$, $|x| \leq 1$

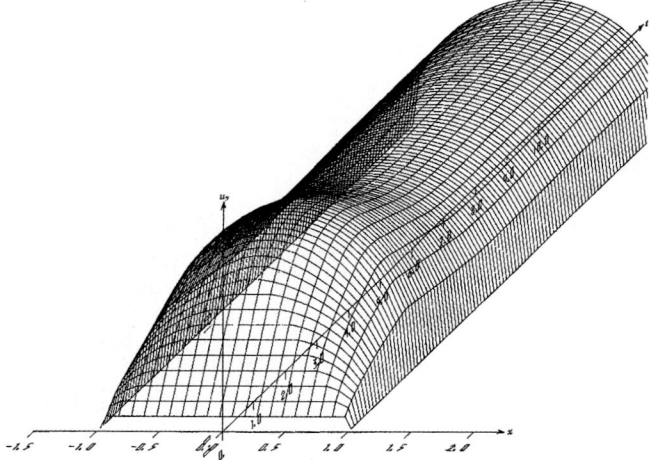

Fig. 22. A plot of the numerical solution for $v = u_y$ for simultaneous nucleation over the whole length of the crack

4.8 Numerical Solution for Nucleation at a Point

The solution for the case where the crack expands from the origin at the wave speed until it fills the region $|x| < 1$. The crack edges stop abruptly at $|x| = 1$. Again we show numerical and early-time analytic solutions in Fig. 23 and Fig. 24.

4.9 Plane Strain

Next we consider the two-dimensional case of plane-strain where u_x and u_z are nonzero but $u_y = 0$. We now must satisfy the simultaneous equations

$$\begin{aligned} \rho \ddot{u}_x - (\lambda + \mu)(u_{x,xx} + u_{z,xz}) - \mu(u_{x,xx} + u_{x,zz}) = 0 \;, \\ \rho \ddot{u}_z - (\lambda + \mu)(u_{x,xz} + u_{z,zz}) - \mu(u_{z,xx} + u_{z,zz}) = 0 \;, \end{aligned} \qquad (42)$$

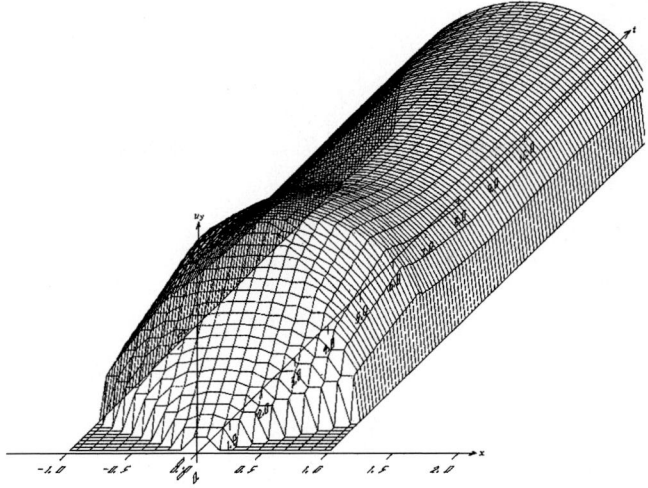

Fig. 23. A plot of $v = u_y$ for nucleation at a point calculated numerically

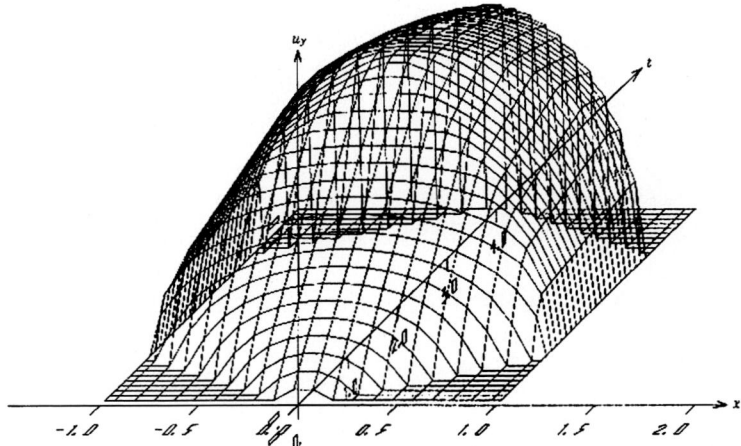

Fig. 24. A plot of $v = u_y$ for nucleation at a point calculated analytically in the region $t \leq 4 - |x|$, $|x| \leq 1$

Writing $\boldsymbol{D}(t,x)$ for the kernel such that on $z = 0$

$$-\frac{\partial}{\partial z}\begin{pmatrix} u_x \\ u_y \end{pmatrix} = \boldsymbol{D}\boldsymbol{u} = \begin{pmatrix} D_{xx} & D_{xz} \\ D_{zx} & D_{zz} \end{pmatrix}\begin{pmatrix} u_x \\ u_y \end{pmatrix} \qquad (43)$$

$$\begin{pmatrix} 1 & 0 \\ 0 & 1 \end{pmatrix}\delta''(t)\delta(x) - \begin{pmatrix} 1 & 0 \\ 0 & \mu \end{pmatrix}\delta(t)\delta''(x)$$
$$+ (1-\mu)\begin{pmatrix} 0 & 1 \\ 1 & 0 \end{pmatrix}\boldsymbol{D}*\delta(t)\delta'(x) - \begin{pmatrix} \mu & 0 \\ 0 & 1 \end{pmatrix}\boldsymbol{D}*\boldsymbol{D} = 0\,. \qquad (44)$$

This quadratic convolution equation for $\boldsymbol{D}$ can be solved numerically by a slight extension of the previously used discretization method and from it the kernel $\boldsymbol{K}$ giving the traction drop $\boldsymbol{\tau}$ computed.

$$\boldsymbol{\tau} = \begin{pmatrix} \tau_x \\ \tau_y \end{pmatrix} = \boldsymbol{K}\boldsymbol{u} = \begin{pmatrix} K_{xx} & K_{xz} \\ K_{zx} & K_{zz} \end{pmatrix} \begin{pmatrix} u_x \\ u_y \end{pmatrix} \quad (45)$$

4.10 The Numerical Scheme

The numerical method is a direct extension of that used in the scalar case:

$$\begin{aligned}
\bar{K}_{xx}(0,0)u_x(i_0,j_0) &= h\tau_x(i_0,j_0) \\
&\quad - \sum_{i=0}^{i_0-1}\sum_j \bar{K}_{xx}(i_0-i, j_0-j) u_x(i,j) \\
&\quad - \sum_{i=0}^{i_0-1}\sum_j \bar{K}_{xz}(i_0-i, j_0-j) u_z(i,j) \,, \\
\bar{K}_{zz}(0,0)u_z(i_0,j) &= h\tau_z(i_0,j_0) \\
&\quad - \sum_{i=0}^{i_0-1}\sum_j \bar{K}_{zx}(i_0-i, j_0-j) u_x(i,j) \\
&\quad - \sum_{i=0}^{i_0-1}\sum_j \bar{K}_{zz}(i_0-i, j_0-j) u_z(i,j) \,.
\end{aligned} \quad (46)$$

4.11 The Numerical Solution

See Figs. 25 and 26.

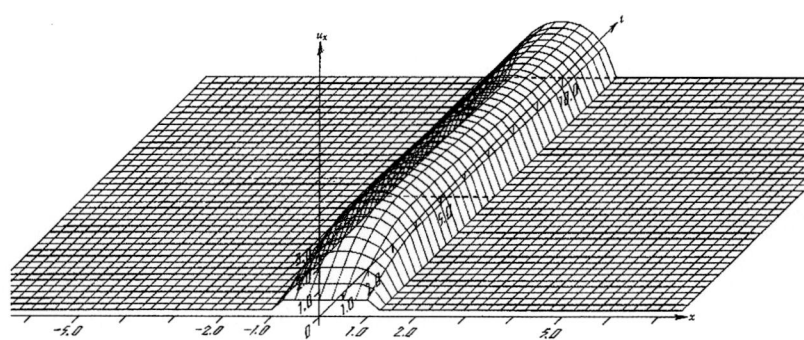

Fig. 25. Here we plot as far as $t = 10$ $u_x(x,t)$ for the plane-strain shear crack instantaneously lubricated over its whole length at $t = 0$

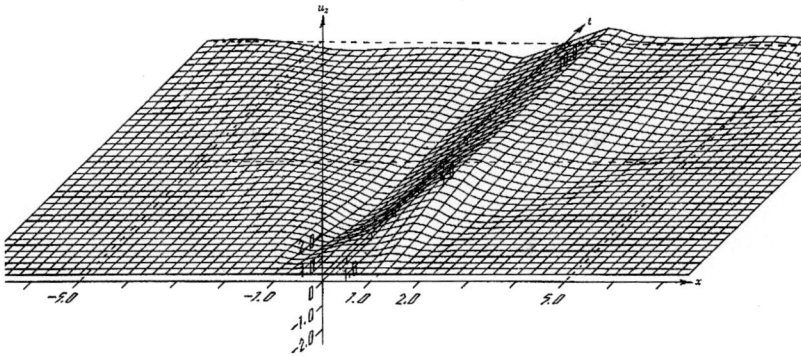

Fig. 26. Here we plot as far as $t = 10$ $u_z(x,t)$ for the plane-strain shear crack instantaneously lubricated over its whole length at $t = 0$

4.12 The Exact Static Solution for Comparison

See Fig. 27.

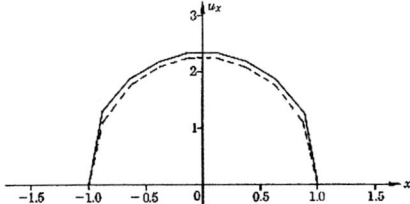

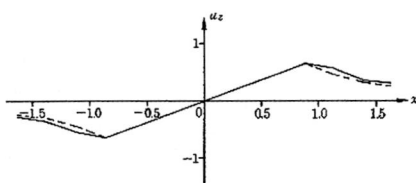

Fig. 27. The exact static solution (*dashed line*) and the dynamical solution computed numerically at $t = 10$ (*full line*)

4.13 Another Numerical Solution

See Figs. 28 and 29.

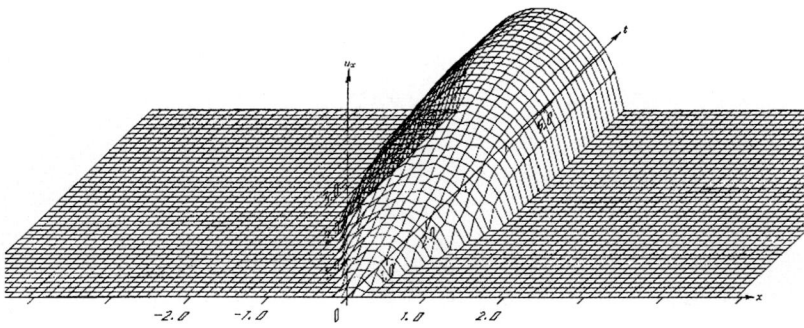

Fig. 28. Here is shown u_x for a crack spreading from a point (the origin) at half the P-wave speed until it fills the region $-1 < x < 1$

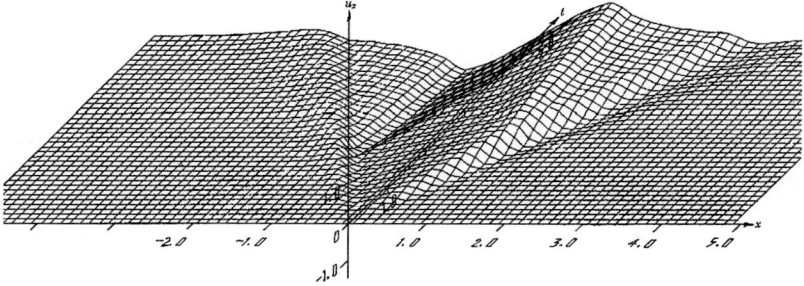

Fig. 29. The normal component of displacement u_z belonging to the u_x of Fig. 28. Notice the rather sharper S pulse than in Fig. 26 due to the "Doppler effect" of the approaching crack edge

Notice in Figs. 26 and 29 the S waves propagating beyond the crack, and that, due to a Doppler effect, the wave is sharper in Fig. 29.

5 A Dynamic Shear Crack with Friction

5.1 The Setup

The following model is a continuum model based on fracture mechanical principles including a simplified fracture criterion adapted to frictional sliding on a preexisting fault rather than true fracture of intact material. This enables us to calculate the propagation of the region of slip as well as the displacements, stress fields, etc. Remarkably we can solve for the complete motion analytically on the fault, although the far fields were calculated by numerically integrating the analytic expressions against the Green's function. We consider the two-dimensional antiplane-strain elastic problem of frictional sliding under joint influence of a driving shear stress and retarding friction.

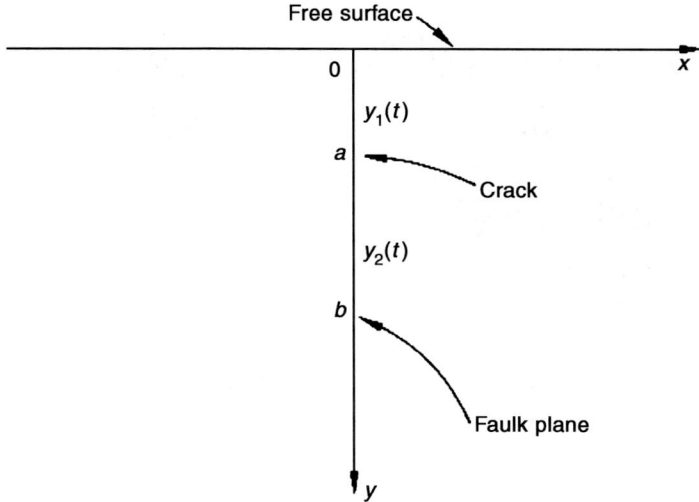

Fig. 30. Here is shown the configuration at some time t after nucleation of the crack at depth a. The crack now occupies the part of the plane $x = 0$ for y between $y_1(t)$ and $y_2(t)$. The level $y = b$ is where the dynamic friction equals the initial stress. For $y < b$ the stress drop is positive, and for $y > b$ it is negative

Let y be vertically down, $y = 0$ the stress-free surface, and $x = 0$ the fault plane (see Fig. 30). Assume u_z is the only nonzero component of displacement and is independent of z. Before motion begins limiting static friction is sufficient to resist the initial shear stress and hold the system in equilibrium. At time $t = a$ we suppose that at the point $y = a$ on the fault plane static friction is broken and that subsequently a region of slip spreads from that point of nucleation in both directions. During the motion the net driving force is the stress drop, i.e. the difference between the initial shear stress and the dynamic friction. Thus, static friction does not enter into the calculation except to set up the initial conditions. We always assume that dynamic friction is less than static friction and both are functions only of position on the fault plane and not of the (non-zero) relative velocity.

5.2 Initial Stress

$$\begin{aligned}
\tau_{yy}^0 &= -\rho g y\,, \\
\tau_{xx}^0 &= \tau_{zz}^0 = -\rho g y - B\,, \\
\tau_{xz}^0 &= \tau_{zx}^0 = A\,.
\end{aligned} \tag{47}$$

5.3 Static Friction

For equilibrium before the earthquake the limiting static friction must be great enough to resist the initial shear stress, which drives the system. So, denoting the coefficient of static friction by γ, in $y > 0$ we must have

$$\gamma(\rho g y + B) \geq A . \tag{48}$$

The constant B was introduced merely to give non-zero friction at zero depth $y = 0$.

5.4 Dynamic Friction

On the other hand, denoting the coefficient of dynamic friction by Γ, for instability leading to an earthquake we require

$$\Gamma(\rho g y + B) \leq A . \tag{49}$$

Hence, defining the stress drop $T(y)$,

$$T(y) = A - \Gamma(\rho g y + B) \geq 0 \tag{50}$$

near the point of nucleation $x = 0, y = a$.

Because the overburden pressure $\rho g y$, increases indefinitely with depth y, T will change sign from positive near $y = 0$ to negative at sufficiently large y. We shall suppose that $T(y)$ is monotone decreasing and $T(b) = 0$ for some $b > 0$.

5.5 The Mathematical Formulation

Let $w(x, y, t)$ be the component of displacement in the z-direction measured from the static state of initial stress.

$$\frac{1}{\beta^2}\frac{\partial^2 w}{\partial t^2} = \frac{\partial^2 w}{\partial x^2} + \frac{\partial^2 w}{\partial y^2}, y > 0, x \neq 0 , \tag{51}$$

where the shear wave speed $\beta = \sqrt{\mu/\rho}$, μ being the shear modulus and ρ the density of the material.

The stress components τ_{zx} and τ_{zy} are given by

$$\tau_{zx} = \tau_{zx}^0 + \mu\frac{\partial w}{\partial x}, \tau_{zy} = \mu\frac{\partial w}{\partial y} , \tag{52}$$

so that on the plane $y = 0$, the free surface of the Earth,

$$\frac{\partial w}{\partial y} = 0 . \tag{53}$$

5.6 Symmetry

Thus if we continue $w(x, y, t)$ to $y < 0$ as an even function of y (53) is satisfied automatically. So, we solve by the method of images (in $y = 0$).

Also, $w(x, y, t)$ is an odd function of x. In the (enlarged) fault plane $x = 0$, $-\infty < y < \infty$ the region of slip (the crack) is $y_1(t) < y < y_2(t)$ and its image $-y_2(t) < y < -y_1(t)$. There $w(0+, y, t) = -w(0-, y, t) \neq 0$, but $w = 0$ elsewhere on $x = 0$, and thus is continuous across the fault. Hence $w(0, y, t) = 0$ on the part of the fault that has not slipped. We shall find it convenient to consider the halfspace $x > 0$, $\infty < y < \infty$. See Fig. 31, which shows the image configuration.

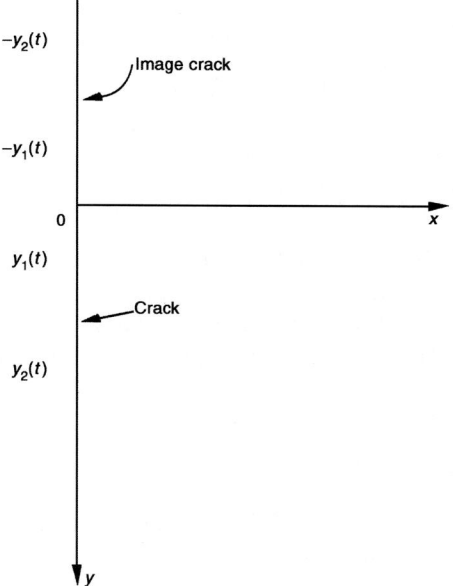

Fig. 31. Antisymmetry in the plane $x = 0$ and the introduction of an image half-space $y < 0$ allows us to consider only the half-space $x > 0$, $-\infty < y < \infty$

5.7 Analysis Confined to the Plane $x = +0$

By using Green's function for the two-dimensional wave equation, and letting $x \downarrow 0$, we get

$$w(+0, y_0, t_0) = \frac{-\beta}{\pi} \int \int_S \frac{w_x(0+, y, t)\, dt\, dy}{\sqrt{\beta^2(t_0 - t)^2 - (y_0 - y)^2}}, \tag{54}$$

where S is the characteristic triangle

$$|y_0 - y| \leq \beta(t_0 - t) \tag{55}$$

and

$$w_x = \partial w/\partial x = -\frac{1}{\mu}T \tag{56}$$

is known on the crack. Elsewhere on the fault plane $w = 0$ but w_x is unknown.

5.8 The Basic Integral Relationships

Let us now define characteristic coordinates

$$\xi = \beta t + y, \; \eta = \beta t - y . \tag{57}$$

Then $y = y_2(t)$ maps into $\xi = \xi_2(\eta)$. Using the formula for $w(0, y_0, t_0) = 0$ for a point beyond the crack tip i.e. for $\xi_0 > \xi_2(\eta)$ expressed in characteristic coordinates, and writing $w(\xi, \eta)$ for $w(0+, y, t)$ when y, t, ξ and η are connected by (57), we obtain

$$\int_0^{\eta_0} \frac{d\eta}{(\eta_0 - \eta)^{1/2}} \int_0^{\xi_0} \frac{w_x(\xi, \eta)\, d\xi}{(\xi_0 - \xi)^{1/2}} = w(0+, y, t) = 0 . \tag{58}$$

Since also (58) holds for any smaller η_0 we have

$$\int_0^{\xi_0} \frac{w_x(\xi, \eta)\, d\xi}{(\xi_0 - \xi)^{1/2}} = 0, \text{ for } \xi_0 > \xi_2(\eta) . \tag{59}$$

This important result will be used repeatedly. It was used first by Kostrov [7] in fracture mechanics.

5.9 The Stress Ahead of the Crack

Next we split the range of integration to get the Abel integral equation

$$\int_{\xi_2(\eta)}^{\xi_0} \frac{w_x(\xi, \eta)\, d\xi}{(\xi_0 - \xi)^{1/2}} = \frac{1}{\mu}\int_0^{\xi_2(\eta)} \frac{T(\xi, \eta)\, d\xi}{(\xi_0 - \xi)^{1/2}}, \text{ for } \xi_0 > \xi_2(\eta) , \tag{60}$$

in which the right side is known. This may be solved for w_x:

$$w_x(\xi, \eta_0) = \frac{1}{\pi\mu(\xi_0 - \xi_2(\eta_0))^{1/2}} \int_0^{\xi_2(\eta_0)} T(\xi, \eta) \frac{(\xi_2(\eta_0) - \xi)^{1/2}}{(\xi_0 - \xi)}\, d\xi . \tag{61}$$

Then the stress singularity just beyond the crack tip is given by

$$w_x(0+, y_0, t_0) = \frac{1}{\pi\mu}\left(\frac{\beta - \dot{y}_2}{2\beta}\right)\int_0^{\xi_2(\eta_0)} \frac{T(\xi, \eta_0)}{(\xi_2(\eta_0) - \xi)^{1/2}}\, d\xi \\ \times \frac{1}{[y_0 - y_2(t_0)]^{1/2}} , \tag{62}$$

showing that $w_x \to \infty$ with $[y_0 - y_2(t_0)]^{-1/2}$ as $y_0 \to y_2(t)$ unless the multiplier of $[y_0 - y_2(t_0)]^{-1/2}$ is zero. From this observation we next calculate the crack-edge trajectory.

5.10 The Crack Edge

Since the stress cannot exceed the static limiting friction, and must therefore be finite, and referring to (62), we obtain the following alternatives for the crack edge

$$\dot{y}_2 = \beta, \text{ or } \int_0^{\xi_2(\eta_0)} \frac{T(\xi, \eta_0)}{(\xi_2(\eta_0) - \xi)^{1/2}} \, d\xi = 0 \, . \tag{63}$$

5.11 The Stopping Locus

Referring to the the important Fig. 32 we find [6] that the crack edge stops, $\dot{y}_2(t) = 0$, at F where the ξ-characteristic through B' meets the crack edge, where B' $(y = -b)$ is the image of the zero-stress-drop point B $(y = b)$.

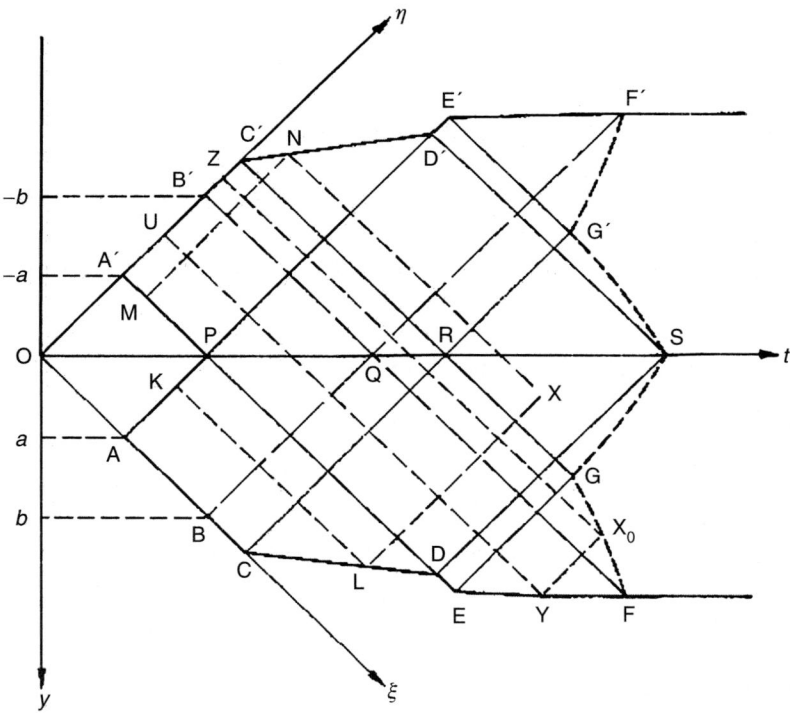

Fig. 32. The crack edge PABCDEEF and its image are shown. FGS is the stopping locus to the right of which the relative displacements on the crack are constant in time. It is interesting to note that F, G, and S are determined by the characteristic lines through certain important points on the crack edge and its image. The light continuous lines together with the stopping locus divide that (t, y)-plane into about 16 regions in which different analytic formulae represent the displacement

We can also find the stopping locus, FGS in Fig. 32, the locus on which $w_t = 0$ for the first time. After $w_t = 0$ it remains zero for greater values of t, since friction would act to inhibit reverse sliding.

5.12 A Simple Example: $T = 1 - y^2$

We non-dimensionalize so that the wave speed $\beta = 1$ and $b = 1$. If we take $T = 1 - y^2$ all the loci that are not characteristics $t \pm y = \text{const.}$ in Fig. 32 are arcs of ellipses in the (y, t)-plane. The whole solution may be calculated analytically for this particular stress drop, and we show plots of the solution for nucleation depths $a = 0$, $a = 0.6$ and the maximum depth $a = 1$.

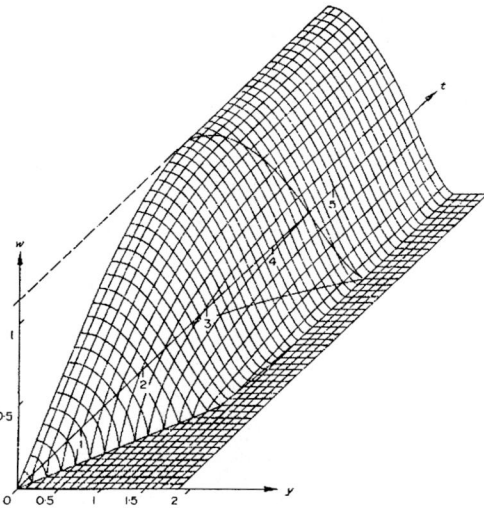

Fig. 33. A plot of $w = w(t, y)$ vertically versus t and y horizontally. The crack edge forms the boundary of the support of w. The stopping locus is drawn on both the surface and the (t, y)-plane. Beyond it w is independent of time, i.e. the motion has stopped. Here the crack nucleates at the surface, $a = 0$

5.13 The Residual Static Stress Drop

The static displacement after the slipping has ceased, i.e. the residual displacement, and the stress drop as $t \to \infty$ on the fault are independent of the nucleation depth a. They are given by (See Figs. 33-36)

$$w(\infty, y) = \begin{cases} \dfrac{32}{81\pi}\sqrt{\dfrac{15}{2}} \left(\dfrac{9}{4} - y^2\right)^{3/2}, & \text{for } 0 \leq y \leq \dfrac{3}{2}, \\ 0, & y > \dfrac{3}{2}. \end{cases} \tag{64}$$

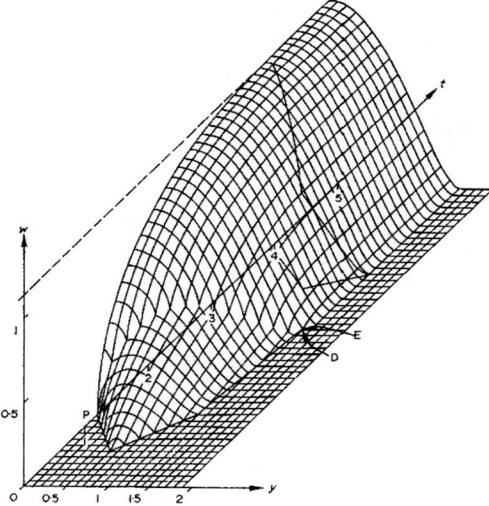

Fig. 34. A plot of $w = w(t, y)$ in the same format as the preceding figure. Here the crack nucleates at depth $a = 0.6$. Notice the sonic jog DE in the crack-edge trajectory caused by the breakout stress wave visible as a discontinuity in the slope of $w = w(t, y)$

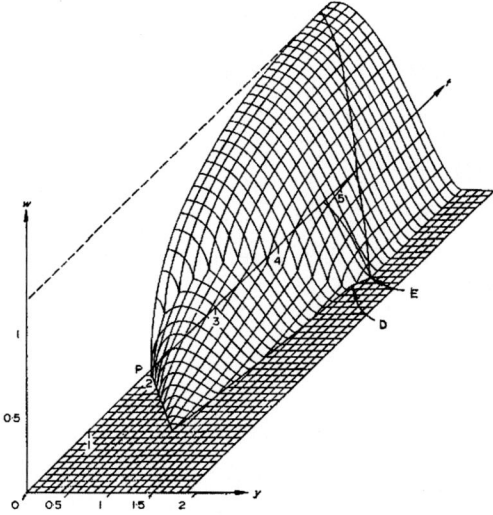

Fig. 35. A plot of $w = w(t, y)$ in the same format as the preceding figures. Here the crack nucleates at the maximum depth $a = 1$. The jog in the crack-edge trajectory now occurs just before stopping

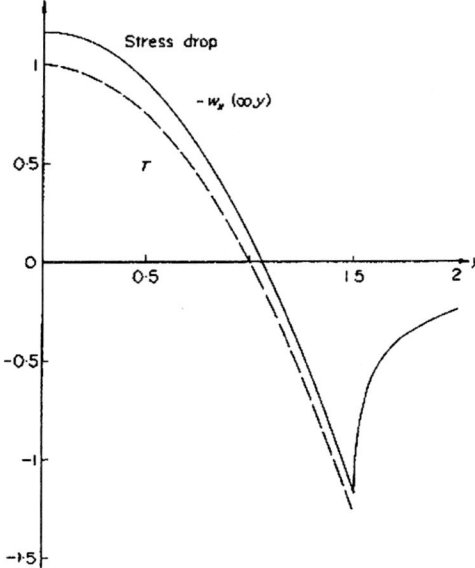

Fig. 36. Here are plotted as functions of y the dynamic stress drop $T = 1 - y^2$ (*dashed line*) which acts before stopping and the final static stress drop $-w_x(\infty, y)$, which is about 16% greater at $y = 0$

$$-\mu w_x(\infty, y) = \begin{cases} \dfrac{32}{27\pi}\sqrt{\dfrac{15}{2}}\left(\dfrac{9}{8} - y^2\right), & \text{for } 0 \leq y \leq \dfrac{3}{2}, \\ \dfrac{32}{27\pi}\sqrt{\dfrac{15}{2}}\left[y\left(y^2 - \dfrac{9}{4}\right)^{1/2} + \dfrac{9}{8} - y^2\right], & y > \dfrac{3}{2}. \end{cases} \quad (65)$$

Notice that the final static displacement is independent of the point $y = a$ of nucleation for this example. See [6].

5.14 Radiated Far-Field Pulses

At large distances the pulses that are radiated in directions making angle θ with the downward vertical are shown in the Figs. 38, 39 and 40. The amplitudes radiated in the various directions are shown in the radiation pattern of Fig. 37. Notice the very pronounced breakout phase, which is a much more energetic arrival than the first motion, especially for takeoff angles near vertically down. This is a Doppler effect of the approaching image crack edge due to the reflection of a strong stress wave when the upgoing crack edge "breaks out" at the free surface. The radiation pattern, however, moderates this effect since it is small near vertical and vanishes when $\theta = 0$.

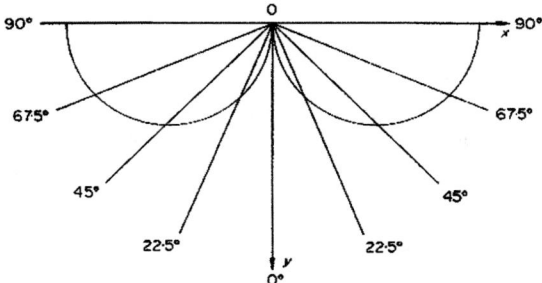

Fig. 37. The radiation pattern proportional to $\sin\theta$, where θ is the angle with the downward vertical

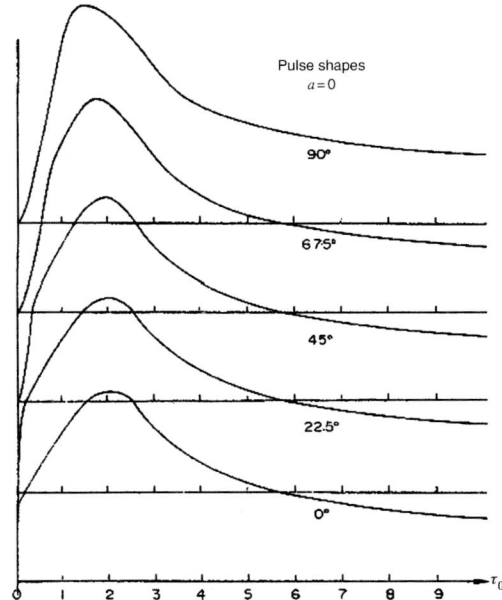

Fig. 38. These are the far-field pulse shapes for depth of nucleation $a = 0$

6 A Model for Repeating Events

We embed the previous model in a viscoelastic halfspace.

6.1 Equations of Motion

$$\rho\ddot{\mathbf{w}} = \nabla\cdot\boldsymbol{\tau} \ ,$$

$$\boldsymbol{\tau}(x,y,t) = \mu \int_{-\infty}^{t} \{1 - \phi[(t-s)/\bar{T}]\} \nabla\dot{\mathbf{w}}(x,y,s)\,\mathrm{d}s \ , \tag{66}$$

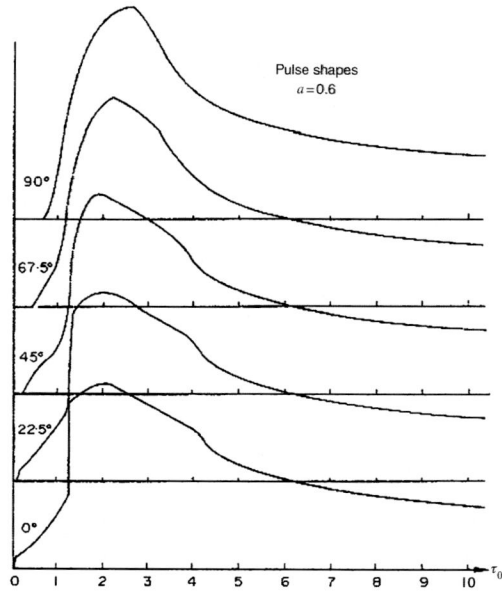

Fig. 39. Far-field pulse shapes for depth of nucleation $a = 0.6$

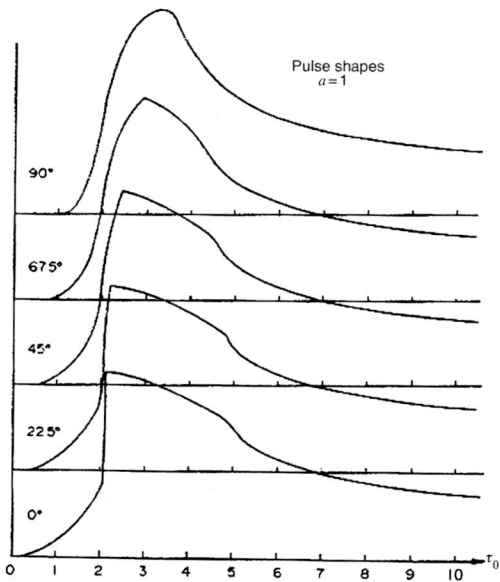

Fig. 40. Far-field pulse shapes for depth of nucleation $a = 1$

where
$$\phi(0) = 0 \text{ and } \Phi = \int_{-0}^{\infty} [1 - \phi(\sigma)] \, d\sigma < \infty. \tag{67}$$

6.2 On the Fault Plane

- Limiting static friction is again great enough to hold equilibrium against the initial traction on the fault plane $x = 0$ except at the point of nucleation of slip.
- Dynamic friction is less than the initial traction in the neighborhood of the point of nucleation.

Let the stress drop be
$$\tau^{\text{drop}}(y) = \tau^0_{zx}(y, 0) - F_{\text{D}}(y) \tag{68}$$

τ^{drop} is positive near the surface, monotone decreasing, and tending to $-\infty$ as $y \to \infty$. Then the driving force is τ^{drop} acting on the region of slip as before.

6.3 Nondimensionalization

Let $y = L$ be the depth at which stress drop τ^D changes sign, and let the time for S-waves to travel a distance L be
$$T = \left(\frac{\rho}{\mu}\right)^{1/2} L = \frac{L}{v_S}. \tag{69}$$

We nondimensionalize by defining
$$y = Ly', \ x = Lx' \ t = Tt', \ s = Ts'. \tag{70}$$

Then τ^D changes sign at $y' = 1$ and, after the elimination of $\boldsymbol{\tau}$ the equation of motion is
$$\ddot{\mathbf{w}}(x', y', t') = \int_{-\infty}^{t'} \{1 - \phi[\epsilon(t' - s')]\} \nabla^2 \dot{\mathbf{w}}(x', y', s') \, ds' \tag{71}$$

all derivatives being with respect to the primed variables y', x', t', which are the natural scaled variables to describe an event. The small parameter ϵ is $T/\bar{T}$. Integrating by parts on the right we get
$$\ddot{\mathbf{w}}(x', y', t') = \nabla^2 \mathbf{w}(x', y', t')$$
$$- \epsilon \int_{-\infty}^{t'} \phi'[\epsilon(t' - s')] \nabla^2 \mathbf{w}(x', y', s') \, ds'. \tag{72}$$

We shall drop the second term on the right during events.

150   R. Burridge

Now change to slow time by setting

$$\bar{t} = \epsilon\,t',\ \bar{s} = \epsilon\,s'\ . \tag{73}$$

Then

$$\epsilon^2 \ddot{\mathbf{w}}(x',y',\bar{t}) = \int_{-\infty}^{\bar{t}} \bar{t}\{1 - \phi[(\bar{t} - \bar{s})]\} \nabla^2 \mathbf{w}(x',y',\bar{s})\,\mathrm{d}\bar{s}\ . \tag{74}$$

where now the dots represent derivatives with respect to $\bar{t}$.

During the long intervals between events $(x', y', \bar{t})$ are appropriate scaled variables and we drop the left-hand side to get

$$0 = \int_{-\infty}^{\bar{t}} \{1 - \phi[(\bar{t} - \bar{s})]\} \nabla^2 \dot{\mathbf{w}}(x',y',\bar{s})\,\mathrm{d}\bar{s}\ , \tag{75}$$

leading to

$$\nabla^2 \dot{\mathbf{w}}(x',y',\bar{t}) = 0\ . \tag{76}$$

6.4 Rate of Strain Between Events

Assume a constant rate of shear at infinity

$$\nabla \dot{w} \to (0, V) \ \text{as}\ x^2 + y^2 \to \infty\ .$$

Then

$$\nabla \dot{w} = (0, V)\ . \tag{77}$$

everywhere and for all times except for short time intervals of duration $O(\epsilon)$ in $\bar{t}$ after events. What changes between events is the stress field which gradually creeps up to trigger the next event.

If the stress at infinity is

$$\boldsymbol{\tau}_\infty = (0, \tau_\infty)\ , \tag{78}$$

Then

$$\tau_\infty = \mu V \bar{T} \int_0^\infty [1 - \phi(\sigma)]\,\mathrm{d}\sigma \tag{79}$$

giving

$$V = \frac{\tau_\infty}{\mu \bar{T} \Phi},\ \Phi = \int_0^\infty [1 - \phi(\sigma)]\,\mathrm{d}\sigma\ , \tag{80}$$

and

$$\nabla \dot{w} = \frac{\tau_\infty}{\mu \bar{T} \Phi}(0, 1)\ . \tag{81}$$

Superposed on $\boldsymbol{\tau}_\infty$ is a stress field decaying at infinity due to the repeated events.

Assume the events repeat periodically with period $\bar{\nu}$. Then viewed on the long $\bar{t}$ time scale, at the time of an event the displacement $(0, w)$ undergoes a sudden step change and suppose that ∇w jumps by $\boldsymbol{f}(x,y)$ at $\bar{t} = n\bar{\nu}$ (n an integer). Then we may write

$$\nabla \dot{w} = \frac{T_\infty}{\mu \bar{T} \Phi}(0,1) + \boldsymbol{f}(x,y) \sum_{n=-\infty}^{\infty} \delta(\bar{t} - n\bar{\nu}) \tag{82}$$

Then the viscoelastic stress-strain relation gives

$$\tau_{xz}(x', y', \bar{t}) = T_\infty - \frac{\mu}{L} f(x', y') \int_{-\infty}^{\bar{t}} [1 - \phi(\bar{t} - \bar{s})]$$
$$\sum_{n=-\infty}^{\infty} \delta(\bar{s} - n\bar{\nu}) \, \mathrm{d}\bar{s} \tag{83}$$
$$= T_\infty - \frac{\mu}{L} f(x', y') \sum_{n=-\infty}^{[\bar{t}/\bar{\nu}]} [1 - \phi(\bar{t} - n\bar{\nu})] \ .$$

We see that the stress drop on the fault plane during an event (equivalent to the residual static stress drop in the previous model) is $(\mu/L)f(y',0)$.

We shall assume that the dynamic friction is such that the dynamic stress drop is quadratic as before:

$$\tau^{\mathrm{drop}}(y') = \tau_0^{\mathrm{drop}}(1 - y'^2) \ . \tag{84}$$

Notice $\tau^{\mathrm{drop}}(0) = \tau_0^{\mathrm{drop}} = F_{\mathrm{drop}} = F_{\mathrm{S}} - F_{\mathrm{D}}$ at the point of nucleation $y' = 0$.

6.5 The Parameters for Repeating Events

According to the previous model the residual stress drop (65) on the region of slip $0 \leq y' \leq 1.5$ is

$$k\tau_0^{\mathrm{drop}}\left(1 - \frac{8}{9}y'^2\right), \ k = \frac{4}{3\pi}\left(\frac{15}{2}\right)^{1/2} \simeq 1.1623 \ . \tag{85}$$

Thus

$$\frac{\mu}{L} f(y', 0) = k\tau_0^{\mathrm{drop}}\left(1 - \frac{8}{9}y'^2\right) \ . \tag{86}$$

Just before an event

$$\tau_{zx}(y', 0) = F_{\mathrm{S}} + a\,y'^2 = T_\infty - \frac{\mu}{L} f(y', 0)\Psi(\bar{\nu}) \ , \tag{87}$$

where

$$\Psi(\bar{\nu}) = \sum_{p=1}^{\infty}[1 - \phi(p\bar{\nu})] \tag{88}$$

So, specializing to $y' = 0$,

$$F_S = \tau_\infty - k\tau_0^{\text{drop}} \Psi(\bar{\nu}), \qquad (89)$$

and, writing $\tau_\infty^{\text{drop}}$ for $\tau_\infty - F_S$, we have

$$\Psi(\bar{\nu}) = \frac{\tau_\infty^{\text{drop}}}{k\, F_{\text{drop}}} = \frac{\tau_\infty - F_S}{k\,(F_S - F_D)}. \qquad (90)$$

This is an equation for $\bar{\nu}$.

Next we consider the coefficient of $y'^{\,2}$ in the expressions for the residual stress.

$$a = \frac{8}{9} k \Psi(\bar{\nu}) \tau_0^{\text{drop}} = \frac{8}{9} \frac{\tau_\infty^{\text{drop}}}{F_{\text{drop}}} \qquad (91)$$

So, if

$$\tau_{zx}(y',0) = F_S + \frac{8}{9} \frac{\tau_\infty^{\text{drop}}}{F_{\text{drop}}} y'^{\,2} \qquad (92)$$

just before an event, and the residual stress drop after an event is

$$k\tau_0^{\text{drop}}\left(1 - \frac{8}{9} y'^2\right), \qquad (93)$$

the events will repeat with period $\bar{\nu}$ given by solving

$$\Psi(\bar{\nu}) = \frac{\tau_\infty^{\text{drop}}}{k\, F_{\text{drop}}}. \qquad (94)$$

In particular if the solid is a Maxwell solid

$$\phi(\sigma) = 1 - e^{-\sigma} \qquad (95)$$

and

$$\Psi(\bar{\nu}) = \sum_1^\infty [1 - \phi(\sigma)] = \frac{e^{-\bar{\nu}}}{1 - e^{-\bar{\nu}}} = \frac{1}{e^{\bar{\nu}} - 1}, \qquad (96)$$

yielding

$$\bar{\nu} = \log\left(1 + \frac{1}{\Psi(\bar{\nu})}\right) = \log\left(1 + \frac{k F_{\text{drop}}}{\tau_\infty^{\text{drop}}}\right). \qquad (97)$$

References

1. R. Burridge and L. Knopoff, Body force equivalents for seismic dislocations, Bull. Seis. Soc. Amer. **54**, 1875–1888 (1964).
2. R. Burridge and L. Knopoff, Model and theoretical seismicity, Bull. Seis. Soc. Amer. **57**, 341–371 (1967).

3. J.M. Carlson, J.S. Langer and B.E. Shaw, Dynamics of earthquake faults, Rev. Mod. Phys. **66**, 657–670, (1994).
4. B.K. Chakrabarti and L.G. Benguigui, *Statistical Physics of Fracture and Breakdown in Disordered Systems*, Oxford Univ. Press, Oxford, pp. 128–148 (1997).
5. R. Burridge, The numerical solution of certain integral equations with nonintegrable kernels arising in the theory of crack propagation and elastic wave diffraction, Phil. Trans. Roy. Soc. **A 265**, 353–381 (1969).
6. R. Burridge and G.S. Halliday, Dynamic shear cracks as models for shallow focus earthquakes, Geophys. J. Roy. Astr. Soc. **25**, 261–283 (1971).
7. B. Kostrov, Unsteady propagation of longitudinal shear cracks, J. Appl. Math. Mech. **30**, 1241–1248 (1966).
8. R. Burridge, A repetitive earthquake source model, J. Geophys. Res. **82**, 1663–1666 (1977).

Geometric Models of Earthquakes

P. Bhattacharyya

Physics Department, Gurudas College, Narkeldanga, Kolkata 700 054 and Centre for Applied Mathematics and Computational Science, Saha Institute of Nuclear Physics, Kolkata -700 064
pratip.bhattacharyya@saha.ac.in

1 Introduction

Earthquakes are caused the movement of adjacent tectonic plates along geological faults (fault dynamics) in the earth's lithosphere. The overall frequency distribution of the magnitude $\mathcal{M}$ of earthquakes, including foreshocks, mainshocks and aftershocks, is said to follow the empirical Gutenberg–Richter law [1, 2]:

$$\log_{10} \text{Nr}(\mathcal{M} > M) = a - bM , \qquad (1)$$

where $\text{Nr}(\mathcal{M} > M)$ denotes the number (the frequency) of earthquakes of magnitude $\mathcal{M}$ that are greater than a certain value M. The constant a is obtained from the total number of earthquakes of all magnitudes: $a = \log_{10} \text{Nr}(\mathcal{M} > 0)$. The coefficient of M is presumed to be universal: $b \approx 1$. In an alternative form, the Gutenberg–Richter law is expressed as a relation for the number (frequency) of earthquakes in which the energy released $\mathcal{E}$ is greater than a certain value E [3, 4]:

$$\log_{10} \text{Nr}(\mathcal{E} > E) \sim E^{-b/\beta} , \qquad (2)$$

where $\beta \approx 3/2$ appears as the coefficient of M in the empirical relation between the values E of energy released in earthquakes and the corresponding values M of their magnitude: $\log_{10} E = \alpha + \beta M$.

A geological fault is created by a fracture in the earth's rock layers (lithosphere) followed by a relative displacement of the two parts along the fracture (shown schematically in Fig. 1). As the two parts of the fault move in opposite directions their surfaces in contact stick and slip intermittently. While the surfaces stick due to friction, stress, and potential energy associated with it, develops due to deformation of the adjacent parts. When the developed stress overcomes the friction, the two surfaces of the fault slip; the potential energy of deformation is thereby released and it causes an earthquake. An earthquake therefore occurs each time there is a slip in the fault. This intermittent stick-slip process is the essential feature of fault dynamics.

156   P. Bhattacharyya

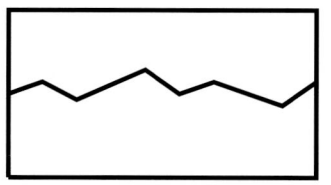 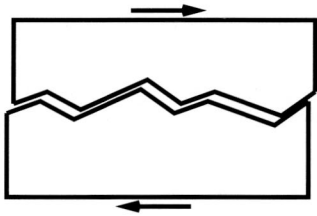

Fig. 1. Schematic diagram of a portion of a geoloical fault

The aim of constructing models of earthquakes is to reproduce the Gutenberg–Richter law by simulating the dynamics of a fault or of a system of interconnected faults. Different classes of models focus on different aspects of the faults, such as geometry, structure, and material properties. One traditional class of models that considers the structure of faults is based on the collective motion of an assembly of locally connected elements that are driven slowly. The first of this kind was the block-spring model introduced by Burridge and Knopoff [5]. The original model and all its variants, for example [6, 7] contain the stick-slip dynamics necessary to produce earthquakes. The underlying principle in this class of models was found to be self-organised criticality [8]. Another traditional class of models is based on the mechanical properties of deformable materials that break under sufficient stress. Fibre bundle models are typical of this class [9].

A relativey new class of models of earthquakes is based on the geometry of faults. The surfaces produced by a fracture in a rock are found to be self-similar fractals, the shape of the one part being complementary to the shape of the other. Since it is created by a fracture in the earth's rock layers, a geological fault, from the point of view of geometry, is composed of a pair of overlapping fractals of complementary shape, the one shifting over the other. The simplest model containing this essential geometric feature of a fault is the Cantor set– the simplest fractal known–shifting uniformly over its complementary set in the unit interval (shown in Fig. 2). However, Chakrabarti and Stinchcombe [10] considered a further simplification: the Cantor set shifting uniformly over its replica (Fig. 3). Though not a realistic model of a geological fault, it is the simplest case of a system of two overlapping fractals, a geological fault being only a much more complicated case. The aim of studying this model is to learn about the time-series and the frequency or probability distribution of the events that a system of overlapping fractals can produce and hence to determine whether fractal-overlap geometry can produce the Gutenberg–Richter law.

It needs to be mentioned that this article (and the papers upon which it is based) neither suggest that a geological fault is a pair of overlapping Cantor sets, nor finds that a pair of overlapping Cantor sets produce the Gutenberg-Richter law. It is only established that a pair of overlapping Cantor

Fig. 2. A finite generation of the Cantor set overlapping on its complement in the unit line segment: the representation of the fractal overlap structure of a geological fault in terms of the simplest fractal

Fig. 3. The Chakrabarti–Stinchcombe model: the Cantor set overlapping on its replica. This is the simplest system of two overlapping fractals

sets and a pair of overlapping fractured rock surfaces (that make a fault) can both produce power law distributions of overlap magnitudes. Their exponents certainly differ, but it provides the scope of obtaining the Gutenberg-Richter exponent by refining the model, for example, by replacing the Cantor sets with more complicated fractals that resemble the surfaces of a real fault.

2 The Cantor Set

In the construction of the Cantor set [11] the initiator (or, the zeroeth generation) is the closed interval $[0, 1]$ represented by a line segment of unit length. The generator is defined by the first stage of the construction (the first generation) which consists of dividing the interval $[0, 1]$ into three parts, then removing the middle open interval $(1/3, 2/3)$. In the next stage (the second generation), the procedure of obtaining the generator is applied to each of the two remaining closed intervals $[0, 1/3]$ and $[2/3, 1]$. In this way (see, e.g., Fig. 4) every generation is obtained by applying the same procedure to each of the closed intervals of the preceding generation. Therefore, the n-th generation contains 2^n closed intervals, represented by 2^n line segments, each of length $1/3^n$. If the construction is continued for infinite number of generations, the remainder set of discrete points is called the Cantor set; the positions of the points in the Cantor set are given by

$$x = \sum_{n=1}^{\infty} A_n/3^n, \quad A_n \in \{0, 2\} \, . \tag{3}$$

It is an exactly self-similar fractal of dimension $\log_b 2 / \log_b 3$.

3 The Simplest Fractal-Overlap Model of a Fault

The geometric model of a fault consists of two dynamically overlapping fractals that represent the two complementary parts of the fault, each shifting relative

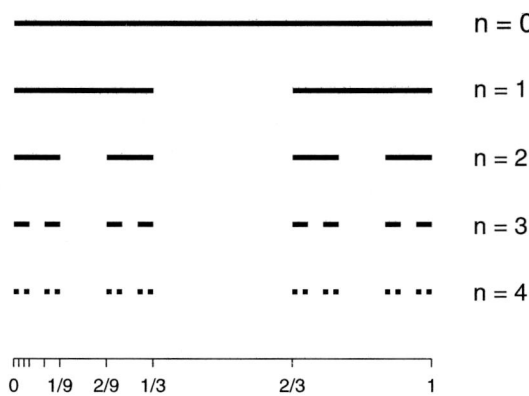

Fig. 4. The "initiator" ($n = 0$) and the first four generations in the construction of the regular Cantor set. The first generation ($n = 1$) provides the "generator". This figure illustrates the process of constructing successive generations described in the text

to the other along the surface of contact. In the simplest case each of the two fractals is the Cantor set [10]. Though it is practically impossible to generate the Cantor set, its properties may be obtained by the examining any finite generation n in its construction procedure. It is also necessary to study a finite generation owing to the fact that the property of self-similarity in all naturally occuring fractals is observed between a lower and an upper cut-off length [11]. The model therefore studied is the n-th generation of the Cantor set overlapping on its replica and its asymptotic properties are obtained by proceeding to the limit $n \to \infty$. The two finite generation sets of line segments schematically represent the profiles of the fractal surfaces of a geological fault. At any instant t of time the magnitude of overlap between the two fault surfaces is measured in this model as the length of overlap between the line segments of the two sets; for the n-th generation sets it is denoted by $Y_n(t)$. The quantity $Y_n(t)$ represents the energy released in an earthquake caused by a slip in the fault at time t whereas the logarithm of $Y_n(t)$ corresponds to the magnitude of the earthquake. The movement of the fault surfaces, each relative to the other, is simulated by shifting one of the sets over the other in one particular direction with uniform velocity. As a simplification, the process of shifting is considered to occur in discrete steps [12]. Assuming that initially the two sets overlap completely, i.e., every line segment of the one set overlaps completely on the corresponding line segment of the other set, the length of a step for the n-th generation is chosen to be $1/3^n$ in order to ensure that each line segment of the one set overlaps completely on a line segment or does not overlap on any of the other set. It is convenient to define time t for this discrete dynamical process as a discrete variable and measure its value as the the number of steps by which one of the sets has shifted from its initial position relative to the other. Further, for the convenience of analysis of the model, periodic boundary conditions are assigned to both the finite generations; the result of open boundary conditions will be deduced from the result of periodic boundary conditions. The overlap magnitude $Y_n(t)$ in the discrete-shift model is given by the number of pairs of overlapping line segments; because of the structure of the Cantor set, Y_n can assume only discrete values that are in geometric progression: $Y_n = 2^{n-k}$, $k = 0, \ldots, n$; the overlap magnitudes are written as 2^{n-k} in order to indicate the generation n of the overlapping sets.

4 Time Series of Overlap Magnitudes

The natural outcome of the fractal-overlap model is the time series of overlap magnitudes. Therefore the first stage in the analysis of the model is the construction of the time series of the overlap magnitudes $Y_n(t)$ for the n-th generation of the Cantor set overlapping on its replica. The time series is constructed by determining the formula for the sequences of constant magnitude overlaps since the complete time series is just the superposition of the sequences of all possible overlap magnitudes. Since the initial condition is chosen as the complete overlap of the two sets, the initial overlap is maximum and its magnitude is given by:

$$Y_n(0) = \max Y_n = 2^n \ . \tag{4}$$

Moreover, owing to periodic boundary conditions, the time-series of the overlaps for the n-th generation repeats itself after every 3^n time-steps as the length of each elementary line segment in the n-th generation of the Cantor set is $1/3^n$:

$$Y_n(t) = Y_n(t + 3^n) \ . \tag{5}$$

Consequently it is only necessary to study the time-series within the period $0 \leq t < 3^n$. Besides, the symmetric structure of all the generations of the Cantor set causes the time-series of the overlaps in the period $0 \leq t < 3^n$ (or, in all periods of the kind $k3^n \leq t < (k+1)3^n$, $k = 0, 1, 2, \ldots$) to be symmetric about the center of the period:

$$Y_n(t) = Y_n(3^n - t), \quad 0 \leq t < 3^n \ . \tag{6}$$

i.e., for every overlap occurring at the time-step t there is an overlap of the same magnitude occurring at the time-step $3^n - t$.

The time series has a recursive structure with a skeleton that is derived from the following observation: After the first 3^{n-1} time-steps of shifting one n-th generation relative to the other, the overlapping region is similar to generation $n - 1$. Thus the next 3^{n-1} time-steps form one period of the time-series of overlaps of two sets of generation $n - 1$. Within this period of 3^{n-1} time-steps, after the first 3^{n-2} time-steps the overlapping region is similar to generation $n-2$ and therefore the next 3^{n-2} time-steps form one period of the time-series of overlaps of two sets of generation $n-2$. As a result of this recursive process a period of the time-series of overlaps of generation-n is a nested structure of the periods of the time-series of the overlaps of all the preceding generations. In general, after $3^{n-1} + 3^{n-2} + \cdots + 3^{n-k}$ time-steps the overlapping region is similar to generation $n-k$, for which, following (4), the magnitude of overlap is given by:

$$Y_n \left(\sum_{r=1}^{k} 3^{n-r} \right) = 2^{n-k}; \quad k = 1, \ldots, n. \tag{7}$$

The sequence of overlaps generated by (7) and the complementary sequence obtained by the symmetry property of (6), along with the initial condition (4), forms the skeleton of the entire time-series. Due to this feature the time series acquires a self-affine profile.

The details of the time series are deduced from further observations: Since the n-th generation of the Cantor set contains 2^n line segments that are arranged by the generator in a self-similar manner, the magnitude of the overlaps, beginning from the maximum, form a geometric progression of descending powers of 2. As the magnitude of the maximum overlap is 2^n (at $t = 0$), the magnitude of the next largest overlap is 2^{n-1}, i.e., a half of the maximum. A pair of nearest line segments form a doublet and the generation-n of the Cantor set has 2^{n-1} such doublets. Within a doublet, each of the two line segments are two steps away from the other. Therefore, if one of the sets is shifted from its initial position by two time-steps relative to the other, only one of the segments of every doublet of the former set will overlap on the other segment of the corresponding doublet of the latter set, thus resulting in an overlap of magnitude 2^{n-1}. An overlap of 2^{n-1} also occurs if quartets formed of pairs of nearest doublets are considered and one of the sets is shifted from its initial position by 2×3 time-steps relative to the other. Similarly, in the case of octets formed of pairs of nearest quartets, a shift of 2×3^2 time-steps is required to produce an overlap of 2^{n-1}. Considering pairs of blocks of 2^{r_1} nearest segments ($r_1 \leq n-1$), an overlap of magnitude 2^{n-1} occurs for a shift of 2×3^{r_1} time-steps:

$$Y_n (2 \times 3^{r_1}) = 2^{n-1}; r_1 = 0, \ldots, n-1. \tag{8}$$

The complementary sequence is obtained as usual by the symmetry property of (6).

Because of the self-similar structure of the Cantor set, the line of argument used to deduce (8) works recursively for overlaps of all consecutive magnitudes. For example, the next overlap magnitude is 2^{n-2}, i.e., a quarter of the maximum. For each time-step t at which an overlap of 2^{n-1} segments occur, there are two subsequences of overlaps of 2^{n-2} segments that are mutually symmetric with respect to t; one of the subsequences precedes t, the other succeeds t. Therefore the sequence of t values at which an overlap of 2^{n-2} segments occurs is determined by the sum of two terms, one from each of two geometric progressions, one nested within the other:

$$Y_n\left(2\left[3^{r_1} \pm 3^{r_2}\right]\right) = 2^{n-2}; \tag{9}$$
$$r_1 = 1, \ldots, n-1;$$
$$r_2 = 0, \ldots, r_1 - 1.$$

The first term belongs to the geometric progression that determines the sequence of t values appearing in equation (8) for the overlaps of 2^{n-1} segments, while the second term belongs to a geometric progression nested within the first. The symmetry property of (6) provides the complementary sequence.

In general the sequence of time-step values at which an overlap of 2^{n-k} segments ($1 \leq k \leq n$) occurs is determined by the sum of k terms, one from each of k geometric progressions, nested in succession:

$$Y_n\left(2\left[3^{r_1} \pm 3^{r_2} \pm \cdots \pm 3^{r_{k-1}} \pm 3^{r_k}\right]\right) = 2^{n-k}; \tag{10}$$
$$k = 1, \ldots, n;$$
$$r_1 = k-1, \ldots, n-1;$$
$$r_2 = k-2, \ldots, r_1 - 1;$$
$$\vdots \quad \vdots$$
$$r_{k-1} = 1, \ldots, r_{k-2} - 1;$$
$$r_k = 0, \ldots, r_{k-1} - 1.$$

For each value of k in the above equation there is a complementary sequence due to the symmetry property of (6). Equation (10) along with (4), (5) and (6) determine the entire time-series of overlaps of the n-th generation of the Cantor set on its replica. Assuming that the initial overlap is given by (4), the time-series is the superposition of the sequences of constant magnitude overlaps given by (10) for all possible overlap magnitudes.

Consider now the special case of unit overlaps, i.e., when the overlap is only on a unit segment. The magnitude of a unit overlap for the n-th generation of the Cantor set is written as 2^{n-n} in order to indicate the generation index explicitly. A unit overlap occurs when $k = n$ in (10). The sequence of t values at which these occur is given by:

$$Y_n\left(2\left[3^{n-1} \pm 3^{n-2} \pm \cdots \pm 3^1 \pm 3^0\right]\right) = 2^{n-n} = 1. \tag{11}$$

The above equation shows 2^{n-1} occurrences of the unit overlap. The same number of unit overlaps also occur in the complementary sequence obtained by using (6). Therefore, in a period of the time-series for the n-th generation, there are altogether 2^n unit overlaps.

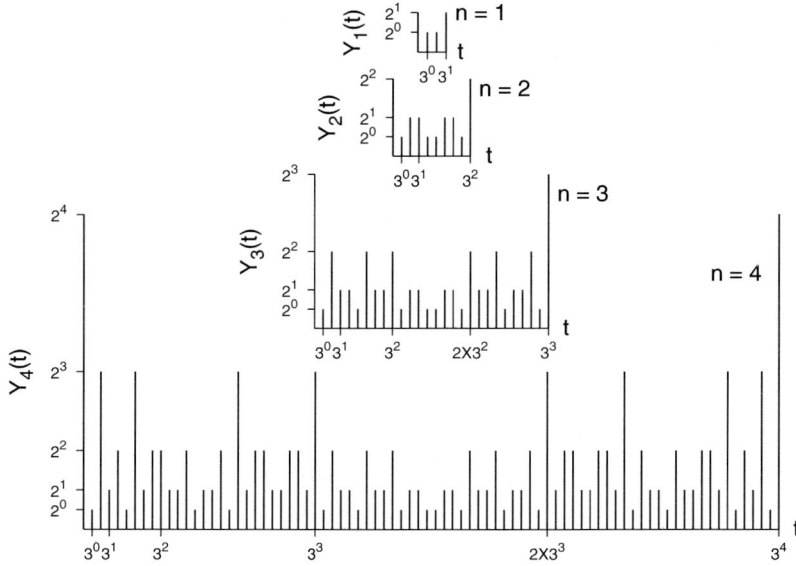

Fig. 5. A period of the time-series of overlap magnitudes for the n-th generation of the Cantor set overlapping on its replica, for the first four generations ($n = 1, 2, 3$ and 4) according to (4), (5), (6) and (10). It shows that a period of the time-series for each generation is nested within a period of the time-series for the next generation. Owing to this recurrence, the sequence of overlap magnitudes in a period of the time-series for the Cantor set ($n = \infty$) forms a self-affine profile

5 Analysis of the Time-Series

The time-series (see, e.g., Fig. 5) is analysed by first determining, using induction, the exact number of overlaps of every magnitude Y_n in a period. The probability of occurrence of an overlap of magnitude Y_n after any arbitrary time-step is thereby obtained. Next the conditional probability of the occurrence of an overlap of magnitude Y'_n is determined if it is given that the preceding overlap in the time-series has magnitude Y_n.

In the following $\text{Nr}(Y_n)$ denotes the frequency of overlaps of magnitude Y_n, i.e., the number of times an overlap of Y_n segments occurs in a period of the time-series for n-th generation of the Cantor set. Equations (4), (8) and (9) and their complementary parts show that:

$$\text{Nr}\left(2^n\right) = 1 , \tag{12}$$

$$\text{Nr}\left(2^{n-1}\right) = 2n \tag{13}$$

and

$$\text{Nr}\left(2^{n-2}\right) = 2\sum_{r_1=1}^{n-1} 2\, r_1 = 2n(n-1) . \tag{14}$$

Similarly the frequency of overlaps of the next few magnitudes are:

$$\text{Nr}\left(2^{n-3}\right) = 2\sum_{r_1=2}^{n-1} 2\sum_{r_2=1}^{r_1-1} 2r_2$$
$$= \frac{4}{3}n(n-1)(n-2), \qquad (15)$$

$$\text{Nr}\left(2^{n-4}\right) = 2\sum_{r_1=3}^{n-1} 2\sum_{r_2=2}^{r_1-1} 2\sum_{r_3=1}^{r_2-1} 2r_3$$
$$= \frac{2}{3}n(n-1)(n-2)(n-3), \qquad (16)$$

$$\text{Nr}\left(2^{n-5}\right) = 2\sum_{r_1=4}^{n-1} 2\sum_{r_2=3}^{r_1-1} 2\sum_{r_3=2}^{r_2-1} 2\sum_{r_4=1}^{r_3-1} 2r_4$$
$$= \frac{4}{15}n(n-1)(n-2)(n-3)(n-4). \qquad (17)$$

In general, from (10), the frequency of overlaps of magnitude $Y_n = 2^{n-k}$ is given by:

$$\text{Nr}\left(2^{n-k}\right) = 2\sum_{r_1=k-1}^{n-1} 2\sum_{r_2=k-2}^{r_1-1} \cdots 2\sum_{r_{k-1}=1}^{r_{k-2}-1} 2r_{k-1}$$
$$= C_k\, n(n-1)(n-2)\cdots(n-k+1)$$
$$= C_k\, \frac{n!}{(n-k)!}. \qquad (18)$$

The value of the constant C_k is determined from the case of unit overlaps in the following way. In the above equation the index k is kept constant and the generation index is chosen to be $n = k$. The result is the frequency of unit overlaps for the k-th generation:

$$\text{Nr}\left(2^{k-k}\right) = C_k\, k!. \qquad (19)$$

On the other hand, the frequency of the unit overlap for the k-th generation is obtained from the sequence defined in (11) by replacing the index n by k:

$$\text{Nr}\left(2^{k-k}\right) = 2^k. \qquad (20)$$

Comparing (19) and (20) gives the value of the constant:

$$C_k = \frac{2^k}{k!}. \qquad (21)$$

Therefore the formula for the frequency of overlaps of magnitude $Y_n = 2^{n-k}$ in the time-series for the n-th generation of the Cantor set is:

$$\text{Nr}\left(2^{n-k}\right) = \frac{2^k}{k!}\frac{n!}{(n-k)!}$$
$$= 2^k \binom{n}{k}, \quad k=0,\ldots,n \qquad (22)$$

where $\binom{n}{k}$ denotes the binomial coefficient. The frequency distribution $\text{Nr}(Y_n)$ obtained in (22) above for the overlap magnitudes $Y_n = 2^{n-k}$, $k = 0,\ldots,n$, is an exact result. Using the binomial theorem, for the entire period:

$$\sum_{k=0}^{n} \text{Nr}\left(2^{n-k}\right) = \sum_{k=0}^{n} 2^k \binom{n}{k} = 3^n, \qquad (23)$$

obviously because there are 3^n time-steps in a period of the time-series.[1]

The time-series of overlap may now be analyzed from the point of view of probability theory. The treatment by probability theory is necessary to determine whether the occurrence of an overlap of a certain magnitude can be predicted from the knowledge of the magnitude of an overlap that has already occurred. Let $\Pr(Y_n)$ denote the probability that after any arbitrary time-step t there is an overlap of Y_n segments. For the general case, $Y_n = 2^{n-k}$, $k = 0,\ldots,n$, it is given by:

$$\Pr\left(2^{n-k}\right) = \frac{\text{Nr}\left(2^{n-k}\right)}{\sum_{k=0}^{n}\text{Nr}\left(2^{n-k}\right)}$$
$$= \frac{2^k}{3^n}\binom{n}{k}$$
$$= \binom{n}{n-k}\left(\frac{1}{3}\right)^{n-k}\left(\frac{2}{3}\right)^k. \qquad (24)$$

The final expression in the above equation has the form of the binomial distribution for n Bernoulli trials with probability $1/3$ for the success of each trial and therefore $\Pr(Y_n)$ stands for the probability of the case where the number of successes is given by $\log_2 Y_n$. Since $\Pr\left(2^{n-k}\right)$ is maximum for $k = \lfloor 2(n+1)/3 \rfloor$ (and also for $k = 2(n+1)/3 - 1$ when $n+1$ is a multiple of 3), the most probable overlap magnitude in general is given by:

$$\widehat{Y}_n = 2^{n-\lfloor 2(n+1)/3 \rfloor} \qquad (25)$$

---

[1] In the model defined by the n-th generation of the Cantor set overlapping on its complement in the unit line segment (the complement of the n-th generation set is obtained by replacing each line segment of length $1/3^n$ in the latter by an empty segment and each empty segment by a line segment) the overlap magnitudes are given by $Y_n = 2^n - 2^k$, $k = 0,\ldots,n$. The time-series of overlaps can be directly derived from (10) by replacing each overlap magnitude 2^{n-k} in the time-series of the original model with an overlap magnitude $2^n - 2^{n-k}$. Consequently the frequency distribution of overlap magnitudes is given by $\text{Nr}\left(2^n - 2^k\right) = 2^{n-k}\binom{n}{k}$.

where the floor function $\lfloor x \rfloor$ is defined as the greatest integer less than or equal to x. Since $\max Y_n = 2^n$, the asymptotic relation for large values of the generation index n is:

$$\widehat{Y}_n \sim (\max Y_n)^{1/3} . \qquad (26)$$

The time series is further analyzed by determining the conditional probability that an overlap of magnitude Y'_n occurs after the time-step $t+1$ if it is known that an overlap of magnitude Y_n has occurred after the previous time-step t, for any arbitrary t. The conditional probability is given by:

$$\Pr(Y'_n \mid Y_n) = \frac{\mathrm{Nr}\,(Y_n, Y'_n)}{\mathrm{Nr}\,(Y_n)} \qquad (27)$$

where $\mathrm{Nr}\,(Y_n, Y'_n)$ is the number of ordered pairs of consecutive overlaps (Y_n, Y'_n) occurring in a period of the time-series for the n-th generation. It follows from (10) that overlaps of magnitude 2^{n-k} in the time-series are immediately succeeded (i.e., after the next time-step) by overlaps of magnitude 2^r, $0 \le r \le k$ only and never by an overlap of magnitude greater than 2^k. For $Y_n = 2^{n-k}$ and $Y'_n = 2^r$, $r = 0, \ldots, k$:

$$\mathrm{Nr}\,(2^{n-k}, 2^r) = \binom{n}{k}\binom{k}{r} . \qquad (28)$$

Therefore, from (22), (27) and (28):

$$\Pr(2^r \mid 2^{n-k}) = \frac{1}{2^k}\binom{k}{r}; \quad \begin{array}{l} k = 0, \ldots, n \\ r = 0, \ldots, k \end{array}$$

$$= \binom{k}{r}\left(\frac{1}{2}\right)^r\left(\frac{1}{2}\right)^{k-r} . \qquad (29)$$

The conditional probability thus follows the binomial distribution for k Bernoulli trials with probability $1/2$ for the success of each trial. Equation (29) shows that the expression of $\Pr(2^r \mid 2^{n-k})$ is independent of the generation index n. Therefore, for fixed k and r, it has the same value for all generations n, provided that $0 \le r \le k \le n$. Since the conditional probability $\Pr(2^r \mid 2^{n-k})$ is maximum when $r = \lfloor (k+1)/2 \rfloor$ (and also for $r = (k+1)/2 - 1$ when $k+1$ is a multiple of 2), the most probable overlap magnitude $\widehat{Y}'_n$ to occur next to an overlap of magnitude $Y_n = 2^{n-k}$ is given by:

$$\widehat{Y}'_n = 2^{\lfloor (k+1)/2 \rfloor} . \qquad (30)$$

For large values of the generation index n and large k, $k \le n$, the above equation gives the following asymptotic relation:

$$\widehat{Y}'_n \sim \sqrt{\frac{\max Y_n}{Y_n}} . \qquad (31)$$

Three special cases of (29) are considered; in each case $Y_n = 2^{n-k}$ and Y'_n is the overlap magnitude occurring next to Y_n in the time-series. In the first case

$$\Pr(Y'_n \leq Y_n \mid Y_n) = \sum_{r=0}^{k} \Pr\left(2^r \mid 2^{n-k}\right)$$
$$= 1 \quad \text{for } 0 \leq k \leq \frac{n}{2}. \quad (32)$$

that implies that, for $k \leq n/2$, an overlap of magnitude $Y_n = 2^{n-k}$ is always followed by an overlap of equal or less magnitude Y'_n. Consequently the case of an overlap of magnitude greater than that of the preceding one (i.e., $Y'_n > Y_n$) appears only for $k > n/2$. In the second case

$$\Pr(Y'_n \geq Y_n \mid Y_n) = \sum_{r=n-k}^{k} \Pr\left(2^r \mid 2^{n-k}\right)$$
$$= \frac{1}{2^k} \sum_{r=n-k}^{k} \binom{k}{r} \quad \text{for } \frac{n}{2} \leq k \leq n. \quad (33)$$

Since the final expression in the above equation involves a partial sum of binomial coefficients, it does not have a closed form, except for the trivial case $k = n$ and a few other special cases. Therefore it must be calculated numerically for specific values of n and k. In the third case, if the magnitude of overlap after a certain time-step is $Y_n = 2^{n-k}$, the probability that an overlap of the same magnitude also occurs after the next time-step is given by:

$$\Pr\left(Y'_n = Y_n \mid Y_n\right) = \Pr\left(2^{n-k} \mid 2^{n-k}\right)$$
$$= \begin{cases} 0 & , 0 \leq k < n/2 \\ \frac{1}{2^k}\binom{k}{n-k} & , n/2 \leq k \leq n \end{cases}. \quad (34)$$

In this way the conditional probability can be determined in various cases where Y'_n is specifically related to Y_n.

6 Emergence of a Power Law

The above analysis of the model, defined by a pair of overlapping n-th generations of the Cantor set with periodic boundary conditions, showed that the probability of occurence of an overlap of magnitude Y_n, which may assume the values $Y_n = 2^{n-k} (k = 0, \ldots, n)$, follows the binomial distribution for the logarithm of the magnitude $\log_2 Y_n = n - k$:

$$\Pr(Y_n) = \binom{n}{\log_2 Y_n} \left(\frac{1}{3}\right)^{\log_2 Y_n} \left(\frac{2}{3}\right)^{n-\log_2 Y_n} \equiv F(\log_2 Y_n). \tag{35}$$

Since the index of the central term (i.e., the term for the most probable event) of the above distribution is $n/3 + \delta$, $-2/3 < \delta < 1/3$, for large values of n the normal approximation to the above binomial distribution near the central term may be written as

$$F(\log_2 Y_n) \sim \frac{1}{\sqrt{n}} \exp\left[-\frac{(\log_2 Y_n)^2}{n}\right], \tag{36}$$

not mentioning the factors that do not depend on Y_n. Now

$$F(\log_2 Y_n)\,\mathrm{d}(\log_2 Y_n) \equiv G(Y_n)\mathrm{d}Y_n \tag{37}$$

where

$$G(Y_n) \sim \frac{1}{Y_n} \exp\left[-\frac{(\log_2 Y_n)^2}{n}\right] \tag{38}$$

is the log-normal distribution of Y_n. As the generation index $n \to \infty$, the normal factor spreads indefinitely (since its width is proportional to $\sqrt{n}$) and becomes a very weak function of Y_n so that it may be considered to be almost constant; thus $G(Y_n)$ asymptotically assumes the form of a simple power law with an exponent that is independent of the fractal dimension of the overlapping Cantor sets:

$$G(Y_n) \sim \frac{1}{Y_n} \text{ for } n \to \infty. \tag{39}$$

This is the Gutenberg–Richter law (2) for the simplest fractal-overlap model of earthquakes. This is qualitatively similar to what is observed in the distribution of real earthquakes: the Gutenberg–Richter power law is found to describe the distribution of earthquakes of small and intermediate energies; however deviations from it are observed for the very small and the very large earthquakes. Similar to the outcome of the simplest fractal overlap model, the distribution of overlap magnitudes has also been observed to follow a power law for more complicated fractals: one random Cantor set overlapping on another, the Sierpinsky gasket and the Sierpinsky carpet overlapping on their respective replica [13], and a fractional Brownian profile overlapping on another [14].

The fact that the fractal-overlap model produces an asymptotic power law distribution of overlaps suggests that the Gutenberg–Richter law owes its origin significantly to the fractal geometry of the faults. Furthermore, since this model contains the geometrical rudiments (i.e., the fractal overlap structure) of geological faults and it produces an asymptotic distribution of overlaps that has qualitative similarity with the Gutenberg–Richter law, it may be an indication that the entire distribution of real earthquake energies is log-normal that is wide enough for the Gutenberg–Richter power law to be observed over a large range of energy values.

References

1. B. Gutenberg and C.F. Richter, Bull. Seismol. Soc. Am. **34**, 185 (1944).
2. B. Gutenberg and C.F. Richter, *Seismicity of the Earth*, Princeton University Press, Princeton (1954).
3. L. Knopoff, Proc. Natl. Acad. Sci. USA **93**, 3756 (1996).
4. L. Knopoff, Proc. Natl. Acad. Sci. USA **97**, 11880 (2000).
5. R. Burridge and L. Knopoff, Bull. Seismol. Soc. Am. **57**, 341 (1967).
6. J.M. Carlson and J.S. Langer, Phys. Rev. Lett. **62**, 2632 (1989); J.M. Carlson and J.S. Langer, Phys. Rev. A **40**, 6470 (1989); J.M. Carlson, J.S. Langer and B.E. Shaw, Rev. Mod. Phys. **66**, 657 (1994).
7. Z. Olami, H.J.S. Feder and K. Christensen, Phys. Rev. Lett. **68**, 1244 (1992).
8. P. Bak, C. Tang and K. Wiesenfeld, Phys. Rev. Lett. **59**, 381 (1987); P. Bak, C. Tang and K. Wiesenfeld, Phys. Rev. A **38**, 364 (1988).
9. P.C. Hemmer, A. Hansen and S. Pradhan, *Rupture Processes in Fibre Bundle Models*, Lect. Notes Phys. **705**, pp. 27–55 (2006).
10. B.K. Chakrabarti and R.B. Stinchcombe, Physica A **270**, 27 (1999).
11. B.B. Mandelbrot, *The Fractal Geometry of Nature* (W.H. Freeman and Company, New York, 1983).
12. P. Bhattacharyya, Physica A **348**, 199 (2005).
13. S. Pradhan, B.K. Chakrabarti, P. Ray and M.K. Dey, Physica Scripta **T106**, 77 (2003).
14. V. De Rubeis, R. Hallgass, V. Loreto, G. Paladin, L. Pietronero and P. Tosi, Phys. Rev. Lett. **76**, 2599 (1996).

Friction, Stick-Slip Motion and Earthquake

H. Matsukawa and T. Saito

Department of Physics and Mathematics, Aoyamagakuin University,
5-10-1 Fuchinobe, Sagamihara, 229-8558, Japan
hm@phys.aoyama.ac.jp

1 Introduction

Friction has close relation to Earthquake [1, 2]. When the friction between two solids shows velocity weakening behavior, stationary motion becomes unstable and stick-slip motion appears, which repeats stopping and fast moving states. Earthquake is a kind of stick-slip motion. Figure 1 shows tectonic plates around Japan. The pacific ocean plate goes down below the land plate and drags the latter. Then the land plate deforms and its restoring force increases with time. When the restoring force becomes larger than the maximum static frictional force between the ocean and land plates the land plate slips. This is the interplate earthquake. Besides this interplate earthquake other types of earthquake exist. We concentrate on the interplate earthquake here, because it can be much stronger than other types of earthquake.

In this report we first introduce the present stage of the investigation of the mechanism of friction. Then we discuss the velocity and waiting time dependence of friction. The friction of rock as well as other systems is well described by the empirical constitutive law proposed by Dietrich [3] and Ruina [4]. The parameters of the law governs the velocity dependence of friction and be measured by experiments [5, 6]. We discuss the relation of the critical depth of the earthquake and the parameters of the rock friction [6, 7]. Then we derive results of the constitutive law from theoretical point of view [1, 8]. Various systems show stick-slip motion [8, 9, 10, 11, 12, 13, 14, 15]. But the behavior of stick-slip motion strongly depends on the system. Some systems show regular periodic stick-slip motion and other systems show irregular one. Earthquake is not a periodic phenomena and its size distribution obeys the power law, the Gutenberg–Richter law [16, 17]. It is known that the Gutenberg–Richter law is reproduced by the Burridge–Knopoff model [18], which is a kind of spring block model between two plates, one of which is driven with a constant velocity [17, 19]. The mechanism of the Gutenberg–Richter law is usually explained in the context of self-organized criticality (SOC) proposed by Bak et al. [20]. We report our study of size dependence of the Burridge–Knopoff model [21]. The

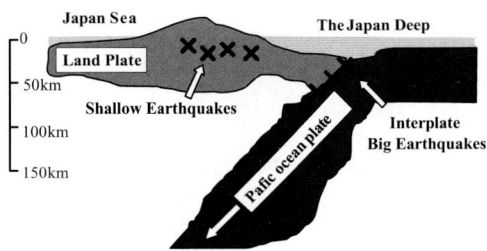

Fig. 1. Tectonic plates around Japan

obtained results indicate that the larger systems show larger deviation from the Gutenberg–Richter law, which is not consistent with the idea of SOC. The relation of the present results with observation is also discussed.

2 Friction

Friction is one of the most familiar physical phenomena in our daily life and has been studied from ancient age. It is well known that a empirical law, which is often called the Amomtons–Coulomb law, holds well for solid friction, which is: (i) frictional force does not depend on the apparent contact area, (ii) frictional force is proportional to the loading force, the proportional coefficient is called frictional coefficient and (iii) kinetic frictional force is smaller than maximum static frictional force and does not depend on sliding velocity [1]. The oldest description of the Amomtons–Coulomb law which exists now was written by da Vinci. The mechanism of friction is, however, not established still now.

The most accepted mechanism is the adhesion theory organized by Bowden and Tabor about 1930's $\sim$ 1940's [22]. As shown in Fig. 2 most of solid surfaces have large roughness from μm scale point of view. Two solid surfaces touch with each other only at actual contact points. The total area of actual contact

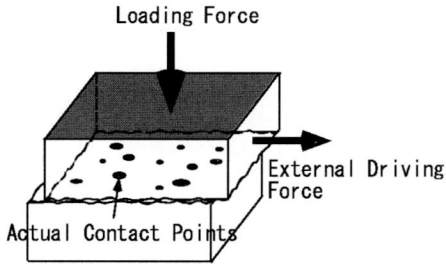

Fig. 2. Actual contact points between two solid surfaces

points A_r is called the actual contact area, which is much smaller than the apparent contact area. Then the pressure at actual contact points is much higher than the elastic limit and reaches yield stress p_c, which is constant. As a result A_r is proportional to the loading force L and given by

$$A_r = L/p_c . \qquad (1)$$

At each actual contact points two solid surfaces adhere each other by intermolecular or interatomic forces. The maximum static frictional force $F_{\text{stat}}^{\max}$ is given by

$$F_{\text{stat}}^{\max} = A_r \times \sigma_s , \qquad (2)$$

where σ_s is the shear stress to cut the adhesion. So we have

$$F_{\text{fric}}^{\max} = \mu_{\text{stat}} \times L , \qquad (3)$$

with the static frictional coefficient,

$$\mu_{\text{stat}} = \sigma_s/p_c . \qquad (4)$$

This is the (simplest version of) adhesion theory, which can explain the 1st and 2nd part of the Amontons-Coulomb law.

The adhesion theory of friction has two important assumptions. One is that the actual contact area is proportional to the loading force. Many experiments support this assumption. But there are still many ideas about its mechanism besides the plastic deformation mentioned above [1]. Greenwood showed that if the height distribution of solid surface obeys the Gaussian distribution the actual contact area is proportional to the loading force for a certain range even within the elastic limit [23]. Moreover if the surface has s self-similar property the actual contact area is exactly proportional to the loading force within the elastic limit. And many actual solid surface has such property approximately.

Another important assumption of the adhesion theory is that at each actual contact points adhesion occurs, which yield the friction. In order to consider the validity of this assumption we discuss the friction between atomically clean two surfaces. To simplify the problem we employ one dimensional model [24, 25]. In Fig. 3a we show one dimensional model of friction between atomically clean surfaces. The lower solid is assumed to be rigid and makes periodic potential to atoms in the upper solid. The solid circles represent original configuration of atoms in upper solid, which interact with each other. In general two atomically clean solid surfaces are incommensurate, that is, the ratio of the lattice constants of two surfaces are irrational. When upper solid moves right each atom in the solid moves right and occupies the position shown by dashed circles. The atoms 1, 4, 6, 7, 9 then gain the potential energy, but the atoms 2, 3, 5, 8 lose it. The energy does not change against the translational motion of upper body as a whole. Then the system has translational invariance and static frictional force vanishes. Such situation is also possible for 2 and 3 dimensional systems and called superlubricity. When the periodic

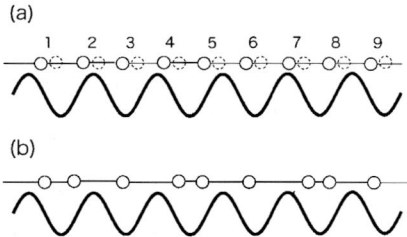

Fig. 3. One dimensional model of friction between atomically clean surfaces. The *lower solid* is assumed to be rigid and makes periodic potential to atoms in the *upper solid*. The atoms in the *upper solid* interact with each other. (**a**) Incommensurate Case. *Solid circles* represent original configuration of atoms in the *upper solid* and the *dashed circles* do that after the translation to the right direction. (**b**) Rearranged configuration

potential made by atoms in the lower solid becomes stronger than a certain critical strength, however, the atoms in the upper solid rearrange their positions in order to gain the potential energy. The rearranged configuration is shown in (b). In this state the system loses the translational invariance and finite static frictional force appears. This is the Aubry's transition of the Frenkel-Kontorova model [26]. Many calculations show two atomically clean surfaces of various materials are in the superlubricity state [27]. In fact Hirano et al. showed that the frictional force between W(011) and Si(001) surfaces is finite when the two surfaces are commensurate, that is, the ratio of two lattice constants along sliding direction is rational, while the frictional force vanishes within experimental error when they are incommensurate [28]. It is to be noted that we discuss here the friction or change of energy against parallel displacement to the surface and does not discuss that against perpendicular displacement to the surface. It is possible that the system has translational invariance against parallel motion but gains contact energy.

Let us return to the problem of friction of usual surfaces, that is, rough surfaces with actual contact points [29]. The interfaces at actual contact points are considered to be relatively clean. When we neglect the randomness at the interfaces the problem is reduced to the friction between atomically clean surfaces and in most cases maximum frictional force vanishes as mentioned above. Next we consider the effect of randomness at the interfaces at actual contact points. The effect of randomness to the pinning has been discussed from many years ago in the context of pinning of Charge Density Wave and Spin Density Wave in low dimensional conductors and that of vortex in type-II superconductors. It has shown that there always exist finite pinning force, that is the static frictional force, for finite randomness. It has also made clear that the pinning has two regimes. One is called strong pinning, in which the interaction between the object and the pinning center is so strong so each pinning center pins the object individually. Another regime is called weak pinning, in which the interaction between them is not so strong so each pinning center can not pin the object. Instead group of pinning centers pins the object collectively. In the latter case the pinning force is much smaller than that of the former. Some calculation show that the pinning by the randomness at actual contact points is in the weak pinning regime and can not yield the frictional force observed in experiments.

Then what is the mechanism of macroscopic friction obeying Amontons-Coulomb law? He et al. proposed a new mechanism for the macroscopic friction [29], in which mobile atoms or molecules between two surfaces play an essential role. The point is that the mobile atoms or molecules can occupy the position where they gain the interaction energies with two surfaces, because they are mobile. As a result the system loses energy against translational motion parallel to the surface and then static frictional force appears. They showed numerically that this mechanism can explain the 1st and 2nd part of the Amontons-Coulomb law. There are much problems, however, to be explained. Further investigations are required.

2.1 Velocity and Waiting Time Dependence of Frictional Force and Earthquake

As described in Amontons–Coulomb law, kinetic frictional force F_{kin} does not depend on sliding velocity in usual condition. This behavior is explained as follows [1]. When the upper solid slides on the lower solid, formation and break of actual contact points occur repeatedly at interface. Once an actual contact point is formed it sticks locally. With increasing time elastic energy increases around that point and finally it breaks and two asperities resulting from the broken contact point slips locally. The elastic energy stored around the point is released and dissipated. Then kinetic frictional force arises. The system repeats local stick-slip motion even though it moves stationary as a whole. The velocity scale which determines the kinetic frictional force is the local slip velocity, which is much greater than the sliding velocity of the moving solid. As a result the total kinetic frictional force does not depend on the sliding velocity of the solid in usual condition.

But many solids actually yield velocity dependent kinetic frictional force. For example the kinetic friction between rocks shows logarithmic velocity dependence, of which sign depends on the condition [5, 6]. The velocity dependence is not so strong and does not appear significantly in usual condition, but plays an essential role in some cases as discussed later. It is also known that the maximum static frictional force increases logarithmically with the waiting time [30, 31]. The waiting time is the duration between the instance when two rocks touch with each other and the instance when we measure the frictional force. Such a waiting time dependence is also observed in frictions of paper, PMMA, metals and so on [8, 31]. This is due to the logarithmic increases of actual contact area, which results from the plastic deformation of actual contact points caused by creep motion of dislocation etc. In fact Dietrich et al.

directly observed such increases of actual contact area by using acrylic plastic and soda glass [32]. The waiting time dependence of static frictional force has close relation to the velocity dependence of kinetic frictional force [8, 30]. As we mentioned just above when one solid slides on another solid with constant velocity local stick-slip motions occur at actual contact points and they rearrange repeatedly. When an actual contact point is formed the waiting time at the contact increases during the stick and then the "static" frictional force of that contact increases. The average sticking time of actual contact points, which we call contact age, is inversely proportional to the sliding velocity. As a result the kinetic frictional force decreases logarithmically with sliding velocity.

This mechanism makes the kinetic frictional force smaller for higher sliding velocity. But the sign of velocity dependence of kinetic frictional force of rock depends on the condition. There is a factor which induces logarithmic increase of kinetic friction with velocity. That is thermal activation process. When two solids slides each other actual contact points are formed and broken. At finite temperature thermal fluctuation helps actual contacts break in the case of low sliding velocity and reduces the kinetic frictional force. In the high sliding velocity regime thermal fluctuation is less effective because the high sliding velocity breaks actual contact by compulsion and the kinetic frictional force increases.

Such behavior of frictional force is well described by empirical constitutive law proposed by Dietrich [3] and Ruina [4]. It is expressed as

$$\mu = \mu_0 + A \ln\left(1 + \frac{V}{V_0}\right) + B \ln\left(1 + \frac{\theta}{\theta_0}\right) . \tag{5}$$

Here μ is the frictional coefficient, V the sliding velocity and θ a state variable. A, B, μ_0, V_0 and θ_0 are certain constants. In the variables of the logarithm functions in the above equation, 1 is added in order to avoid the divergence of the function. We focus on the case when it can be neglected. The time evolution of the state variable θ is given by

$$\frac{d\theta}{dt} = 1 - \frac{\theta V}{D_c} , \tag{6}$$

where D_c is a constant with the dimension of length. The solution of above differential equation is given by

$$\theta(t) = \int_0^t \exp\left\{-\frac{x(t) - x(t')}{D_c}\right\} dt' , \tag{7}$$

where the time is measured from the beginning of the contact of two solids. As seen from the above equation $\theta(t)$ is the contact age and D_c is a memory length. The system forget the memory in the duration that $x(t) - x(t')$ becomes greater than D_c. In the static case θ increases with time linearly according to (6). Then we have a static frictional coefficient increasing logarithmically with time from (5). In the stationary sliding state the state variable does not change. Then we get $\theta = D_c/V$ from (6). By substituting this relation to (5) we obtain the velocity dependence of the frictional coefficient as

$$\mu = \text{const.} + (A - B)\ln(V) \ . \tag{8}$$

The sign of $(A - B)$ determines the sign of the velocity dependence of kinetic frictional force [6, 7]. The frictional force shows velocity increasing behavior for the case of $(A - B) > 0$ and velocity weakening behavior for $(A - B) < 0$. Tse and Rice determined the coefficients A and B as a function of temperature for granite based on the measurement of Stesky [5] and showed that $(A - B) < 0$ below $300°C$ and $(A - B) > 0$ above $300°C$ [6]. The velocity weakening behavior of frictional force makes stationary motion unstable and causes stick-slip motion. Granite is main composition of tectonic plate. The stick-slip motion of tectonic plate at fault is earthquake. It is considered that the change of the sign of $(A - B)$ at $300°C$ determines the depth limit of tectonic earthquakes [6, 7].

2.2 Mechanism of Velocity and Waiting Time Dependence of Frictional Force

Here we consider the mechanism of velocity and waiting time dependence of frictional force in more detail theoretically. To simplify the problem we neglect the internal degrees of freedom and randomness of two solid and consider a single particle pushed by spring on 1 dimensional periodic potential as shown in Fig. 4 [8]. The single particle represent the center of gravity degrees of freedom of sliding solid and the periodic potential does the pinning potential, of which period corresponds to the mean separation of actual contact points. The Langevin equation of motion of the system is expressed as

$$M\ddot{x} + \Gamma\dot{x} = -\frac{\partial U_{\text{tot}}}{\partial x} + R(t) \ , \tag{9}$$

where M is the mass of the single particle, x is its coordinate, Γ the dissipation coefficient, U_{tot} the total potential acting on the particle and $R(t)$ thermal fluctuation force, which satisfies the fluctuation dissipation theorem,

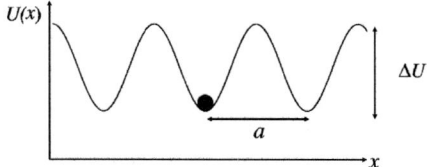

Fig. 4. A single particle on a periodic potential

$$\Gamma = \frac{1}{2k_B T} \int_{-\infty}^{\infty} \langle R(0)R(t) \rangle dt \, . \tag{10}$$

U_{tot} is given as

$$U_{\text{tot}} = U(x) + V_{\text{ext}} \, . \tag{11}$$

Here $U(x)$ is a periodic pinning potential with period a and amplitude ΔU shown in Fig. 4 and

$$V_{\text{ext}} = \frac{k}{2} \{x - x_0(t)\}^2 \, , \tag{12}$$

is the elastic energy of the pushing spring with spring constant k. $x - x_0(t)$ is the distortion of the spring.

Consider the case of finite spring distortion, in which U_{tot} has a shape shown in Fig. 5, and creep regime, where $Nk_B T$ is much less than ΔU. Here N is the total number of actual contact points. In this case the particle can move by thermal activation process and its time evolution is expressed as

$$\dot{x} = a \left\{ \frac{1}{\tau_{\rightarrow}} - \frac{1}{\tau_{\leftarrow}} \right\} \, , \tag{13}$$

where

$$\tau_{\rightleftarrows} = \frac{\omega_0}{2\pi} \exp \left[\frac{-\Delta U_{\rightleftarrows}}{N k_{\rm B} T} \right] \, . \tag{14}$$

Here $\Delta U_{\rightarrow}$ and $\Delta U_{\leftarrow}$ indicate the energy barriers to the right and left directions, respectively. From (13, 14) $\dot{x}$ is expressed as

$$\dot{x} \simeq \frac{\omega_0 a}{2\pi} 2 \sinh \left\{ \frac{ka(x_0 - x)}{2Nk_B T} \right\} \exp \left[\frac{-\Delta U_0}{N K_{\rm B} T} \right] \, . \tag{15}$$

We consider the low temperature regime where thermal activation process over the higher barrier can be neglected. The force by spring is equal to the external force F_{ext}, which coincides with the frictional force F_{fric} in the stationary state. Then we obtain

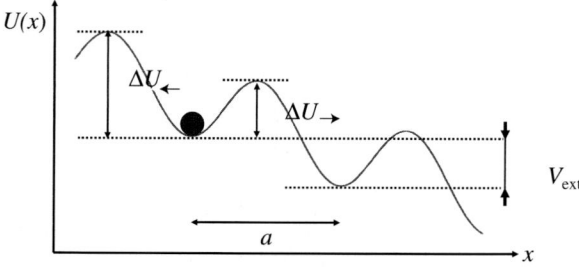

Fig. 5. A single particle pushed by the spring on a periodic potential

$$F_{\text{fric}} = \frac{2}{a}\left\{\Delta U_0 + Nk_B T \ln\left[\frac{2\pi V}{\omega_0 a}\right]\right\}. \tag{16}$$

Here V is the sliding velocity. When ΔU_0 is considered to be constant the above equation gives the kinetic frictional force of which velocity dependence is logarithmic.

Next we consider the effect of time growth of actual contact area caused by the plastic deformation of actual contact points [1]. Such deformation results from thermal creep of dislocation etc. under normal stress and the calculation just above is also applicable to this problem. In (16) $\frac{2}{a}\Delta U_0$ corresponds to the maximum static frictional force for the case of friction and to the yield stress σ_c for the case of plastic deformation at absolute zero. $\Delta U_0/N$ corresponds to the pinning energy. Therefore in the case of plastic deformation we have

$$\sigma = \sigma_c\left\{1 + \frac{k_B T}{u_0}\ln\left[-\frac{\dot\epsilon}{\dot\epsilon_0}\right]\right\}, \tag{17}$$

where σ is the stress, u_0 the pinning energy, $\dot\epsilon$ strain rate and $\dot\epsilon_0$ a certain constant. Let us consider the time evolution of an actual contact point, which has cylindrical shape with area of the base S and height ℓ. The volume of the contact $\ell(t) \times S(t)$ is assumed to be constant. Then

$$\dot\epsilon = \frac{\dot\ell}{\ell} = -\frac{\dot S}{S}, \tag{18}$$

If we write

$$S(t) = S_0 + s(t), \tag{19}$$

where $S_0 = S(t=0)$, we obtain from (18, 19)

$$\dot\epsilon = -\frac{\dot s/S_0}{1 + s/S_0}$$
$$= -\frac{\dot\xi}{1+\xi}, \tag{20}$$

where $\xi = s(t)/S_0$. The stress is expressed as

$$\sigma = \frac{L}{S}$$
$$= \frac{L/S_0}{1 + s/S_0}. \tag{21}$$

Because L/S_0 is equal to the yield stress at absolute zero σ_c, $\frac{\sigma}{\sigma_c}$ is given by

$$\frac{\sigma}{\sigma_c} = \frac{1}{1+\xi}. \tag{22}$$

By substituting (17, 20) into the above equation we obtain the following equation,

$$\frac{k_B T}{u_0} \ln\left[\frac{\dot{\xi}/\dot{\epsilon}_0}{1+\xi}\right] = -\frac{\xi}{1+\xi} . \tag{23}$$

Now we focus on the low temperature regime, take only leading order term of $k_B T$ and get

$$\frac{k_B T}{u_0} \ln\left[\dot{\xi}/\dot{\epsilon}_0\right] = -\xi . \tag{24}$$

The solution of the above equation is easily obtained as

$$\xi = \frac{k_B T}{u_0} \ln\left[1 + t/\tau\right] , \tag{25}$$

where

$$\tau = \frac{k_B T}{u_0 \dot{\epsilon}_0} . \tag{26}$$

Then we obtain the time evolution of the area of an actual contact point as

$$S(t) = S(0) + \frac{k_B T}{u_0} \ln\left[1 + t/\tau\right] . \tag{27}$$

The area of an actual contact point increases with time logarithmically then the actual contact area also increases logarithmically in the static case. In the stationary sliding state with sliding velocity V, the waiting time t in the above equation should be replaced by the contact age D_c/V. We have

$$A_r(V) = A_r(0) + \frac{N k_B T}{u_0} \ln\left[1 + D_c/(V\tau)\right] . \tag{28}$$

ΔU_0 in (16) is considered to be proportional to the actual contact area $A_r(V)$. Then we obtain the empirical relation (8) from (16, 28).

3 Stick-Slip Motion

Earthquake is a kind of stick-slip motion as mentioned in Sect. 1. There are many systems which show stick-slip motion from microscopic [9, 10, 11, 12] to macroscopic [8, 13, 14, 15] scales and many types of stick-slip motion appear. Earthquake in not a periodic phenomena and its size distribution obeys the power law, the Gutenberg–Richter law [16, 17]. It is well known that at faults gauge exists which is a kind of rock waste as shown in Fig. 6 [2]. So the earthquake is a kind of stick-slip motion which occurs between two surfaces with interpositions. We have many frictional systems with interpositions. Typical interpositions are lubricants in machines. We introduce liquid lubricant between two sliding surfaces in machines and reduce the friction and wear. When machines increase their accuracy or become small, such as nano-machines, the

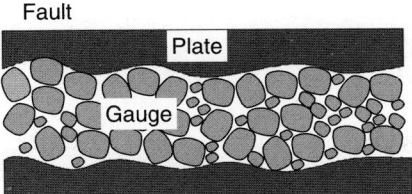

Fig. 6. Gauge and plates at fault

separation between two sliding surfaces and then the lubricant layer become narrow. It is shown by experiments employing surface force apparatus that when the lubricant layer becomes less than about 5 molecular layer periodic stick-slip motion appears, that means the existence of finite maximum static frictional force [11, 12]. Above a certain critical sliding velocity stick-slip motion disappears and stationary motion appears. The mechanism of these phenomena are explained as shown in Fig. 7. Because lubricant molecules are confined into small space and affected by the periodic potential made by two solid surfaces the lubricant solidifies and yields finite static frictional force even above the bulk melting temperature. Then stick occurs. The external force is applied to the upper solid and the elastic energy of the system increases with time. At certain time the solidified lubricant molecules melt by the stored elastic energy. Then the system slips. After slip the stored elastic energy is released and the lubricant molecules solidify again. By repeating this process the system shows periodic stick-slip motion. When the sliding velocity increases above a certain critical velocity the melting lubricant molecules does not have enough time to solidify. As a result the lubricant keeps liquid state then stationary motion appears in the high velocity regime.

Let us consider macroscopic systems. Nasuno et al. examined the frictional phenomena of granular systems between two plates [13]. Figure 8 shows their

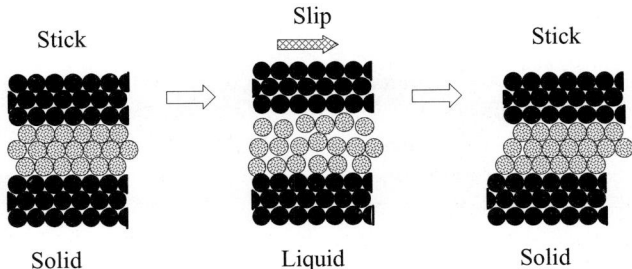

Fig. 7. Periodic stick-slip motion appearing in the system with confined lubricant molecules. The *dark circles* represent molecules or atoms in two *solids* and *gray circles* represent@those of lubricant

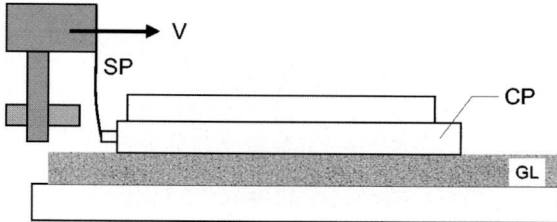

Fig. 8. Experimental apparatus of Nasuno et al. [13]. Granular layer is covered by plate, which is driven by a leaf spring

experimental apparatus. Cover plate on the granular layer is pushed by a leaf spring. Its origin is moved with constant velocity V. When they use spherical smooth glass particles with uniform radius as granular particles, the system shows periodic stick-slip motion in the low velocity regime. When the velocity is increased the amplitude of stick-slip motion decreases and the system shows continuous transition to the stationary motion above a certain critical velocity. In the case of clean art sand consisting of rough particles with diameter $100 \sim 600\,\mu\mathrm{m}$ they observed irregular stick-slip motion, but the amplitude distribution has two peaks and does not show power law.

Granular systems, however, shows power law behavior in other case. Dalton et al. put granular particles between two fixed cylinders and covered them by annual plate as shown in Fig. 9 [14]. The annual plate is rotated by torsion spring. They employed tapioca particles with large distribution of size as granular particles and observed stick-slip motion. The amplitude distribution of stick-slip shows power law behavior in this experiment.

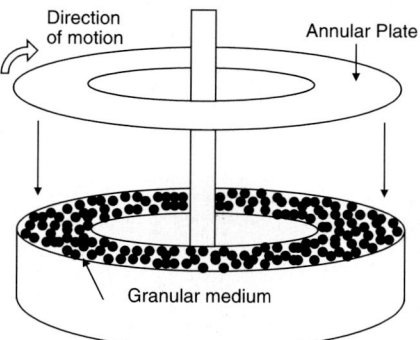

Fig. 9. Experimental apparatus of Dalton et al. [14]. Granular particles are put between two fixed cylinders and covered by annual plate. The annual plate is rotated by torsion spring

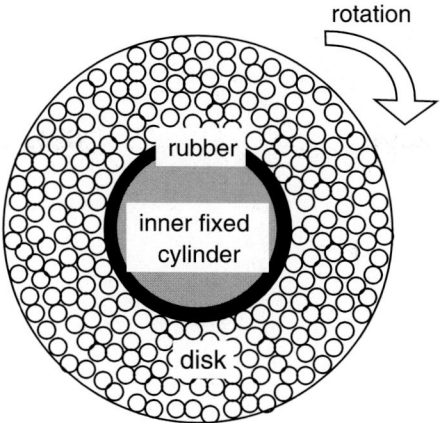

Fig. 10. Experimental apparatus of Hirata [15]. Disks are putted between two cylinders, inner one is fixed and outer one is rotated

The distribution of size or randomness of the shape of interpositions is not a necessary condition for the appearance of power law behavior of stick-slip motion. Hirata putted disks with uniform radius between two cylinders, inner one is fixed and outer one is rotated [15]. See Fig. 10. The observed stick-slip motion shows size distribution which obeys power law.

We observed in this subsection many sliding systems with interpositions. Some systems show regular stick slip motion and others do irregular one. Some of the latter show power law behavior in the size distribution of stick-slip motion. The condition for the appearance of power law is not clear. One scenario of the appearance of power law behavior is self-organized criticality (SOC) proposed by Bak et al. [20]. They proposed that some systems approach to their critical states by themselves under repeated perturbation, because the critical states are marginally stable states and then the systems reach the critical states at first starting from the unstable states under repeated perturbation. They showed that a simple sand pile model yields power law behavior of avalanches. But it is clear that SOC scenario can not explain the condition for the appearance of power law behavior discussed above. The validity of SOC scenario is also discussed later.

4 Numerical Study of the Burridge–Knopoff Model

In this section we report results of our numerical simulation of the Burridge–Knopoff model with special emphasis on size dependence [21].

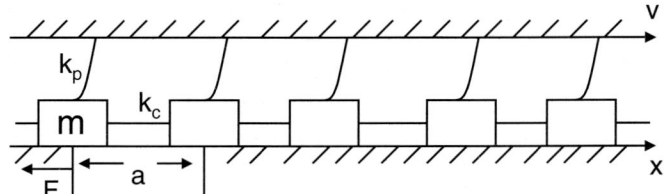

Fig. 11. The Burridge–Knopoff Model

4.1 Model

As mentioned above earthquake is a nonperiodic stick-slip motion and its size distribution obeys the power law, the Gutenberg–Richter law [16, 17]. It is known that the model proposed by Burridge and Knopoff [18], which is a kind of spring block model, can reproduce the Gutenberg–Richter law [17, 19]. The model is shown in Fig. 11. The upper plate moves with constant loading velocity v. The block of mass m is connected to the upper plate by linear spring with spring constant k_p. The block is also connected to each other by linear spring with spring constant k_c, of which natural length is a. Frictional force acts between the lower plate and each block. Its velocity dependence is given by hand in order to produce earthquake. One of the necessary condition is velocity weakening behavior as discussed before. In this section we show results of our numerical simulation of the Burridge–Knopoff model with special emphasis on its size dependence.

The equation of motion of the model is expressed as

$$m\ddot{x}_i = k_p(vt - x_i) + k_c(x_{i+1} + x_{i-1} - 2x_i) - \Phi(\dot{x}_i) \ . \tag{29}$$

Here x_i is the position of i-th block measured from the natural position of the spring connected with the upper plate at $t = 0$ and $\Phi(\dot{x}_i)$ is the frictional force. In order to make the equation dimensionless we introduce unit of time $\sqrt{m/k_p}$ and unit of length $\Phi_{\text{stat}}^{\text{max}}/k_p$, where $\Phi_{\text{stat}}^{\text{max}}$ is the maximum static frictional force, and measure the time and length by these units. Then the normalized equation of motion is given as

$$\ddot{x}_i = (vt - x_i) + l^2(x_{i+1} + x_{i-1} - 2x_i) - \phi(\dot{x}_i) \ , \tag{30}$$

where $l = \sqrt{k_c/k_p}$ is the dimensionless stiffness parameter and x_i, t, v and $\phi(\dot{x}_i)$ are normalized dimensionless quantities. In the actual earthquake the loading velocity is about $1 \sim 10$ cm per year. This is negligible velocity compared to the velocity of fault at earthquake. Therefore we take a limit of $v \to 0$. We follow the work by Carlson et al. [19] and adopt the functional form of $\phi(\dot{x}_i)$ as

$$\phi(\dot{x}_i) = \frac{1 - \sigma}{1 + 2\alpha\dot{x}_i/(1 - \sigma)} \ , \tag{31}$$

for $\dot{x}_i > 0$, and the maximum static frictional force to the loading direction is unity and that to opposite direction is $-\infty$. That is, every block never move to the opposite direction to the loading direction. σ is the difference between the maximum static frictional force and the kinetic frictional force in the limit of zero velocity. Finite value of σ ensures the limit of zero loading velocity and robustness of the numerical results against the magnitude of time mesh introduced in the calculation. α is the measure of velocity weakening behavior of kinetic frictional force. In the calculation we employed the following values, $l = \sqrt{60}, a = 1, \sigma = 0.01$ and $\alpha = 4$. The initial position of each block has a uniform distribution of the width 0.01 around the natural position of the spring connecting with the plate. First we calculate the total force of the spring acting on each block and search its maximum value, $f_{\max}$. If $f_{\max}$ is less than unity, we put the time t forward as $(1 - f_{\max})/v$. If $f_{\max}$ is equal to or greater than unity, earthquake occurs and each block moves according to the equation of motion (30). During the earthquake the position of upper plate is kept constant. In each run we observed 300000 events of earthquake. First 100000 events are abandoned to avoid the transient phenomena. In the calculation of size distribution of earthquake we calculate two distributions by employing data between the 100001st and 200000th events and by employing data between the 200001st and 300000th events and compare them in order to check that the observation time is long enough to obtain stationary distribution.

4.2 Numerical Results

In Fig. 12 we show the displacement of center of gravity coordinate of blocks as a function of time for the system with 100 blocks. Clear stick-slip motion is observed. The solid and dashed straight lines are tries to connect the lower and upper corners of the displacement-time relation, respectively. If the lower corners are on the straight line we can predict the time of the occurrence of earthquake, so the system is called time predictable. On the other hand if the upper corners are on the straight line we can predict the size of the earthquake, so the system is called size predictable. As seen from the figure the system shows time and size predictable behavior in certain duration but does not in longer duration. Figure 13 shows the displacement of the blocks of the system with 8000 blocks. In this system we can not find any predictability. The predictability of the model depends on the size of the model.

Next we discuss the size distribution of earthquake obtained from our numerical calculation. As mentioned above the size distribution of the earthquake obeys the power law, that is, $\rho(m) \simeq p \times 10^{-\beta m}$, where $\rho(m)dm$ is the number of earthquake in the range $m \sim m + dm$, m is the magnitude of earthquake and p is a certain constant. The magnitude is the strength of earthquake in logarithmic scale, so the above relation gives the power law distribution. It is also often claimed, however, that characteristic earthquake exist for each fault or region, which has characteristic large magnitude and

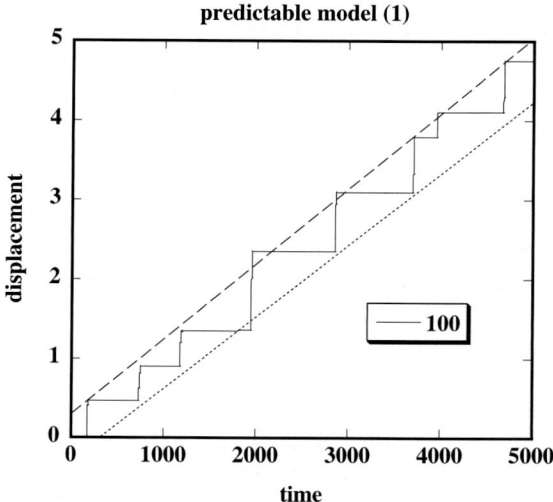

Fig. 12. Displacement of center of gravity coordinate of blocks as a function of time of the system with 100 blocks

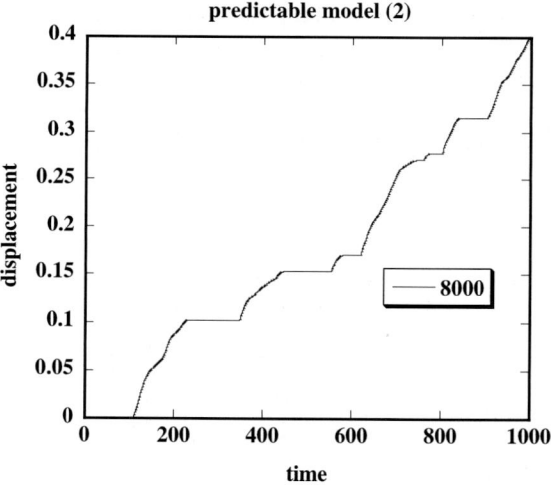

Fig. 13. Displacement of center of gravity coordinate of blocks as a function of time of the system with 8000 blocks

may has characteristic period at that fault or region. In fact Carlson and Langer observed the power law behavior in not a large magnitude regime and also observed a peak at certain large magnitude in the size distribution of earthquake in the simulation of the Burridge–Knopoff model [17]. The latter may correspond to the characteristic earthquake.

In the present study we define the magnitude m as,

$$m = \log_{10}\left[\sum_i \Delta x_i\right], \qquad (32)$$

where Δx_i is the displacement of i-th block during one event.

Figures 14 and 15 show the size distribution of earthquake for systems with various numbers of blocks. As seen from the figures the regime where power law behavior is observed becomes narrower with increasing number of blocks and the peak at large magnitude around $m \simeq 2.5$ is pronounced, which may correspond to the characteristic earthquake.

In order to clarify how the power law behavior depends on the number of blocks quantitatively, we follow the work by Stirling et al. [33] and calculate the frequency ratio as a function of the number of blocks in the system, which is the ratio of predicted number of earthquake by employing the power law and the actual number of earthquake. The larger value of the ratio means the less@validity of the power law. The actual method of the calculation of the frequency ratio is as follows. First we evaluated the power of the power law of each system by fitting the data shown in Figs. 14 and 15 in the range where the law holds well, that is $m = -1.7 \sim -1.0$. Then we evaluated the frequency of earthquake with magnitude $m = -1.5$ predicted by the power law with the power obtained above and actual frequency of earthquake with magnitude $m = 1.0$. The frequency ratio is obtained by the above frequency divided by the actual frequency of the earthquake with magnitude $m = -1.5$. As shown by the figure the frequency ratio increases with the number of blocks.

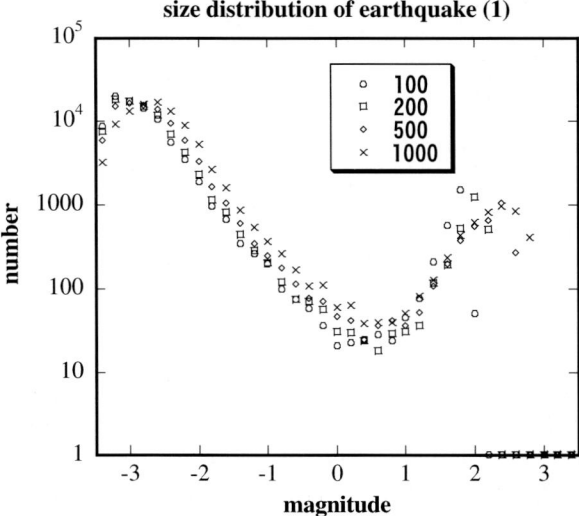

Fig. 14. Size distribution of earthquake for systems with 100, 200, 500, 1000 blocks

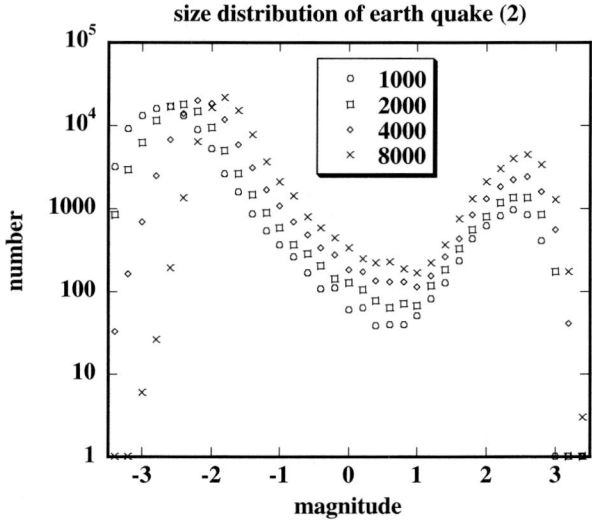

Fig. 15. Size distribution of earthquake for systems with 1000, 2000, 4000, 8000 blocks

Therefore we can conclude that the deviation from the power law behavior becomes larger for larger systems.

5 Summary and Discussion

In this report we first discussed the mechanism of friction, which is not established still now. We then invesigated velocity and waiting time dependence of frictional force. After the introduction of constitutive law of friction, we discussed that the velocity dependence of granite changes from velocity weakening to strengthening behaviors above a certain temperature, which is considered to determine the critical depth of the occurrence of tectonic earthquake. Then we derived results of the constitutive law of friction theoretically. Earthquake is a kind of stick-slip motion of the system with interpositions. There are many systems which have interpositions and show stick-slip motion. The behavior of stick-slip motion is, however, depends on the system and we can not clarify the condition that determines the behavior. The one of the interesting feature of earthquake is that its size distribution obeys the power law, the Gutenberg–Richter law. It is known that the Burridge–Knopoff model, a kind of spring-block model, reproduces the behavior. We investigated the size dependence of the model and showed that the deviation from the power law becomes larger for larger systems.

Let us discuss the relation between the power law behavior of the size distribution of earthquake and self-organized criticality (SOC) [20]. As mentioned before one possible explanation of the power law behavior is SOC. If SOC scenario is true, however, the power law holds better for larger system because true critical state exists only in infinite system and the deviation from the critical state grows for smaller systems. This is not consistent with the results obtained in the previous section. There is other system which shows power law behavior only in small systems. Yoshioka performed sand pile avalanche experiments [34]. He observed that the power law behavior holds only in small systems and large deviation from the power law appears in the larger systems than a certain critical size.

The number of blocks in the Burridge–Knopoff model is considered to be proportional to the length of the fault. Then the present result means that the Gutenberg–Richter law holds better in shorter faults and the deviation from the law increases for longer faults. Stirling et al. examined the size distribution of earthquake in each fault [33]. They evaluated the frequency ratio in the previous section and obtained the similar results with Fig. 16, that is, the deviation from the Gutenberg–Richter law becomes larger for longer faults.

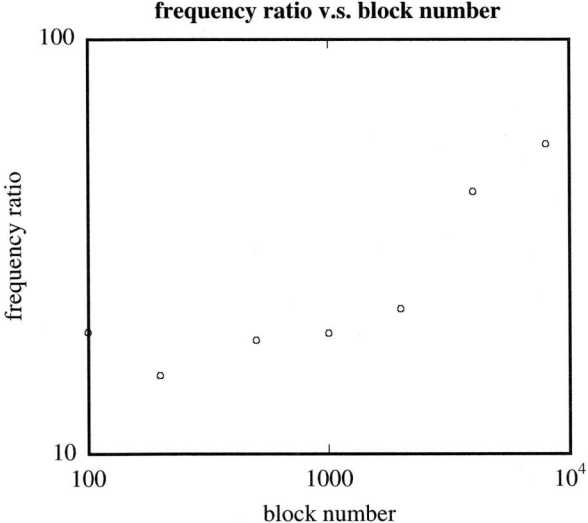

Fig. 16. Frequency ratio as a function of number of blocks

In these systems, the Burridge–Knopoff model, the sand pile avalanche system and earthquake system, the power law may results from the nature of the system that has finite degrees of freedom and not from criticality. Of course the power law behavior appears only in the limited range in finite systems. We may have to investigate a scenario that explain the limited power law behavior in finite systems.

Acknowledgements

One of the author (H.M.) would like to express his sincere thanks to Prof. B. Chakrabarti and local support members of moe05 for the organization of the nice conference and kind hospitality. He also thanks to Mr. K. Murakami for his contribution in early stage of the study in Sect. 4.

References

1. B.N. J. Persson, *Sliding Friction, Physical Principles and Applications*, 2nd edn., Springer, Berlin, Heidelberg, New York (2000).
2. C.C. Scholz, *The Mechanics of Earthquake and Faulting*, Cambridge Univ Press, Cambridge (1990).
3. J.H. Dieterich, J. Geophys. Res. **84**, 2161 (1979).
4. A.L. Ruina, J. Geophys. Res. **88**, 10359 (1983).
5. R. Stesky, Can. J. Earth. Sci. **15**, 361 (1978).
6. S. Tse and J. Rice, J. Geophys. Res. **91**, 8452 (1986).
7. W.F. Brace and J.D. Byerlee, Science **168**, 1573 (1970).
8. F. Heslot, T. Baumberger, B. Perrin, B. Caroli and C. Caroli, Phys. Rev. **E49**, 4973 (1994).
9. C.M. Mate, G.M. McCelland, R. Erlandsson and S. Chaing, Phys. Rev. Lett. **59**, 1942 (1987).
10. K. Matsushita, H. Matsukawa and N. Sasaki, Solid State Commun. **136**, 51 (2005).
11. J.N. Israelachvili, Surf. Sci. Rpt. **14**, 109 (1992).
12. H. Yoshizawa, P. McGuiggan and J. Israelachvili, Science **259**, 1305 (1993); H. Yoshizawa, Y.-L. Chen and J. Israelachvili, J. Chem. Phys. **97**, 4128 (1993); H. Yoshizawa and J. Israelachvili, J. Chem. Phys. **97** 11300 (1993).
13. S. Nasuno, A. Kudrolli and J.P. Gollub, Phys. Rev. Lett. **79**, 949 (1997); S. Nasuno, A. Kudrolli, A. Bak and J.P. Gollub, Phys. Rev. **E58**, 2161 (1998).
14. F. Daltion and D. Corcoran, Phys. Rev. **E63**, 061312 (2001).
15. T. Hirata, J. Phys. Soc. Jpn. **68**, 3195 (1999).
16. B. Gutenberg and C.F. Richter, Ann. Geofis. **9**, 1 (1956).
17. J.N. Carlson and J.S. Langer, Phys. Rev. Lett. **62**, 2632 (1989).
18. R. Burridge and L. Knopoff, Bull. Seismol. Soc. Am. **57**, 3411 (1967).
19. J.N. Carlson, J.S. Langer, B.E. Shaw and C. Tang, Phys. Rev. **A44**, 884 (1991).
20. P. Bak, C. Tang and K. Wiesenfeld, Phys. Rev. Lett. **59**, 381 (1987); P. Bak and C. Tang, J. Geophys. Res. **94**, 15635 (1989).
21. T. Saito and H. Matsukawa, *Friction, Stick-Slip Motion and Earthquake,* Lect. Notes in Phys. **705**, pp. 169–189 (2006).
22. F.P. Bowden and D. Tabor, *The Friciton and Lubricaiton of Solids*, Oxford University Press, Oxford (1950; paperback edition 1986).
23. J.A. Greenwood, In *Fundamentals of Friction: Macroscopic and Microscopic Process*, Ed. I.L. Singer and H.M. Pollock, Kluwer, Dordrecht (1992) pp. 37–57.
24. H. Matsukawa and H. Fukuyama, Phys. Rev. **B49**, 17286 (1994); T. Kawaguchi and H. Matsukawa, Phys. Rev. **B56**, 13932 (1997), **B58**, 15866 (1998).

25. J. Ringlein and M.O. Robbins, Am. J. Phys. **72**, 884 (2004).
26. M. Peyrard and S. Aubry, J. Phys. **C16**, 1593 (1983).
27. M. Hirano and K. Shinjo, Phys. Rev. **B41**, 11837 (1990).
28. M. Hirano. K. Shinjo, R. Kaneko and Y. Murata, Phys. Rev. Lett. **78**, 1448 (1997).
29. G. He, M.H. Müsser and M.O. Robbins, Science **284**, 50 (1999); M.H. Müsser, L. Wenning and M.O. Robbins, Phys. Rev. Lett. **86**, 1295 (2001).
30. J.H. Dieterich, Pure and Appl. Geophys. **116**, 790 (1979).
31. T. Baumberger, Solid State Commun. **102**, 175 (1997).
32. J.H. Dieterich and B.D. Kilgore, Pure and Appl. Geophys. **143**, 283 (1994); Tectonophysics **256**, 219 (1996).
33. M.W. Stirling, S.G. Wesnousky and K. Shimazaki, J. Geophys. Res. **124**, 833 (1996).
34. N. Yoshioka, Earth Planets Space **55**, 283 (2003).

Statistical Features of Earthquake Temporal Occurrence

Á. Corral

Departament de Física, Facultat de Ciències, Universitat Autònoma de Barcelona, 08193 Bellaterra, Spain
Alvaro.Corral@uab.es

The physics of an earthquake is a subject with many unknowns. It is true that we have a good understanding of the propagation of seismic waves through the Earth and that given a large set of seismographic records we are able to reconstruct a posteriori the history of the fault rupture (the origin of the waves). However, when we consider the physical processes which lead to the initiation of a rupture with a subsequent slip and its growth through a fault system to give rise to an earthquake, then our knowledge is really limited. Not only the friction law and the rupture evolution rules are largely unknown, but the role of many other processes such as plasticity, fluid migration, chemical reactions, etc., and the couplings between them, remain unclear [1, 2].

On the other hand, one may wonder about *the physics of many earthquakes*. How do the collective properties of the set defined by all earthquakes in a given region, or better, in the whole world, emerge from the physics of individual earthquakes? How does seismicity, which is the structure formed by all earthquakes, depend on its elementary constituents – the earthquakes? And which are these properties? Which kind of dynamical process does seismicity constitute? It may be that these collective properties are largely independent on the physics of the individual earthquakes, in the same way that many of the properties of a gas or a solid do not depend on the constitution of its elementary units – the atoms (for a broad range of temperatures it doesn't matter if we have atoms, with its complicated quantum structure, or microscopic marbles). It is natural then to consider that the physics of many earthquakes has to be studied with a different approach than the physics of one earthquake, and in this sense we can consider the use of statistical physics not only appropriate but necessary to understand the collective properties of earthquakes.

Here, we provide a summary of recent work on the statistics of the temporal properties of seismicity, considering the phenomenon as a whole and with the goal of looking for general laws. We show the fulfillment of a scaling law for recurrence-time distributions, which becomes universal for stationary seismicity and for aftershock sequences which are transformed into stationary processes by means of a nonlinear rescaling of time. The existence of a decreasing power-law regime in the distributions has paradoxical consequences on the time evolution of the earthquake hazard and on the expected time of occurrence of an incoming event, as we will see. On the other hand, the scaling law for recurrence times is equivalent to the invariance of seismicity under renormalization-group-like transformations, for which the role of correlations between recurrence times and magnitudes is essential. Finally, we relate the recurrence-time densities studied here with the method previously introduced by Bak et al. [3].

1 The Gutenberg–Richter Law and the Omori Law

Traditionally, the knowledge of seismicity has been limited to a few phenomenological laws, the most important being the Gutenberg–Richter (GR) law and the Omori law. The GR law determines that, for a certain region, the number of earthquakes in a long period of time decreases exponentially with the magnitude; to be concrete, $N(M_c) \propto 10^{-bM_c}$, where $N(M_c)$ is the number of earthquakes with magnitude M greater or equal than a threshold value M_c, and the b-value is a constant usually close to one [4, 5, 6, 7].

If we introduce the seismic rate, $r(t, M_c)$, defined as the number of earthquakes with $M \geq M_c$ per unit time in a time interval around t, then, the GR relation can be expressed in terms of the mean seismic rate, $R(M_c)$, as

$$R(M_c) \equiv \langle r(t, M_c) \rangle = \frac{1}{T} \int_0^T r(t, M_c) dt = \frac{N(M_c)}{T} = R_0 10^{-bM_c}, \quad (1)$$

where T is the total time under consideration and R_0 is the (hypothetical) mean rate in the region for $M_c = 0$ (its dependence, as well as that of other parameters, on the region selected for study is implicit and will not be indicated when it is superfluous). In fact, the GR law must be understood as a probabilistic law, and then we conclude that earthquake magnitude follows an exponential distribution, this is, $\text{Prob}[M \geq M_c] = N(M_c)/\mathcal{N} \propto e^{-\ln 10\, bM_c}$, with $\mathcal{N}$ the total number of earthquakes, of any magnitude. Due to the properties of the exponential distribution, the derivative of $\text{Prob}[M \geq M_c]$, which is the probability density (with a minus sign), is also an exponential.

In terms of the seismic moment or of the dissipated energy, which are increasing exponential functions of the magnitude, the GR law transforms into a power-law distribution, the usual signature of scale invariance. This means that earthquakes have no characteristic size of occurrence, if we take the seismic moment or the energy as more appropriate measures of earthquake size than the magnitude [4].

The Omori law (in its modified form) states that after a strong earthquake, which is called *mainshock*, the seismic rate for events with $M \geq M_c$ in a certain region around the mainshock increases abruptly and then decays in time essentially as a power law; more precisely,

$$r(t, M_c) = \frac{r_0(M_c)}{(1+t/c)^p} \qquad (2)$$

where t is the time measured from the mainshock, $r_0(M_c)$ is the maximum rate for $M \geq M_c$, which coincides with the rate immediately after the mainshock, i.e., $r(t=0, M_c) = r_0(M_c)$, c is a short-time constant (of the order of hours or a few days), which describes the deviation from a pure power law right after the mainshock, and the exponent p is usually close to 1. In fact, c depends to a certain degree on M_c and, together with $r_0(M_c)$ and p, depends also on the mainshock magnitude [7, 8, 9].

Nowadays it has been confirmed that the Omori law does not only apply to strong earthquakes, but to any earthquake, with a productivity factor (r_0) for small earthquakes which is orders of magnitude smaller than for large events. In this way, the classification of earthquakes in mainshocks and aftershocks turns out to be only relative, as we will have a cascade process in which aftershocks become also mainshocks of secondary sequences and so on. When an aftershock happens to have a magnitude larger than the mainshock a change of roles occur: the mainshock is considered a foreshock and the aftershock becomes the mainshock. Also, the triggering of strong aftershocks may cause that the overall seismic rate departs significantly from the Omori law, as it happens in earthquake swarms.

In any case, the Omori law illustrates clearly the temporal clustering of earthquakes, for which events (aftershocks) tend to gather close (in time) to a strong event (the mainshock), becoming more dilute as time from the mainshocks grows. In addition, the fact that the seismic rate decays essentially as a power law means that the relaxation process has no characteristic time, in opposition to the usual situation in physics (think for instance in radioactive decay). Finally, the Omori law has a probabilistic interpretation, as an all-return-time distribution, measuring the probability that earthquakes occur at a time t after a mainshock.

2 Recurrence-Time Distributions and Scaling Laws

One can go beyond the GR law and the Omori law and wonder about the temporal properties of individual earthquakes (from a statistical point of view), in particular about the time interval between consecutive earthquakes. In this case, it is necessary to assume that earthquakes are point events in time, or at least that their temporal properties are well described by their initiation time. In contrast to the previous approaches, this perspective has been much less studied and no general law has been proposed; rather, the situation is confusing in the literature, where claims range from nearly-periodic behavior for large earthquakes to totally random occurrence of mainshocks (see the citations at [10, 11]). Furthermore, it can be argued that the times between consecutive earthquakes depend strongly on the selection of the coordinates of the region under study and the range of magnitudes selected (which change the sequence of events) and therefore one is dealing with an ill-defined variable. We will see that the existence of universal properties for these times invalidates this objection.

Following the point of view of Bak et al., we have addressed this problem by considering seismicity as a phenomenon on its own. In this way, we will not separate events into different kinds (foreshocks, mainshocks, aftershocks, or microearthquakes, etc.), nor divide the crust into provinces with different tectonic properties, but will place all events and regions on the same footing; in other words, we wonder about the very nature of seismicity as a whole, from a complex-system perspective, in opposition to a reductionist approach [3, 12].

This exposition will concentrate on the temporal properties of seismicity, and their dependence with space and magnitude, but equally important are the spatial properties. It turns out that all the aspects of seismicity are closely related to each other and one cannot study them separately. Although all the events are important, as they are the elementary constituents of seismicity, we will need to consider windows of observation in space, time, and magnitude; of course, this is due to the incompleteness of seismic records but also to the fact that the variation of the quantities we measure with the range of magnitudes selected or with the size of the spatial region under study will allow us to establish self-similar properties for seismicity.

Let us select an arbitrary region of the Earth, a temporal period, and a minimum magnitude M_c, in such a way that only events in this space-time-magnitude window are taken into account. We can consider the resulting events as a point process in time, disregarding the magnitude and the spatial degrees of freedom (this is not arbitrary, as M_c and the size of the region will be systematically varied later on), in this way we can order the events in time, from $i = 1$ to $N(M_c)$ and characterize each one only by its occurrence time, t_i. From here we can define the recurrence time τ (also called waiting time, interevent time, interoccurrence time, etc.) as the time interval between consecutive events, i.e., $\tau_i \equiv t_i - t_{i-1}$. The mean recurrence time, $\langle \tau(M_c) \rangle$, is obviously given by the inverse of the rate, $R^{-1}(M_c)$; however, as the recurrence time is broadly distributed, the mean alone is a poor characterization of the process and it is inevitable to work with the probability distribution of recurrence times. So, we compute the recurrence-time probability density as

$$D(\tau; M_c) = \frac{\text{Prob}[\tau < \text{ recurrence time} \leq \tau + d\tau]}{d\tau}, \qquad (3)$$

where $d\tau$ has to be small enough to allow D to represent a continuous function but large enough to contain enough data to be statistically significant (note that the spatial dependence of D is not indicated explicitly).

2.1 Scaling Laws for Recurrence-Time Distributions

We can illustrate this procedure with the waveform cross-correlation catalog of Southern California obtained by Shearer et al.[1] [13] for the years 1984–2002, containing 26700 events with $M \geq 2.5$ (84209 events with $M \geq 2$). The recurrence-time probability densities for several values of M_c are shown in Fig. 1 (left). First, one can see that τ ranges from seconds to more than 100 days (in fact, we have restricted our analysis to recurrence-times greater than one minute; shorter times do not follow the same trend than the rest, probably due to the incompleteness of the records in that time scale). Also, the different distributions look very similar in shape, although the ranges are different (obviously, the larger M_c, the smaller the number of events $N(M_c)$, and the larger the mean time between them).

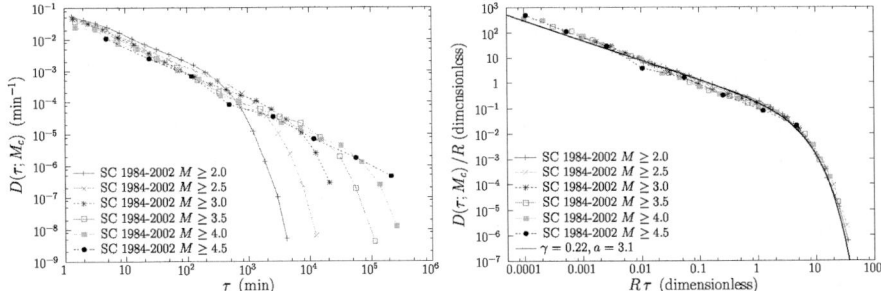

Fig. 1. (*Left*) Probability densities of recurrence times in Southern California (SC) for the period 1984–2002, for several M_c values. (*Right*) The same probability densities rescaled by their rate. The data collapse illustrates the fulfillment of a scaling law. The continuous line is a gamma fit

Figure 1 (right) shows the same distributions but rescaled by the mean rate, as a function of the rescaled recurrence time, i.e., $D(\tau; M_c)/R(M_c)$ versus $R(M_c)\tau$. In this case all the distributions collapse onto a single curve f and we can establish the fulfillment of a *scaling law* [14],

$$D(\tau; M_c) = R(M_c) f(R(M_c)\tau) \ . \tag{4}$$

where f is the scaling function, and corresponds to the recurrence-time density in the hypothetical case $R(M_c) = 1$. Note that we could have arrived to a similar equation by scaling arguments, but there would be no reason for the function f to be independent on M_c. Only imposing the self-similarity of the process in time-magnitude can lead to the fact that f does not depend on M_c and therefore to the fact that f is a scaling function. As $R(M_c)$ verifies the GR law, the scaling law can be written

---

[1] Available at http://www.data.scec.org/ftp/catalogs/SHLK/

$$D(\tau; M_c) = 10^{-bM_c}\tilde{f}(10^{-bM_c}\tau) \,. \qquad (5)$$

The GR law can be calculated from the scaling law; just calculate the mean recurrence time, $\langle \tau(M_c) \rangle = \int_0^\infty \tau D(\tau; M_c) d\tau = 10^{bM_c} \int_0^\infty z\tilde{f}(z)dz \propto 10^{bM_c}$, and as $\langle \tau(M_c) \rangle$ is the inverse of the mean rate, then, $R(M_c) \propto 10^{-bM_c}$. But the scaling law does not only include the GR law, it goes one step further, as it implies that the GR law is fulfilled *at any time*, if times are properly selected; indeed, events separated by recurrence times τ' for $M \geq M_c'$ and τ for $M \geq M_c$ occur at a GR ratio, $10^{-b(M_c'-M_c)}$, if and only if the ratio of the recurrence times is given by $10^{b(M_c'-M_c)}$. Notice that the only requirement for the GR law to be fulfilled (for a long period of time) is that $D(\tau; M_c)$ has a mean that verifies the GR law, i.e., $\langle \tau(M_c) \rangle = R^{-1}(M_c) = R_0 10^{bM_c}$; therefore, the fulfillment of the GR law at any time is a new feature of seismicity.

To make it more concrete, we can count the number of events in Southern California with $M \geq 3$ coming after a recurrence time $\tau = 100$ hours and compare with the number of events with $M \geq 4$ after the same recurrence time; then the ratio of these numbers has nothing to do with the GR relation. However, if for $M \geq 4$ we select events with $\tau = 1000$ hours (the b-value in the GR law is very close to 1 in Southern California) then, the number of these events is about 1/10 of the number of events with $M \geq 3$ and $\tau = 100$ hours, the same proportion as when we consider all events (no matter the value of τ). This could be somehow analogous to the well-known law of corresponding states in condensed-matter physics: two pairs of consecutive earthquakes in different magnitude windows would be in "corresponding states" if their rescaled recurrence times are the same.

2.2 Relation with the Omori Law

In general, as seismicity is not stationary, the scaling function f will change with the spatio-temporal window of observation. In the case of Omori aftershock sequences, the scaling function, and therefore the distribution of recurrence times, is related to the Omori law, as we now see. Let us assume, just for simplicity, that the aftershock sequence can be modeled as a nonhomogeneous Poisson process (also called nonstationary Poisson process, this is a Poisson process but with a time-variable rate, in such a way that at any instant the probability of occurrence, per unit time, is not constant but is independent on the occurrence of other events); in this case the rate of occurrence will be given by the Omori law, (2). Then, the recurrence-time density is a temporal mixture of Poisson processes, which have a density $D(\tau|r(M_c)) = re^{-r\tau}$, so,

$$D(\tau; M_c) = \frac{1}{\mu}\int_{r_m}^{r_0} rD(\tau|r)\rho(r; M_c)dr = \frac{1}{\mu}\int_{r_m}^{r_0} r^2 e^{-r\tau}\rho(r; M_c)dr \,, \qquad (6)$$

where $\rho(r; M_c)$ is the density of rates, μ is a normalization factor that turns out to be the mean value of r, $\mu = \langle r(M_c) \rangle = \int r\rho(M_c)dr$ and $r_0(M_c)$ and

$r_m(M_c)$ the maximum and minimum rate, respectively, assuming $r_0 \gg r_m$; the factor r appears because the probability of a given $D(\tau|r)$ to contribute to $D(\tau; M_c)$ is proportional to r.

The density of rates can be obtained by the projection of $r(t; M_c)$ onto the r axis, turning out to be,

$$\rho(r; M_c) \propto \left|\frac{dr}{dt}\right|^{-1} \Rightarrow \rho(r; M_c) = \frac{C}{r^{1+1/p}} \quad \text{for} \quad r_m \leq r \leq r_0 \qquad (7)$$

with C just a constant (depending on M_c) that can be obtained from normalization. Substituting, we get

$$D(\tau; M_c) = \frac{C}{\mu} \int_{r_m}^{r_0} r^{1-1/p} e^{-r\tau} dr = \frac{C[\Gamma(2-1/p, r_m\tau) - \Gamma(2-1/p, r_0\tau)]}{\mu \tau^{2-1/p}}, \qquad (8)$$

with $\Gamma(\alpha, z) \equiv \int_z^\infty z^{\alpha-1} e^{-z} dz$ the incomplete gamma function (note that $\Gamma(1, z) = e^{-z}$). It is clear that for intermediate recurrence times, $1/r_0 \ll \tau \ll 1/r_m$, we get a power law of exponent $2 - 1/p$ for the recurrence time density,

$$D(\tau; M_c) \simeq \frac{C\Gamma(2-1/p)}{\mu \tau^{2-1/p}}, \qquad (9)$$

with $\Gamma(\alpha)$ the usual (complete) gamma function. This power-law behavior has been derived before by Senshu and by Utsu for nonhomogeneous Poisson processes [7], but our procedure can be easily extended beyond this case, just defining a different $D(\tau|r)$, for which the value of the recurrence-time exponent $2 - 1/p$ is still valid. Notice that the value of this exponent is close to one if the p-value is close to one, but both exponents are only equal if $p = 1$, in any other case we have $2 - 1/p < p$, which means that in general $D(\tau; M_c)$ decays more slowly than $r(t; M_c)$. If we consider large recurrence times, $r_m\tau \gg 1$, we can use the asymptotic expansion $\Gamma(\alpha, z) \to z^{\alpha-1} e^{-z} + \cdots$ for $z \to \infty$ [15], to get

$$D(\tau; M_c) \simeq \frac{C r_m^{1-1/p}}{\mu} \frac{e^{-r_m\tau}}{\tau}, \qquad (10)$$

which in the limit we are working is essentially an exponential decay.

Although the equations derived here for a nonhomogeneous Poisson process with Omori rate reproduce well the recurrence-time distribution of aftershock sequences, $D(\tau; M_c)$ [16], the choice of an exponential form for $D(\tau|r)$ is not justified, as we will see in the next sections. Nevertheless, for the moment we are only interested in the form of $D(\tau; M_c)$.

2.3 Gamma Fit of the Scaling Function

The fact that the density $D(\tau; M_c)$ for a nonhomogeneous Poisson-Omori sequence is a power law for intermediate times and follows (10) for long times suggests that a simpler parameterization of the distribution can be obtained by the combination of both behaviors; in the case of the scaling function f, which must follow the same distribution as D (but with mean equal to one), we can write

$$f(\theta) \propto \frac{e^{-\theta/a}}{\theta^{2-1/p}(1+\theta)^{1/p-1}}.$$

However, as both power laws of θ are very similar and in the long time limit it is the exponential alone what is really important, we can simplify even further and use the gamma distribution to model f; so,

$$f(\theta) = \frac{C}{a\Gamma(\gamma)} \left(\frac{a}{\theta}\right)^{1-\gamma} e^{-\theta/a}, \qquad (11)$$

where θ plays the role of a dimensionless recurrence time, $\theta \equiv R\tau$, a is a dimensionless scale parameter, and C is a correction to normalization due to the fact that the gamma distribution may not be valid for very short times; this will allow the shape parameter γ not to be restricted to the case $\gamma > 0$, the usual condition for the gamma distribution (nevertheless, if $\gamma \leq 0$ the factor $\Gamma(\gamma)$ is inappropriate for normalization). As f is introduced in such a way that the mean of θ is $\langle\theta\rangle = 1$, the parameters are not independent; for instance, for $C = 1$, $\langle\theta\rangle = \gamma a$ and in consequence $a = 1/\gamma$. So, essentially, we only have one parameter to fit, γ, to characterize the process. In the case of Omori sequences, $1 - \gamma = 2 - 1/p$ and $a = R/r_m, \Rightarrow \gamma \simeq r_m/R$, but we will see that the gamma distribution has a wider applicability than just Omori sequences.

A fit of the gamma distribution to the rescaled distribution for Southern-California, shown in Fig. 1(right), yields the parameter values $\gamma \simeq 0.22$ and $a \simeq 3$; this yields a power-law exponent for small and intermediate times $1 - \gamma \simeq 0.78$ and allows to calculate a p-value $p = (1+\gamma)^{-1} \simeq 0.82$, which can be interpreted as an average for Southern California, and a minimum rate $r_m \simeq R/3$. Of course, with our resolution we only can establish $1-\gamma \simeq p \simeq 0.8$.

2.4 Universal Scaling Law for Stationary Seismicity

We have mentioned the nonstationary character of seismicity and that in consequence the scaling function f depends on the window of observation. A more robust, universal law can be established if we restrict our study to stationary seismicity. By stationary seismicity we mean in fact homogeneity in time, which implies that the statistical properties of the process do not depend on the time window of observation, in particular, the mean rate must be practically constant in time.

It is obvious that an aftershock sequence following the Omori law (with $p > 0$) is not stationary, but observational evidence shows that in other cases seismicity can be well described by a stationary process, for example worldwide seismicity for the last 30 years (for which there are reasonably good data) or regional seismicity in between large aftershock sequences. It should be clear that considering stationary seismicity has nothing to do with declustering (the removal of aftershocks from data). We simply consider periods of time for which no aftershock sequence dominates in the spatial region selected for study, but many smaller sequences may be hidden in the data, intertwined in such a way to give rise to an overall stationary seismic rate.

The total number of earthquakes in Southern-California (from Shearer et al.'s catalog) as a function of time since 1984 is displayed in Fig. 2. Clearly, the behavior of the number of earthquakes in time is nonlinear, with episodic abrupt increments which correspond to large aftershock sequences, following the trend prescribed by the Omori law, $N(M_c, t) = N(M_c, 0) + \int_0^t r_0(M_c)/(1+t'/c)^p dt'$. However, there exist some periods which follow a linear increase of $N(M_c, t)$ versus t; in particular, we have chosen for analysis the intervals (in years, with decimal notation) 1984–1986.5, 1990.3–1992.1, 1994.6–1995.6, 1996.1–1996.5, 1997–1997.6, 1997.75–1998.15, 1998.25–1999.35, 2000.55–2000.8, 2000.9–2001.25, 2001.6–2002, and 2002.5–2003. These intervals comprise a total time span of 9.25 years and contain 6072 events for $M \geq 2.5$, corresponding to a mean rate $R(2.5) = 1.7$ earthquakes/day. Note from the figure that not only the rate of occurrence is nearly constant for each interval, but different intervals have similar values of the rate.

We will study all these stationary periods together, in order to improve the statistics. The probability densities of the recurrence times are calculated from all the periods and the corresponding rescaled distributions appear in Fig. 3 (left). The good quality of the data collapse indicates the validity of a scaling law of the type of (4), although the scaling function f is clearly different than

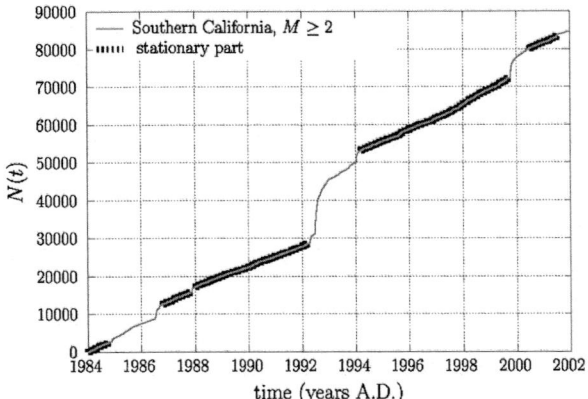

Fig. 2. Accumulated number of earthquakes in Southern California as a function of time. Some stationary or nearly stationary periods mentioned in the text are specially marked, see Subsect. 5.3

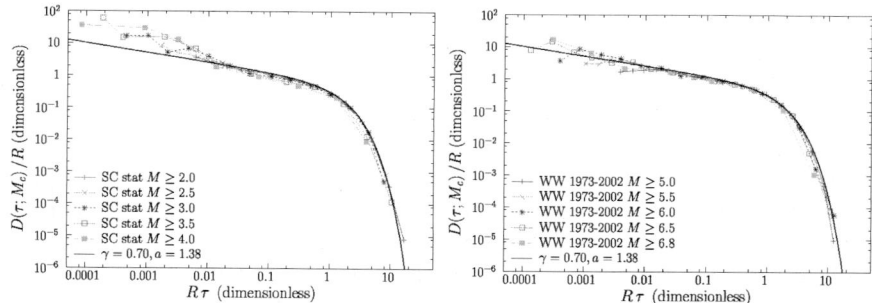

Fig. 3. Rescaled recurrence-time probability densities for the stationary periods explained in the text for Southern California (*left*) and for worldwide seismicity (*right*). The *solid line* is the same function in both cases, showing the universal character of the scaling law fulfilled

the one for the whole time period 1984–2002, in particular, the power-law is much flatter, which is an indication that the clustering degree is smaller in this case, in comparison, but still exists. Nevertheless, it is remarkable that this kind of clustering is different than the clustering of aftershock sequences, as in this case we are dealing with a stationary process. The figure shows also a plot of the scaling function f parameterized with a gamma distribution with $\gamma = 0.7$ and $a = 1.38$, which indeed implies a power-law exponent $1 - \gamma = 0.3$.

We now present the results for recurrence times in worldwide scale, using the NEIC-PDE worldwide catalog (National Earthquake Information Center, Preliminary Determination of Epicenters[2]) which covers the period 1973–2002 and yields 46055 events with $M \geq 5$. In this case the total number of earthquakes grows linearly in time, which confirms the stationarity of worldwide seismicity. The corresponding rescaled recurrence-time probability densities are shown in Fig. 3 (right), together with the scaling function used in the previous case (i.e., Southern-California stationary seismicity). The collapse of the data onto a single curve is again an indication of the validity of a scaling law, and the fact that this curve is well fit by the same scaling function than in the Southern-California stationary case is a sign of *universality*. We use the term universality with the usual meaning in statistical physics, in which it refers to very different systems (gases or magnetic solids, or in our case seismic occurrence in quite diverse tectonic environments) sharing the same quantitative properties.

In fact, the universality of the scaling law for recurrence-time distributions in the stationary case has been tested for several other regions, namely, Japan, Spain, New Zealand, New Madrid (USA), and Great Britain, with magnitude values ranging from $M \geq 1.5$ to $M \geq 7.5$ (which is equivalent to a factor 10^9

---

[2] Available at http://wwwneic.cr.usgs.gov/neis/epic/epic_global.html

in the minimum dissipated energy), and for spatial areas as small as $0.16° \simeq$ 20 km [14, 17].

2.5 Universal Scaling Law for Omori Sequences

We now return to nonstationary seismicity to show how the universal scaling law for recurrence times applies there. For this purpose, let us consider the Landers earthquake, with magnitude $M = 7.3$, the largest event in Southern California in the last decades, taking place in 1992, June 28, at 34.12°N, 116.26°W. After the earthquake, seismicity in Southern California followed the usual behavior when large shallow events happen: a sudden enormous increase in the number of earthquakes and a consequent slow decay in time, in good agreement with the Omori law.

The previous universal results for stationary seismicity can be generalized in the nonstationary case by replacing the mean seismic rate $R(M_c)$ by the "instantaneous" seismic rate $r(t, M_c)$ as the scaling factor in (4). Then, in order to obtain the rescaled, dimensionless recurrence time θ, we will rescale each recurrence time as

$$\theta_i \equiv r(t_i; M_c)\tau_i \,. \tag{12}$$

This means that it is the instantaneous rate of occurrence which sets the time scale.

First, we examine the complete seismicity with $M \geq M_c$ for a square region (in a space in which longitude and latitude are considered as rectangular coordinates), the region containing the Landers event (but not centered on it); that is, as in previous sections, we will not separate aftershocks from the rest of events. As expected, after a few days from the mainshock, the seismic rate decreases as a pure power law, which is equivalent to take $t \gg c$ in (2), so,

$$r(t; M_c) = r_0 \left(\frac{c}{t}\right)^p,$$

which lasts until the rate reaches the background seismic level. This form for $r(t; M_c)$ is fit to the measured seismic rate, see Fig. 4 (right). One advantage of analyzing the pure power-law regime only, rather than the whole sequence using the modified Omori law, (2), is that it is believed that the deviations from power-law behavior for short times are due to the incompleteness of the catalogs after strong events; therefore, in this way we avoid the problem of incompleteness.

Next, using the results of the fit rather than the direct measurement of $r(t; M_c)$ we calculate

$$\theta_i = r_0 \tau_i \left(\frac{c}{t_i}\right)^p \,. \tag{13}$$

In fact, this rescaling could be replaced by $\theta_i = r(t_{i-1})\tau_i$ or by $\theta_i = r(t_{i-1} + \tau_i/2)\tau_i$, with no noticeable difference in the results, as the rate varies very slowly at the scale of the recurrence time.

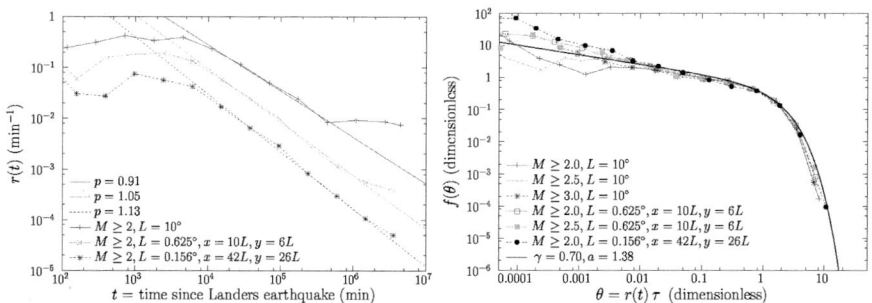

Fig. 4. (*Left*) Seismic rate as a function of the time elapsed since the Landers earthquake for regions of different size L including the event, using the SCSN catalog. Only events with $M \geq 2$ are considered. The *straight lines* correspond to power laws, with exponents given by the p-value. (*Right*) Recurrence-time probability densities for the power-law regime of the decay of the rate after the Landers event, rescaled at each time by the rate. The *solid line* represents the universal scaling function in terms of a gamma distribution

The probability densities of the rescaled recurrence times θ_i obtained in this way are displayed in Fig. 4 (right), showing a slow power-law decay followed by a faster decay, in surprising agreement, not only qualitative but also quantitative, with the results for stationary seismicity, in such a way that the universal scaling function for the stationary case is still valid [10, 14]. Therefore, as in that case, the power-law regime in the density is a sign of clustering, but as the primary clustering structure of the sequence has been removed by the rescaling with Omori rate, this implies the existence of a secondary clustering structure inside the main sequence, due to the fact that any large aftershock may generate its own aftershocks [18, 19]. What is remarkable is that this structure seems to be identical to the one corresponding to stationary seismicity.

An important consequence of this is that the time behavior of seismicity depends on just one variable: the seismic rate. Another implication is the fact that aftershock sequences cannot be described as a nonhomogeneous Poisson process, as in that case one should obtain an exponential distribution for $f(\theta)$. The use of the nonhomegeneous Poisson process previously in this work must be understood only as a first approximation to justify the use of the gamma fit for the distribution of recurrence times in an Omori sequence. Nevertheless, we will see in the next sections that the generalization of the nonhomogeneous Poisson process taking into account the results explained here leads to similar conclusions for the time distribution in the sequence.

The rescaling of the recurrence times with the seismic rate $r(t, M_c)$ can be applied also to the occurrence times t_i, in order to transform the Omori sequence (or in general any sequence with a time-variable rate) into a stationary sequence. For this purpose we define the accumulated rescaled recurrence time Θ, defined as $\Theta_i = \theta_1 + \theta_2 + \cdots + \theta_i$, which plays the role of a stationary occurrence time, in the same way that θ_i plays the role of a stationary recurrence time, in general. This allows the complete comparison between stationary seismicity and aftershock sequences.

3 The Paradox of the Decreasing Hazard Rate and the Increasing Time Until the Next Earthquake

Other functions, in addition to the probability density, are suitable for describing the general properties of recurrence times. Although from a mathematical point of view the functions we are going to introduce are fully equivalent to the probability density, they show much clearly some interesting temporal features of seismicity.

3.1 Decreasing of the Hazard Rate

Let us consider first the *hazard rate*, $\lambda(\tau; M_c)$, defined for a certain region and for $M \geq M_c$ as the probability per unit time of an immediate earthquake given that there has been a period τ without activity [20],

$$\lambda(\tau; M_c) \equiv \frac{\text{Prob}[\tau < \tau' \leq \tau + d\tau \,|\, \tau' > \tau]}{d\tau} = \frac{D(\tau; M_c)}{S(\tau; M_c)},$$

where τ' is a generic label for the recurrence time, while τ refers to a particular value of the same quantity, the symbol $|$ denotes conditional probability, and $S(\tau; M_c)$ is the survivor function, $S(\tau; M_c) \equiv \text{Prob}[\tau' > \tau] = \int_\tau^\infty D(\tau'; M_c) d\tau'$. Introducing the scaling law (4) for D in the definitions it is immediate to obtain that both $S(\tau; M_c)$ and $\lambda(\tau; M_c)$ verify also scaling relations, $S(\tau; M_c) = g(R\tau)$ and $\lambda(\tau; M_c) = Rh(R\tau)$. If we make use of the gamma parameterization (11), with $C \simeq 1$, we get for the scaling function h,

$$h(\theta) = \frac{1}{a} \left(\frac{a}{\theta}\right)^{1-\gamma} \frac{e^{-\theta/a}}{\Gamma(\gamma, \theta/a)}.$$

For short recurrence times this function diverges as a power law,

$$h(\theta) \simeq \frac{1}{\Gamma(\gamma) a^\gamma \theta^{1-\gamma}}$$

(in fact, in this limit the hazard rate becomes undistinguishable from the probability density). On the other hand, $h(\theta)$ tends as a power law to the value $1/a$ as $\theta \to \infty$; indeed, making use of the expansion $\Gamma(\gamma, z) \to z^{\gamma-1} e^{-z}[1 - (1-\gamma)/z + (1-\gamma)(2-\gamma)/z^2 + \cdots]$ [15], we get

$$h(\theta) = \frac{1}{a} \left[1 + \frac{a(1-\gamma)}{\theta} + \cdots\right].$$

The overall behavior for $\gamma < 1$ is that $h(\theta)$ decreases monotonically as θ increases; so, contrary to common belief and certainly counterintuitively, these calculations allow us to predict that the hazard does not increase with the elapsed time since the last earthquake, but just the opposite, it decreases up to an asymptotic value that corresponds to a Poisson process of rate R/a. This means that although the hazard rate decreases, it never reaches the zero value, and sooner or later a new earthquake will strike.

If we compare the hazard rate for $\gamma < 1$ with that of a Poisson process with the same mean (given by $\gamma = a = 1$, and rate R), we see that for short recurrence times the hazard rate is well above the Poisson value, implying that at any instant the probability of having an earthquake is higher than in the Poisson case. In contrast, for long times the probability is below the Poisson value, by a factor $1/a$. This is precisely the most direct characterization of clustering in time, for which we can say that events tend to attract each other, being closer in short time scales and more separated in long time scales (in comparison with the Poisson process).

In conclusion, we predict that seismicity is clustered independently on the scale of observation; in the case of stationary seismicity this clustering is much less trivial than the clustering due to the increasing of the rate in aftershock sequences. Taking advantage of the self-similarity implied by the scaling law, we could extrapolate the clustering behavior to the largest events worldwide ($M \geq 7.5$) over relatively small spatial scales (hundreds of kilometers) and we would obtain a behavior akin to the long-term clustering observed by other means [21].

3.2 Increasing of the Residual Time Until the Next Earthquake

Let us introduce now the *expected residual recurrence time*, $\epsilon(\tau_0; M_c)$, which provides the expected time till the next earthquake, given that a period τ_0 without earthquakes (in the spatial area and range of magnitudes considered) has elapsed [20],

$$\epsilon(\tau_0; M_c) \equiv \langle \tau - \tau_0 \,|\, \tau > \tau_0 \rangle = \frac{1}{S(\tau_0; M_c)} \int_{\tau_0}^{\infty} (\tau - \tau_0) D(\tau; M_c) d\tau \, .$$

where | denotes that the mean is calculated only when the condition $\tau > \tau_0$ is fulfilled. Again, the scaling law for D implies a scaling form for this function, which is $\epsilon(\tau_0) = e(R\tau_0)/R$, and introducing the gamma parameterization we get for the scaling function

$$e(\theta) = a \frac{\Gamma(\gamma+1, \theta/a)}{\Gamma(\gamma, \theta/a)} - \theta = a \left[\gamma + \left(\frac{\theta}{a}\right)^\gamma \frac{e^{-\theta/a}}{\Gamma(\gamma, \theta/a)} \right] - \theta,$$

making use of the relation $\Gamma(\gamma+1, z) = \gamma \Gamma(\gamma, z) + z^\gamma e^{-z}$. For short times we obtain

$$e(\theta) = a\gamma + \frac{a}{\Gamma(\gamma)} \left(\frac{\theta}{a}\right)^{\gamma} - \theta + \cdots ;$$

remember that the unconditional mean is $\langle\theta\rangle = \gamma a = 1$, precisely the value obtained for $\theta = 0$. For long times $e(\theta)$ reaches, again as a power law, an asymptotic value equal to a, i.e.,

$$e(\theta) = a \left[1 - \frac{a(1-\gamma)}{\theta} + \cdots \right].$$

The global behavior of $e(\theta)$ is monotonically increasing as a function of θ if $\gamma < 1$. Therefore, the residual time until the next earthquake should grow with the elapsed time since the last one. Notice the counterintuitive behavior that this represents: if we decompose the recurrence time τ as $\tau = \tau_0 + \tau_f$, with τ_f the residual time to the next event, the increase of τ_0 implies the increase of the mean value of τ_f, but the mean value of τ is kept fixed. In fact, this is fully equivalent to the previously reported decreasing-hazard phenomenon and just a more dramatic version of the classical waiting-time paradox [22, 23, 24].

This result seems indeed paradoxical for any time process, as we naturally expect that the residual recurrence (or waiting) time decreases as time increases; think for instance that you are waiting for the subway: you are confident that the next train is approaching; or when you celebrate your birthday, your expected residual lifetime decreases (at any time, in fact). Of course, for an expert statistician the case of earthquakes is not paradoxical, but only counterintuitive, and he or she can provide the counterexamples of newborns (mainly in underdeveloped countries) or of private companies, which become healthier or more solid as time passes and therefore their expected residual lifetime increases with time. These counterintuitive behaviors can be referred to as a phenomenon of *negative aging*.

Nevertheless, for the concrete case of earthquakes the increasing of the expected residual recurrence time is still paradoxical, since one naively expects that the longer the time one has been waiting for an earthquake, the closer it will be, due to the fact that as time passes stress increases on the faults and the next earthquake becomes more likely. Nevertheless, note that our approach does not deal with individual faults but with two-dimensional, extended regions, and in this case the evolution of the stress is not so clear. It is worth mentioning that, as far as the author knows, no conclusive study of this kind has been performed for observational data in individual faults, the difficulty on associating earthquakes to faults is one the major problems here.

3.3 Direct Empirical Evidence

Our predictions for earthquake recurrence times follow the line initiated by other authors. Davis et al. [25], pointed out that when a lognormal distribution is a priori assumed for the recurrence times, the expected residual time increases with the elapsed time. However, the increase there was associated to the update of the distribution parameters as the time since the last earthquake (which was taken into account in the estimation) increased, and not to an intrinsic property of the distribution. Sornette and Knopoff [26] showed that the increase (or decrease) depends completely on the election of the distribution, and studied the properties of a number of them. We now will see that the observational data provide direct and clear evidence in favor of the picture of an incoming earthquake which is moving away in time.

Indeed, in order to rule out the possibility that these paradoxical predictions are an artifact introduced by the gamma parameterization, we must contrast them with real seismicity; in fact, both the hazard rate and the expected residual recurrence time can be directly measured from the catalogs, with no assumption about their functional form. Their definitions provide a simple way to estimate these functions, and in this way we have applied these definitions to the recurrence-time data [17]. From the results displayed in Fig. 5 it is apparent that in all cases the hazard rate decreases with time whereas the expected residual recurrence time increases, as we have predicted. Although both quantities are well approximated by the proposed universal scaling functions, we emphasize that their behavior does not depend on any modeling of the process and in particular is independent on the gamma parameterization. Moreover, the fact that $\epsilon(\tau_0)$ is far from being constant at large times means that the time evolution is not properly described by a Poisson process, even in the long-time limit.

We conclude stating that the contents of this section can be summarized in this simple sentence: *the longer since the last earthquake, the lowest the hazard for a new one*, which is fully equivalent to this one (although less shocking): *the longer since the last earthquake, the longer the expected time till the next*. Moreover, this happens in a self-similar way, thanks to the scaling laws which are fulfilled.

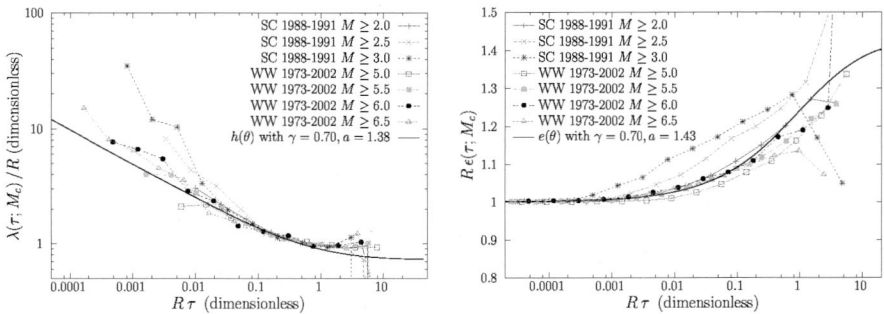

Fig. 5. Rescaled hazard rate (*left*) and rescaled expected residual recurrence time (*right*) as a function of time for Southern California, 1988–1991 (nearly stationary period) and for worldwide seismicity, 1973–2002. The observational data agrees with the scaling functions derived from the gamma distribution. In the right plot the parameter a is not free, but $a = 1/\gamma$ to enforce $e(0) = 1$

4 Scaling Law Fulfillment as Invariance Under a Renormalization-Group Transformation

It is interesting to realize that the scaling law for the recurrence-time distribution, (4), implies the invariance of the distribution under a renormalization-group transformation. Let us investigate deeper the meaning of the scaling analysis we have performed and its relation with the renormalization group.

Figure 6 displays the magnitude M versus the occurrence time t of all worldwide earthquakes with $M \geq M_c$ for different periods of time and M_c values. The top of the figure is for earthquakes with $M \geq 5$ for the year 1990. If we rise the threshold up to $M_c = 6$ we get the results shown in Fig. 6 (medium). Obviously, as there are less earthquakes in this case, the distribution of recurrence times (time interval between consecutive "spikes" in the plot) becomes broader with respect to the previous case, as we know. The rising of the threshold can be viewed as a mathematical transformation of the seismicity point process, which is referred to as *thinning* in the context of stochastic processes [27] and is also equivalent to the common *decimation* performed for spin systems in renormalization-group transformations [28, 29, 30]; the term decimation is indeed appropriate as only one tenth of the events survive this transformation, due to the fulfillment of the GR law with $b = 1$.

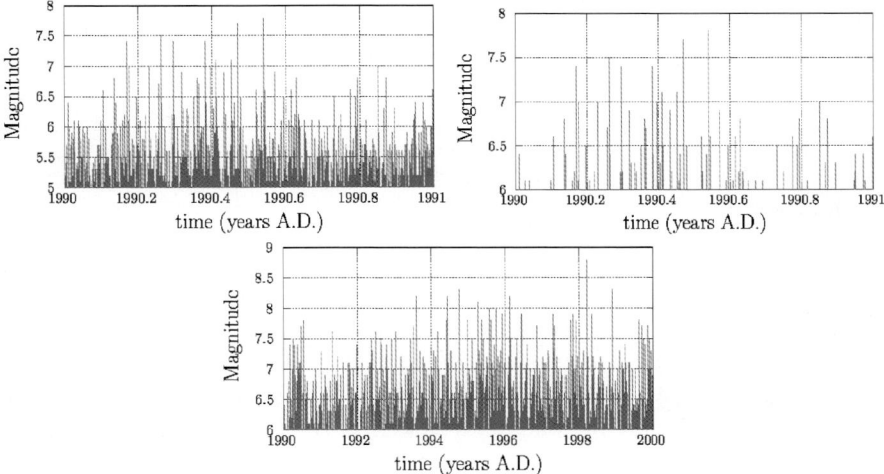

Fig. 6. Magnitude versus time of occurrence of worldwide earthquakes for several magnitude-time windows. The rising of the magnitude threshold from 5 to 6 illustrates the thinning or decimation process characteristic of the first step of a renormalization-group transformation. The second step is given by the extension (rescaling) of the time axis from one year (1990) to 10 years (1990–2000). Notice the similarity between the first plot and the last one, which is due to the invariance of seismicity under this transformation

Figure 6 (bottom) shows the same as Fig. 6 (medium) but for ten years, 1990–1999, and represents a scale transformation of seismicity (also as in the renormalization group), contracting the time axis by a factor 10 to compensate for the previous decimation. The similarity between Fig. 6 (top) and 6 (bottom) is apparent, and is confirmed when the probability densities of the corresponding recurrence times are calculated and rescaled following (4), see again Fig. 3 (right).

4.1 Simple Model to Renormalize

A simple model may illustrate these ideas [31]. Let us assume that seismicity could be described as a time process for which each recurrence time τ_i (which separates event $i - 1$ and i) only depends on M_{i-1}, the magnitude of the last event that has occurred before event i. Any other dependences are ignored, and in particular the values of the magnitudes are generated independently from the rest of the process. It is possible to shown that for this process the recurrence-time density for events above M'_c, $D(\tau; M'_c)$, can be related to the recurrence-time density for events above M_c conditioned to $M_{pre} \geq M'_c$ or to $M_{pre} < M'_c$, which we denote $D(\tau | M_{pre} \geq M'_c; M_c)$ and $D(\tau | M_{pre} < M'_c; M_c)$, respectively, where M_{pre} refers to the magnitude of the event immediately previous to the recurrence time, and it is assumed that $M'_c > M_c$. The relation turns out to be

$$\begin{aligned} D(\tau; M'_c) &= pD(\tau | M_{pre} \geq M'_c; M_c) + qpD(\tau | M_{pre} \geq M'_c; M_c) \\ &* D(\tau | M_{pre} < M'_c; M_c) + q^2 p D(\tau | M_{pre} \geq M'_c; M_c) \\ &* D(\tau | M_{pre} < M'_c; M_c) * D(\tau | M_{pre} < M'_c; M_c) + \cdots \\ &= pD(\tau | M_{pre} \geq M'_c; M_c) * \sum_{k=0}^{\infty} q^k [D(\tau | M_{pre} < M'_c; M_c)]^{*k} \end{aligned} \quad (14)$$

where $*$ denotes the convolution product and p is the probability that an earthquake is above M'_c, given that it is above M_c, i.e.,

$$p \equiv \text{Prob}[M \geq M'_c | M \geq M_c] = 10^{-b(M'_c - M_c)}, \quad (15)$$

using the GR law, whereas $q \equiv \text{Prob}[M < M'_c | M \geq M_c] = 1 - p$. Equation (14) enumerates the number of ways in which two consecutive events for $M \geq M'_c$ may be separated by a recurrence time τ; these are the number of events with $M < M'_c$ in between, each one contributing with a probability q, and then the time τ between the two events is in fact a $(k + 1)$-th return-time for the process with $M \geq M_c$; from here and the independence between recurrence times the convolutions arise.

Let us translate (14) to Laplace space, by using $F(s) \equiv \int_0^{\infty} e^{-s\tau} F(\tau) d\tau$; then, the convolutions turn out to be simple products, i.e.,

$$D(s; M'_c) = pD(s | M_{pre} \geq M'_c; M_c) \sum_{k=0}^{\infty} q^k [D(s | M_{pre} < M'_c; M_c)]^k . \quad (16)$$

As $qD(s|M_{pre} < M'_c; M_c) < 1$ the series can be summed, yielding

$$D(s; M'_c) = \frac{pD(s|M_{pre} \geq M'_c; M_c)}{1 - D(s; M_c) + pD(s|M_{pre} \geq M'_c; M_c)}, \quad (17)$$

using that $D(s; M_c) = pD(s|M_{pre} \geq M'_c; M_c) + qD(s|M_{pre} < M'_c; M_c)$. We have obtained an equation for the transformation of the recurrence-time probability density under the thinning or decimation caused by the raising of the magnitude threshold from M_c to M'_c. The second part in the process is the simple rescaling of the distributions, to make them have the same mean and comparable with each other; we obtain this by removing the effect of the decreasing of the rate, which, due to thinning, is proportional to p, so,

$$D(\tau; M'_c) \to p^{-1} D(p^{-1}\tau; M'_c), \quad (18)$$

and in Laplace space,

$$D(s; M'_c) \to D(ps; M'_c). \quad (19)$$

Finally, the renormalization-group transformation T is obtained by combining the decimation with the scale transformation,

$$\mathsf{T}[D(s; M_c)] = \frac{pD(ps|M_{pre} \geq M'_c; M_c)}{1 - D(ps; M_c) + pD(ps|M_{pre} \geq M'_c; M_c)}. \quad (20)$$

A third step which is usual in renormalization-group transformations is the renormalization of the field, M in this case, but as we are only interested in recurrence times it will not be necessary here. The fixed points of the renormalization-group transformation are obtained by the solutions of the fixed-point equation

$$\mathsf{T}[D(s; M_c)] = D(s; M_c). \quad (21)$$

This equation is equivalent to the scaling law for the recurrence-time densities, (4), the only difference is that now it is expressed in Laplace space, as we are not able to provide the form of the operator T in real space.

4.2 Renormalization-Group Invariance of the Poisson Process

We can get some understanding of the transformation T by considering first the simplest possible case, that in which there are no correlations in the process; so we have to break the statistical dependence between the magnitude and the subsequent recurrence time. This means that

$$D(\tau|M_{pre} \geq M'_c; M_c) = D(\tau|M_{pre} < M'_c; M_c) = D(\tau; M_c) \equiv D_0(\tau; M_c) \quad (22)$$

and then the renormalization transformation turns out to be

$$\mathsf{T}[D_0(s; M_c)] = \frac{pD_0(ps; M_c)}{1 - qD_0(ps; M_c)}. \quad (23)$$

if we introduce $\omega \equiv ps$ and substitute $p = \omega/s$ and $q = 1 - \omega/s$ in the fixed-point equation $\mathsf{T} D_0(s; M_c) = D_0(s; M_c)$, we get, separating variables and equaling to an arbitrary constant k

$$\frac{1}{sD_0(s; M_c)} - \frac{1}{s} = \frac{1}{\omega D_0(\omega; M_c)} - \frac{1}{\omega} \equiv k \, ; \qquad (24)$$

due to the fact that p and s are independent variables and so are s and ω. The solution is then

$$D_0(s; M_c) = (1 + ks)^{-1} \, , \qquad (25)$$

which is the Laplace transform of an exponential distribution,

$$D_0(\tau; M_c) = k^{-1} e^{-\tau/k} \, . \qquad (26)$$

The dependence on M_c enters by means of k, as $k = \langle \tau(M_c) \rangle$; in the case of seismicity the GR law holds and $k = R^{-1}(M_c) = R_0^{-1} 10^{bM_c}$.

Summarizing, we have shown that the only process without correlations which is invariant under a renormalization-group transformation of the kind we are dealing with is the Poisson process. This means that if one considers as a model of seismicity a renewal process (i.e., independent identically distributed return times) with uncorrelated magnitudes, then the recurrence-time distributions will not verify a scaling law when the threshold M_c is raised, except if $D(\tau; M_c)$ is an exponential (which constitutes the trivial case of a Poisson process).

Even further, the Poisson process is not only a fixed point of the transformation, but a stable one (or attractor) for a thinning transformation in which events are randomly removed from the process (random thinning). If magnitudes are assigned to any event independently of any other variable (other magnitudes or recurrence times) the decimation of events after the risen of the threshold M_c is equivalent to a random thinning, and therefore the resulting process must converge to a Poisson process, under certain conditions [27].

The fact that for real seismicity the scaling function f is not an exponential tells us that our renormalization-group transformation is not performing a random thinning; this means that the magnitudes are not assigned independently on the rest of the process and therefore there exists correlations in seismicity. This of course is not new, but let us stress that correlations are fundamental for the existence of the scaling law (4): the only way to depart from the trivial Poisson process is to consider correlations between recurrence times and magnitudes in the process. This is the motivation for the model explained in this section, for which we have chosen the simplest form of correlations between magnitudes and subsequent recurrence times. In fact, Molchan has shown that even for this correlated model the Poisson process is the only possible fixed point, implying that this type of correlations are too weak and one needs a stronger dependence of the recurrence times on history to depart from the Poisson case. After all, this is not surprising, as we know from the study of equilibrium critical phenomena that in order to flow away from trivial fixed points, long-range correlations are necessary. Therefore, the problem of finding a model of correlations in seismicity yielding a nontrivial recurrence-time scaling law is open.

5 Correlations in Seismicity

In the preceding section we have argued that the existence of a scaling law for recurrence time distributions is inextricably linked with the existence of correlations in the process, in such a way that correlations determine the form of the recurrence-time distribution. In consequence, an in-depth investigation of correlations in seismicity is necessary.

Our analysis will be based in the conditional probability density; for instance, for the recurrence time we have,

$$D(\tau|X) \equiv \frac{\text{Prob}[\tau < \text{recurrence time} \leq \tau + d\tau \mid X]}{d\tau},$$

where $|X$ means that the probability is only computed for the cases in which the condition X is fulfilled. If it turns out to be that $D(\tau|X)$ is undistinguishable from the unconditional density, $D(\tau)$, then, the recurrence time is independent on the condition X; on the contrary, if both distributions turn out to be significantly different, this means that the recurrence time depends on the condition X and we could define a correlation coefficient to account for this dependence, although in general we might be dealing with a nonlinear correlation.

Moreover, as we will compare values of the variables in different times (for example, the dependence of the recurrence time τ_i on the value of the preceding recurrence time, τ_{i-1}, for all i), we introduce a slight modification in the notation, particularly with respect the previous section, including the subindices denoting the ordering of the events in the probability distributions. Further, in order to avoid complications in the notation, we will drop the dependence of the conditional density on M_c when unnecessary.

5.1 Correlations Between Recurrence Times

Let us start with the temporal sequence of occurrences, for which we obtain the conditional distributions $D(\tau_i|\tau_a \leq \tau_{i-1} < \tau_b)$; in particular we distinguish two cases; short preceding recurrence times, $D(\tau_i|\tau_{i-1} < \tau_b)$, where τ_b is small, and long preceding recurrences, $D(\tau_i|\tau_{i-1} \geq \tau_a)$, with τ_a large. The results, both for worldwide seismicity and for Southern-California stationary seismicity, turn out to be practically the same, see Fig. 6 of [11]. For short τ_{i-1}, a relative increase in the number of short τ_i and a decrease of long τ_i is obtained, in comparison with the unconditional distribution, which leads to a steeper power-law decay of the conditional density for short and intermediate times. In the opposite case, long τ_{i-1}'s imply a decrease in the number of short τ_i and an increase in the longer ones, in such a way that a flatter power-law exists here. In any case, the behavior for long τ_i is exponential. So, short τ_{i-1}'s imply an average reduction of τ_i and the opposite for long τ_{i-1}'s, and then both variables are positively correlated.

This behavior corresponds to a clustering of events, in which short recurrence times tend to be close to each other, forming clusters of events, while longer times tend also to be next each other. This clustering effect is different from the clustering reported in previous sections, associated to the non-exponential nature of the recurrence-time distribution, but is similar, in some sense, to the usual clustering of aftershock sequences, as these sequences also show this kind of correlations, although mainly due to the time-variable rate.

In fact, the case of nonstationary seismicity was studied by Livina et al. for Southern California [32, 33], with the same qualitative behavior. These authors explain their results in terms of the persistence of the recurrence time, which is a concept equivalent to the kind of clustering we have described. The results could be also similar to the long-term persistence observed in climate records.

The effect of correlations can be described in terms of a scaling law, which constitutes a generalization of the scaling law for (unconditioned) recurrence time distributions, (4). In this way, we can write the conditional recurrence-time distribution in terms of a scaling function which depends on two variables, $R\tau_i$ and $R\tau_a$, or $R\tau_i$ and $R\tau_b$, see [33]. Further, the study of correlations between recurrence times can be studied beyond consecutive events, i.e., we can measure the distribution of τ_i conditioned to τ_{i-2}, or τ_{i-3}, etc. The results for these distributions show no qualitative difference, at least up to $i-10$, in comparison with what we have explained for $i-1$.

The main results of this subsection can be summarized in one single sentence, reflecting the positive correlation between recurrence times: *the shortest the time between the two last earthquakes, the shortest the recurrence of the next one*, on average.

5.2 Correlations Between Recurrence Time and Magnitude

If magnitudes are taken into account, there are two main types of correlations with the recurrence times. First, we consider how the magnitude of one event influences the recurrence time of a future event, in particular the next one, measuring $D(\tau_i|M_{i-1} \geq M'_c)$. The results for the case of worldwide seismicity and for Southern California (in a stationary case) are again similar and show a clear (negative) correlation between M_{i-1} and τ_i [34].

Figure 7 (left) shows, for Southern-California in the stationary case, how larger values of the preceding magnitudes, given by $M_{i-1} \geq M'_c$, lead to a relative increase in the number of short τ_i and a decrease in long τ_i, implying that M_{i-1} and τ_i are anticorrelated. For the cases for which the statistics is better, the densities show the behavior typical of the gamma distribution,

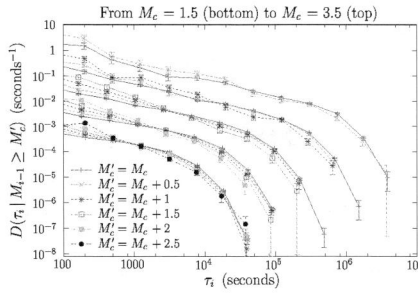 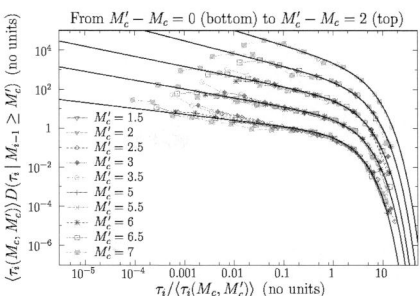

Fig. 7. (*Left*) Recurrence-time distributions for Southern-California conditioned to the value of the preceding magnitude, for the stationary period comprised between 1994–1999. Each set of curves, which have been shifted up for clarity sake, corresponds to a value of M_c, which is, from *bottom* to *top*: 1.5, 2, 2.5, 3, 3.5. (*Right*) Rescaled recurrence-time distributions conditioned to the value of the preceding magnitude, both for worldwide and for Southern-California stationary seismicity. Each set of data, shifted up again for clarity sake, corresponds this time to a different value of $M'_c - M_c$, these being 0, 0.5, 1, 1.5, and 2, from *bottom* to *top*, and is fit by a gamma distribution, with (decreasing) power-law exponents 0.30, 0.45, 0.52, 0.65, and 0.77, respectively

with a power law that becomes steeper for larger M'_c; the different values of the power-law exponent are given at the figure caption.

Remarkably, this behavior can be described by a scaling law for which the scaling function depends now on the difference between the threshold magnitude for the $i-1$ event, M'_c, and the threshold for i, M_c, i.e.,

$$D(\tau_i | M_{i-1} \geq M'_c; M_c) = R(M_c, M'_c) f(R(M_c, M'_c)\tau_i, M'_c - M_c)$$

with $R(M_c, M'_c) \equiv 1/\langle \tau(M_c, M'_c) \rangle$ and $\langle \tau_i(M_c, M'_c) \rangle$ the mean of the distribution $D(\tau_i | M_{i-1} \geq M'_c; M_c)$. The original scaling law for unconditioned distributions, (4), is recovered taking the case $M'_c = M_c$. Figure 7 (right) illustrates this scaling law both for worldwide and for Southern-California stationary seismicity.

The second kind of correlations deals with how the recurrence time to one event τ_i influences its magnitude M_i, or, equivalently, how the recurrence time to one event depends on the magnitude of this event. For this purpose, we measure $D(\tau_i | M_i \geq M'_c)$; the results for Southern-California stationary seismicity are shown in Fig. 8 (left). It is clear that in most of their range the distributions are nearly identical, and when some difference is present this is inside the uncertainty given by the error bars. However, there is one exception: very short times, $\tau_i \simeq 3$ min, seem to be favored by larger magnitudes, or, in other words, short times lead to larger events; nevertheless, due to the short value of the time involve, we can ignore this effect. Then, the recurrence time and the magnitude after it can be considered as independent from a

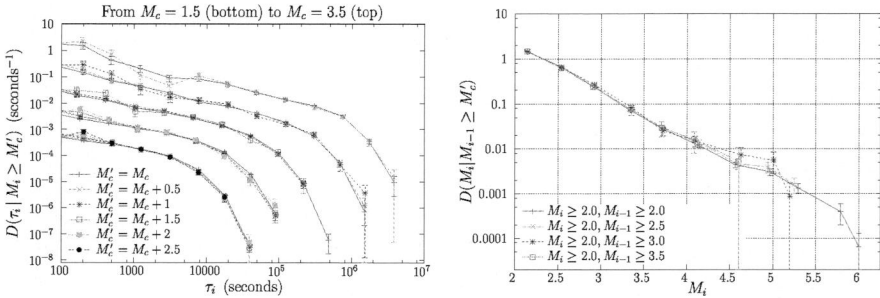

Fig. 8. (*Left*) Recurrence-time distributions for Southern-California conditioned to the value of the magnitude of the incoming event, for the stationary period comprised between 1994–1999. The curves have been shifted up and each set corresponds to a value of M_c, which ranges from 1.5 to 3.5 in steps of 0.5, from *bottom* to *top*. (*right*) Distribution of magnitudes conditioned to the value of the preceding magnitude for Southern California for the set of stationary periods indicated at the text. Only events separated by $\tau \geq 30$ min have been considered

statistical point of view, with our present resolution (it might be that a very weak dependence is hidden in the error bars of the distributions).

Further, as in the previous subsection, we have gone several more steps backwards in time, measuring conditional distributions up to $D(\tau_i|M_{i-10} \geq M'_c)$, and also we have extended the conditional distributions to the future, measuring $D(\tau_i|M_j \geq M'_c)$ with $j > i$. The results are not qualitatively different than what is described previously, with the first type of distributions dependent on M'_c and the second type independent. From here we can conclude the dependence of the recurrence time and the independence of the magnitude with the sequence of previous recurrence times, at least with our present statistics and resolution.

Two sentences may serve to summarize the behavior of seismicity described here. First, *the bigger the size of an earthquake, the shortest the time till next*, due to the anticorrelation between magnitudes and forward recurrence times. Note that in the case of stationary seismicity this result is not trivially derived from the law of aftershock productivity. Second, the belief that *the longer the recurrence time for an earthquake, the bigger its size*, is false, as the magnitude is uncorrelated with the previous recurrence times. This shows clearly the time irreversibility of seismicity.

5.3 Correlations Between Magnitudes

Although not directly related with the temporal properties, we study the correlations between consecutive magnitudes, M_{i-1} and M_i, by means of the distribution $D(M_i|M_{i-1} \geq M'_c)$. As the analysis of the 1994–1999 period for Southern California did not provide enough statistics for the largest events, we considered a set of stationary periods, these being: Jan 1, 1984 – Oct 15, 1984; Oct 15, 1986 – Oct 15, 1987; Jan 1, 1988 – Mar 15, 1992; Mar 15, 1994 – Sep 15, 1999; and Jul 1, 2000 – Jul 1, 2001; all of them visible in Fig. 2.

Again we find that both worldwide seismicity and stationary Southern-California seismicity share the same properties, but with a divergence for short times. Figure 10 of [11] shows the distributions corresponding to the two regions, and whereas for the worldwide case the differences in the distributions for different M'_c are compatible with their error bars, for the California case there is a systematic deviation, implying a possible correlation.

In order to find the origin of this discrepancy we include an extra condition, which is to restrict the events to the case of large enough recurrence times, so we impose $\tau_i \geq 30$ min. In this case, the differences in Californian distributions become no significant, see Fig. 8 (right), which means that the significant correlations between consecutive magnitudes are restricted to short recurrence times [34]. Therefore, we conclude that the Gutenberg–Richter law is valid independently of the value of the preceding magnitude, provided that short times are not considered. This is in agreement with the usual assumption in the ETAS model, in which magnitudes are generated from the Gutenberg–Richter distribution with total independence of the rest of the process [18, 19]. Of course, this independence is established within the errors associated to our finite sample. It could be that the dependence between the magnitudes is weak enough for that the changes in distribution are not larger than the uncertainty. With our analysis, only a much larger data set could unmask this hypothetical dependence.

The deviations for short times may be an artifact due to the incompleteness of earthquake catalogs at short time scales, for which small events are not recorded. Helmstetter et al. [35] propose a formula for the magnitude of completeness in Southern California as a function of the elapsed time since a mainshock and its magnitude; applying it to our stationary periods, for which the larger earthquakes have magnitudes ranging from 5 to 6, we obtain that a time of about 5 hours is necessary in order that the magnitude of completeness reaches a value below 2 after a mainshock of magnitude 6. After mainshocks of magnitude 5.5 and 5 this time reduces to about 1 hour and 15 min, extrapolating Helmstetter et al.'s results. In any case, it is perfectly possible that large mainshocks (not necessarily the preceding event) induce the loss of small events in the record and are the responsible of the deviations from the Gutenberg–Richter law at small magnitudes for short times. If an additional physical mechanism is behind this behavior, this is a question that cannot be answered with this kind of analysis.

As in the previous subsection, we have performed measurements of the conditional distributions for worldwide seismicity involving different M_i and M_j, separated up to 10 events, with no significant variations in the distributions, as expected. This confirms the independence of the magnitude M_i with its own history.

These results, together with those of the previous subsection allow to state that *an earthquake does not "know" how big is going to be* (at least from the information recorded at the catalogs, disregarding spatial structure, and with our present resolution) [11, 34].

5.4 Correlations Between Recurrence Times and Distances

Up to now we have considered seismicity as a point process in time, marked by the magnitude. If, in addition to this, we take into account the spatial degrees of freedom, these new variables allow to study other types of correlations. Of outstanding importance will be the distances between earthquakes, and specially the distances between consecutive earthquakes, which we may call jumps.

The correlations between jumps and recurrence times have been investigated in [36], and they show a curious behavior. There are two kinds of recurrence time distributions conditioned to the distance, one for short distances, which can be represented by a gamma distribution with a decaying power law of exponent around 0.8, and the distribution for long distances, which is an exponential. It is clear then that in one case we are dealing with aftershocks and in the other with Poissonian events. The particularity of these distributions is that they are independent on the distances, provided that the set of values of the distances are short, or long. For worldwide earthquakes, the difference between short and long distances is around 2° (200 km), whereas for Southern California this value is 0.1°, approximately.

We may note that the (unconditional) distribution of recurrence times is then a mixture of these two kind of conditional distributions, and therefore, the existence of a universal recurrence-time distribution is a consequence of a constant proportion of short and long distances in seismicity, or of aftershocks and uncorrelated events.

6 Bak et al's Unified Scaling Law

Bak, Christensen, Danon, and Scanlon introduced a different way to study recurrence times in earthquakes [3]. They divided the region of Southern California into approximately equally-sized squared subregions (when longitude and latitude are taken as rectangular coordinates), and computed the series of recurrence times for each subregion. The main difference with the procedure explained in the previous sections is that Bak et al. included all the series of recurrence times into a unique recurrence-time distribution, performing therefore a mixing of the distributions for all subregions. As seismic rate displays large variations in space (compare the rates of occurrence in Tokyo and in Moscow, and the same happens at smaller scales) Bak et al.'s procedure leads to a very broad distribution of recurrence times.

It was found that the recurrence-time densities defined in this way, $\mathcal{D}(\tau; M_c, \ell)$, for different magnitude thresholds M_c and different linear size ℓ of the subregions, verify the following scaling law (see Fig. 9 (left)),

$$\mathcal{D}(\tau; M_c, \ell) = \mathcal{R} F(\mathcal{R}\tau),$$

which was named *unified scaling law*, where F is the scaling function, showing a power-law decay with exponent close to 1 for small recurrence times and a different power-law decay for long times, with exponent around 2.2 [37], whereas $\mathcal{R}(M_c, \ell)$ is the spatial average of the mean seismic rate, i.e., the average of $R_{xy}(M_c, \ell)$ for all the regions with seismic activity (labeled by xy), so, $\mathcal{R} = \sum_{xy} R_{xy}/n$, where n is the number of such regions. Note that $\mathcal{R}$ is the inverse of the mean of $\mathcal{D}$. From the GR law for each region, $R_{xy} = R_{xy0} 10^{-bM_c}$ and from the fractal scaling of n with ℓ, $n = (\mathcal{L}/\ell)^{d_f}$, we get, $\mathcal{R} = R_0 (\ell/\mathcal{L})^{d_f} 10^{-bM_c}$, with $R_0(\mathcal{L}) = \sum_{xy} R_{xy0}(\ell)$ and $\mathcal{L}$ a rough measure of the linear size of the total area under study. Therefore we can write the scaling law as

$$\mathcal{D}(\tau; M_c, \ell) = \ell^{d_f} 10^{-bM_c} \tilde{F}(\ell^{d_f} 10^{-bM_c} \tau),$$

which relates the recurrence-time density, defined in the Bak et al.'s way, with the GR law and with the fractal distribution of epicenters, and from here the name of unified scaling law. Molchan and Kronrod have studied this law in the framework of multifractals [38].

Later it was found that the unified scaling law holds beyond the case of Southern California, for instance for Japan, Spain, New Zealand, New Madrid (USA), or Iceland, as well as worldwide [39, 40]. However, it turned out that the scaling function is not universal, as there are differences for different regions, mainly in the crossover between short and long times, although the value of the long-time power-law exponent seems to be in all cases 2.2, and therefore universal, see Fig. 9 (left). The deviations from the hyperbolic-like behavior (exponent close to one) for very short times have also been studied [40].

It is clear that the short-time exponent must be related (but not identical!) to the Omori p-value; on the other hand, the long-time exponent is a consequence of a power law distribution of seismic rates in space, as we now show [41]. Therefore, with the purpose of understanding the relation of Bak et al.'s results with the rest of this work, let us generalize the nonhomogeneous Poisson-Omori process previously introduced, in order to include the universal scaling law for Omori sequences. We have explained that these sequences can be characterized by an r-dependent recurrence-time probability density of the form $D(\tau|r) \propto r^\gamma \tau^{\gamma-1} e^{-r\tau/a}$ (note that this includes the nonhomogeneous Poisson process, given by $\gamma = a = 1$, but for real field data $\gamma \simeq 0.7$). We expect that, for a given spatial area, this is valid not only for Omori sequences but also for a general time-varying rate; then the overall probability density of the recurrence times, independently of r, is given by the mixing of all $D(\tau|r)$ [37],

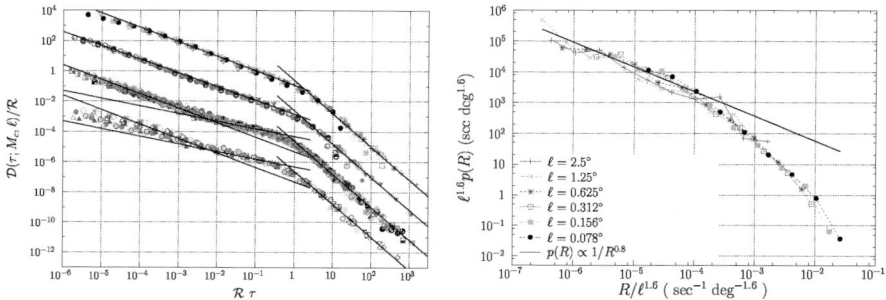

Fig. 9. (*Left*) Mixed recurrence-time densities (defined in the way of Bak et al.) and rescaled by $\mathcal{R}$, for several values of ℓ and M_c. The different sets of curves correspond, from top to bottom, to (i) Southern California, 1984–2001; (ii) Northern California, 1985–2003; (iii) Stationary seismicity: Southern California, 1988–1991; worldwide, 1973–2002; Japan, 1995–1998; and Spain, 1993–1997; (iv) Stationary seismicity: New Zealand, 1996–2001, and New Madrid, 1975–2002; A total of 84 distributions are shown, ℓ ranging from 0.039° to 45°, and $1.5 \leq M_c \leq 6$. The distributions are shifted to the bottom for clarity sake. All the left tails are fit by a (decreasing) power-law with exponent 2.2, the right part of the distributions are fit by a power law with exponent 0.95 or 0.9, see [39]. (*Right*) Distribution of mean rates of occurrence for events with $M \geq 2$ in Southern-California, 1984–2001 (averaging 1984–1992 and 1993–2001), using diverse values of ℓ. The distributions are rescaled by $L^{1.6}$ and the straight line is a power law fit which turns out to be $\propto 1/R^{0.8}$.

$$D(\tau|r_m) = \frac{1}{\mu} \int_{r_m}^{r_0} r D(\tau|r) \rho(r) dr, \qquad (27)$$

in fact, this is just (6); we recall that $\rho(r)$ is the density of rates, μ is the mean rate, r_0 is the maximum rate, and r_m is the minimum rate, related to the background seismicity level. Note that we have emphasized the dependence of the resulting distribution on r_m.

Let us consider that the distribution of rates comes essentially from Omori sequences, then, as we already know, $\rho(r) = C/r^{1+1/p}$. The analysis is simplified for $\gamma = 1/p$, although the conclusions will be of general validity, so, in this case,

$$D(\tau|r_m) \propto \frac{C}{\mu} \frac{(e^{-r_m \tau/a} - e^{-r_0 \tau/a})}{\tau^{2-1/p}}, \qquad (28)$$

where, in the same way as for a nonhomegeneous Poisson process, the minimum rate r_m determines the exponential tail of $D(\tau|r_m)$ for large τ, which is preceded by a decreasing power law with exponent $2 - 1/p$ if $r_0 \gg r_m$.

Up to now we have arrived to a slightly different, more convenient variation of the distribution corresponding to a nonhomogeneous Poisson process. Next step is to take into account the spatial degrees of freedom, fundamental in Bak et al.'s approach. In fact, as we have explained, their approach performs a mixing of recurrence times coming from different spatial areas (or subregions), which are characterized by disparate seismic rates. In particular, each area will have a different r_m, depending on its background seismicity level. As the minimum rate is difficult to measure (it depends on the size of the time intervals selected), we assume that the minimum rate r_m is somehow proportional to the mean rate of the sequence μ, which in turn is in correspondence with the mean rate in the area, R. This spatial heterogeneity of seismicity can be well described by a power-law probability density of mean rates R, $p(R) \propto 1/R^{1-\alpha}$, with $\alpha \simeq 0.2$, see Fig. 9 (right) and [37]; then,

$$p(r_m) \propto 1/r_m^{1-\alpha}$$

and therefore the recurrence-time probability density comes from the mixing,

$$\mathcal{D}(\tau) \propto \int_{r_{mm}}^{r_{mM}} r_m D(\tau|r_m) p(r_m) dr_m \qquad (29)$$

where r_m varies between r_{mm} and r_{mM}. Integration, taking into account that C/μ depends on r_m, leads, for $r_{mm}\tau \ll 1 \ll r_{mM}\tau$, to

$$\mathcal{D}(\tau) \propto 1/\tau^{2+\alpha} \qquad (30)$$

In this way the power law for long times, reflects the spatial distribution of rates. The universal value of the exponent $2+\alpha$ [39], would imply the universality of seismicity spatial heterogeneities. In consequence, Bak et al.'s unified scaling law provides a way to measure these properties. Further, (30) shows that the change of exponent in $\mathcal{D}(\tau)$ appears for τ larger than $1/r_{mM}$, which corresponds, for the area of highest seismicity, to the mean of events that are in the tail of the Omori sequence, or in background seismicity, and therefore at the onset of correlation with the mainshock. It is in this sense that the change of exponent separates events with different correlation. On the other hand, the power law for short times is not affected by the spatial mixing and therefore $\mathcal{D} \propto 1/\tau^{2-1/p}$.

7 Conclusions

We hope we have convinced the reader about the interest to study of the temporal features of seismicity. The research of the author was illuminated by the ideas and philosophy of the late Per Bak.

References

1. *Earthquake Science and Seismic Risk Reduction*, Ed. F. Mulargia, R. J. Geller, Kluwer, Dordrecht (2003)
2. J. B. Rundle, D. L. Turcotte, R. Shcherbakov, W. Klein, C. Sammis, Rev. Geophys. **41**, 1019 (2003)

3. P. Bak, K. Christensen, L. Danon, T. Scanlon, Phys. Rev. Lett. **88**, 178501 (2002)
4. Y. Y. Kagan, Physica D **77**, 160 (1994)
5. D. L. Turcotte, *Fractals and Chaos in Geology and Geophysics*, Cambridge University Press, Cambridge (1997)
6. T. Utsu, Pure Appl. Geophys. **155**, 509 (1999)
7. T. Utsu, Statistical Features of Seismicity. In: *International Handbook of Earthquake and Engineering Seismology, Part A*, Ed. W.H. K. Lee, H. Kanamori, P.C. Jennings, C. Kisslinger, pp. 719–732, Academic Press, Amsterdam (2002)
8. P. A. Reasenberg, L. M. Jones: Science **243**, 1173 (1989)
9. T. Utsu, Y. Ogata, R.S. Matsu'ura, J. Phys. Earth **43**, 1 (1995)
10. A. Corral, Nonlinear Proc. Geophys. **12**, 89 (2005)
11. A. Corral, Tectonophys., accepted (2006)
12. P. Bak, *How Nature Works: The Science of Self-Organized Criticality* (Copernicus, New York, 1996)
13. P. Shearer, E. Hauksson, G. Lin, D. Kilb, Eos Trans. AGU **84** (46): Fall Meet. Suppl., Abstract S21D–0326 (2003)http://www.data.scec.org/ftp/ catalogs/SHLK/
14. A. Corral, Phys. Rev. Lett. **92**, 108501 (2004)
15. *Handbook of Mathematical Functions*, Ed. M. Abramowitz, I.A. Stegun, Dover, New York (1965)
16. R. Shcherbakov, G. Yakovlev, D.L. Turcotte, J.B. Rundle, Phys. Rev. Lett. **95**, 218501 (2005)
17. A. Corral, Phys. Rev. E **71**, 017101 (2005)
18. Y. Ogata, Pure Appl. Geophys. **155**, 471 (1999)
19. A. Helmstetter, D. Sornette, Phys. Rev. E **66**, 061104 (2002)
20. J.D. Kalbfleisch, R.L. Prentice, *The Statistical Analysis of Failure Time Data*, Wiley, New York (1980)
21. Y.Y. Kagan, D.D. Jackson, Geophys. J. Int. **104**, 117 (1991)
22. W. Feller: *An Introduction to Probability Theory and Its Applications*, 2nd edn., vol 2, Wiley, New York (1971)
23. G.J. Székely, *Paradoxes in Probability Theory and Mathematical Statistics*, Reidel, Dordrecht (1986)
24. M. Schroeder, *Fractals, Chaos, Power Laws*, Freeman, New York (1991)
25. P.M. Davis, D.D. Jackson, Y.Y. Kagan, Bull. Seismol. Soc. Am. **79**, 1439 (1989)
26. D. Sornette, L. Knopoff, Bull. Seismol. Soc. Am. **87**, 789 (1997)
27. D.J. Daley, D. Vere-Jones, *An Introduction to the Theory of Point Processes*, Springer, New York (1988)
28. L.P. Kadanoff, *Statistical Physics: Statics, Dynamics and Renormalization*, World Scientific, Singapore (2000)
29. W.D. McComb, *Renormalization Methods*, Clarendon Press, Oxford (2004)
30. K. Christensen, N.R. Moloney, *Complexity and Criticality*, Imperial College Press, London (2005)
31. A. Corral, Phys. Rev. Lett. **95**, 028501 (2005)
32. V. Livina, S. Tuzov, S. Havlin, A. Bunde, Physica A **348**, 591 (2005)
33. V.N. Livina, S. Havlin, A. Bunde, Phys. Rev. Lett. **95**, 208501 (2005)
34. A. Corral, Phys. Rev. Lett. **95**, 159801 (2005)
35. A. Helmstetter, Y.Y. Kagan, D.D. Jackson, Bull. Seismol. Soc. Am. **96**, 90 (2006)

36. A. Corral, Universal earthquake-occurrence jumps, correlations with time, and anomalous diffusion. Submitted (2006)
37. A. Corral, Phys. Rev. E **68**, 035102 (2003)
38. G. Molchan, T. Kronrod, Geophys. J. Int. **162**, 899 (2005)
39. A. Corral, Physica A **340**, 590 (2004)
40. J. Davidsen, C. Goltz, Geophys. Res. Lett. **31**, L21612 (2004)
41. A. Corral, K. Christensen, Phys. Rev. Lett. **96**, accepted (2006)

Spatiotemporal Correlations of Earthquakes

H. Kawamura

Department of Earth and Space Science, Faculty of Science, Osaka University,
Toyonaka 560-0043, Japan
kawamura@ess.sci.osaka-u.ac.jp

Statistical properties of earthquakes are studied both by the analysis of real earthquake catalog of Japan and by numerical computer simulations of the spring-block model in both one and two dimensions. Particular attention is paid to the spatiotemporal correlations of earthquakes, e.g., the recurrence-time distribution or the time evolution before and after the mainshock of seismic distribution functions, including the magnitude distribution and the spatial seismic distribution. Certain eminent features of the spatiotemporal correlations, e.g., foreshocks, aftershocks, swarms and doughnut-like seismic pattern, are discussed.

1 Introduction

Although earthquakes are obviously complex phenomena, the basic physical picture of earthquakes seems to have been well established now: Earthquake is a stick-slip frictional instability of a fault driven by steady motions of tectonic plates [1, 2]. Although it remains to be extremely difficult at the present stage to say something really credible for each individual earthquake event, if one collects many events and take average over these events, a clear tendency often shows up there. Thus, it is sometimes possible to say something credible for the *average* or *statistical* properties of earthquakes, or more precisely, sets of earthquakes.

In this article, we wish to review some of our recent studies on the statistical properties of earthquakes. Our study of earthquakes is motivated by the following three issues.

Critical versus Characteristic

It has long been known empirically that certain power-laws often appear in the statistical properties of earthquakes, e.g., the Gutenberg–Richter (GR) law for the magnitude distribution of earthquakes, or the Omori law for the time evolution of the frequency of aftershocks [2]. Power-law means that there is no characteristic scale in the underlying physical phenomenon. In statistical physics, one of the most widely recognized occasion of the appearance of a power-law is critical phenomena associated with a thermodynamic second-order phase transition. Indeed, inspired by this analogy, Bak and collaborators introduced the concept of the "self-organized criticality (SOC)" into earthquakes [3, 4]. According to this picture, The Earth's crust is always in the critical state generated dynamically, and power-laws associated with statistical properties of earthquakes are regarded as realizations of the intrinsic critical nature of earthquakes. Indeed, the SOC idea gives a natural explanation of the scale-invariant power-law behaviors frequently observed in earthquakes, including the GR law and the Omori law.

In contrast to such an SOC view of earthquakes, an opposite view has also been common in earthquake studies, a view which regards earthquakes as "characteristic". In this view, earthquakes are supposed to possess their own characteristic scales, e.g., a characteristic energy scale or a characteristic time scale.

Thus, whether an earthquake is critical or characteristic remains to be one of central issues of modern earthquake studies.

Possible Precursory Phenomena – Spatiotemporal Correlations of Earthquakes

In conjunction with earthquake prediction, possible precursory phenomena associated with large earthquakes have special importance. If one takes a statistical approach, a natural quantity to be examined might be spatiotemporal correlations of seismicity, i.e., how earthquakes correlate in space and time. If we could identify the property in which a clear anomaly is observed preceding the large event, it might be useful for earthquake prediction. In the present article, we wish to investigate among others various types of spatiotemporal correlation functions of earthquakes.

We note that, in 1980's, Kagan and Knopoff performed a pioneering study of the spatial moment structure of earthquake epicenters for various seismic catalogs via two-, three-, and four-point moment functions, to find that the spatial distribution of earthquakes has a scale-invariant pattern [5].

The Constitutive Relation and the Nature of Stick-slip Dynamics

Since earthquakes can be regarded as a stick-slip frictional instability of a pre-existing fault, the statistical properties of earthquakes should be governed by the physical law of rock friction [1, 2]. Unfortunately, our present understanding of physics of friction is still poor. We do not have precise knowledge of the constitutive relation governing the stick-slip dynamics at earthquake faults. The difficulty lies partly in the fact that a complete microscopic theory of friction is still not available, but also in the fact that the length and time scales relevant to earthquakes are so large that the applicability of laboratory experiments of rock friction is not necessarily clear.

A question of fundamental importance in earthquake studies might be how the properties of earthquakes depend on the constitutive relation and the material parameters characterizing earthquake faults, including the elastic properties of the crust and the constitutive parameters characterizing the friction force.

To answer this question and to get deeper insight into the physical mechanism governing the stick-slip process of earthquakes, a proper modeling of an earthquake might be an important step. In fact, earthquake models of various levels of simplifications have been proposed in geophysics and statistical physics, and their statical properties have been extensively studied mainly by means of numerical computer simulations. One of the standard model is the so-called spring-block model originally proposed by Burridge and Knopoff (BK model) [6], which we will employ in the present particle. Yet, our present understanding of the question how the earthquake properties depend on the constitutive relation and the material parameters characterizing earthquake faults remains far from satisfactory.

In the present article, in order to approach the three goals mentioned above, we take two complementary approaches: In one, we perform numerical computer simulations based on the spring-block model to clarify the spatiotemporal correlations of seismic events. Both the one-dimensional (1D) BK model [7, 8] and the two-dimensional (2D) BK model [9] are studied. In the other, we analyze the earthquake catalog of Japan to examine the spatiotemporal correlations of real seismicity [10]. We then compare the results of numerical model simulation and the analysis of real earthquake catalog, hoping that such a comparison might give us useful information about the nature of earthquakes.

The following part of the article is organized as follows. In Sect. 2, we introduce the model employed in our numerical computer simulations, and explain some the details of the simulations. We also introduce the earthquake catalog used in our analysis of real seismicity of Japan. Then, in Sect. 3, we report on the results of our analysis of the spatiotemporal correlations of real earthquakes based on the earthquake catalog of Japan, together with the results of our numerical simulations of the 1D and 2D BK models. The statistical properties examined in this section include; (i) the magnitude distribution, (ii) the local recurrence-time distribution, (iii) the global recurrence-time distribution, (iv) the time evolution of the spatial seismic distribution before the mainshock, (v) the time evolution of the spatial seismic distribution after the mainshock, and (vi) the time-resolved magnitude distribution before and after the mainshock. Finally, Sect. 4 is devoted to summary and discussion of our results.

2 The Model, the Simulation and the Catalog

In this section, we introduce the spring-block model which we will use in our numerical computer simulation, and explain some of the details of the simulation. We also introduce the seismic catalog of Japan which we will use in our analysis of the spatiotemporal correlations of real earthquakes. The results of our numerical computer simulations and the analysis of seismic catalog of Japan will be presented in the following Sect. 3.

2.1 The Spring-Block Model of Earthquakes

The spring-block model of earthquakes was originally proposed by Burridge and Knopoff [6]. In this model, an earthquake fault is simulated by an assembly of blocks, each of which is connected via the elastic springs to the neighboring blocks and to the moving plate. All blocks are subject to the friction force, the source of the nonlinearity in the model, which eventually realizes an earthquake-like frictional instability. The model contains several parameters representing, e.g., the elastic properties of the crust and the frictional properties of faults.

In the 1D BK model, the equation of motion for the i-th block can be written as

$$m\ddot{U}_i = k_p(\nu' t' - U_i) + k_c(U_{i+1} - 2U_i + U_{i-1}) - \Phi_i \,, \tag{1}$$

where t' is the time, U_i is the displacement of the i-th block, ν' is the loading rate representing the speed of the plate, and Φ_i is the friction force working at the block i.

In order to make the equation dimensionless, we measure the time t' in units of the characteristic frequency $\omega = \sqrt{k_p/m}$ and the displacement U_i in units of the length $L = \Phi_0/k_p$, Φ_0 being the static friction. Then, the equation of motion can be written in the dimensionless form as

$$\ddot{u}_i = \nu t - u_i + l^2(u_{i+1} - 2u_i + u_{i-1}) - \phi_i \,, \tag{2}$$

where $t = t'\omega$ is the dimensionless time, $u_i \equiv U_i/L$ is the dimensionless displacement of the i-th block, $l \equiv \sqrt{k_c/k_p}$ is the dimensionless stiffness parameter, $\nu = \nu'/(L\omega)$ is the dimensionless loading rate, and $\phi_i \equiv \Phi_i/\Phi_0$ is the dimensionless friction force working at the block i.

The form of the friction force ϕ is specified by the constitutive relation. As mentioned, this part is a vitally important, yet largely ambiguous part in the proper description of earthquakes. In order for the model to exhibit a dynamical instability corresponding to an earthquake, it is essential that the friction force ϕ possesses a frictional *weakening* property, i.e., the friction should become weaker as the block slides.

As the simplest form of the friction force, we assume here the form used by Carlson et al., which represents the velocity-weakening friction force [11, 12].

Namely, the friction force $\phi(\dot{u})$ is assumed to be a single-valued function of the velocity $\dot{u}_i$, i.e., ϕ_i, gets smaller as $\dot{u}_i$ increases,

$$\phi(\dot{u}) = \begin{cases} (-\infty, 1], & \text{for } \dot{u}_i \leq 0, \\ \dfrac{1-\sigma}{1+2\alpha\dot{u}_i/(1-\sigma)}, & \text{for } \dot{u}_i > 0, \end{cases} \qquad (3)$$

where its maximum value corresponding to the static friction has been normalized to unity. As noted above, this normalization condition $\phi(\dot{u}=0)=1$ has been utilized to set the length unit L. The back-slip is inhibited by imposing an infinitely large friction for $\dot{u}_i < 0$, i.e., $\phi(\dot{u} < 0) = -\infty$.

In this velocity-weakening constitutive relation, the friction force is characterized by the two parameters, σ and α. The former, σ, represents an instantaneous drop of the friction force at the onset of the slip, while the latter, α, represents the rate of the friction force getting weaker on increasing the sliding velocity. In our simulation, we regard σ to be small, and fix $\sigma = 0.01$.

It should be emphasized again that, although the above Carlson-Langer velocity-weakening friction is rather simple and has been widely used in numerical simulations, the real constitutive relation might not be so simple with features possibly different from the simplest velocity-weakening one. Indeed, there have been several other proposals for the law of rock friction, e.g., the slip-weakening friction force [1, 2, 13, 14] or the rate- and state-dependent friction force [1, 2, 15, 16, 17, 18, 19]. In this article, we leave the study of these different constitutive relations to other references or to future studies, and assume the simplest velocity-weakening friction force given above.

We also assume the loading rate ν to be infinitesimally small, and put $\nu = 0$ during an earthquake event, a very good approximation for real faults [11, 12]. Taking this limit ensures that the interval time during successive earthquake events can be measured in units of ν^{-1} irrespective of particular values of ν. Taking the $\nu \to 0$ limit also ensures that, during an ongoing event, no other event takes place at a distant place, independently of this ongoing event.

The extension of the 1D BK model to 2D is rather straightforward. In 2D, the blocks are considered to be arranged in the form of a square array connected with the springs of the spring constant k_c [20]. We consider the isotropic and uniform case where k_c is uniform everywhere in the array independent of the spatial directions. All blocks are connected to the moving plate via the springs of the spring constant k_p, and are also subject to the velocity-weakening friction force defined above. The plate is driven along the x-direction with a constant rate ν. It is assumed that all blocks move along the x-direction only, i.e., the displacement of the block at the position (i_x, i_y) is assumed to be given by $\mathbf{u}(i_x, i_y) = (u_x(i_x, i_y), 0)$.

Since the early study by Burridge and Knopoff [6], the properties of this BK model has been studied by numerical simulations.

Carlson, Langer and collaborators performed a pioneering study of the statistical properties of the 1D BK model quite extensively [11, 12, 20, 21, 22], with particular attention to the magnitude distribution of earthquake events. It was observed that, while smaller events persistently obeyed the GR law, i.e., staying critical or near-critical, larger events exhibited a significant deviation from the GR law, being off-critical or "characteristic" [11, 20, 21, 22, 23]. Shaw, Carlson and Langer studied the same model by examining the spatiotemporal patterns of seismic events preceding large events, observing that the seismic activity accelerates as the large event approaches [24].

The BK model was also extended in several ways, e.g., taking account of the effect of the viscosity [25, 26, 27], modifying the form of the friction force [25, 27], or taking account of the long-range interactions [28]. The 2D version of the BK model was also analyzed by Carlson [20] and by Myers et al. [14].

The study of statistical properties of earthquakes was promoted in early nineties, inspired by the work by P. Bak and collaborators who emphasized the concept of "self-organized criticality (SOC)" [3, 4]. The SOC idea was developed mainly on the basis of the cellular-automaton versions of the earthquake model [3, 4, 29, 30, 31, 32, 33, 34, 35]. The statistical properties of these cellular-automaton models were also studied quite extensively, often interpreted within the SOC framework. These models apparently reproduce several fundamental features of earthquakes such as the GR law, the Omori law of aftershocks, the existence of foreshocks, etc. We note that, although many of these cellular-automaton models were originally introduced to mimic the spring-block model, their statistical properties are not always identical with the original spring-block model. Furthermore, as compared with the spring-block model, the cellular-automaton models are much more simplified so that the model does not have enough room to represent various material properties of the earthquake fault in a physically appealing way. Thus, the spring-block model has an advantage over the cellular-automaton models that the dependence on the material parameters, including the constitutive and elastic properties, are more explicit.

2.2 The Numerical Simulation

As already mentioned, numerical model simulation is a quite useful tool for our purposes: First, generating huge number of events required to attain the high precision for discussing the statistical properties of earthquakes is easy to achieve in numerical simulations whereas it is often difficult to achieve in real earthquakes, especially for larger ones. Second, various material parameters characterizing earthquake faults are extremely difficult to control in real faults, whereas they are easy to control in model simulations.

We solve the equation of motion (2) by using the Runge-Kutta method of the fourth order. The width of the time discretization Δt is taken to be $\Delta t \nu = 10^{-6}$. We have checked that the statistical properties given below are unchanged even if we take the smaller Δt. Total number of 10^7 events are generated in each run, which are used to perform various averagings. In calculating the observables, initial 10^4 events are discarded as transients.

In order to eliminate the possible finite-size effects, the total number of blocks N are taken to be large. In 1D, we set $N = 800$, imposing the periodic boundary condition. The size dependence of the results was examined in [8] with varying N in the range $800 \leq N \leq 6400$. In 2D, we set our lattice size 160×80 with the periodic boundary condition on the longer side, and the free boundary condition on the shorter side.

We study the properties of the model, with varying the frictional parameter α and the elastic parameter l. In this article, attention is paid to the dependence on the parameter α, since the parameter α, which represents the extent of the frictional weakening, turns out to affect the result most significantly [8].

2.3 The Seismic Catalog of Japan

In order to compare the simulation data on the 1D and 2D BK models with the corresponding data for real earthquakes, we analyze in the following section the seismic catalog of Japan provided by Japan University Network Earthquake Catalog (JUNEC) available at http://kea.eri.u-tokyo.ac.jp/CATALOG/junec/monthly.html. The catalog covers earthquakes which occurred in Japan area during July 1985 and December 1998. Total of 199,446 events are recorded in the catalog. As an example, we show in Fig. 1, a seismicity map of Japan generated from the JUNEC catalog, where all large earthquakes of their magnitudes greater than five, which occurred in Japan area during 1985–1998, are mapped out [36].

3 Statistical Properties of Earthquakes

In this section, we show the results of our numerical simulations of the 1D and 2D BK models, and compare them with the results of our analysis of the seismic catalog of Japan (JUNEC catalog). We study several observables, i.e., (i) the magnitude distribution, (ii) the local recurrence-time distribution, (iii) the global recurrence-time distribution, (iv) the time evolution of the spatial seismic distribution before the mainshock, (v) the time evolution of the spatial seismic distribution after the mainshock, and (vi) the time-resolved magnitude distribution before and after the mainshock. We show these results consecutively below.

3.1 The Magnitude Distribution

In Fig. 2 we show the magnitude distribution $R(m)$ of earthquakes of Japan generated from the JUNEC catalog, where $R(m)\mathrm{d}m$ represents the rate of events with their magnitudes in the range $[m, m + \mathrm{d}m]$. The data lie on a straight line fairly well for $m > 3$, obeying the GR law. The slope of the straight line gives the power-law exponent about $b \simeq 0.9$.

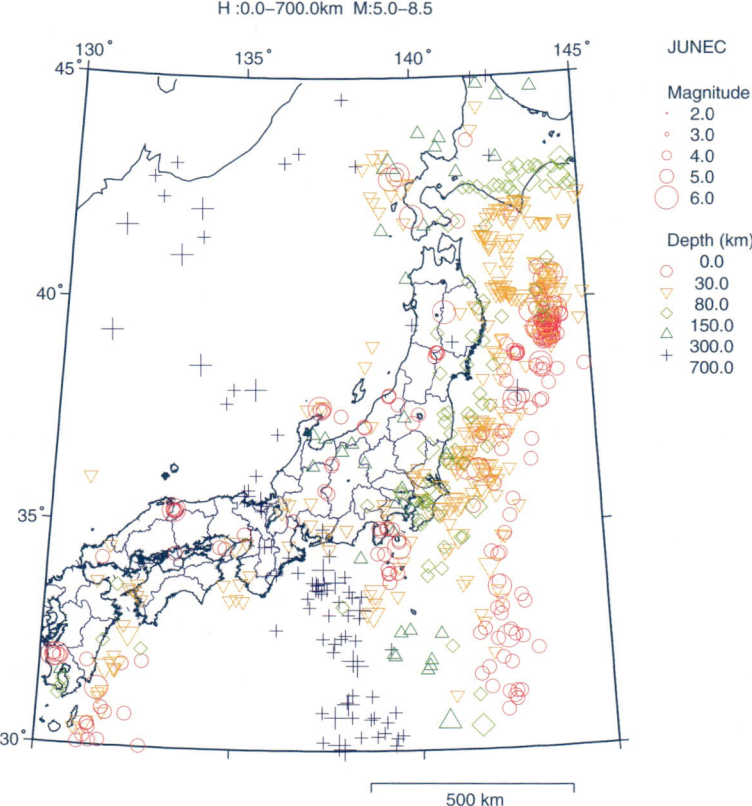

Fig. 1. A seismicity map of Japan generated from the JUNEC catalog. Large earthquakes of their magnitudes greater than five, which occurred in Japan area during 1985–1998, are mapped out by using the program developed by H. Tsuruoka [36]

In Fig. 3(a), we show the magnitude distribution $R(\mu)$ of earthquakes calculated from our numerical simulation of the 1D BK model. The parameter α is varied in the range $0.25 \leq \alpha \leq 10$, while the elastic parameter l is fixed to $l = 3$.

In the BK model, the magnitude of an event, μ, is defined as the logarithm of the moment M_0, i.e.,

$$\mu = \ln M_0, \quad M_0 = \sum_i \Delta u_i, \tag{4}$$

where Δu_i is the total displacement of the i-th block during a given event and the sum is taken over all blocks involved in the event [11]. It should be noticed that the absolute numerics of the magnitude value of the BK model μ has no direct quantitative correspondence to the magnitude m of real earthquakes.

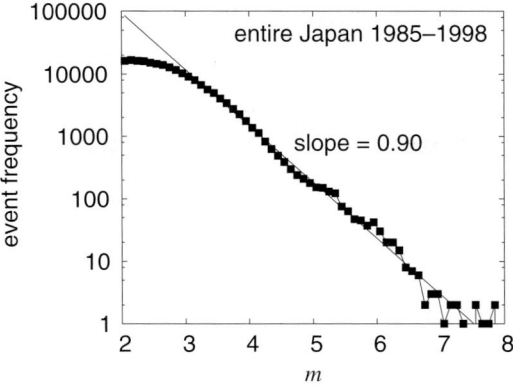

Fig. 2. The magnitude distribution of earthquakes in Japan generated from the JUNEC catalog. The data for $m > 3$ lie on a *straight line* with the GR exponent $b \simeq 0.9$

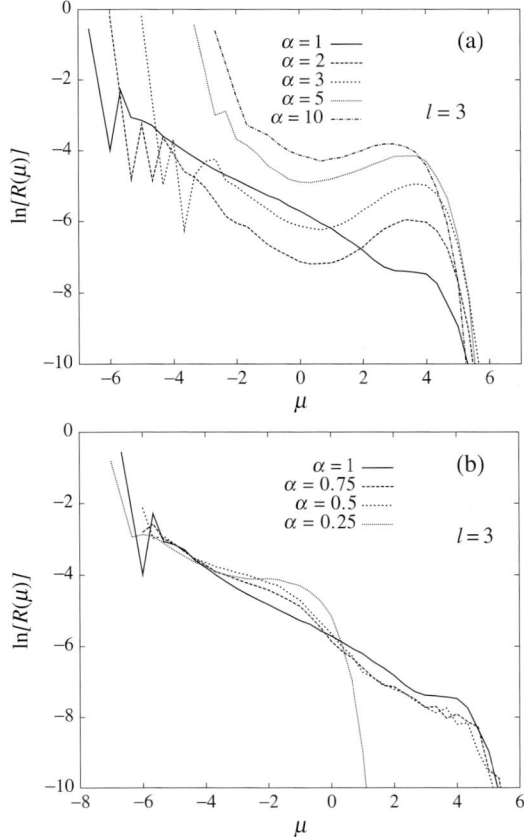

Fig. 3. The magnitude distribution of seismic events calculated for the 1D BK model, for the range of larger $\alpha \geq 1$ (**a**), and for the range of smaller $\alpha \leq 1$ (**b**)

As can be seen from Fig. 3, the form of the calculated magnitude distribution depends on the α-value considerably. The data for $\alpha = 1$ lie on a straight line fairly well, apparently satisfying the GR law. The value of the exponent B describing the GR-like power-law behavior, $\propto 10^{-B}$, is estimated to be $B \simeq 0.50$ corresponding to $b \simeq 0.75$, which is slightly smaller than the b-value observed for the JUNEC catalog $b \simeq 0.9$. Remember the relation $b = \frac{3}{2}B$.

On the other hand, the data for larger α, i.e., the ones for $\alpha \geq 2$ deviate from the GR law at larger magnitudes, exhibiting a pronounced peak structure, while the power-law feature still remains for smaller magnitudes. These features of the magnitude distribution are consistent with the earlier observation of Carlson and Langer [11, 21]. It means that, while smaller events exhibit self-similar critical properties, larger events tend to exhibit off-critical or characteristic properties, much more so as the velocity-weakening tendency of the friction is increased. The observed peak structure gives us a criterion to distinguish large and small events. Below, we regard events with their magnitudes μ greater than $\mu_c = 3$ as large events of the 1D BK model, $\mu_c = 3$ being close to the peak position of the magnitude distribution of Fig. 3(a). In an earthquake with $\mu = 3$, the mean number of moving blocks are about 76 ($\alpha = 1$) and 60 ($\alpha = 2, 3$).

As can be seen from Fig. 3(b), the data at smaller $\alpha < 1$ exhibit considerably different behaviors from those for $\alpha > 1$. Large events are suppressed here. For $\alpha = 0.25$, in particular, all events consist almost exclusively of small events only. This result might be consistent with the earlier observation which suggested that the smaller value of $\alpha < 1$ tended to cause a creeping-like behavior without a large event [21]. In particular, Vasconcelos showed that a single block system exhibited a "first-order transition" at $\alpha = 0.5$ from a stick-slip to a creep [37], whereas this discontinuous transition becomes apparently continuous in many-block system [38, 39]. Since we are mostly interested in large seismic events here, we concentrate in the following on the parameter range $\alpha \geq 1$. In contrast to the parameter α, the magnitude distribution turns out to be less sensitive to the stiffness parameter l. Further details of the l-dependence is given in [8].

In Fig. 4, we show the magnitude distribution $R(\mu)$ of the 2D BK model with varying the α value. At larger magnitudes, a deviation from the GR law is observed for any value of α, i.e., the calculated magnitude distribution exhibits a peak structure at larger magnitude irrespective of the α-value, suggesting that larger earthquakes tend to be characteristic. From the observed peak structure, we regard events with their magnitudes μ greater than $\mu_c = 5$ as large events of the 2D BK model.

While the BK model tends to reproduce the GR law for earthquakes of smaller magnitudes, its relevance to the GR law observed in real seimicity is not entirely clear. It has been suggested that the GR b-value might be related to the fractal dimension of the fault interface [40, 41, 42]. More recently,

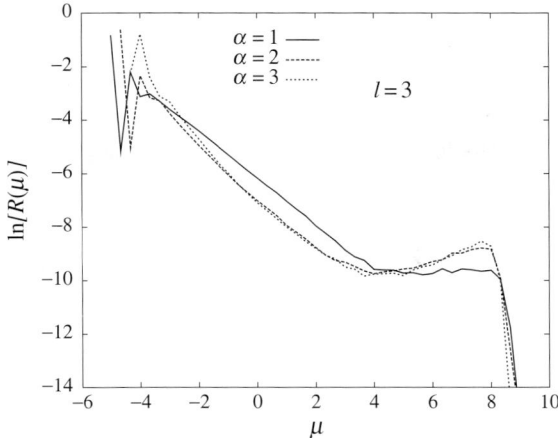

Fig. 4. The magnitude distribution of seismic events calculated for the 2D BK model

Chakrabarti and collaborators proposed an earthquake model in which the magnitude distribution of earthquakes is related to the contact area distribution between the two fractal surfaces of the plates [43]. In these approaches, the intrinsic nonuniformity at earthquake faults plays an essential role in realizing the GR law and the critical nature of earthquakes. Whether this is really true, as well as its possible relation to the BK model where the nonuniformity is apparently absent, or at least not explicit, needs to be examined further.

3.2 The Local Recurrence-Time Distribution

A question of general interest may be how large earthquakes repeat in time, do they occur near periodically or irregularly? One may ask this question either locally, i.e., for a given finite area on the fault, or globally, i.e., for an entire fault system. The picture of characteristic earthquake presumes the existence of a characteristic recurrence time. In this case, the distribution of the recurrence time of large earthquakes, T, is expected to exhibit a peak structure at such a characteristic time scale. If the SOC concept applies to large earthquakes, by contrast, such a peak structure would not show up.

In the upper panel of Fig. 5, we show the distribution of the local recurrence time T of large earthquakes of Japan with $m \geq m_c = 4$, calculated from the JUNEC catalog. In defining the recurrence time locally, the subsequent large event is counted when a large event with $m \geq m_c = 4$ occurs with its epicenter lying within a circle of radius 30 km centered at the epicenter of the previous large event. The mean recurrence time $\bar{T}$ is then estimated to be 148 days. In the middle and lower panels of Fig. 5, the same data are re-plotted on a double-logarithmic scale [middle panel], and on a

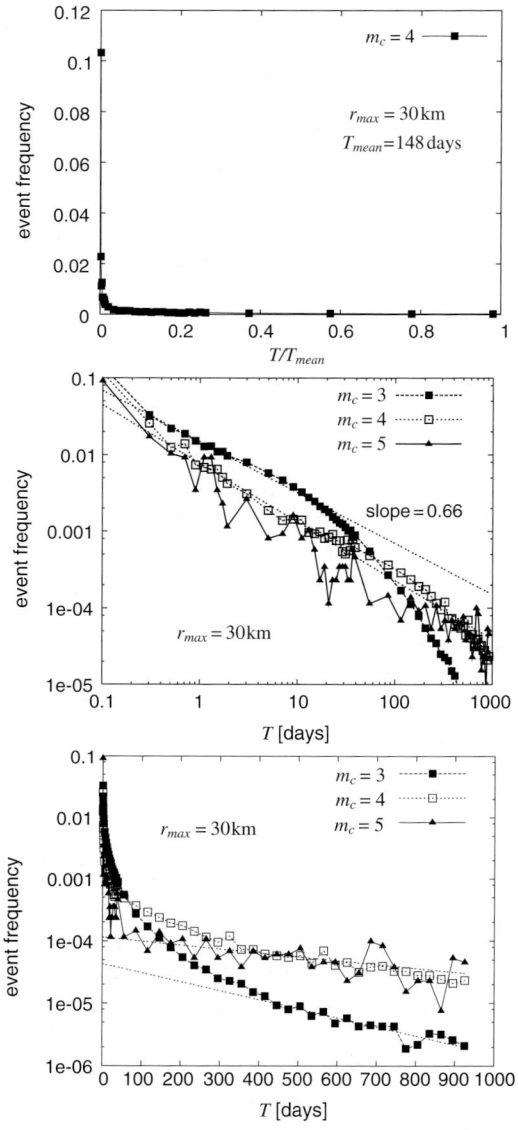

Fig. 5. The local recurrence-time distribution of large earthquakes in Japan generated from the JUNEC catalog. The distribution is given on a bare scale for large earthquakes with $m_c = 4$ [*upper figure*], on a double-logarithmic scale with $m_c = 3, 4$ and 5 [*middle figure*], and on a semi-logarithmic scale with $m_c = 3, 4$ and 5 [*lower figure*]

semi-logarithmic scale [lower panel], with the magnitude threshold $m_c = 3, 4$ and 5. One sees from these figures that the local recurrence-time distribution exhibits power-law-like critical features at shorter times, which seems to cross over to a faster exponential-like decay at longer times. The time range in which the data obey a power-law tends to get longer for larger earthquakes. The power-law exponent estimated for earthquakes of $m_c = 5$ in the time range $T < 1000$ days is about $\frac{2}{3}$. This value is a bit smaller than the standard Omori exponent $\simeq 1$.

In the upper panel of Fig. 6, we show the distribution of the local recurrence time T of large earthquakes with $\mu \geq \mu_c = 3$, calculated for the 1D BK model. In the insets, the same data including the tail part are re-plotted on a semi-logarithmic scale. In defining the recurrence time locally in the BK model, the subsequent large event is counted when a large event occurs with its epicenter in the region within 30 blocks from the epicenter of the previous large event. The mean recurrence time $\bar{T}$ is then estimated to be $\bar{T}\nu = 1.47$, 1.12, and 1.13 for $\alpha = 1$, 2 and 3, respectively.

As can be seen from Fig. 6, the tail of the distribution is exponential at longer $T > \bar{T}$ for all values of α, while the form of the distribution at shorter $T < \bar{T}$ is non-exponential and differs between for $\alpha = 1$ and for $\alpha = 2$ and 3. For $\alpha = 2$ and 3, the distribution has an eminent peak at around $\bar{T}\nu \simeq 0.5$, not far from the mean recurrence time. This means the existence of a characteristic recurrence time, suggesting the near-periodic recurrence of large events. This characteristic behavior is in sharp contrast to the critical behavior without any peak structure observed in the JUNEC data. It should also be mentioned, however, that there were reports in the literature of a near-periodic recurrence of large events at several real faults [2, 44].

For $\alpha = 1$, by contrast, the peak located close to the mean $\bar{T}$ is hardly discernible. Instead, the distribution has a pronounced peak at a shorter time $\bar{T}\nu \simeq 0.10$, just after the previous large event. In other words, large events for $\alpha = 1$ tend to occur as "twins". This has also been confirmed by our analysis of the time record of large events. In fact, as shown in [8], a large event for the case of $\alpha = 1$ often occurs as a "unilateral earthquake" where the rupture propagates only in one direction, hardly propagating in the other direction. When a large earthquake occurs in the form of such a unilateral earthquake, further loading due to the plate motion tends to trigger the subsequent large event in the opposite direction, causing a twin-like event. This naturally explains the small-T peak observed in Fig. 6 for $\alpha = 1$.

In the lower panel of Fig. 6, the local recurrence-time distribution of large events is shown for the cases of $\alpha = 1$, with varying the magnitude threshold as $\mu_c = 2$, 3 and 4. As can be seen form this figure, the form of the distribution for $\alpha = 1$ largely changes with the threshold value μ_c. Interestingly, in the case of $\mu_c = 4$, the distribution has *two* distinct peaks, one corresponding to the twin-like event and the other to the near-periodic event. Thus, even in the case of $\alpha = 1$ where the critical features are apparently dominant

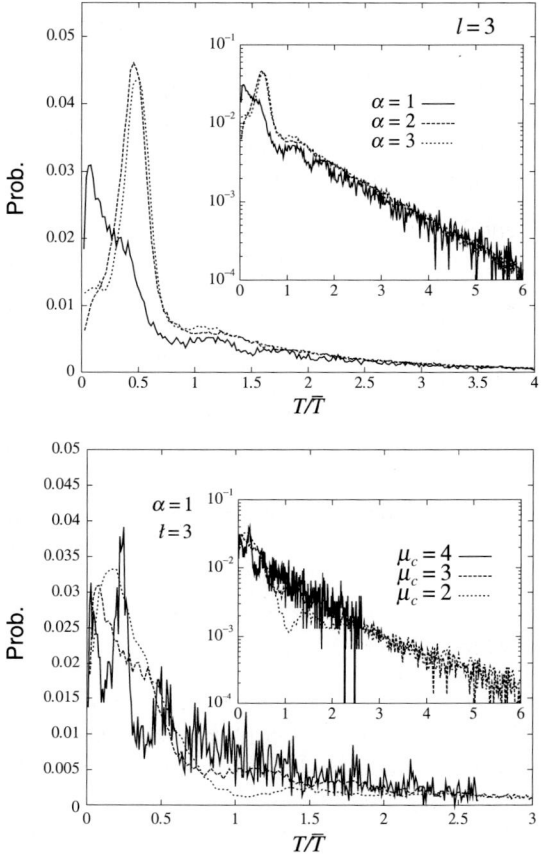

Fig. 6. The local recurrence-time distribution of large events of the 1D BK model. In the *upper figure*, the distribution is given for $\alpha = 1, 2$ and 3 with fixing $\mu_c = 3$, whereas, in the *lower figure*, the distribution is given with varying the magnitude threshold as $\mu_c = 2, 3$ and 4 with fixing $\alpha = 1$

for smaller thresholds $\mu_c = 2$ and 3, features of a characteristic earthquake becomes increasingly eminent when one looks at very large events.

In Fig. 7, the local recurrence-time distribution is shown for the 2D BK model. In defining the recurrence time locally in the 2D BK model, a subsequent large event is counted when a large event occurs with its epicenter lying within a circle of its radius 5 blocks centered at the epicenter of the previous large event. The mean recurrence time $\bar{T}$ is then estimated to be $\bar{T}\nu = 1.47$, 1.12, and 1.13 for $\alpha = 1, 2$ and 3, respectively. The behavior of the 2D model is similar to the 1D model, with enhanced characteristic features. The peak structure of the distribution is very prominent even for $\alpha = 1$.

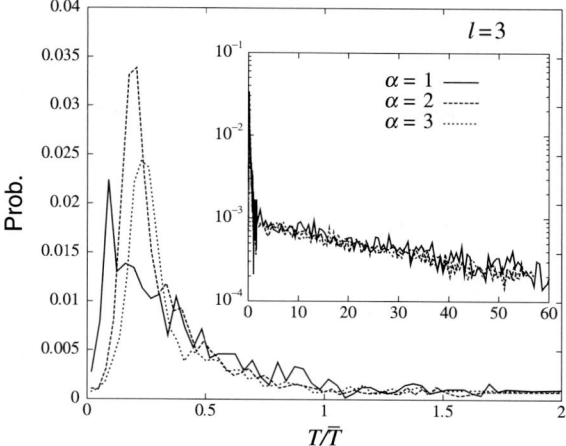

Fig. 7. The local recurrence-time distribution of large events of $\mu > \mu_c = 5$ of the 2D BK model

3.3 The Global Recurrence-Time Distribution

In this subsection, we examine the global recurrence-time distribution. By the word "global", we consider the situation where the region in which one identifies the next event is sufficiently wide, much wider than the typical size of the rupture zone of large events.

From the JUNEC catalog, we construct in the upper panel of Fig. 8 such *global* recurrence-time distribution for entire Japan for large earthquakes with $m_c = 4$. The mean recurrence time $\bar{T}$ is then estimated to be 0.73 days. In the middle and lower panels of Fig. 8, the same data are re-plotted on a double-logarithmic scale [middle], and on a semi-logarithmic scale [lower], with the magnitude threshold $m_c = 3, 4$ and 5. One sees from these figures that the global recurrence-time distribution of the JUNEC data is very much similar to the corresponding local recurrence-time distribution shown in Fig. 5: It exhibits power-law-like critical features at shorter times, which crosses over to a faster exponential-like decay at longer times. The time range in which the data obey a power-law gets longer for larger earthquakes. The exponent describing the power-law regime is roughly about $\frac{2}{3}$, which is not far from the corresponding exponent of the local recurrence-time distribution.

Figure 9 exhibits the *global* recurrence-time distribution of large events with $\mu_c = 3$ calculated for the 1D BK model. As can clearly be seen from the figure, in the BK model, the form of the distribution takes a different form from the local one: The peak structure seen in the local distribution no longer exists here. Furthermore, the form of the distribution tail at larger T is not a simple exponential, faster than exponential: See a curvature of the data in the inset of Fig. 9. These features of the global recurrence-time distribution

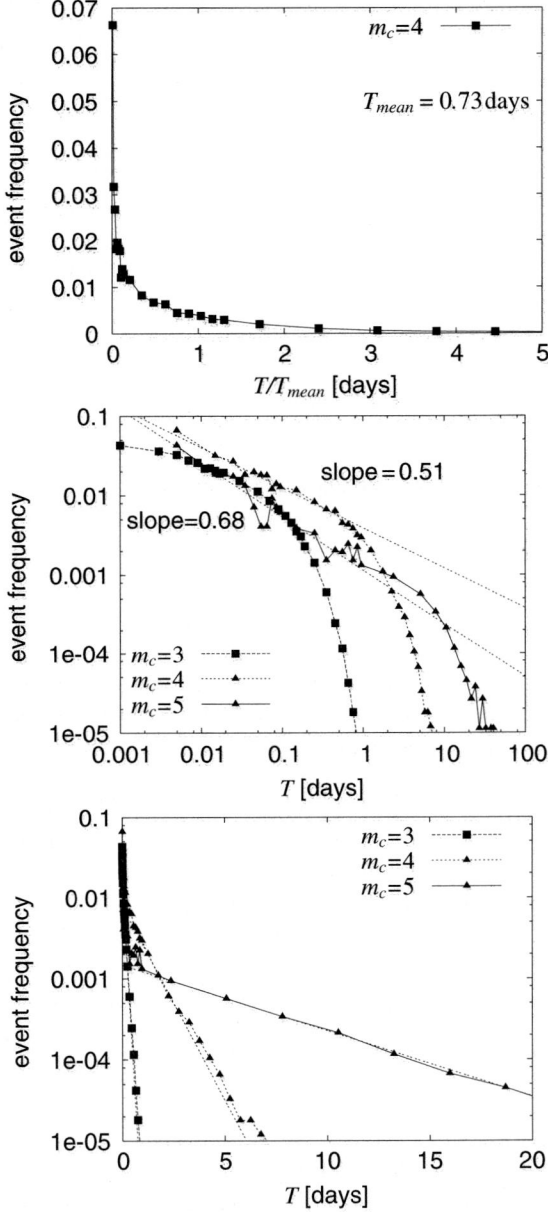

Fig. 8. The global recurrence-time distribution of large earthquakes in entire Japan generated from the JUNEC catalog. The distribution is given on a bare scale for large earthquakes with $m_c = 4$ [*upper figure*], on a double-logarithmic scale with varying $m_c = 3, 4$ and 5 [*middle figure*], and on a semi-logarithmic scale with varying $m_c = 3, 4$ and 5 [*lower figure*]

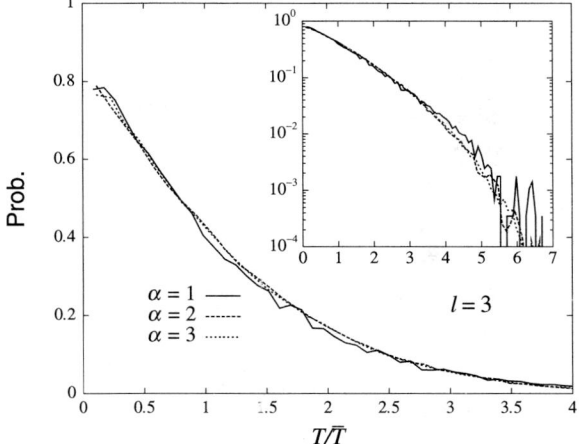

Fig. 9. The global recurrence-time distribution of large events of the 1D BK model. The distribution is given for $\alpha = 1, 2$ and 3 with fixing the magnitude threshold $\mu_c = 3$.

turn out to be rather robust against the change of the parameter values such as α and l, as long as the system size is taken to be sufficiently large.

The observation that the local and the global recurrence-time distributions exhibit mutually different behaviors means that the form of the distribution depends on the length scale of measurements. Such scale-dependent features of the recurrence-time distribution of the BK model are in apparent contrast with the scale-invariant power-law features of the recurrence-time distributions observed in the JUNEC data shown in Figs. 5 and 8, and the ones reported for some of real faults [45, 46]. We note that essentially the same behavior as the one of the 1D BK model is also observed in the 2D BK model (the data not shown here).

3.4 Spatiotemporal Seismic Correlations before the Mainshock

In this subsection, we study the spatiotemporal correlations of earthquake events before the mainshock.

We begin with the JUNEC catalog. First, we define somewhat arbitrarily the mainshock as an event whose magnitude m is greater than $m_c = 5$, and pay attention to the frequency of earthquake events preceding the mainshock, particularly its time and distance dependence. Both the time t and the distance r are measured relative to the time and the position of the subsequent mainshock. The distance between the two events is measured here as the distance between their epicenters. The event frequency is normalized by the factor r associated with the area element of the polar coordinate.

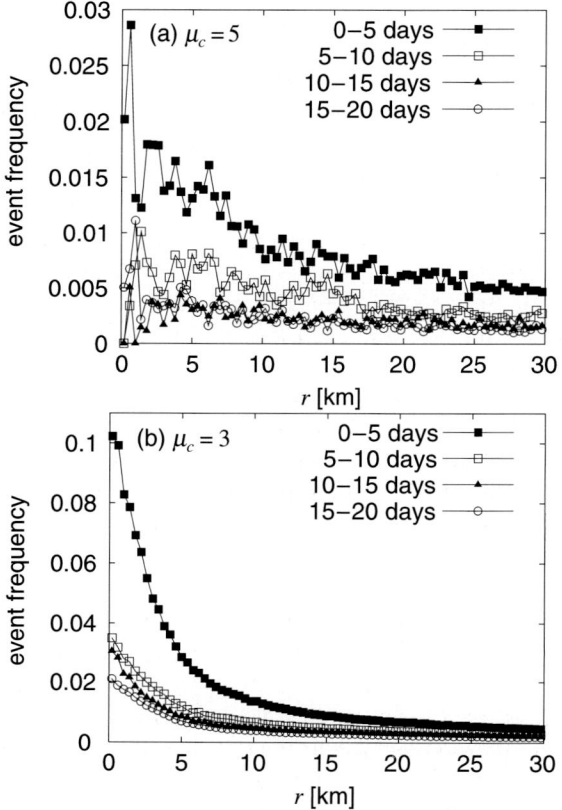

Fig. 10. The time evolution of the event frequency of arbitrary magnitude preceding the mainshock, plotted as a function of the distance from the epicenter of the upcoming mainshock. The magnitude threshold for the mainshock is $m_c = 5$ (**a**) and $m_c = 3$ (**b**). The data are generated from the JUNEC catalog, while the contribution from the two special regions, i.e., the Izu region and the Ebino region, has been omitted from the data: See the text for further details

Figure 10 exhibits the event frequency plotted as a function of r, the distance from the upcoming mainshock, for four time periods preceding the mainshock each containing five days. The data show the time evolution of the spatial pattern of event number irrespective of its magnitude, which occur within 30 km from the epicenter of the upcoming mainshock and during the last 20 days toward the mainshock. The data have been averaged over the mainshocks contained in the data set taken over entire Japan, but *omitting the contributions of the two special narrow regions, i.e., Izu and Ebino*. In these two regions, active earthquake "swarms" occurred during the period, which lead to significantly different behavior in the spatiotemporal correlations (we return to this point later in this subsection). In the data set, total of 990 mainshocks are included, each mainshock accompanying about 8.7 preceding events on average in the time/space range shown in the figure. As can clearly be seen from Fig. 10(a), there is a tendency that the seismic activity accelerates as the mainshock approaches, and this tendency is more pronounced in a closer vicinity of the epicenter of the upcoming mainshock. The data demonstrate that the mainshock accompanies *foreshocks*, at least in the statistical sense.

It should be emphasized here that the evidence of foreshock activity becomes clear only after averaging over a large number of mainshocks [48] and [49]. Note that, in the time/space region of interest, each single mainshock accompanies on average less than ten foreshocks only, a too small number to say anything definite.

One may wonder if certain qualitative features of the spatiotemporal correlations shown in Fig. 10(a) might change depending on the region, the depth and the magnitude-threshold of the mainshock. Examination of the JUNEC data, however, reveals that the qualitative features of Fig. 10(a) are rather robust against these parameters. The average number of foreshocks depends somewhat on each region in Japan: For example, the mean numbers of foreshocks in the space/time region of Fig. 10(a) are 9.4 for the northern part of Japan (40°N-latitude), 14.7 for the middle part (36°N–40°N) and 3.6 for the southern part (–36°N). Nevertheless, the qualitative features of the spatiotemporal correlations turn out to be more or less common.

In Fig. 10(b), we show the similar spatiotemporal seismic correlations as in Fig. 10(a), but now reducing the magnitude threshold of the mainshock from $m_c = 5$ to $m_c = 3$. By this definition, the total number of mainshocks is increased to 53,835 so that the better statistics is expected. Indeed, the data shown in Fig. 10(b) are far less erratic than those in Fig. 10(a), reflecting the improvement of the statistics. Nevertheless, the qualitative features remain almost the same as in Fig. 10(a).

In order to clarify the relation with the inverse Omori law which describes the time evolution of the frequency of foreshocks, we show in Fig. 11 the time dependence of the frequency of foreshocks before the mainshock on a double-logarithmic scale: In the upper panel, the data with the distance range $r_{max} = 30$ km are shown with varying the magnitude threshold as $m_c = 3, 4$ and 5, while in the lower panel, the data with $m_c = 5$ are shown with varying the distance range as $r_{max} = 5, 30$ and 300 km. Except for the case of very large distance range $r_{max} = 300$ km, the inverse Omori exponent comes around 0.5.

Now, we wish to turn to the spatiotemporal correlations in the two special regions, the contribution of which have intentionally been omitted in our analysis of Figs. 10 and 11. In fact, these special regions are associated with "swarms". If there occurs an earthquake swarm which contains in itself a few large events which can be regarded as mainshocks, they make a significant contribution to the above spatiotemporal correlations because a huge number of small events are contained in swarms. The two swarms we discarded in our analysis of Figs. 10 and 11 are, (i) the Izu earthquake swarm, and (ii) the

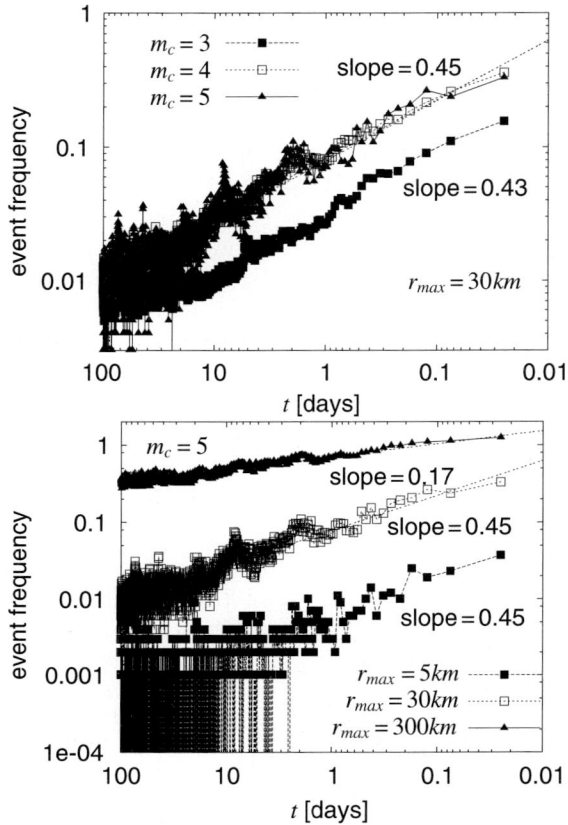

Fig. 11. The frequency of seismic events of arbitrary magnitude preceding the mainshock, plotted versus the time until the mainshock on a double-logarithmic scale. The data are generated from the JUNEC catalog, where the contribution from the two special regions, i.e., the Izu region and the Ebino region, has been omitted: See the text for further details. In the *upper figure*, the magnitude threshold of the mainshock is varied as $m_c = 3, 4$ and 5 with fixing the distance range of observation $r_{max} = 30$ km, whereas, in the *lower figure*, the distance range of observation is varied as $r_{max} = 5, 30$ and 300 km with fixing the magnitude threshold $m_c = 5$

Ebino earthquake swarm (Ebino is located close to the Kirishima volcanoes in southern Kyusyu). Indeed, if the contribution of these two swarm regions were not separated in our analysis of Figs. 10 and 11, the resulting distribution function would look considerably different.

In Fig. 12, we show the spatiotemporal seismic correlations associated with mainshocks with $m_c = 5$ in the Izu swarm region [34.8°N–35.1°N latitude; 139.0°E–139.5°E longitude] (a), and in the Ebino swarm region [31.8°N–32.0°N; 130.2°E–130.6°E] (b), respectively. The total numbers of mainshocks

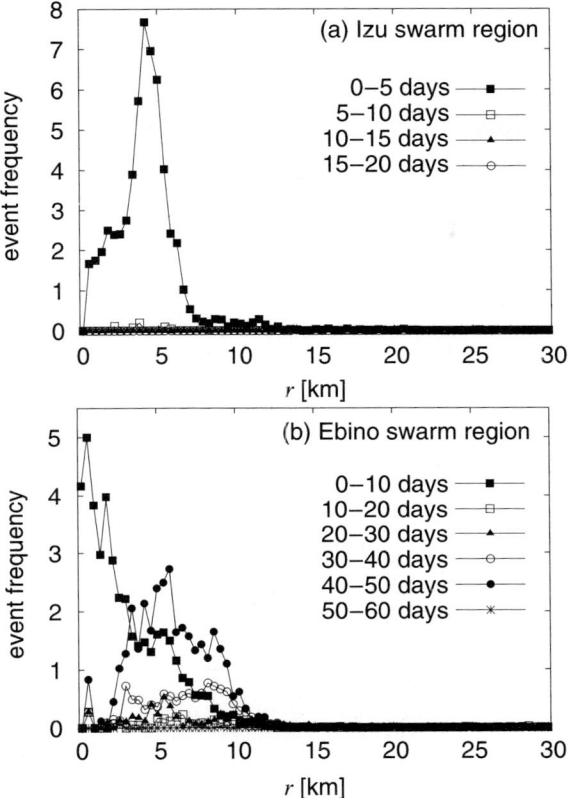

Fig. 12. The time evolution of the event frequency of arbitrary magnitude preceding the mainshock, plotted versus the distance from the epicenter of the upcoming mainshock. The data are generated from the JUNEC catalog, for the Izu earthquake swarm region [34.8°N–35.1°N latitude; 139.0°E–139.5°E longitude] (**a**), and for the Ebino earthquake swarm region [31.8°N–32.0°N; 130.2°E–130.6°E] (**b**), both during the period 1985–1998. The magnitude threshold of the mainshock is $m > m_c = 5$

are 4 (Izu) and 6 (Ebino) here. In either case, the average number of foreshocks per mainshock is very large. In the case of Izu, it is about 256 within the range of 30 km and 20 days from the mainshock, which is an order of magnitude larger than the number associated with swarm-unrelated mainshocks.

As can immediately be seen from Fig. 12, the spatiotemporal pattern of these regions are quite different from the one in other regions shown in Fig. 10, which suggests that the property of swarm-related earthquakes might differ qualitatively from that of swarm-unrelated earthquakes. For example, foreshocks in the Izu swarm region suddenly accelerate on the onset time of about a week, with the center of activity about 5 km away from the mainshock position, the associated spatial correlation function exhibiting a pronounced peak at about $r \simeq 5$ km. This seismic pattern is not dissimilar to the doughnut-like seismic pattern discussed by Mogi as occurring preceding the mainshock (Mogi doughnut) [47]. However, since the doughnut-like seismic pattern as observed in Fig. 12 is not observed in our analysis of more generic case of the wider region, at least in a statistically significant manner, it is likely to be related to the specialty of Izu (and Ebino) region. Swarm activity is considered to be related to the activity of underground magma or water. Hence, some of the statistical properties of swarm-related earthquakes might be better discriminated from those of more general swarm-unrelated earthquakes.

We examine next the corresponding spatiotemporal correlations of the BK model. In the BK model, the distance from the epicenter is measured in units of blocks. Figure 13 represents the time evolution of the spatial distribution of seismicity before the mainshock with $\mu > \mu_c = 3$ for $\alpha = 1$. Very much similar behaviors are observed for other values of α. Similarly to the JUNEC data, preceding the mainshock, there is a tendency of the frequency of smaller events to be enhanced at and around the epicenter of the upcoming mainshock. This was also observed previously by Shaw et al. [24].

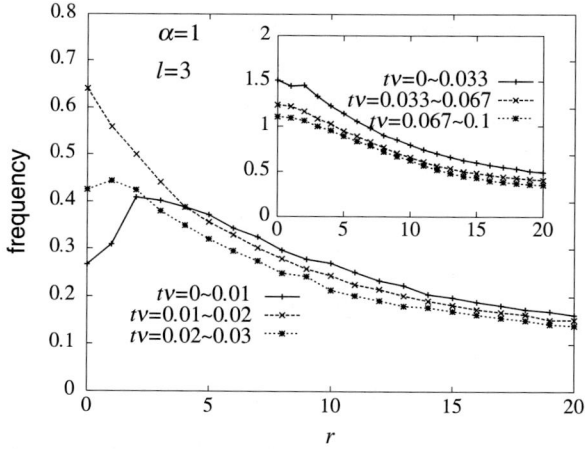

Fig. 13. The time evolution of the spatial seismic distribution function before the mainshock of $\mu > \mu_c = 3$, calculated for the 1D BK model. The inset represents a similar plot with longer time intervals

Interestingly, however, as the mainshock becomes imminent, the frequency of smaller events is suppressed in a close vicinity of the epicenter of the upcoming mainshock, though it continues to be enhanced in the surroundings (Mogi doughnut) [2, 47]. We note that the quiescence observed here occurs only in a close vicinity of the epicenter of the mainshock, within one or two blocks from the epicenter, and only at a time close to the mainshock. The time scale for the appearance of the doughnuts-like quiescence depends on the α-value: The time scale of the onset of the doughnut-like quiescence tends to be longer for larger α. It is not clear at the present stage whether the doughnut-like quiescence as observed here in the BK model has any relevance to the one observed in Fig. 12 for certain earthquake swarms.

As was discussed in detail in [8], in the BK model, the size of the "hole" of the doughnut-like quiescence as well as its onset time scale have no correlation with the magnitude of the upcoming event. In other words, the doughnut-like quiescence is not peculiar to large events in the present model. This means that, by monitoring the onset of the "hole" in the seismic pattern of the BK model, one can certainly deduce the time and the position of the upcoming event, but unfortunately, cannot tell about its magnitude. Yet, one might get some information about the magnitude of the upcoming event, not from the size and the onset time of the "hole", but from the size of the "ring" surrounding the "hole". Thus, we show in Fig. 14 the spatial correlation functions before the mainshock in the time range $0 \leq t\nu \leq 0.001$ for the case of $\alpha = 2$, with varying the magnitude range of the upcoming event. In the figure, the direction in which the rupture propagates farther in the upcoming event is always taken to be the positive direction $r > 0$, whereas the direction in which the rupture propagates less is taken to be the negative direction $r < 0$. As can be seen from the figure, although the size of the "hole" around the

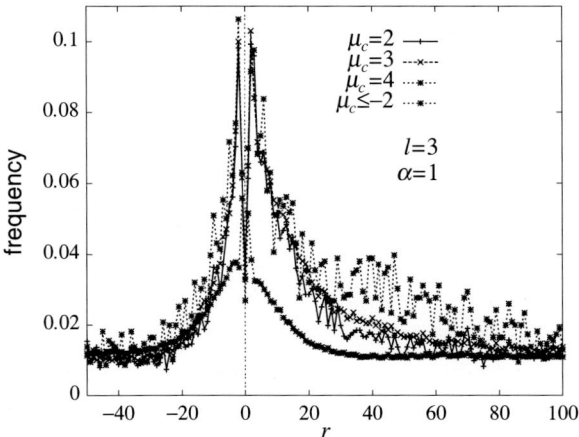

Fig. 14. The frequency of seismic events preceding the events of various magnitude range plotted versus r, the distance from the epicenter of the upcoming mainshock. The *curves* correspond to the large events of $\mu > \mu_c = 2, 3$ and 4, and to the smaller events of $\mu < -2$. The direction in which the rupture propagates farther in the upcoming mainshock is always taken to be the positive direction $r > 0$, whereas the direction in which the rupture propagates less is taken to be the negative direction $r < 0$. The parameters are taken to be $\alpha = 2$ and $l = 3$, the time range being $0 \leq t\nu \leq 0.001$ before the mainshock

origin $r = 0$ has no correlation with the magnitude of the upcoming event as mentioned above, the size of the region of the active seismicity surrounding this "hole" is well correlated with the size and the direction of the rupture of the upcoming event. This coincidence might enable one to deduce the position and the size of the upcoming event by monitoring the pattern of foreshocks, although it is still difficult to give a pinpoint prediction of the time of the upcoming mainshock. We note that such a correlation between the size of the seismically active region and the magnitude of the upcoming event was observed in the BK model in [50], and was examined in real earthquake data as well [51].

As shown in Fig. 15, a very much similar behavior including the doughnut-like quiescence is observed also in the 2D BK model. Note that the event frequency has been normalized here by the measure factor r associated with the area element of the polar coordinate. The time scale of the onset of the doughnut-like quiescence seems to be longer in 2D than in 1D.

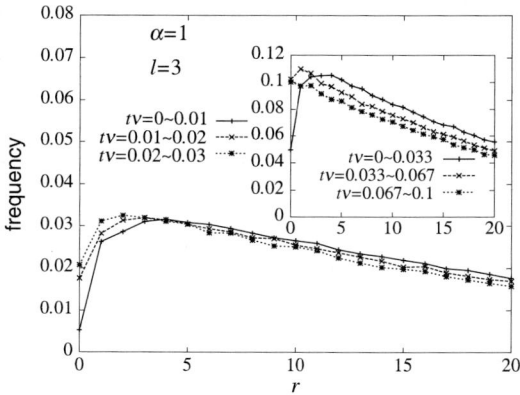

Fig. 15. The time evolution of the spatial seismic distribution function before the mainshock of $\mu > \mu_c = 5$, calculated for the 2D BK model. The inset represents a similar plot with longer time intervals

3.5 Spatiotemporal Seismic Correlations after the Mainshock

In this subsection, we examine the spatiotemporal correlations of earthquake events *after* the mainshock. The time evolution of the spatial seismic pattern calculated from the JUNEC catalog are shown in Fig. 16, with the magnitude threshold for the mainshock $m_c = 5$ (a), and $m_c = 3$ (b). As can be seen from these figures, aftershock activity is clearly observed. The rate of the aftershock activity is highest just at the epicenter of the mainshock, not in the surroundings. It is sometimes mentioned in the literature that the aftershock

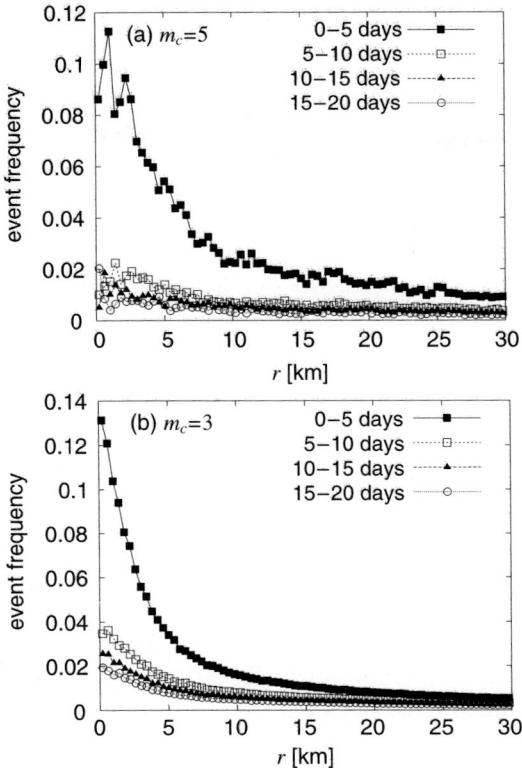

Fig. 16. The time evolution of the event frequency of arbitrary magnitude after the mainshock, plotted versus the distance from the epicenter of the mainshock. The magnitude threshold for the mainshock is $m_c = 5$ (**a**), and $m_c = 3$ (**b**). The data are generated from the JUNEC catalog

activity is highest near the edge of the rupture zone of the mainshock (see, e.g., [2]). However, this is not the case here.

In order to further clarify the relation with the Omori law, we show in Fig. 17 the time dependence of the frequency of aftershocks after the mainshock on a double-logarithmic scale: In the upper panel, the data with the distance range $r_{max} = 30$ km are shown with varying the magnitude threshold as $m_c = 3, 4$ and 5, while, in the lower panel, the data with $m_c = 5$ are shown with varying the distance range as $r_{max} = 5, 30$ and 300 km. In all cases analyzed, a straight-line behavior corresponding to the power-law decay is observed (Omori law). In contrast to the corresponding quantity before the mainshock, the slope of the straight line, i.e., the value of the Omori exponent, seems to depend on the distance range r_{max} and the magnitude of the mainshock m_c. As the distance range r_{max} gets smaller and the magnitude of the mainshock m_c gets larger, the Omori exponent tends to get larger. If

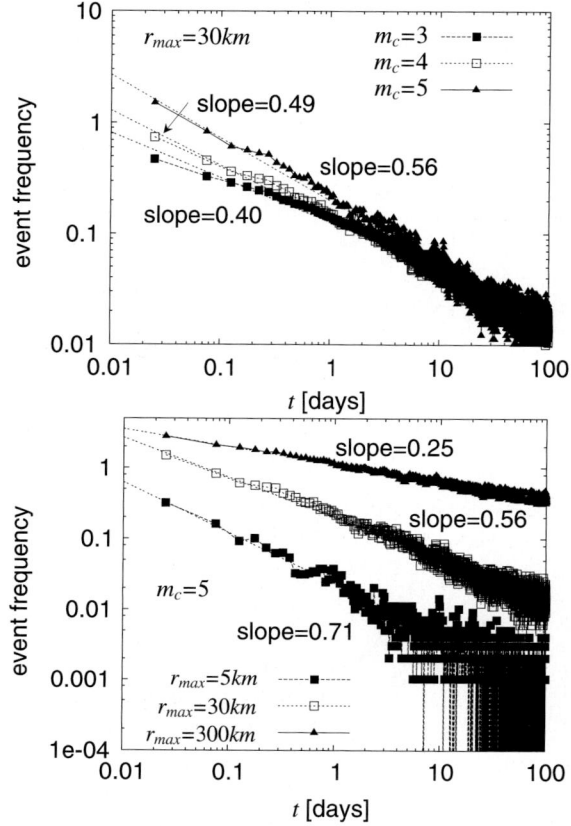

Fig. 17. The frequency of seismic events of arbitrary magnitude after the mainshock, plotted versus the time after the mainshock on a double-logarithmic scale. The data are generated from the JUNEC catalog. In the *upper figure*, the magnitude threshold of the mainshock is varied as $m_c = 3, 4$ and 5 with fixing the distance range of observation $r_{max} = 30$ km, whereas, in the *lower figure*, the distance range of observation is varied as $r_{max} = 5$, 30 and 300 km with fixing the magnitude threshold $m_c = 5$

one looks at the range $r_{max} = 5$ km for mainshocks greater than $m_c = 5$, the Omori exponent is about 0.71.

The corresponding spatiotemporal correlations after the mainshock are calculated for the 1D BK model, and the results are shown in Fig. 18 for the cases of $\alpha = 1$. As can be seen from the figure, aftershock activity is clearly observed with the maximum rate occurring at the epicenter of the mainshock. In contrast to the JUNEC data, the r-dependence of the spatial seismic distribution is non-monotonic, aftershock activity being suppressed in the surrounding region where the rupture was largest in the mainshock. The

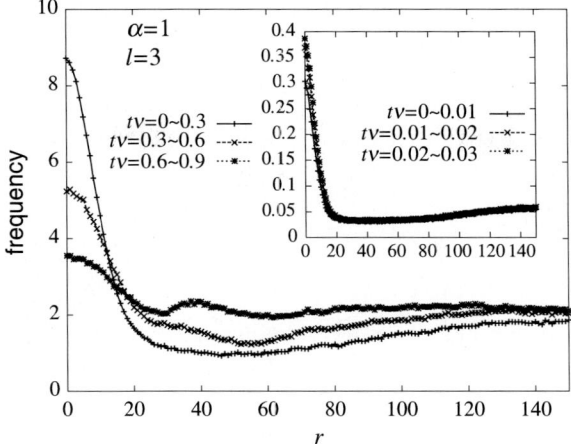

Fig. 18. The time evolution of the spatial seismic distribution function after the mainshock of $\mu > \mu_c = 3$, calculated for the 1D BK model. The inset represents a similar plot with longer time intervals

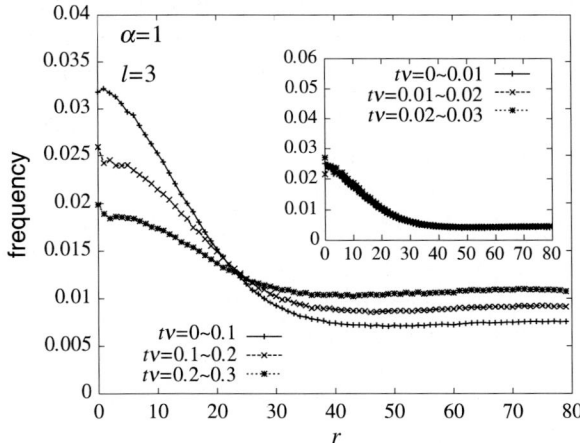

Fig. 19. The time evolution of the spatial seismic distribution function after the mainshock of $\mu > \mu_c = 5$, calculated for the 2D BK model. The inset represents a similar plot with longer time intervals

seismicity near the epicenter is kept almost constant in time for some period after the mainshock, say, in the time range $t\nu < 0.03$, which is in apparent contrast to the power-law decay as expected from the Omori law: See the insets of Fig. 18. At longer time scales, the seismicity near the epicenter seems to decay gradually, although the decay observed here is not a power-law decay. A very much similar behavior is observed also for the 2D BK model: See Fig. 19.

Hence, aftershocks obeying the Omori law is not realized in the BK model, as already reported [21]. This is in apparent contrast to the observation for real faults.

Such an absence of aftershocks in the BK model might give a hint to the physical origin of aftershocks obeying the Omori law, e.g., they may be driven by the slow chemical process at the fault, or by the elastoplaciticity associated with the ascenosphere, etc., which are not taken into account in the BK model.

3.6 The Time-Dependent Magnitude Distribution

As an other signature of precursory phenomena, we examine a "time-resolved" local magnitude distribution for several time periods before the large event. Figure 20 represents such a local magnitude distribution before large earthquakes, calculated from the JUNEC catalog, i.e., the distribution of seismicity in the vicinity $r_{max} = 30$ km of the epicenter of the mainshock with its magnitude greater than $m_c = 4$. It can clearly be seen from the figure that the GR law persists even just before the mainshock, whereas, preceding the mainshock, the GR exponent b gets smaller compared with the space- and time-averaged b-value. For example, 100 days before the mainshock, the b-value becomes about 0.60, considerably smaller than the averaged value $b \simeq 0.88$. Such a decrease of the b-value was also reported in the literature [52, 53].

For comparison, we show the corresponding local magnitude distributions before the mainshock of the BK models both in 1D (Fig. 21) and in 2D (Fig. 22). The parameter α is taken to be $\alpha = 1$. In 1D, only events with their epicenters within 30 blocks from the epicenter of the upcoming mainshock are

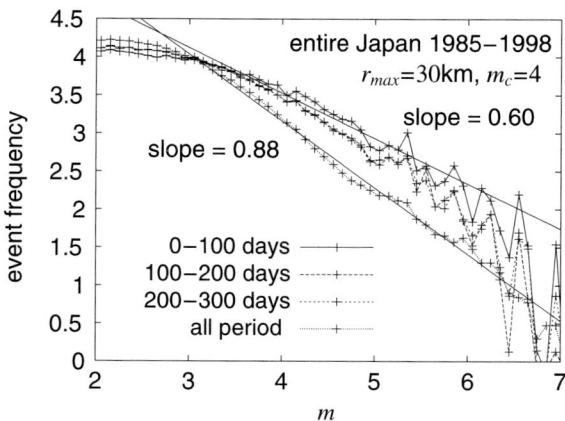

Fig. 20. The time-resolved local magnitude distribution of earthquakes in Japan before the mainshock of $m > m_c = 4$, generated from the JUNEC catalog. Preceding the mainshock, the b-value decreases considerably from the long-time average value

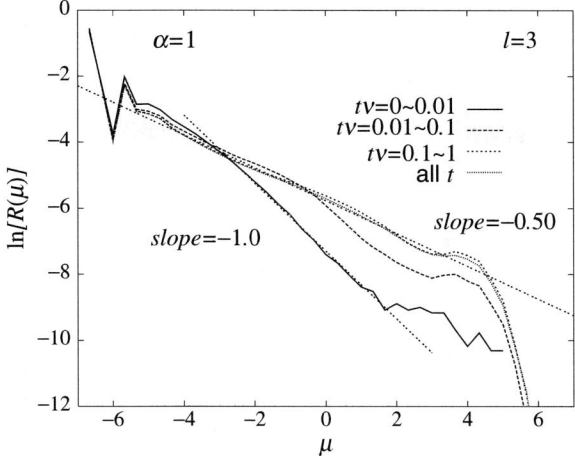

Fig. 21. The time-resolved local magnitude distribution of the 1D BK model before the mainshock of $\mu > \mu_c = 3$

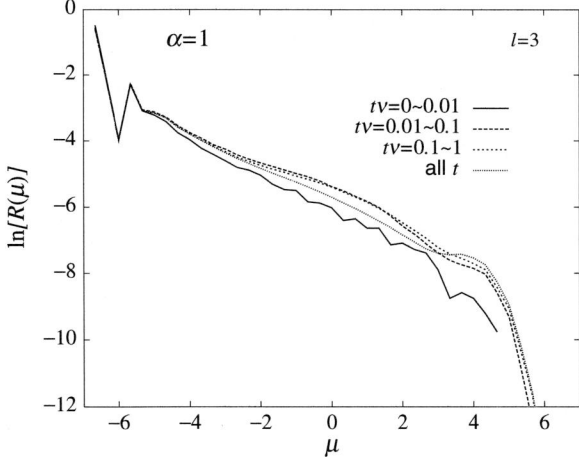

Fig. 22. The time-resolved local magnitude distribution of the 2D BK model before the mainshock of $\mu > \mu_c = 5$

counted, while, in 2D, only events with their epicenters lying in a circle of its radius 5 blocks centered at the epicenter of the upcoming mainshock are counted. As can be seen from the figures, as the mainshock approaches, the form of the magnitude distribution changes significantly. In 1D, the apparent B-value describing the power-law regime tends to *increase* as the mainshock approaches, from the time-averaged value $B \simeq 0.50$ ($b \simeq 0.75$) to the value $B \simeq 1.0$ ($b \simeq 1.5$) just before the mainshock: It is almost doubled. This tendency is opposite to what we have just found for the JUNEC catalog and several other observations for real faults [52, 53]. However, a similar increase of the apparent B-value preceding the mainshock was reported for some of real faults [54]. For the case of larger $\alpha > 1$ (the data not shown here), the change of the B-value preceding the mainshock is still appreciable, though in a less pronounced manner. More complicated behavior is observed in 2D. As the mainshock approaches, the apparent B-value describing the power-law regime *slightly increases first, and then, increases* just before the mainshock.

Next, we analyze similar time-resolved local magnitude distributions, but *after* the large event. Figure 23 represents such a local magnitude distribution after the large event calculated from the JUNEC catalog. In this case, the deviation from the averaged distribution is relatively small as compared with the one observed before the mainshock, although there seems to be a tendency that the observed b-value decreases slightly after the mainshock.

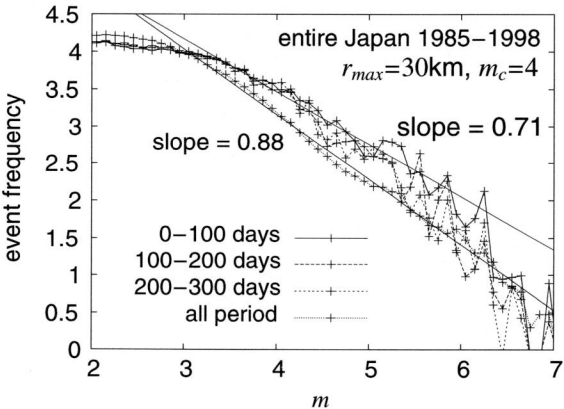

Fig. 23. The time-resolved local magnitude distribution of earthquakes in Japan after the mainshock of $m > m_c = 4$, generated from the JUNEC catalog

For comparison, we show the corresponding local magnitude distributions after the mainshock for the BK models in 1D (Fig. 24) and in 2D (Fig. 25). The parameter α is taken to be $\alpha = 1$. As can be seen from the figures, the form of the magnitude distribution changes only little after the mainshock except that the weight of large events is decreased appreciably, particularly in 2D.

4 Summary and Discussion

In summary, we studied the spatiotemporal correlations of earthquakes both by the analysis of real earthquake catalog of Japan and by numerical computer

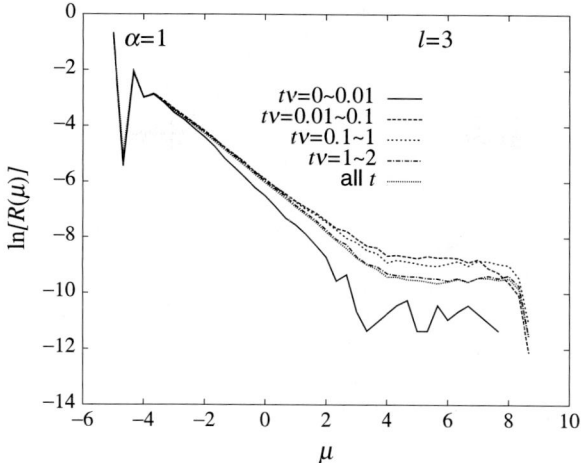

Fig. 24. The time-resolved local magnitude distribution of the 1D BK model after the mainshock of $\mu > \mu_c = 3$

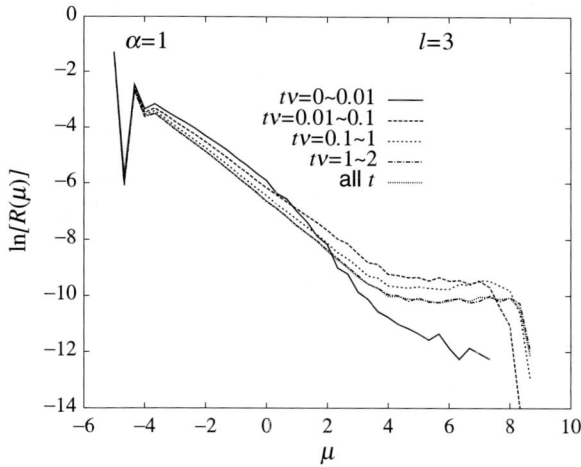

Fig. 25. The time-resolved local magnitude distribution of the 2D BK model after the mainshock of $\mu > \mu_c = 5$

simulations of the spring-block model in 1D and 2D. Particular attention was paid to the magnitude distribution, the recurrence-time distribution, the time evolution of the spatial distribution of seismicity before and after the mainshock, and the time evolution of the magnitude distribution before and after the mainshock. Certain eminent features of the spatiotemporal correlations, including foreshocks, aftershocks, swarms and doughnut-like seismic pattern, were discussed in some detail.

In our numerical simulations of the BK model, particular attention was paid to the issue how the statistical properties of earthquakes depend on the frictional properties of earthquake faults. We have found that when the extent of the velocity-weakening property is suppressed, the system tends to be more critical, while, as the velocity-weakening property is enhanced, the system tends to be more off-critical with enhanced features of characteristic earthquakes.

Overall, the BK model tends to exhibit more characteristic or off-critical statistical properties, particularly for large earthquakes, than the real seismicity which exhibits much more critical statistical properties. This discrepancy between the model and the real seismicity has been recognized for some time now, but its true cause has remained to be unclear.

First, we need to recognize that the earthquake catalog is taken not for a single fault, but over many faults. There exists a suspicion that, even if the property of a single individual fault is more or less characteristic, the property obtained after averaging over many faults each of which has different characteristics, becomes apparently critical as a whole. If this is really the case, the real observation is not necessarily inconsistent with the observation for the BK model, since the BK model deals with the property of a single uniform fault. There is a claim that the extent of the criticality of earthquakes might depend on the type of earthquake faults, i.e., a matured fault with relatively regular fault zone behaves more characteristic, while an inmatured fault with relatively irregular fault zone behaves more critical [55]. Difficulty in testing such a hypothesis is that the statistical accuracy of events for a single fault is rather limited, particularly for large events. We need to be careful because, when the event number is not sufficient, an apparent deviation from the criticality might well arise simply due to the insufficient statistics, pretending a characteristic earthquake.

Other possibility is that, since smaller earthquakes are more or less critical even in the BK model, in real seismicity, the critical behavior might be limited to moderately large earthquakes which are contained in enough number in the earthquake catalog, while very large earthquakes, which are very few in number in the catalog, might be more or less characteristic. Anyway, the question of either critical or characteristic earthquakes is one of major fundamental questions left in earthquake studies.

The BK model was found to exhibit several intriguing precursory phenomena associated with large events: Preceding the mainshock, the frequency of smaller events is gradually enhanced, whereas just before the mainshock it is suppressed only in a close vicinity of the epicenter of the upcoming mainshock (the Mogi doughnut). The apparent B-value of the magnitude distribution increases significantly preceding the mainshock. On the other hand, the Omori law of aftershocks is not observed in the BK model.

Some of these precursory phenomena observed in the BK model are also observed in real earthquake catalog, but some of them are not. For example, the enhancement of foreshock activity is observed in common both in the BK model and in the JUNEC catalog. By contrast, the doughnut-like quiescence generically observed in the BK model is not observed in standard earthquakes in the JUNEC catalog, although it is observed in certain earthquake swarms.

Here, in order to make a further link between the BK model and the real world, it might be of some interest to estimate various time and length scales involved in the BK model. For this, we need to estimate the units of time and length of the BK model in terms of real-world earthquakes. Concerning the time unit ω^{-1}, we estimate it via the rise time of large earthquakes, $\sim \pi/\omega$, which is typically about 10 seconds. This gives an estimate of $\omega^{-1} \sim 3$ sec. Concerning the length unit L, we estimate it making use of the fact that the typical displacement in large events of our simulation is of order one L unit, which in real-world large earthquakes is typically 5 meters. Then, we get $L \sim 5$ meters. Since the loading rate ν' associated with the real plate motion is typically 5 cm/year, the dimensionless loading rate $\nu = N/(L\omega)$ is estimated to be $\nu \sim 10^{-9}$. If we remember that the typical mean recurrence time of large events in our simulation is about one unit of ν^{-1}, the mean recurrence time of our simulation corresponds to 100 years in real world.

In our simulation of the BK model, the doughnut-like quiescence was observed before the mainshock at the time scale of, say, $t\nu \lesssim 10^{-2}$. This time scale corresponds to about 1 year. In our simulation, the doughnut-like quiescence was observed in the region only within a few blocks from the epicenter of the mainshock. To give the corresponding real-world estimate, we need the real-world estimate of our block size a. In the BK model, the length scale a is entirely independent of the length scale L, and has to be determined independently. We estimate a via the typical velocity of the rupture propagation, $la\omega$, which is about 3 km/sec in real earthquakes. From this relation, we get $a \sim 3$ km. The length scale associated with the doughnut-like quiescence is then estimated to be $3 \sim 6$ km in radius. If we remember that the rupture size of large events in our simulation with $\mu = \mu_c = 3$ was about 60 blocks, the size of the rupture size of large earthquakes of our simulation corresponds to 180 km and more. This is comparable to the size of the rupture zone of real earthquakes of their magnitude eight. Hence, the large events in the BK model might correspond to exceptionally large earthquakes in real seismicity, which might explain the reason, at least partially, why the deviation from the GR law observed in the BK model at larger magnitudes is hardly observed in real seismicity. In real faults, the possible maximum size of the rupture zone might be limited by the fault geometry, i.e., by the boundary of a fault.

Of course, it is not a trivial matter at all how faithfully the statistical properties as observed for the BK model represent those of real earthquakes. We should be careful not to put too much meaning to the quantitative estimates given above.

As an other precursory effect, the change of the apparent B-value is observed in the BK model, i.e., an increase in 1D or an initial decrease followed by a subsequent increase in 2D. In the JUNEC catalog, the B-value turns out to decrease prior to the mainshock consistently with several other observations for real faults. Meanwhile, an increase of the B-value, which is similar to the one observed in the 1D BK model, was reported in some real faults. Thus, to elucidate the detailed mechanism behind the change of the B-value preceding the large event is an interesting open question.

One thing seems to be clear: Much needs to be done before we understand the true nature of earthquakes. We hope that further progress in statistical-physical approach to earthquakes, combined with the ones in other types of approaches from various branches of science, would eventually promote our fuller understanding of earthquakes.

Acknowledgements

The author is thankful to Mr. T. Mori and Mr. A. Ohmura for their collaboration and discussion. He is also thankful to Prof. B. Chakrabarti for organizing the *Models of Earthquakes: Physics Approaches* conference and for giving me an opportunity of presenting a talk there and writing this article.

References

1. C.H. Scholz, Nature **391**, 3411 (1998).
2. C.H. Scholz, *The Mechanics of Earthquakes and Faulting* (Cambridge Univ. Press, 1990).
3. P. Bak, C. Tang and K. Wiesenfeld, Phys. Rev., Lett. **59**, 381 (1987).
4. P. Bak and C. Tang, J. Geophys. Res. **94**, 15635 (1989).
5. Y.Y. Kagan, in this book.
6. R. Burridge and L. Knopoff, Bull. Seismol. Soc. Am. **57** (1967) 3411.
7. T. Mori and H. Kawamura, Phys. Rev. Lett. **94**, 058501 (2005).
8. T. Mori and H. Kawamura, J. Geophys. Res., in *press* (physics/0504218).
9. T. Mori and H. Kawamura, unpublished.
10. H. Kawamura and T. Mori, unpublished.
11. J.M. Carlson, J.S. Langer, B.E. Shaw and C. Tang, Phys. Rev. A**44**, 884 (1991).
12. J.M. Carlson, J.S. Langer and B.E. Shaw, Rev. Mod. Phys. **66**, 657 (1994).
13. B.E. Shaw, J. Geophys. Res. **100**, 18239 (1995).
14. C.R. Myers, B.E. Shaw and J.S. Langer, Phys. Rev. Lett. **77**, 972 (1996).
15. J. Dietrich, J. Geophys. Res. **84**, 2161 (1979).
16. A. Ruina, J. Geophys. Res. **88**, 10359 (1983).
17. S.T. Tse and J.R. Rice, J. Geophys. Res. **91**, 9452 (1986).
18. N. Kato, J. Geophys. Res. **109**, B12306 (2004).
19. A. Ohmura and H. Kawamura, unpublished.
20. J.M. Carlson, Phys. Rev. A**44**, 6226 (1991).
21. J.M. Carlson and J.S. Langer, Phys. Rev. Lett. **62**, 2632 (1989); Phys. Rev. A**40**, 6470 (1989).
22. J.M. Carlson, J. Geophys. Res. **96**, 4255 (1991).
23. J. Schmittbuhl, J.-P. Vilotte and S. Roux, J. Geophys. Res. **101**, 27741 (1996).
24. B.E. Shaw, J.M. Carlson and J.S. Langer, J. Geophys. Res. **97**, 479 (1992).

25. C.R. Myers and J.S. Langer, Phys. Rev. E**47**, 3048 (1993).
26. B.E. Shaw, Geophys. Res. Lett. **21**, 1983 (1994).
27. R. De and G. Ananthakrisna, Europhys. Lett. **66**, 715 (2004).
28. J. Xia, H. Gould, W. Klein and J.B. Rundle, Phys. Rev. Lett. **95**, 248501 (2005); cond-mat/0601679.
29. H. Nakanishi, Phys. Rev. A**41**, 7086 (1990).
30. K. Ito and M. Matsuzaki, J. Geophys. Res. **95**, 6853 (1990).
31. S.R. Brown, C.H. Scholz and J.B. Rundle, Geophys. Res. Lett. **18**, 215 (1991).
32. Z. Olami, H.J. Feder and K. Christensen, Phys. Rev. Lett. **68**, 1244 (1992).
33. S. Hergarten and H. Neugebauer, Phys. Rev. E**61**, 2382 (2000).
34. S. Hainzl, G. Zöller and J. Kurths, J. Geophys. Res. **104**, 7243 (1999); Geophys. Res. Lett. **27**, 597 (2000).
35. A. Helmstetter, S. Hergarten and D. Sornette, Phys. Rev. Lett. **88**, 238501 (2002); Phys. Rev. E**70**, 046120 (2004).
36. H. Tsuruoka, Abstract of the annual meeting of the Seismological Society of Japan, p04 (1997).
37. G.L. Vasconcelos, Phys. Rev. Lett. **76**, 4865 (1996).
38. M.S. Vieira, G.L. Vasconcelos and S.R. Nagel, Phys. Rev. E**47**, R2221 (1993).
39. I. Clancy and D. Corcoran, Phys. Rev. E**71**, 046124 (2005).
40. T.C. Hanks, J. Geophys. Res. **84**, 2235 (1979).
41. D.J. Andrews, J. Geophys. Res. **85**, 3867 (1980).
42. K. Aki, in *Earthquake Prediction, An International Review*, (ed. D.W. Simpson and P.G. Richards, American Geophysical Union, Washington D.C.) p.566 (1981).
43. B.K. Chakrabarti and R.B. Stinchcombe, Physica, A**270**, 27 (1990); P. Bhattacharyya, Physica, A**348**, 199 (2005); P. Bhattacharyya, A. Chatterjee and B.K. Chakrabarti, physics/0510038.
44. S.P. Nishenko and R. Buland, Bull. Seismol. Soc. Am. **77**, 1382 (1987).
45. P. Bak, K. Christensen, L. Danon and T. Scanlon, Phys. Rev. Lett. **88**, 178501 (2002).
46. A. Corral, Phys. Rev. Lett. **92**, 108501 (2004); Phys. Rev. E**68**, 035102 (2003).
47. K. Mogi, Bull. Earthquake Res. Inst. Univ. Tokyo, **47**, 395 (1969); Pure Appl. Geophys. **117**, 1172 (1979).
48. B. C. Papazachos, Techtonophysics 28, 213 (1975).
49. L. M. Jones and P. Molner, Nature 262, 677 (1976).
50. S.L. Pepke, J.M. Carlson and B.E. Shaw, J. Geophys. Res. **99**, 6769 (1994).
51. V.G. Kossobokov and J.M. Carlson, J. Geophys. Res. **100**, 6431 (1995).
52. S. Suehiro, T. Asada and M. Ohtake, Papers Meteorol. Geophys. **15**, 71 (1964).
53. S.C. Jaume and L.R. Sykes, Pure Appl. Geophys. **155**, 279 (1999).
54. W.D. Smith, Nature **289**, 136 (1981).
55. S.G. Wesnousky, Nature **335**, 340 (1988); M.W. Stirling, S.G. Wesnousky and K. Shimazaki, J. Goephys. Res. **124**, 833 (1996).

Space-time Combined Correlation Between Earthquakes and a New, Self-Consistent Definition of Aftershocks

V. De Rubeis[1], V. Loreto[2], L. Pietronero[2] and P. Tosi[1]

[1] Istituto Nazionale di Geofisica e Vulcanologia, Roma, Italy
    derubeis@ingv.it and tosi@ingv.it
[2] "La Sapienza" University, Physics Department, and INFM, Center for Statistical Mechanics and Complexity, Roma, Italy
    pietronero@roma1.infn.it and loreto@roma1.infn.it

1 Introduction

Seismicity is recognized to be a complex natural phenomenon either in space, time and energy domains: earthquakes occur as a sudden energy release after a strongly variable time period of stress accumulation, in locations not deterministically defined, with magnitude range spanning over several orders. But seismicity is certainly not a pure random process: spatial locations of events clearly display correlations with tectonic structures at all scales (from plates borders to small faults settings); on the other hand time evolution is clearly linked with strongest shocks occurrence and energy distribution displays hierarchical features. Although it is still not possible to propose deterministic models for earthquakes, well established statistical relations constrain seismicity under very specific and intriguing relations.

In the time domain a main shock is followed by aftershocks with a decay rate that follows the modified Omori law [30]:

$$n(t) = \frac{A}{(t+c)^p} \qquad (1)$$

where $n(t)$ is the daily number of events after the time t, A and c may vary for each sequence and $p \simeq 1$. This law accounts for temporal rate evolution of events after a main shock. It should be noted that relation (1) is adequate for representing general characteristics of shocks decay in time but, going into detail, the behavior of $n(t)$ is not so simple due to the presence of secondary aftershocks [31]. In other words the modified Omori law is the first approximation of a deeply structured process where aftershocks produces secondary aftershocks and so on.

Space distribution of seismicity also displays non-trivial features for which in the last few years new tools and approaches have been devised. Interestingly it has been recognized that seismicity displays scale invariant features both for aftershocks and for complete catalogues [11, 13, 28]. On a large scale space distributions are constrained by limits imposed by the most significant tectonic interfaces as plate limits, large faults systems or volcanic structures [15]. A quantification of these features is possible through the notion of fractal dimension of epicenters : a background like distribution is recognized to have a fractal dimension D_s near topological dimension of the embedding space, while sequences display much lower D_s values. Very different D_s values have been measured both in different places and as time varying feature of an area [3].

Seismogenic sources demonstrate their resistance limit to stress loading by generating earthquakes; stress is redistributed to neighboring space and this cause the aftershock sequence to develop in space. Stress is likely redistributed because the triggering of a sequence leads to a new dynamical equilibrium. This interpretation raises a certain number of open issues. One of them concerns the triggering of seismicity. Short range triggering (at a distance of the order of seismic source dimension) may be sufficiently justified by stress changes induced by the main shock. Long range triggering is a more controversial topic which has been discussed both in the framework of mechanical (physical) modeling and of purely statistical approaches. Experimentally, from the analysis of real data, some authors have recognized long range triggering [10], other statistical approaches consider long distance correlations among events as useful assumption to de-cluster a catalog [6]. Others have proposed physical explanations [8] including Coulomb stress modifications [17, 21, 22, 25, 26]. Interesting modeling schemes have been proposed, using, for instance, cellular automata to model the behavior of the crust, typically considered in a critical state [1]. The advantage of assuming a critical state for the crust is that at long distances a trigger effect can be obtained by a relatively small amount of stress: rupture is achieved by a preexistent load on a fault surface, plus a relatively small portion of far away produced stress sufficient to cause the event. This new event can itself produce a new stress diffusion and so on. This chain of stress-strain transfer raises the question of whether a self-consistent relation between space and time should be invoked in order to explain the complex triggering phenomena. This relation can be expressed as a sort of triggering velocity which can produce a mechanism of stress diffusion. Into this frame short range aftershocks and long range triggering could be unified under the same mechanism [14]. Several authors have attempted to find space-time influence ranges after large main shocks into specific regions and on world wide catalogues. Generally influence regions of the order of 100 km from main shocks are likely to occur [4, 23] and in [18] it has been found that for the largest earthquakes of last century, very long range correlations exist as well as a gap around 300 km, only partially explained with a directional effect due to source geometry. In [14] the authors focused their interest on the stress diffusion at global level. They found that there is not significant triggering at distances larger than 100–150 km. On the other hand the mean triggering distance increases very slowly with time, if compared with a normal diffusion process. They did not make any a priori assumption about the role of each event, considering every earthquake – regardless its magnitude – as a possible triggering event of following seismicity. Other papers [19] focus on the space-time relation of seismicity in three very different seismic catalogues (different for space-time range domains), pointing out that the two dimension domains (space and time) should not be considered separately, as suggested by [24] since spatial correlation structures are evolving on time.

Despite the tremendous effort deployed in this direction, several questions remain open. Is long range triggering a normal process or should it be more considered an exception? What is the influence range of an event and how does it depend on magnitude? Is the hierarchical distinction of seismic events into fore-main-after shocks related to real geophysical reasons? In order to address these questions it appears evident how important is to consider the seismic process under both its spatial and temporal domains in a combined analysis. In this work we first investigate scale invariance of world seismicity for space and time separately. We then apply a novel combined method of analysis [29] suitable for point processes, based on space-time correlations among earthquakes.

2 Scale Invariance in Space

Seismicity shows peculiar space distributions. Depending on the scale we can find, at larger ranges, close relations with plate limits, large fault systems, volcanic structures. Despite these clear patterns, at smaller scales randomness plays a major role increasing substantially the degree of uncertainty. For these reasons, the concept of background seismicity has been postulated in addition to spatial patterns related to sequences, with the aim of better fitting time and spatial behaviors. Moreover spatial distribution of earthquakes during a sequence has received a wide interest over the years; modeling has become more sophisticated being more related to source geometry and stress increasing-decreasing patterns. Nevertheless, even though the high level of quality reached in modeling, the understanding of spatial distributions of earthquakes remains problematic. Recently statistical approaches have been enriched with the consideration of scale invariance of events localization. Several authors have put in evidence this feature and they have quantified it through the definition of a spatial fractal dimension D_s [12, 15]. In specific areas D_s has been studied as a time variable parameter, to point out relations with the occurrence of highest magnitude events [3]: for Central Italy background seismicity has shown to have high D_s values (close to embedding space value) while during sequences D_s values are sensibly lower: interestingly the lowering begins shortly before the largest events [28], moreover when its value was reaching again the background activity level, it was possible to give a definition of time length of the sequence based on its spatial clustering activity. Spatial fractal dimension of seismicity has been shown to be not homogeneous, leading to the quantification of its multifractal spectrum [11, 13].

A frequently used and computationally efficient method to analyze scale invariance in seismicity is the correlation integral [5]. It is defined as:

$$C(l) = \frac{2}{N(N-1)} \sum_{i=1}^{N-1} \sum_{j=i+1}^{N} \Theta\left(l - \|\mathbf{x}_i - \mathbf{x}_j\|\right), \qquad (2)$$

where l is a measure of distance (in a suitable metric) in the space considered, N is the total number of events, $\mathbf{x}$ is the coordinate vector and Θ is the Heaviside step function. If (2) scales as a power law, $C(l) \propto l^D$, the correlation dimension D can be defined by

$$d(l) = \frac{\delta \log C(l)}{\delta \log l} \quad D = \lim_{l \to 0} d(l) \qquad (3)$$

Experimentally self similarity can be found by plotting the local slope $d(l)$ of $\log C(l)$ versus $\log l$ [16].

We first consider the application of the correlation integral calculation to a global seismic catalog. Data are retrieved from National Earthquake Information Center, U.S.G.S., and events are recorded in the time period ranging from 1973 to 2004, with magnitudes $m_b > 5.0$. Time period and minimum magnitude choice were driven by completeness criteria.

The correlation integral for earthquakes in a three-dimensional space is defined as (following (2):

$$C(r) = \frac{2}{N(N-1)} \sum_{i=1}^{N-1} \sum_{j=i+1}^{N} \Theta\left(r - \|\mathbf{x}_i - \mathbf{x}_j\|\right), \qquad (4)$$

where $\mathbf{x}$ is the vector of space coordinates. $\log C(r)$ behavior as function of $\log r$ is shown in Fig. 1. Distances between all seismic hypocenter pairs are measured in three-dimensions connecting two points with a straight line: this imply that two hypocenters, localized at two antipodal points near the Earth surface and at a depth of 20 km, are separated by a distance equivalent to the Earth diameter minus 40 km.

An interesting pattern is evidenced in Fig. 1. Shorter ranges (3 km < r < 30 km) are defined by an unique slope giving a relatively high value of the correlation dimension ($D \simeq 1.9$). From 30 km to 300 km the slope is varying giving a not uniquely defined D value, as evidenced from the local slope graph which depicts a decreasing value of $d(r)$. For distances over 300 km the slope is again constant, giving a value of $D \simeq 1.2$. Several factors contribute to such values over all ranges; at shorter distances certainly localization error influences the values, forcing D values to be higher, but, in general, seismotectonics

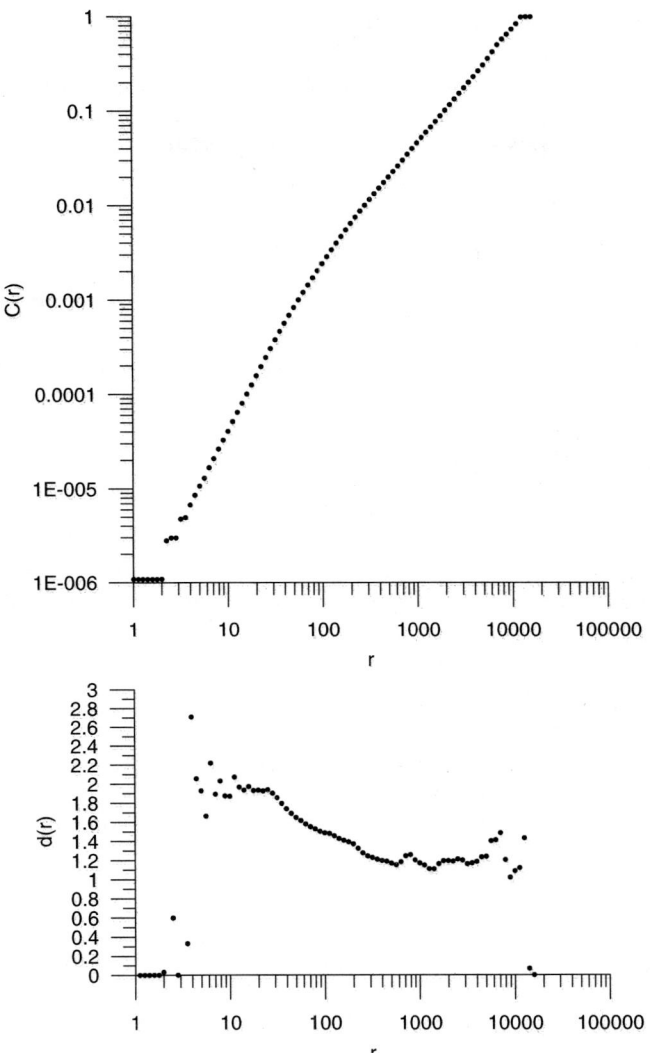

Fig. 1. Correlation integral of global seismicity in space (*upper panel*), r is the distance in km between hypocenters. In the *lower panel* are shown the local slopes

plays the most important role. For such short ranges $D \simeq 2$ may correspond to a distribution of events over a fault plane: if error had played a greater influence, we would have observed $D \simeq 3$. Highest ranges, corresponding to values of D around 1.2, suggest a linear setting of seismicity, probably due to large faults and/or plates boundaries, that are linear at such scales.

3 Scale Invariance in Time

Correlation integral (2) can be applied to earthquakes time occurrence too. In this case we have:

$$C(\tau) = \frac{2}{N(N-1)} \sum_{i=1}^{N-1} \sum_{j=i+1}^{N} \Theta\left(\tau - |t_i - t_j|\right) \qquad (5)$$

where t_i is the occurrence time of earthquake i.

The behavior of (5) is shown in Fig. 2 either in bi-logarithmic and local slope plot. The pattern is very different if compared with the spatial correlation integral. In this case the correlation integral (5) shows a very constant and stable pattern expressed by a constant local slope and a well defined value of D over a wide range of times. With least square fit on the log-log plot, in the time range 24 min – 3000 days, we obtain $D_E = 0.94$ a value near the embedding dimension ($D = 1$). Local slope values are constantly very high in the same time range. The interpretation of this behavior is based on the evidence that temporal distribution of seismicity can be considered nearly random. This result could seem to be in contrast with Omori law which states that, in time, seismicity rate after a mainshock follows a decreasing power law. The paradox is only apparent, because while the rate can vary as a power law, time occurrence remains random: these are two different aspects of the seismic process evolving in time. To clarify this point, we show in Fig. 3 the evolution of seismic rate of a simulated sequence following the Omori's law. In the figure we report the log-log plot of the correlation integral (5) calculated from the same sequence: it is evident the high value of the time fractal dimension, indicating random time occurrence.

4 Space-Time Combined Correlation Integral

The results presented so far have been explained in terms of seismotectonic interpretations, but some apparent contradictions arise. Seismicity is mainly the expression of sequences composed by mainshocks (sometimes preceded by foreshocks) and aftershocks, usually well recognized both in space and time. Moreover there are triggering processes acting at short distances and probably at larger distances too: triggering is a strong deterministic factor which should lower the random content of seismic occurrences in space and time. Nevertheless all these features seem to leave little or nothing influence into the separate space and time analysis, carried on through the correlation integral approach. We believe that the main reason for this lies in the fact that space and time occurrences of seismicity are deeply interconnected: correlation integral, calculated on space distribution alone, mix together all time occurrences and the opposite happens for time correlation integral. This leads to a blurring of combined space-time features. A sequence is a perfect

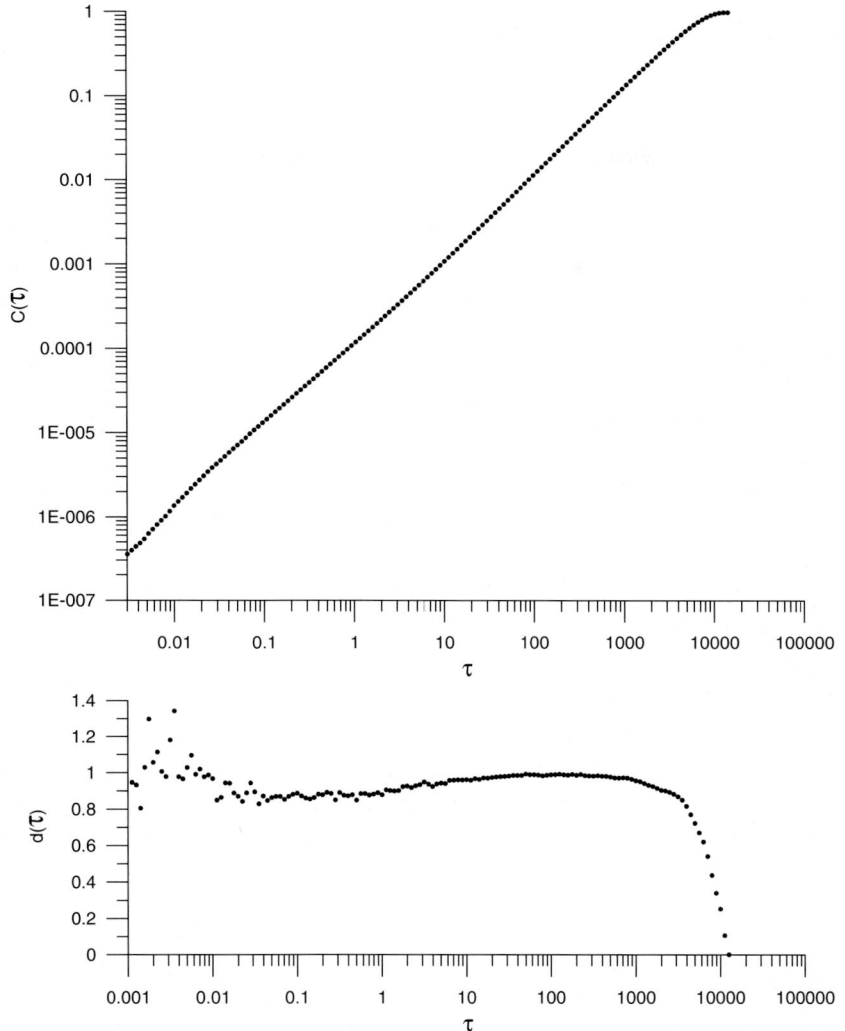

Fig. 2. Correlation integral of global seismicity in time (*upper panel*), τ is the time interval in days. In the *lower panel* are shown the local slopes

example of such interconnections: foreshocks, whenever present, may show a sort of critical activation around a seismogenic area; mainshock delineates the main activated fault; immediately after, through stress redistribution driven by source geometry, the sequence spread out around the mainshock, producing peculiar space-time patterns. The intrinsic self similar character of seismicity have to be reflected into space, time and magnitude events distribution: one

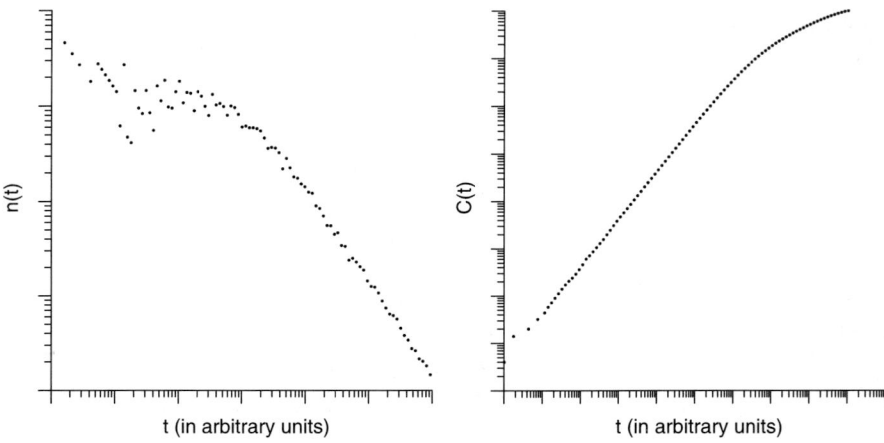

Fig. 3. Seismic rate $n(t)$ of a synthetic seismic sequence following the Omori law (1)(*left panel*); correlation integral of the same sequence: fractal dimension calculated on log-linear portion of the curve shows a random behavior ($D = 0.96$) (*right panel*)

strong reason for this is due to the fractal distribution of faults [12] and to their interactions.

It is with this spirit that we introduce here a new approach leading to a self consistent analysis of both spatial and temporal correlations. We define the space-time combined correlation integral as:

$$C_c(r,\tau) = \frac{2}{N(N-1)} \sum_{i=1}^{N-1} \sum_{j=i+1}^{N} \left(\Theta\left(r - \|\mathbf{x}_i - \mathbf{x}_j\|\right) \cdot \Theta\left(\tau - \|t_i - t_j\|\right)\right), \quad (6)$$

with an obvious use of the notation. It is worth noticing that (6) includes (5) when $r = r_{\max}$ and (4) when $\tau = \tau_{\max}$. Combined correlation integral (6) constitutes in fact a true generalization of time and space cases taken separately, including them as particular end point cases. Moreover it allows combined correlation analysis in cases when there is a set of non comparable dimensions. In our case there are three spatial coordinates and one temporal. The visualization of $C_c(r,\tau)$ involves displaying a three dimensional surface (Fig. 4), instead of a two dimensional plot as for (2) applied to space and time separately. From $C_c(r,\tau)$ we define the time correlation dimension D_t and its local slope d_t as:

$$d_t(r,\tau) = \frac{\partial \log C_c(r,\tau)}{\partial \log \tau} \quad D_t(\tau) = \lim_{\tau \to 0} d_t(r,\tau), \quad (7)$$

and the space correlation dimension D_s with its local slope d_s as:

$$d_s(r,\tau) = \frac{\partial \log C_c(r,\tau)}{\partial \log r} \quad D_s(r) = \lim_{r \to 0} d_s(r,\tau). \quad (8)$$

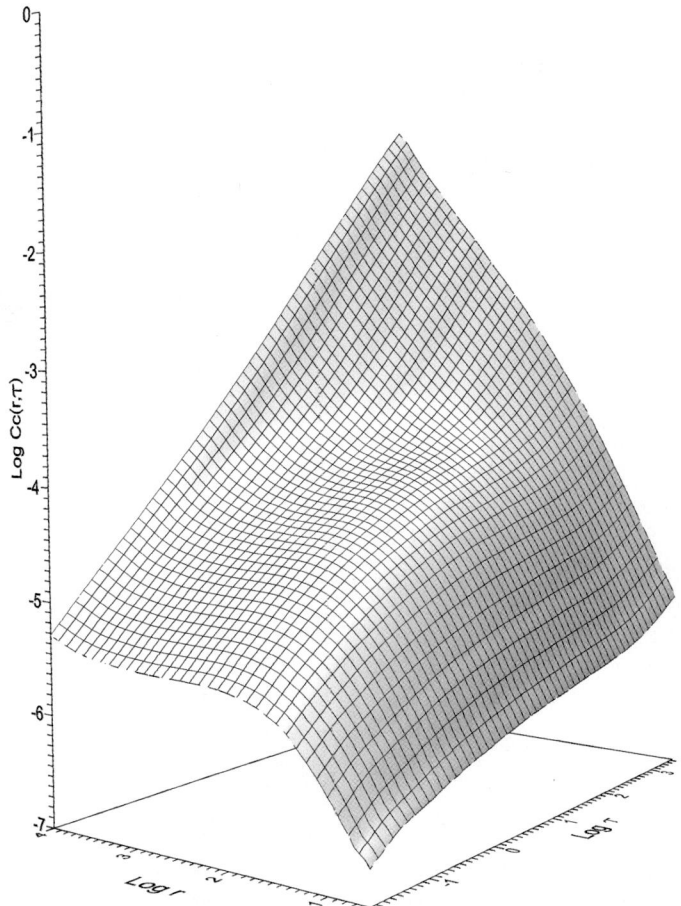

Fig. 4. Space-time combined correlation integral $C_c(r,\tau)$. For r equal to the maximum distance in the catalog the C_c surface reduces to the plot in Fig. 1, while for τ equal to the whole time data span, C_c come down to the plot in Fig. 2

For a random distribution of earthquakes both in space and time – therefore without any combined space-time correlations – d_t and d_s would be constant for all r and τ, equal to their respective embedding dimension values. For events characterized by their specific d_t and d_s values, but still no combined space-time correlations, d_t will be equal for all r and the same to d_s for all τ.

We now first analyze the behavior of $d_t(r,\tau)$ (7) and later on of $d_s(r,\tau)$ (8) for the global seismic catalog. The two plots represent the local partial derivative of $C_c(r,\tau)$ (6), displayed in Fig. 4, that will appear as superimposed gray contour lines.

5 Time Combined Correlation Dimension

The plot of $d_t(r,\tau)$ (7) is shown in Fig. 5 with $\log \tau$ as abscissa and $\log r$ as ordinate. A clear pattern emerges whose most evident features are as follows. A zone with high value of $d_t(r,\tau)$ is well separated from a time clustering domain ranging from 0.1 to 1000 days, at distances up to 200–300 km. The clustering feature is revealed by the low values of $d_t(r,\tau) < 0.4$. The boundary between these two main regions is not sharp and it is not at a constant r value: this means that the limit of the clustering region depends in a non-trivial way on space and time.

It is now important to clarify the relations among the value of correlation dimension, the notion of clustering and of random-causal relationship. Values of the correlation dimension range between 0 and the value of the embedding dimension, which for time is 1. Moreover correlation dimension is a direct measure of clustering: having a set with the correlation dimension equal to the embedding space means that this set covers it uniformly; conversely a set with a low correlation dimension is concentrated into particular patterns, the limit case being $d_t(r,\tau) = 0$ where all the elements of the set are concentrated on a point (which has the topological dimension equal to 0). This topological relation is also deeply related with the randomness content of the set.

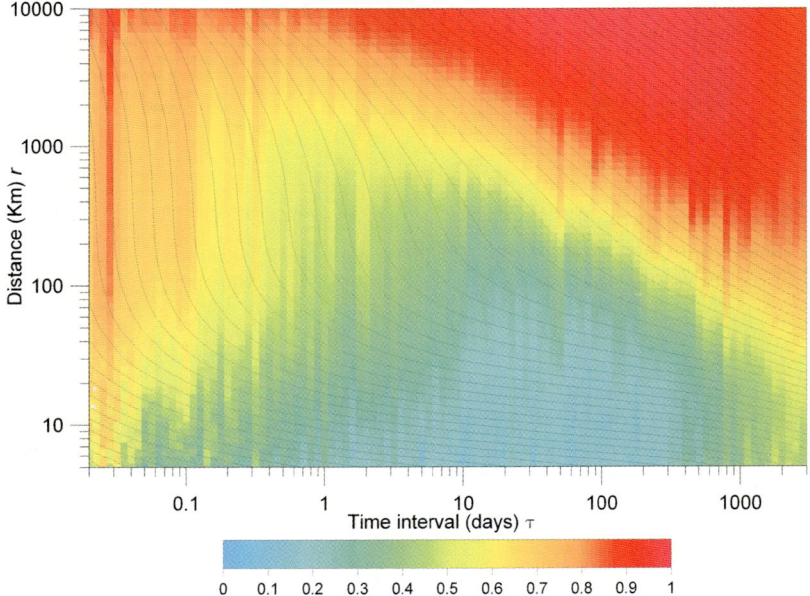

Fig. 5. Local slopes of time correlation dimension $d_t(r,\tau)$ (*colored shaded contour*) for global seismicity. In *dark contour lines* is represented the space-time combined correlation integral $C_c(r,\tau)$

A random distribution of points will cover uniformly the whole embedding space, with no preference for particular positions or patterns. This results in a statistically complete and homogeneous covering of the embedding space.

Low $d_t(r,\tau)$ values significantly support the hypothesis that, inside specific space and time intervals limits, earthquakes are correlated. This feature, being limited to small values of r and τ, tends to disappear for increasing space and time intervals values. In order to check the validity of this experimental result, we have applied the same analysis to the global catalog after a reshuffling procedure. Reshuffling consists in randomly mixing the time occurrences of all events, maintaining their hypocentral coordinates (see Table 1). This procedure does not change the statistical properties of the data set separately for time and space, but it destroys their mutual connections. The result shows that all patterns vanish (Fig. 6), evidencing constant high values of d_t at all distances and time intervals. This behavior of the reshuffled catalog reflects the intrinsic absence of space-time connections and it duplicates, over all distance ranges, the behavior of largest space interval, as shown when sole time correlation integral (5) was applied (Fig. 2). In Fig. 6 a not colored area appears, indicating that, for short space and time intervals, the necessary minimum amount of earthquakes data pairs is not met, making not reliable the calculation of $d_t(r,\tau)$. This is due to the complete breakup of seismic sequences operated by reshuffling, so that it is unlikely to find earthquake couples at small distances and short time interval.

The comparison between Fig. 5 and Fig. 6 demonstrates two main results: first the low $d_t(r,\tau)$ values area is not trivial, but reflects truly space-time connections of seismic events at short space and time ranges; second this feature is due to the presence of seismic sequences, which operates typically

Table 1. Example of reshuffling procedure. Upper part: original catalog. Lower part: shuffled catalog

Year	Mount	Day	Hour	Min	Sec	Lat	Long
1996	06	13	06	57	58.25	−20.416	−178.310
1997	02	08	12	39	41.85	1.657	97.951
1998	02	25	09	12	26.49	40.876	143.834
1999	03	12	20	42	35.06	−19.292	−68.847
2000	08	09	00	08	41.81	−15.693	167.986
Year	Mount	Day	Hour	Min	Sec	Lat	Long
1996	06	13	06	57	58.25	−19.292	−68.847
1997	02	08	12	39	41.85	−20.416	−178.310
1998	02	25	09	12	26.49	−15.693	167.986
1999	03	12	20	42	35.06	1.657	97.951
2000	08	09	00	08	41.81	40.876	143.834

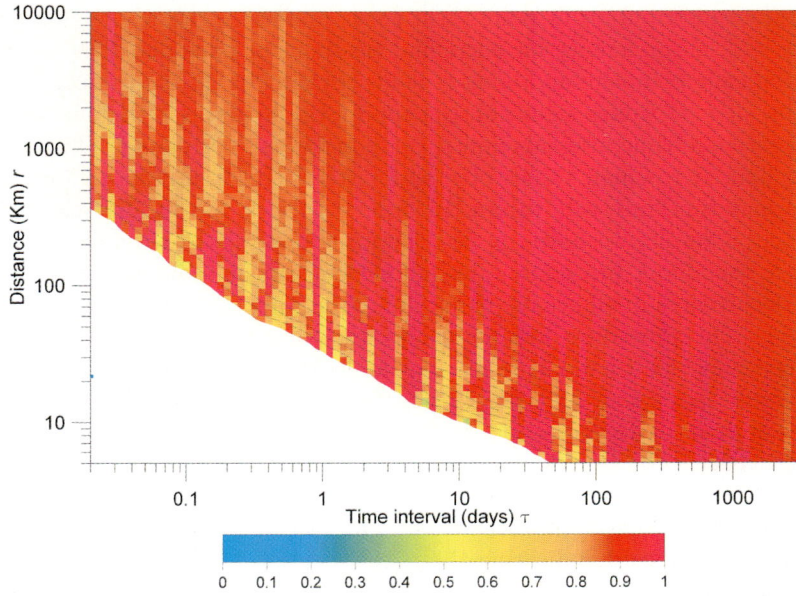

Fig. 6. Local slopes of time correlation dimension $d_t(r,\tau)$ (*colored shaded contour*) and space-time combined correlation integral $C_c(r,\tau)$ (*dark contour lines*) for reshuffled catalog

inside this ranges. A very interesting consequence of this result is that we can pose space and time limits to sequence evolution. Typically a sequence is defined as the set of events triggered by the main shock. Spatially the fault plane, activated through the mainshock, is successively filled with the majority of aftershocks: in fact one standard way to delineate the seismogenic fault, after a mainshock occurrence, is to localize as much as possible the aftershocks with the highest precision, in order to arrive, through exact latitude, longitude and, particularly, depth, to the delineation of the fault plane. This procedure became a standard after the development of local seismic networks, integrated by sophisticated relocation procedures. In time, a sequence is traditionally delineated by the persistence of the Omori's law. Such definitions are simplistic and could be considered valid as a first order approximation. Certainly events over the main seismogenic fault and lasting until the validity of the Omori law, pertain to the seismic sequence. The real problem is introduced by the intrinsic self similar character of seismicity: the main seismogenic fault is not isolated because an entire network of faults may be interested by stress redistribution. As previously stated, faults are often distributed and sized in a fractal pattern; for seismicity a direct consequence of this is an intrinsic scale invariance either for space, time and magnitude. It becomes problematic, inside this framework, to establish limits of each sequence, as the true set of all events related among themselves. Here we find the validity of our result: we delineate through time combined correlation integral a temporal connection among events and we find that the space-time limit of this relation is dynamically varying.

Figure 5 shows that the space-time boundary defining the transition from low $d_t(r,\tau)$ values to a random domain is gradual. We decided to exclude, from correlation ranges, events that are certainly out from any kind of correlation. To achieve this we have to exclude events that gives a value of $d_t(r,\tau)$ high enough to assure randomness. To this purpose we simulated a random global catalog, composed with the same number of events of the real one, but with time and space occurrence randomly distributed inside their real limits. For the space we considered a uniform distribution of points on a sphere, with random depth into the range 0–700 km. For time we assigned to each event a random occurrence inside the time span of the catalog. As expected all simulated $d_t(r,\tau)$ values are very near 1, with average $\bar{d}_t(r,\tau) = 0.99$ and standard deviation $\sigma = 0.10$. Having $\bar{d}_t(r,\tau) - 2\sigma = 0.79$ we set $d_t(r,\tau) = 0.8$ as the lower limit of random behavior. The plot of this limit, as extrapolated from $d_t(r,\tau)$ applied to real global data, is shown as function of r and τ, in the range 1–2000 days, on Fig. 7. The behavior can be well approximated by a straight line fit in a log-log scale:

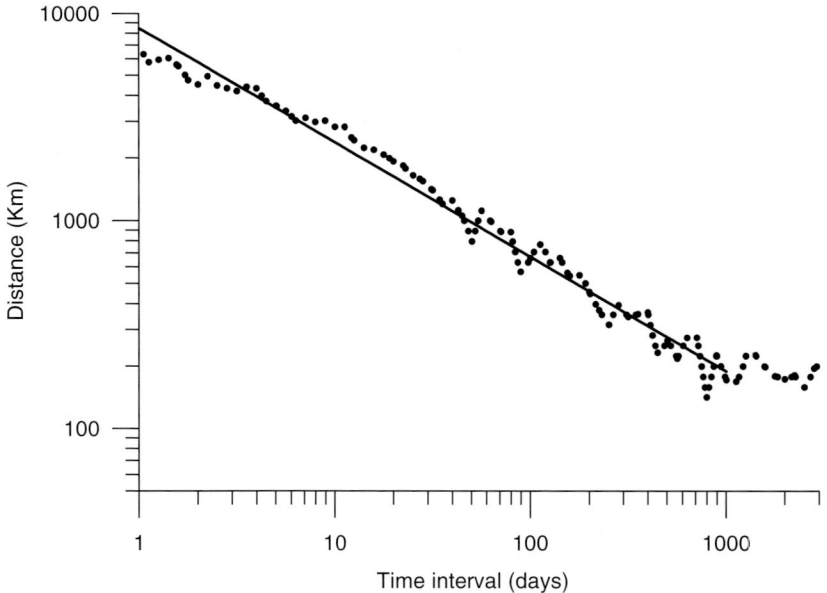

Fig. 7. Time intervals τ and inter-distances r for which $d_t = 0.8$ representing the limit between time clustering (*lower left*) and time randomness (*above right*, see Fig. 5). In the time range 1–1000 days least square fit of points in log-log scale results in the line of equation $\log r = -0.55 \log \tau + 3.9$

$$\log R_I = -0.55 \log \tau + 3.9 \ . \tag{9}$$

This results points to the existence of an influence area (of radius R_I) which shrinks over time. This results is quite robust since the slope of (9) changes little if we repeat the same procedure with any threshold value in the interval $0.6 \leq d_t(r,\tau) \leq 0.9$.

It is interesting to verify whether the space-time boundary between correlated and random features depends on the magnitude of the main shock : larger earthquakes should extend their influence over larger distances and for longer times. We want to check the persistence of the linear fit (9) and how its parameters may change, allowing the magnitude of the ith main shocks (6), the "triggering" event to be band limited into the ranges 5.0–5.5, 5.5–6.0, 6.0–6.5 and 6.5–7.0, while the jth "triggered" events maintain the catalog threshold for all different tests. In Fig. 8 are reported all limits calculated as for (9): it is evident that increasing the ith main shock magnitude the line migrates toward longer distance and time intervals. The least squares slopes of the lines are quite similar to the slope of basic magnitude fit (9).

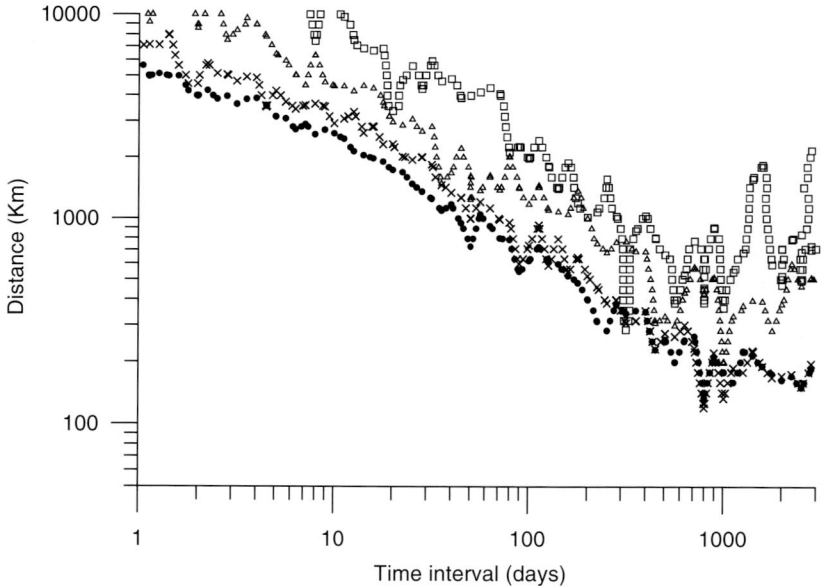

Fig. 8. Variation of the points corresponding to $d_t = 0.8$ with respect to the magnitude of the reference event. The points marking the limit between time clustering and randomness are magnitude-dependent ($5.0 \leq m_b < 5.5$ dots; $5.5 \leq m_b < 6.0$ crosses; $6.0 \leq m_b < 6.5$ triangles; $6.5 \leq m_b < 7.0$ squares)

6 Space Combined Correlation Dimension

Similarly to d_t, we calculated d_s (8) from space-time combined correlation integral C_c (Fig. 9). The pattern is very different if compared with Fig. 5. In space domain, having considered source depth too, d_s values may vary from 0 to 3, being hypocenters points embedded into a 3D space. Actually this does not happen for two intrinsic reasons: first, global seismicity is depth limited, being forced by brittle rheology of the crust into few hundreds of kilometers: this dimension is small if compared to the whole latitude and longitude extension; second, seismicity tend to be located – and hence driven – by planar patterns mainly due to seismogenic faults, or linear patterns as plate boundaries. The effects of these geophysical factors give the result of a lowering of the real dimension of hypocenters, which turns out to be not larger than 2. Despite these limitations, the pattern of d_s values remains well differentiated, denoting peculiar space-time regions. At a first approximation we can distinguish three main regions: a $1 < d_s < 2$ value band present for all times (from small portion of a day to 3000 days) and limited to shorter distances (10–20 km at shortest time intervals, to 30 km for time intervals of about 200 days and increasing until 200–300 km for the longest time intervals); a low d_s values zone (sensibly less than 1.0) present at shortest time intervals

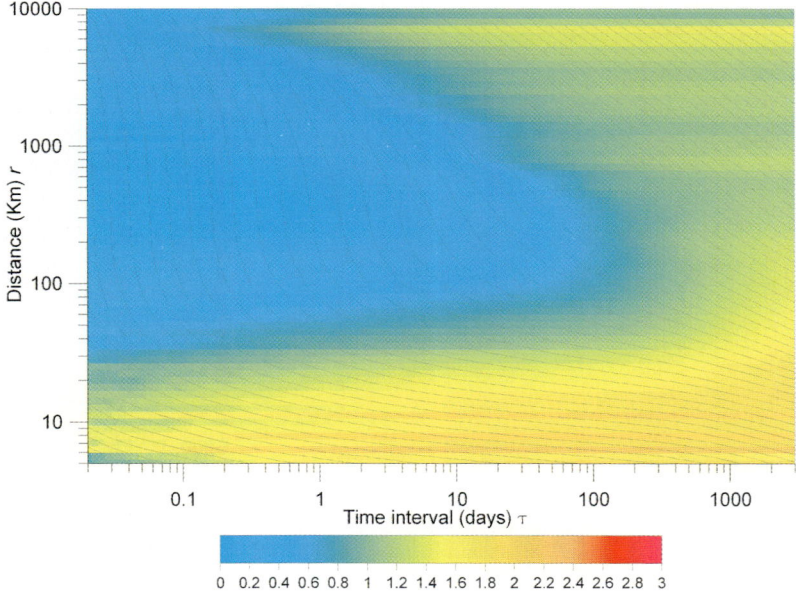

Fig. 9. Local slopes of space correlation dimension $d_s(r, \tau)$ (*colored shaded contour*) for global seismicity. In *dark contour lines* is represented the space-time combined correlation integral $C_c(r, \tau)$

(till around 100 days) and at distances from few tens of km to few thousands; finally an intermediate d_s values ($1.0 < d_s < 1.5$) for distances larger than 100 km and long time intervals.

It appears evident that the pattern of d_s values is much more articulated than the simple spatial fractal dimension calculated following (4). In fact the resulting two-slopes behavior (Fig. 1) is represented now as the particular endpoint case at the right end of the plot in Fig. 9.

To check whether the plot of the space correlation dimension could be a trivial result, we have applied the same procedure on the reshuffled catalog. The resulting plot (Fig. 10) shows that the non-combined statistical properties of data are not sufficient to produce the clustering domains, but a real connection between space and time is needed. Similarly to the case of d_t, the d_s behavior of the reshuffled catalog, for which there was no difference of values at different distances, has the same behavior (Fig. 1) for all time intervals, being nothing more that a random under-sampling of the whole data set.

Highest d_s values in Fig. 9 ($1 < d_s < 2$), confined at shortest spatial distances, may reflect the role played by seismogenic faults (planar structures), where the main cloud of seismic events are embedded. It is important to note that, at such short distances, the localization error of hypocenters may play an important role. Specifically it may increase the fractal spatial dimension: in

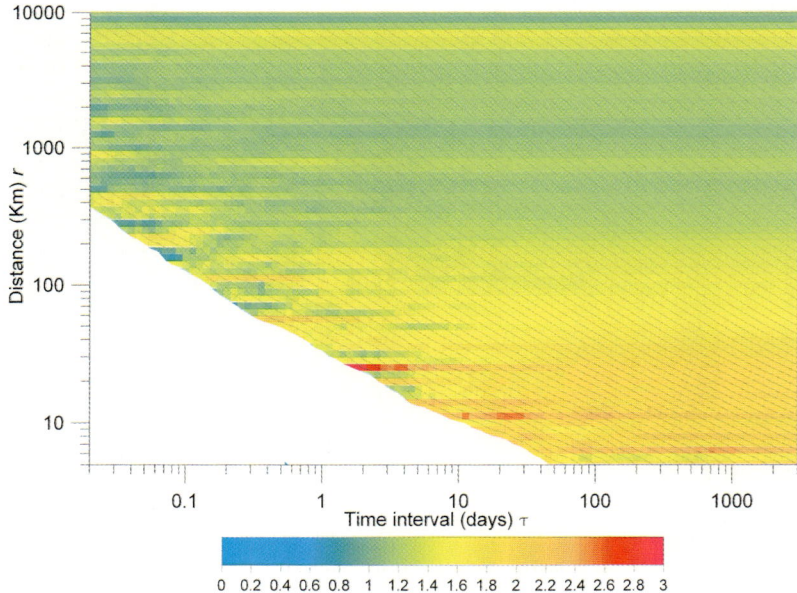

Fig. 10. Local slopes of space correlation dimension $d_s(r, \tau)$ (*colored shaded contour*) and space-time combined correlation integral $C_c(r, \tau)$ (*dark contour lines*) for reshuffled catalog

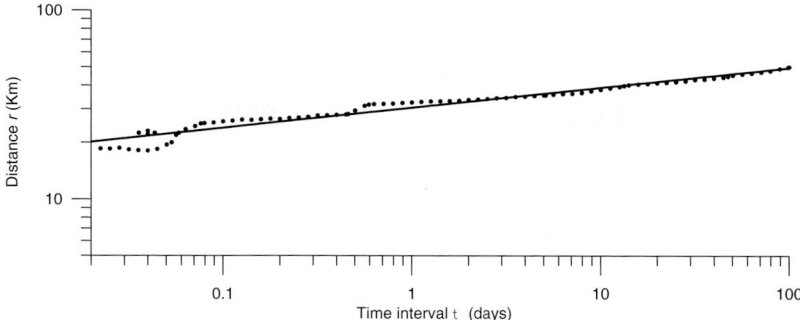

Fig. 11. Time intervals τ and inter-distances r for which $d_s = 1.0$ representing the limit between space clustering (*above*) and higher values of d_s (*lower*, see Fig. 9) corresponding to dimension of a line up to that of a plane. In the time range from 29 minutes to 100 days least square fit of points in log-log scale results in the line of equation $\log r = 0.11 \log \tau + 1.5$

fact it has been experimentally proved that noise increases the fractal dimension of a data set [2]. But a factor emerges indicating that this factor is not so determinant: the distance limit of this high valued d_s zone is not constant over the time intervals, but it increases slowly. Localization error of seismic events can be considered constant for the time period covered by the data catalog at these high magnitude level: an increasing of the range can be explained by a real seismic process. Moreover spatial localization noise must reflect its influence on d_s values, pushing them near the embedding space dimension, which, having considered depth, is 3. Focusing the attention on the limit of this high d_s value zone, we set $d_s = 1$ as the limit value between high d_s and a space clustering region (Fig. 11). A linear fit in the log-log plot well describes the power law increase. Least squares fit inside the time interval between 30 minutes to 100 days gives:

$$\log R_S = 0.11 \log \tau + 1.5 \ . \tag{10}$$

We can interpret the short distances domain bounded by the distance R_S from the reference event given by this relation as the result of the distribution of hypocenters inside every seismic zone during a sequence: this is supported by the short spatial distance ranges and time intervals covered, typical those of a developing sequence. The low positive slope expressed by (10) may indicate a sub-diffusion of events as the sequence evolves in time. This finding is in agreement with the accepted migration of aftershocks away from a main shock based on the interpretation of uniformly spatially distributed events as aftershocks. Many authors have described this migration in terms of a law $\bar{l}(t) \sim t^H$, where $\bar{l}(t)$ is the mean distance between main event and aftershocks occurring after time t, with an exponent $H < 0.5$ corresponding to a sub-diffusive process often observed for local situations [9, 14, 19, 27].

The lowest d_s value zone should be considered in connection with the intermediate $1.0 < d_s < 1.3$ one: in fact these two zones together cover all time intervals for spatial inter-distances bigger than circa 30km. d_s values around 1 imply a linear pattern: it is reasonable if we consider that seismicity is placed on active tectonic lineaments, as big faults and plate borders. Lower d_s values imply a spatial clustering, which is typical when seismicity is considered as a collection of individual sequences, each composed by the main shock and its cloud of aftershocks.

The whole space-time pattern of d_s delineates an evolution of spatial localization of hypocenters. During all time intervals seismicity is localized on shorter distances with a dimension around 2, which is typical of planar or sub-planar structures: this well reflects the typical behavior of a sequence at distances of the order of the source fault. Inside the time range of 100 days at longer distances seismicity is distributed in clusters: considering the low d_s values of this space-time zone, we can state that the clustering is strong. Over longer time intervals clusters evolve filling linear structures typical of regional or continental sizes. We can note that the pattern previously evidenced, with sole spatial correlation dimension analysis (Fig. 1), reflects the end stage of spatial evolution through time of seismic sequences, which appears to be strongly time conditioned: all patterns of shorter time intervals, now clearly evidenced, were completely hidden, and hence lost, through the sole spatial analysis.

7 Discussion and Conclusion

The combined space-time correlation integral analysis strongly enhances the comprehension of seismic interactions, showing how important is the interplay between space and time dimensions. Like some previously cited authors, we do not separate seismicity into main and aftershocks. We consider medium-high energy global seismicity, where every event can be viewed as a potential mainshock. Our method operates a sort of stacking of every possible sequence. We can thus interpret the obtained results as a representation of an ideal averaged single sequence. We hence describe our findings in a organized space-time representation, in order to give an unitary view of the process.

We can define a new kind of seismic sequence over which we can quantify spatial and temporal domains. As for the time occurrence aspects we can define a clustering sequence, where events are time clustered, inside a defined space-time window. As result each event can be part of several clustering sequences, maintaining relations with events happened before and after its time occurrence. Long spatial distances, reached by clustering relation shortly after the occurrence of an event, give a support to long range interaction.

As for the spatial occurrence aspects we delineated a near zone filled by hypocenters of following events. The limit of this area slowly increases with time in good agreement with results found by several authors

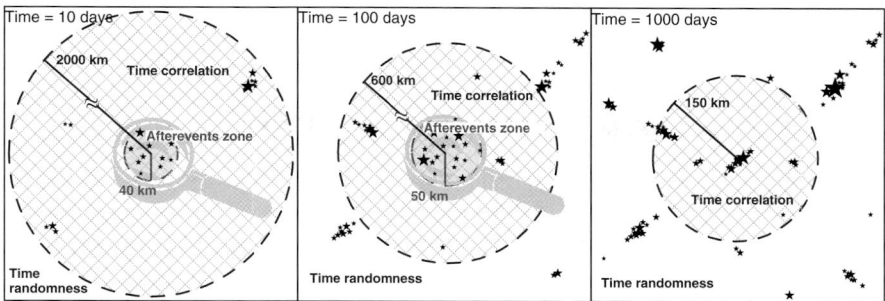

Fig. 12. Cartoon showing the main steps of space-time evolution of a standard sequence. *Left panel*: after 10 days from the main event seismic events are time correlated until a spatial range of 2000 km from the mainshock; space randomness is around an area of 40 km in radius. *Center panel*: after 100 days from the main event time correlation area has shrunken to 600 km from the mainshock; space randomness area has slowly increased to 50 km. *Right panel*: after 1000 days from the main event time correlation area has reduced to 150 km

[7, 9, 14, 19, 20, 27] who found, using other methods, a migration of aftershocks away from a main shock in a sub-diffusive fashion. At regional scales seismicity appears in spatial clusters for short to medium time period, after which it tends to fill linear structures like big fault lines and plate boundaries.

We summarize both space-time results in Fig. 12, where we recognize a "near field" domain, evidenced by the behavior of the space correlation dimension, and a "far field" domain, defined by the behavior of the time correlation dimension. Both domain ranges evolve in time: the first one slowly increasing and the latter quickly shrinking. An interpretation of this scenario, in terms of stress transfer mechanisms activated inside a sequence, is possible.

A given earthquake induces different kinds of stress transfer mechanisms that can generally be classified as co-seismic and post-seismic. The first group is based on the elastic properties of the crust and can be either static or dynamic. Post-seismic stress transfer (sometimes referred to as quasi-static) is associated with the slow viscous relaxation of the lower part of the crust and the upper part of the mantle. The debate is wide open as to which one of these mechanisms is principally responsible for triggering of earthquakes. It is generally accepted that stress changes $d\sigma$ decay over distance s as the power-law $d\sigma \propto |s|^{-a}$ with the exponent a dependent on the specific mechanisms of stress transfer and on the lithosphere rheology. A general distinction can be made between "near field" behavior, occurring at distances from the triggering event of the order of the fault length (and more generally of the order of the size of the seismogenic structure), and "far field" or long-range behavior. Many factors drive stress transfer and seismicity in the near field such as: source mechanisms, pre-existing weakness zones, heterogeneity of the fault plane and fluid migration. When focusing on the far field, i.e. on what happens in a region much larger than the fault length, many of these factors integrate out and a general statistical description appears possible. This longer range indicates an ever changing stress field that tends to weaken over time, either by earthquake occurrences or by aseismic creep (slow slips not generating elastic waves) and, more generally, by all the mechanisms falling under the denomination of stress leakage. Depending on the distance from the main event, an alteration in stress can statistically affect the failure occurrence probability and, hence, the seismicity rate. Greater the stress change detected, greater should be the seismicity rate change. Consequently, for very small stress changes the triggering/shadowing effect is negligible and suggests the existence of some sort of elastic threshold, below which the stress change should not be able to affect the seismicity rate. Such a lower cut-off could be identified, for instance, with the level of "tidal stress" that is induced by the distortion of the earth caused by the pull of the sun and moon. Typical values of tidal stress changes are in the order of 0.01 bars, and do not directly influence seismicity. Adoption of this threshold allows identification of a length scale $R(t)$, defined as the distance from the main event's epicenter for which the stress change falls below the lower cut-off. We interpret this length scale $R(t)$ as the radius of the region causally connected to the main event shortly after the event itself. Since the level of stress drops over time one should expect that $R(t)$ decreases over time. This is what can be observed if we interpret $R(t)$ and t as the previously defined R_I and τ (9).

In summary, we have introduced a new statistical tool, the combined space-time correlation integral, which allows us to perform a simultaneous and self-consistent investigation of the spatial and temporal correlation properties of earthquakes. This tool leads to the discovery, visualization and deep analysis of the complex interrelationships existing between the spatial distribution of epicenters and their occurrence in time. Three main results emerged: the comparison between space and time correlations, performed on the worldwide seismicity catalog and the corresponding reshuffled catalog, strongly suggests that earthquakes do interact spatially and temporally. From the study of time clustering, a new universal relation linking the influence area R_I of an earthquake and the time elapsed since its occurrence is discovered. Finally, analysis of the space clustering reveals the existence of a region where events are uniformly distributed in space. The size of this region increases slowly with time, supporting existing theories on aftershock diffusion. The ensemble of our results set the basis for further validation on both worldwide and local scales, as well as for the introduction of suitable modeling schemes.

Beyond relevance in seismology it is worth stressing how the combined correlation integral C_c could be applied in other contexts, where understanding the interplay of spatial and temporal correlations is crucial for the correct interpretation of phenomena such as solar flares, acoustic emissions, dynamical systems theory, and the dynamics of extended systems in physics and biology.

References

1. P. Bak & C. Tang, J. Geophys. Res., **94**, 15635 (1989).
2. A. Ben-Mizrachi, I. Procaccia & P. Grassberger, Phys. Rev. A, **29**, 975 (1984).
3. V. De Rubeis, P. Dimitriu, E. Papadimitriu & P. Tosi, Geophys. Res. Lett., **20**, 1911 (1993).
4. P. Gasperini & F. Mulargia, Bull. Seismol. Soc. Am., **79**, 973 (1989).
5. P. Grassberger & I. Procaccia, Physica D, **9**, 189 (1983).
6. C. Godano, P. Tosi, V. De Rubeis & P. Augliera, Geophys. J. Int., **136**, 99 (1999).
7. C. Godano & F. Pingue, Geophys. Res. Lett., **32**, 18302 (2005).
8. R.A. Harris, J. Geophys. Res., **103**, 24347 (1998).
9. A. Helmstetter, G. Ouillon & D. Sornette, J. Geophys. Res., **108**, 2483 (2003).
10. D.P. Hill et al, Science, **260**, 1617 (1993).
11. T. Hirabayashi, K. Ito & T. Yoshii, Pure Appl. Geophys., **138**, 591 (1992).
12. T. Hirata, PAGEOPH, **131**, 157 (1989).
13. T. Hirata & M. Imoto, Geophys. J. Int., **107**, 155 (1991).
14. M. Huc & I.G. Main, J. Geophys. Res., **108**(B7), 2324 (2003).
15. Y.Y. Kagan & L. Knopoff, Geophys. J.R. Astron. Soc., **62**, 303 (1980).
16. H. Kantz & T. Schreibre, *Nonlinear Time Series Analysis*, Cambridge University Press, Cambridge (1997).
17. G.C. P. King, R.S. Stein & J. Lin, Bull. Seismol. Soc. Am., **84**, 935 (1994).
18. C. Lomnitz, Bull. Seismol. Soc. Am., **86**, 293 (1996).
19. D. Marsan, C.J. Bean, S. Steacy & J. McCloskey, J. Geophys. Res., **105**, 28081 (2000).
20. D. Marsan & C. Bean, Geophys. J. Int., **154**, 179 (2003).
21. W. Marzocchi, J. Selva, A. Piersanti $ E. Boschi, J. Geophys. Res., **108**(B11), 2538 (2003).
22. D. Melini, E. Casarotti, A. Piersanti & E. Boschi, Earth and Planetary Science Letters, **204**, 363 (2002).
23. P.A. Reasenberg, J. Geophys. Res., **104**, 4755 (1999).
24. B.E. Shaw, Geophys. Res. Lett., **20**, 907 (1993).
25. R.S. Stein, G.C. P. King & J. Lin, Science, **265**, 1432 (1994).
26. R.S. Stein, Nature, **402**, 605 (1999).
27. F. Tajima & H. Kanamori, Phys. Earth Planet. Int., **40**, 77 (1985).
28. P. Tosi, Annali di Geofisica, **41**, 215 (1998).
29. P. Tosi, De Rubeis V., Loreto V. & L. Pietronero, Annals of Geophysics, **47**, 1849 (2004).
30. T. Utsu, Y. Ogata & S. Matsu'ura, J. Phys. Earth, **43**, 1 (1995).
31. T. Utsu, Journal of the Faculty of Sciences, Hokkaido University, Ser. VII, III, 4 (1970).

Short-Term Prediction of Medium- and Large-Size Earthquakes Based on Markov and Extended Self-Similarity Analysis of Seismic Data

M.R.R. Tabar[1,2,a], M. Sahimi[3,b], F. Ghasemi[1,c],
K. Kaviani[4,d], M. Allamehzadeh[5,e], J. Peinke[6,f] M. Mokhtari[5,g],
M. Vesaghi[1,h], M.D. Niry[1,i], A. Bahraminasab[7,k],
S. Tabatabai[8,l], S. Fayazbakhsh[1,m] and M. Akbari[5,n]

[1] Department of Physics, Sharif University of Technology, P.O. Box 11365-9161, Tehran, Iran
   [c]f_ghasemi@mehr.sharif.edu, [h]vesaghi@sharif.edu,
   [i]mdniry@mehr.sharif.edu, [m]S_Fayazbakhsh@mehr.sharif.edu
[2] CNRS UMR 6202, Observatoire de la Côte d'Azur, BP 4229, 06304 Nice Cedex 4, France
   [a]rahimitabar@sharif.edu
[3] Department of Chemical Engineering and Material Science, University of Southern California Los Angeles, CA 90089, USA
   [b]moe@iran.usc.edu
[4] Department of Physics, Az-zahra University, P.O. Box 19834, Tehran, Iran
   [d]kaviani@scintist.com
[5] Department of Seismology, The International Institute of Earthquake Engineering and Seismology, IIEES, P.O. Box 19531, Tehran, Iran
   [e]Mallam@iiees.ac.ir, [g]Mokhtari@iiees.ac.ir, [n]mary.Akbari@gmail.com
[6] Carl von Ossietzky University, Institute of Physics, 26111 Oldendurg, Germany
   [f]peinke@uni-olden-burg.de
[7] ICTP, Strada Costiera 11, 34100 Trieste, Italy
   [k]abahrami@ictp.it
[8] Institute of Geophysics, University of Tehran, Iran
   [l]S.Tabatai@tabagroup.com

We propose a novel method for analyzing precursory seismic data before an earthquake that treats them as a Markov process and distinguishes the background noise from real fluctuations due to an earthquake. A short time (on the order of several hours) before an earthquake the Markov time scale t_M increases sharply, hence providing an alarm for an impending earthquake. To distinguish a false alarm from a reliable one, we compute a second quantity, T_1, based on the concept of extended self-similarity of the data. T_1 also changes strongly before an earthquake occurs. An alarm is accepted if *both* t_M and

T_1 indicate it *simultaneously*. Calibrating the method with the data for one region provides a tool for predicting an impending earthquake within that region. Our analysis of the data for a large number of earthquakes indicate an essentially *zero* rate of failure for the method.

1 Introduction

Earthquakes are complex phenomena to analyze. The interaction between the heterogeneous morphology of rock and the mechanism by which earthquakes occur gives rise to distinct characteristics in different parts of the world. Seismic data as time series exhibit complex patterns, as they encode features of the events that have occured over extended periods of time, as well as information on the disordered morphology of rock and its deformation during the time that the events were occuring. It is for such reasons that seismic records appear seemingly chaotic.

Published reports indicate the existence of precursory anomalies preceding earthquakes. The reported anomalies take on many different forms, and contain aspects of seismic wave propagation in rock, and its chemical, hydrological, and electromagnetic properties. The spatio-temporal patterns of seismicity, such as anomalous bursts of aftershocks, quiescence or accelerated seismicity, are thought to indicate a state of progressive damage within the rock that prepares the stage for a large earthquake. Numerous papers have reported that large events are preceded by anomalous trends of seismic activity both in time and space. Several reports also indicate that seismic activity increases as an inverse power of the time to the main event (sometimes referred to as an inverse Omori law for relatively short time spans), while others document a quiescence, or even contest the existence of such anomalies at all [1, 2, 3]. If such anomalies can be analyzed and understood, then one might be able to forecast future large events.

There are two schools of thought on the length of the time period over which the anomalies occur and accumulate. One school believes that the anomalies occur within days to weeks before the main shock, but probably not much earlier [4], and that the spatial precursory patterns develop at short distances from impending large earthquakes. Proponents of this school look for the precursory patterns in the immediate vicinity of the mainshock, i.e., within distances from the epicenter that are on the order of, or somewhat larger than, the length of the main shock rupture.

The second school believes that the anomalies may occur up to *decades* before large earthquakes, and at distances much larger than the length of the main shock rupture, a concept closely linked to the theory of critical phenomena [1, 2, 3] which was advocated [1, 2, 5, 6] as early as 1964 with a report [5] on the pre-monitory increase in the total area of the ruptures in the earthquake sources in a medium magnitude range, documenting the existence of long-range correlations in the precursors (over 10 seismic source lengths) with worldwide similarity. More recently, Knopoff et al. [7] reported on the existence of long-range spatial correlations in the increase of medium-range magnitude seismicity prior to large earthquakes in California.

Beginning in the late 1970s, models of rock rupture and their relation with critical phenomena and earthquakes were pursued. Vere-Jones [8] pioneered this approach. Hence, a method for the analysis of the data was introduced that, for certain values of its parameters, led to a power law (typical of critical phenomena) for the system's time-to-failure. Allègre et al. [9] proposed a percolation model of damage/rupture prior to an earthquake, emphasizing the multiscale nature of rupture prior to a critical point which was similar to a percolation threshold [10, 11]. Their model was actually just a rephrasing of the real-space renormalization group approach to the percolation model developed by Reynolds et al. [12]. Similar ideas were also explored in a hierarchical model of rupture by Smalley et al. [13]. Sahimi and co-workers [14, 15, 16] proposed a connection between percolation, the spatial distribution of earthquakes' hypocenters, and rock's fracture/fault networks.

Sornette and Sornette [17] proposed an observable consequence of the critical point model of Allègre et al. [9] with the goal of verifying the proposed scaling laws of rupture. Almost simultaneously, but following apparently an independent line of thought, Voight [18] introduced the idea of a time-to-failure analysis in the form of an empirical second-order nonlinear differential equation, which for certain values of the parameters would lead to a time-to-failure power law, in the form of an inverse Omori law. This failure was used and tested later for predicting volcanic eruptions. Then, Sykes and Jaumé [19] performed the first empirical study to quantify with a specific law an acceleration of seismicity prior to large earthquakes. They used an exponential law to describe the acceleration, and did not use or discuss the concept of a critical earthquake. Bufe and Varnes [20] re-introduced a time-to-failure power law to model the observed accelerated seismicity quantified by the so-called cumulative Benioff strain. Their justification of the power law was a mechanical model of material damage. They neither referred to nor discussed the concept of a critical earthquake.

Sornette and Sammis [21] were the first to reinterpret the work of Bufe and Varnes [20], and all the previous ones reporting accelerated seismicity, within a model in which the occurence of large earthquakes is viewed as a critical point phenomenon in the sense of the statistical physics framework of critical phase transitions. Their model generalized significantly the previous works in that the proposed critical point theory did not rely on an irreversible fracture process, but invoked a more general *self-organization* of the stress field prior to large earthquakes. Moreover, using insights from the critical phenomena, Sornette and Sammis [21] generalized the power-law description of the accelerated seismicity by considering *complex* scaling exponents which result in *log-periodic* corrections to the scaling [21, 22, 23, 24, 25, 26]. Such a generalized power law with log-periodic corrections was shown [27] to describe the increase in the energy that rock releases as it undergoes fracturing. These ideas were further developed by Huang et al. [28]. Empirical evidence for these concepts was provided by Bowman et al. [29], who showed that large earthquakes in California with magnitudes larger than 6.5 are systematically preceded by a power-law acceleration of seismic activity in time over several decades, in a spatial domain about 10–20 times larger than the impending rupture length (i.e., of the order of a few hundred kilometers). The large event can, therefore, be viewed as a temporal singularity in the seismic history time series. Such a theoretical framework implies that a large event results from the collective behavior and accumulation of many previous smaller-sized events. Similar analysis was reported by Brehm and Braile [30] for other earthquakes.

The work of Ouillon and Sornette [31] on mining-induced seismicity, and Johansen and Sornette [32] in laboratory experiments, made similar conclusions on systems of very different scales, in good agreement with the scale-invariant phenomenology, reminiscent of systems undergoing a second-order critical phase transition. In this picture, the system is subjected to an increasing external mechanical solicitation. As the external stress increases, micro-ruptures occur within the medium which locally redistribute stress, creating stress fluctuations within the rock. As damage accumulates, fluctuations interfere and become more and more spatially and temporally correlated, i.e., there are more and more, larger and larger domains that are significantly stressed and, therefore, larger and larger events can occur at smaller and smaller time intervals. The accelerating spatial smoothing of the stress field fluctuations eventually culminates in a rupture with a size on the order of the system's size. This is the final rupture of laboratory samples, or earthquakes breaking through the entire seismo-tectonic domain to which they belong. This concept was verified in numerical experiments led by Mora et al. [33, 34], who showed that the correlation length of the stress field fluctuations increases significantly before a large shock occurrs in a discrete numerical model.

More recently, Bowman and King [35] argued, based on empirical data, that, in a large domain including the impending major event, and similar to the critical domain proposed in Bowman et al. [29], the maximum size of natural earthquakes increases with time up to the main shock. If one assumes that the maximum rupture length at a given time is given by (or related to) the stress field correlation length, then the work of Bowman and King [35] shows that the correlation length increases before a large rupture. Note that Keilis-Borok and co-workers [2] have also repeatedly used the concept of a critical point, albeit in a broader and looser sense than its restricted meaning used in the statistical physics of phase transitions.

So far we have discussed the case in which the stress rate is imposed on a system. The problem is, however, more complex when the strain rate is imposed. In this case, the system may not evolve towards a critical point. A possible unifying view point between the two cases is to study whether the dissipation of energy by the deteriorating system slows down or accelerates. The answer to this question depends on the nature of the external loading (an imposed stress, rather than strain, rate), the evolution of the system and how the resulting evolving mechanical characteristics of the system interacts with the external loading conditions. For a constant applied stress rate, the dissipated energy rate diverges in general in a finite time leading to a critical behavior. For a constant strain rate, on the other hand, the answer depends on the damage law. For the Earth crust, the situation is in between the ideal constant strain and constant stress loading states. The critical point may then emerge as a mode of localization of a global input of energy to the system.

The critical point approach leads also to an alternative physical picture of the so-called seismic cycle. From the beginning of the cycle, small earthquakes accumulate and modify the stress field within the Earth crust, making it correlated over larger and larger scales. When the correlation length reaches the size of the local seismo-tectonic domain, a very large rupture may occur which, together with its early aftershocks, destroys correlations at all spatial scales. This is the end of the seismic cycle, and the beginning of a new one, leading to the next large event. As earthquakes are distributed in size according to the Gutenberg–Richter law, small to medium-size events are negligible in the energetic balance of the tectonic system, which is dominated by the largest final event. However, they are seismo-active in the sense that their occurrence prepares that of the largest one. The opposite view of the seismic cycle is to consider that it is the displacements of the large-scale tectonic plate which dominates the preparation of the largest events, which can be modelled to first order as a simple stick-slip phenomenon. In that case, all the smaller-size events would be seismo-passive, in the sense that they would reflect only the boundary loading conditions acting on isolated faults without much correlations from one event to the other.

Summarizing, in the critical point approach to earthquakes, as the stress on rock increases, micro-ruptures develop that redistribute the stress and generate fluctuations in it. As damage accumulates, the fluctuations become spatially and temporally correlated, resulting in a larger number of significantly-stressed large domains. The correlations accelerate the spatial smoothing of the fluctuations, culminating in a rupture with a size on the order of the system's size, and representing the final state in which earthquakes occur. Numerical and empirical evidence for this picture indicates that, similar to critical phenomena, the correlation length of the stress-field fluctuations increases significantly before a large earthquake.

Notwithstanding the advances that have been made, the concept of a critical earthquake concept remains only a working hypothesis: from an empirical point of view, the reported analyses possess significant deficiencies and a full statistical analysis establishing the confidence level of the hypothesis remains to be performed. In this vain, Zoller et al. [36] and Zoller and Hainzl [37, 38] recently performed novel and systematic spatio-temporal tests of the critical point hypothesis for large earthquakes based on the quantification of the predictive power of both the predicted accelerating moment release and the growth of the spatial correlation length, hence providing fresh support to the concept.

In order to prove, or refute, the notion that a boundary between tectonic plates is a truly critical system, one must check the existence or absence of a build-up of cooperativity prior to a large event in terms of cumulative (Benioff) strain. This means that one should make direct measurements of the stress field and its evolution in space and time in the region in which a large earthquake is expected, compute its spatial correlation function, deduce the spatial correlation length, and show that it increases with time as a power-law which defines a singularity when the earthquake occurs. Unfortunately, such a procedure is, at present, far beyond our technical observational abilities. It is generally believed that large earthquakes nucleate at a depth of about 10–15 km, implying that the stress field and the correlations would have to be measured at such depths in order to yield an unambiguous signature of what is happening. Moreover, the tensorial stress field would have to be measured with a high enough resolution in order to provide evidence of a clear increase of the correlation length. Such measurements are clearly out of reach at present.

A predictive theory of earthquakes should be able to forecast, (1) *when* and (2) *where* they occur in a wide enough region. It should also be able to (3) distinguish a false alarm from a reliable one. In this paper we propose a completely new method for predicting earthquakes which possesses the three features. The method estimates the Markov time scale (MTS) t_M of a seismic time series – the time over which the data can be represented by a Markov process [39, 40, 41, 42, 43, 44, 45]. As the seismic data evolve with the time, so also does t_M. We show that the time evolutioon of t_M provides an effective alarm a short time before earthquakes. The method distinguishes abnormal variations of t_M *before* the arrival of the P-waves, hence providing enough of a warning for triggering a damage/death-avoiding response *prior to* the arrival of the more damaging S-waves. To distinguish a false alarm from a real one, we describe a second new method of analyzing seismic data which provides a complementary method for predicting when an earthquake may happen. An alarm for an earthquake is then accepted when *both* methods indicate *simultaneously* that an earthquake is about to happen.

The paper is organized as follows. In the next section we introduce two different methods for estimating the Markov time scale of seismic data. In section III we show how the concept of the extended self-similarity (ESS) may be used for analyzing seismic time series. A key quantity deduced from the ESS analysis is a time scale T_1 which is able to detect the change in the correlations in the data, even with a small number of data points. Section IV presents the results of the analysis of seismic data for several earthquakes.

2 Analysis of Seismic Time Series as Markov Process

We have developed two methods for analyzing seismic time series as Markov processes and estimating their Markov time scale (MTS) t_M. In what follows we describe the two methods.

Method 1: In this method one first checks whether the seismic data follow a Markov chain and, if so, measures the MTS $t_M(t)$. As is well-known, a given dynamic process with a degree of stochasticity may have a finite or an infinite MTS t_M – the minimum time interval over which the data can be considered as a Markov process. To estimate t_M, we note that a complete characterization of the statistical properties of stochastic fluctuations of a quantity $x(t)$ requires the numerical evaluation of the joint probability distribution function (PDF) $P_n(x_1, t_1; \ldots; x_n, t_n)$ for an arbitrary n, the number of the data points in the time series $x(t)$. If $x(t)$ is a Markov process, an important simplification is made as P_n, the n-point joint PDF, is generated by the *product* of the conditional probabilities $P(x_{i+1}, t_{i+1}|x_i, t_i)$, for $i = 1, \ldots, n-1$. A necessary condition for $x(t)$ to be a Markov process is that the Chapman-Kolmogorov (CK) equation,

$$P(x_3, t_3|x_1, t_1) = \int d(x_2)\, P(x_3, t_3|x_2, t_2)\, P(x_2, t_2|x_1, t_1)\,, \tag{1}$$

should hold for any value of t_2 in the interval $t_3 < t_2 < t_1$. Hence, one should check the validity of the CK equation for various x_1 by comparing the directly-computed conditional probability distributions $P(x_3, t_3|x_1, t_1)$ with the ones computed according to right side of (1). The simplest way of determining t_M for stationary or homogeneous data is the numerical computation of the quantity, $S = |P(x_3, t_3|x_1, t_1) - \int dx_2 P(x_3, t_3|x_2, t_2)\, P(x_2, t_2|x_1, t_1)|$, for given x_1 and x_3, in terms of, for example, $t_2 - t_1$ (taking into account the possible numerical errors in estimating S). Then, $t_M = t_2 - t_1$ for that value of $t_2 - t_1$ for which S vanishes or is nearly zero (achieves a minimum).

Method 2: In second method the MTS is estimated using the least square test. The exact mathematical definition of the Markov process is given by [46], by

$$P(x_k, x_k|x_{k-1}, t_{k-1}; \cdots; x_1, t_1; x_0, t_0)$$
$$= P(x_k, t_k|x_{k-1}, t_{k-1})\,. \tag{2}$$

Intuitively, the physical interpretation of a Markov process is that, it is a process with no memory; it "forgets its past." In other words, only the most nearby conditioning, say (x_k, t_k), is relevant to the probability of finding the system at a particular state x_k at time t_k. Thus, the ability to predict the value of $x(t)$ at time t is not enhanced by knowing its values in steps prior to the the most recent one. Therefore, an important simplification that can be made for a Markov process is that, a conditional multivariate joint PDF can be written in terms of the products of the simple two-parameter conditional PDFs [46] as (3)

$$P(x_k, t_k; x_{k-1}, t_{k-1}; \cdots ; x_1, t_1 | x_0, t_0)$$
$$= \prod_{i=1}^{k} P(x_i, t_i | x_{i-1}, t_{i-1}) . \tag{3}$$

Here, we use the least square method to estimate the MTS of the fluctuations in the seismic data. Testing (3) for large values of k is beyond the present computational capability. However, for $k = 3$, where we have three points, the relation,

$$P(x_3, t_3 | x_2, t_2; x_1, t_1) = P(x_3, t_3 | x_2, t_2) , \tag{4}$$

should hold for any value of t_2 in the interval $t_1 < t_2 < t_3$.

A process is Markov if (4) is satisfied for a *certain* time separation, with the MTS t_M being, $t_M = t_3 - t_2$. To measure the MTS we use a fundamental theory of probability that allows us to write any three point PDF in terms of conditional probability functions as,

$$P(x_3, t_3; x_2, t_2; x_1, t_1)$$
$$= P(x_3, t_3 | x_2, t_2; x_1, t_1) P(x_2, t_2; x_1, t_1) . \tag{5}$$

Using the Markov Processes' properties, and substituting (5), we obtain,

$$P_{Markov}(x_3, t_3; x_2, t_2; x_1, t_1)$$
$$= P(x_3, t_3 | x_2, t_2) P(x_2, t_2; x_1, t_1) . \tag{6}$$

We then compute the deviations of P_{Markov} from that given by (5), using the least square method:

$$\chi^2 = \int dx_3 dx_2 dx_1$$
$$\times \frac{[P(x_3, t_3; x_2, t_2; x_1, t_1) - P_{Markov}(x_3, t_3; x_2, t_2; x_1, t_1)]^2}{\sigma^2 + \sigma_{Markov}^2} \tag{7}$$

where $\sigma^2 + \sigma_{\text{Markov}}^2$ are the variance of the terms in the nominator. One should then plot the reduced χ^2, $\chi_\nu^2 = \frac{\chi^2}{\mathcal{N}}$, ($\mathcal{N}$ is the number of degrees of freedom), as a function of time scale $t_3 - t_2$. The MTS is that value of $t_3 - t_2$ at which χ_ν^2 is minimum.

Our analysis of seismic data (see below) indicates that the average t_M for the *uncorrected* background seismic time series is much smaller than that for data close to an impending earthquake. Thus, at a certain time before an earthquake, t_M rises significantly and provides an alarm for the earthquake. As we show below, the alert time t_a is on the order of hours, and depends on the earthquake's magnitude M and the epicenter's distance d from the data-collecting station(s).

The sharp rise in t_M at the moment of alarm is, in some sense, similar to the increase in the correlation length ξ of the stress-field fluctuations in the critical phenomena theories of earthquake, since t_M is also the time over which the events leading to an earthquake are correlated. Therefore, just as the correlation length ξ increases as the catastrophic rupture develops, so also does t_M. However, whereas it is exceedingly difficult to directly measure ξ, t_M is computed rather readily. Moreover, whereas ξ is defined for the *entire* rupturing system over long times, t_M is computed *online* (in real time), hence reflecting the correlations of the most recent events that are presumably most relevant to an impending earthquake.

3 The Extended Self-Similarity of Seismic Data

To distinguish a false alarm that might be indicated by t_M from a true one, we have developed a second time-dependent function, which is compute based on the extended self-similarity (ESS) of the seismic time series [47, 48]. This concept is particularly useful if the time series for seismic data fluctuations (or other types of time series) do not, as is often the case, exhibit scaling over a broad interval. In such cases, the time interval in which the structure function of the time series, i.e.,

$$S_q(\tau) = \langle |x(t+\tau) - x(t)|^q \rangle , \qquad (8)$$

behaves as

$$S_q(\tau) \sim \tau^{\xi_q} , \qquad (9)$$

is small, in which case the existence of scale invariance in the data can be questioned. In Fig. 1 the logarithmic plot of the structure function $S_{0.1}(\tau)$ versus τ for a typical set of data with small scaling region is shown. In such cases, instead of rejecting outright the existence of scale invariance, one must first explore the possibility of the data being scale invariant via the concept of ESS. The ESS is a powerful tool for checking multifractal properties of data, and has been used extensively in research on turbulent flows. Thus, when analyzing the seismic time series, one can, in addition to the τ-dependence of the structure function, compute a generalized form of scaling using the ESS concept. In many cases, when the structure functions $S_q(\tau)$ are plotted against a structure function of a specific order, say $S_3(\tau)$, an extended scaling regime is found according to [47, 48],

$$S_q(\tau) \sim S_3(\tau)^{\zeta_q} . \qquad (10)$$

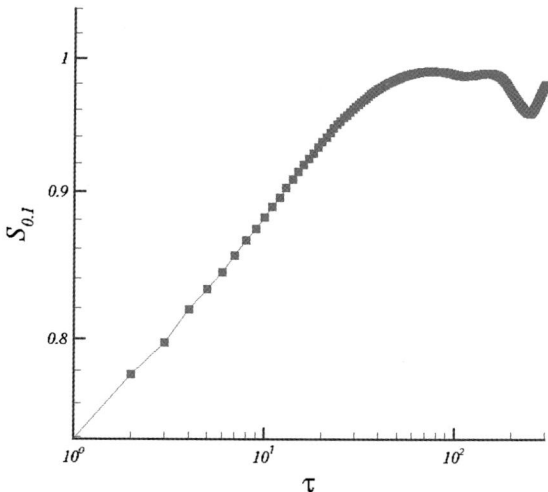

Fig. 1. The structure function $S_{0.1}$ against τ for a typical seismic time series, indicating that the scaling region is small and less than one order of magnitude variations in τ

Clearly, meaningful results are restricted to the regime where S_3 is *monotonic*. For any Gaussian process the exponents ζ_q follow a simple equation,

$$\zeta_q = \frac{1}{3}q. \qquad (11)$$

Therefore, systematic deviation from the simple scaling relation, (11), should be interpreted as deviation from monofractality. An additional remarkable property of the ESS is that it holds rather well even in situations when the ordinary scaling does not exit, or cannot be detected due to small scaling range, which is the case for the data analyzed here. In Fig. 2 we plot the behavior structure function S_q for $q = 2 - 6$ verses the the third moment. It is evident the scaling region is at least two order of magnitude.

It is well-known that the moments with $q < 1$ and $q > 1$ are related, respectively, to the frequent and rare events in the time series [47, 48]. Thus, for the seismic time series one may also be interested in the frequent events in signal. In Fig. 3 we show the results for the moment $q = 0.1$ against a third-order structure function for two types of synthetics data with scaling exponent $\alpha = 0.9$ and $\alpha = 1$. The exponent α is related to the exponent of spectral density β, i.e., $S(f) \sim 1/f^\beta$, via, $\beta = 2\alpha - 1$. As shown in Fig. 3, the interesting feature is that the starting point of $S_{0.1}(\tau)$ versus $S_3(\tau)$ is different for the data for different types of correlation exponents. To determine the distance form the origin, we define [47, 48],

$$T(\tau = 1) = [S_q^2(\tau = 1) + S_3^2(\tau = 1)]^{1/2}. \qquad (12)$$

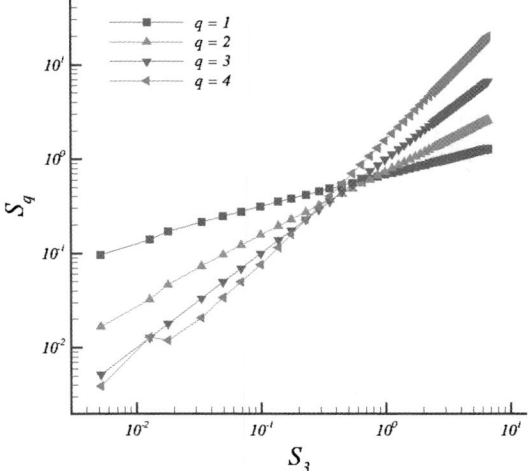

Fig. 2. Generalized scaling analysis of a seismic time series. The structure functions S_q are displayed versus S_3 in the log-log scale.

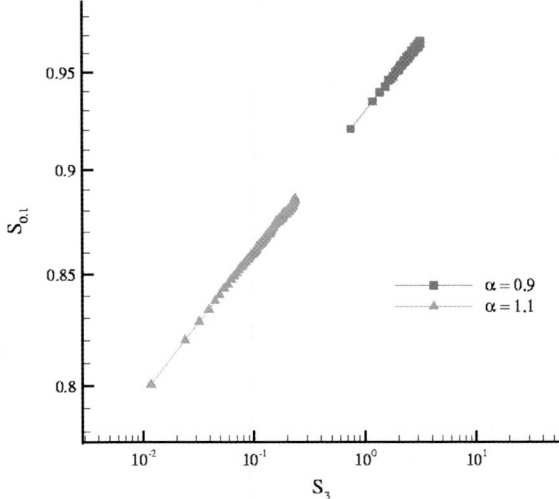

Fig. 3. The structure function $S_{0.1}$ against $S_3(\tau)$ for two type of time series with different scaling exponents.

Our analysis indicates that since prior to an earthquake the number of frequent events (development of cracks that join up) suddenly rises, one also obtains a corresponding sudden change in S_p with $p < 1$ (we use p = 0.1). Close to an earthquake the function $T_1(t)$, also estimated online, suddenly changes and provides a second alert. Its utility is not only due to the fact that it provides a means of distinguishing a false alarm from a true one indicated by t_M, but also that it is estimated very accurately even with very few data points, say 50, hence enabling online analysis of the data collected over intervals of about 1 second. Thus, even with few data points, the method can detect the change of correlations in the incoming data. For example, for the correlated synthetic data with a spectral density $1/f^{2\alpha-1}$, one obtains, $T_1 = -7(\alpha - 1)$.

4 Test of the Method

To test the method described above, we first generated two synthetic data sets with equal averages and variances (zero and unity, respectively), and mixed them together, as shown in Fig. 4, by replacing the last 50 data points of the first set with the first 50 points of the second. As Fig. 4 indicates, the Markov time scale t_M and T_1 are able to determine the time at which the two data sets are mixed.

Now we utilize the method for analyzing seismic time series. We have analyzed the data for *vertical ground velocity* $V_z(t)$ for 173 earthquakes with magnitudes $3.2 \leq M \leq 6.3$ that occurred in Iran between 28°N and 40°N latitude, and 47°E and 62.5°E longitude, between January 3 and July 26, 2004. Recorded by 14 stations, the data can be accessed at http://www.iiees.ac.ir/bank/bank_2004.html. The frequency was 40 Hz for 2 of the stations and 50 Hz for the rest. The vertical ground velocity data were analyzed because, with the method described above, they provide relatively long (on the order of several hours) and, hence, useful alarms for the impending earthquakes. Fourty (discrete) data points/second are recorded in the broad-band seismogram for the vertical ground velocity $x(t) \equiv V_z$. To analyze such data and provide alarms for the area for which the data are analyzed, we proceed as follows.

(1) The data are analyzed in order to check whether they follow a Markov chain [the directly-computed $P(x_3, t_3 | x_1, t_1)$ must be equal to the right side of (2)].
(2) The MTS $t_M(t)$ of the data are estimated by the above two methods. When using method 1 described above, for long-enough data series (10^3 data points or more) the function $t_M(t)$ is computed as the point when $S \to 0$, but for shorter series the minimum in S provides estimates of $t_M(t)$. We also utilize method 2 described above to estimate t_M. To carry out such computations, we use 1000 data points in each window for calculating t_M and move 200 data points at each step inside and outside of the window. This means that for the stations with frequency 50, we estimate a t_M every 4 seconds.
(3) $T_1(t)$ is computed for the same data. To compute $S_q(\tau)$ (we used $q = 1/10$) the data $x(t)$ are normalized by their standard deviation, hence making T_1 dimensionless. We calculate the time series T_1 with 200 data points. In order to obtain an unambiguous alert from $T_1(t)$, we sometimes calculate the quantity T_1 for the series, $y(t_i) = x(t_i) - x(t_{i-1})$.

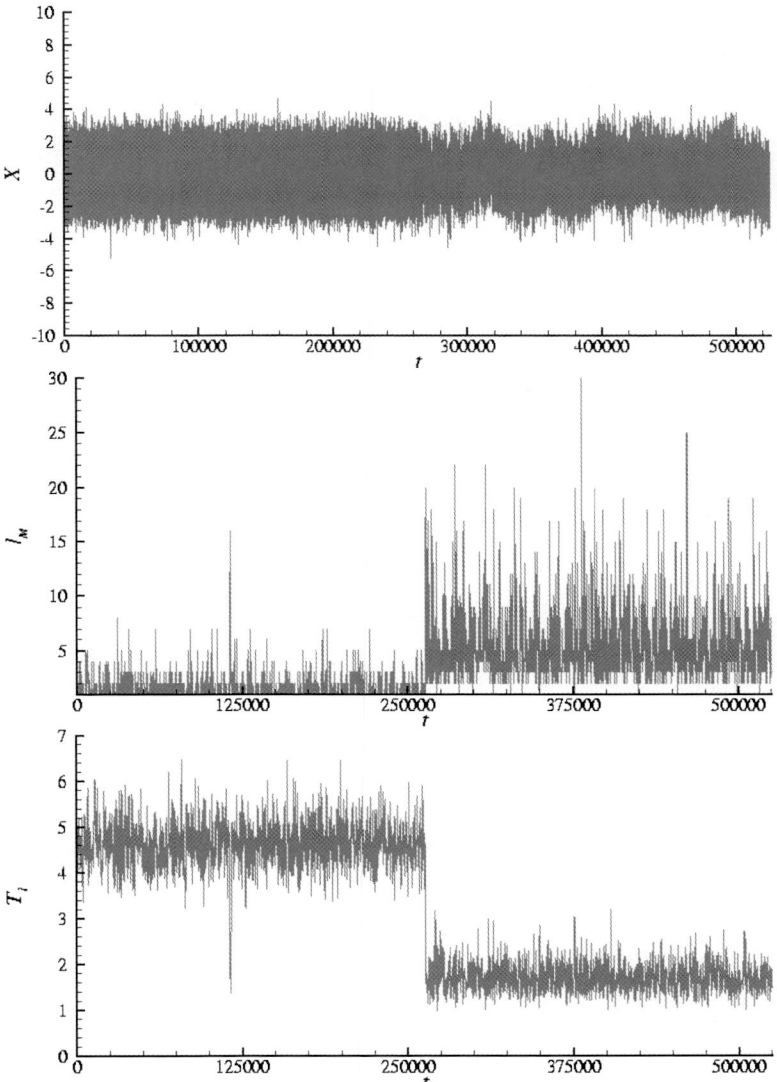

Fig. 4. Mixing of two synthetic time series with different correlation exponent (*top*); plots of the corresponding t_M (*middle*), and time variations of T_1 (*bottom*). As seen, t_M and T_1 distinguish the two types of data

(4) Steps (1)–(3) are repeated for a large number of previously-occurred earthquakes of size M at a distance d from the station, referred to as (M, d) earthquakes. Earthquakes with $M < M_c$ and $d > d_c$ are of no practical importance and are ignored.

(5) Define the thresholds t_{Mc} and T_{1c} such that for $t_M > t_{Mc}$ and $T_1 > T_{1c}$ one has an alert for an earthquake ($M > M_c, d < d_c$). If t_{Mc} and T_{1c} are too large no alert is obtained, whereas one may receive useless alerts if they are too small. By comparing the data for all the earthquakes with $M > M_c$ registered in a given station, t_{Mc} and T_{1c} for the earthquakes are estimated.

(6) Real-time data analysis is performed to compute the function $t_M(t)$ and $T_1(t)$. An alarm is turned on if $t_M > t_{Mc}$ and $T_1 > T_{1c}$ *simultaneously*. When the alarm is turned on, it indicates that an earthquake of magnitude $M \geq M_c$ at a distance $d \leq d_c$ is about to occur. The procedure can be carried out for *any* station. The larger the amount of data, the more precise the alarm will be.

Figure 5 presents $T_1(t)$ and $t_M(t)$ for a $M = 5.7$ earthquake that occurred on March 13, 2005 at 03:31:21 am in Saravan at (27.37N, 62.11E, depth 33) in southern Iran. The data were collected at Zahedan station (near Zahedan, Iran) at a distance of ∼150 km from the epicenter. The earthquake catalogue in the internet address given above indicates that, for several days before the main event, there was no foreshock in that region. As Fig. 5 indicates, T_1 and t_M provided a five hour alarm for the Saravan earthquake. Since the data used for computing t_M and T_1 were, respectively, in strings of 1000 and 200 points, there is no effect of the events *before* they were collected and, hence, the patterns in Fig. 5 reflect the events taking place in the time period in which the data were collected. The thresholds used are, $t_{Mc} = 10$ and $T_{1c} = 4$.

Figure 6 presents $T_1(t)$ and $t_M(t)$ for a $M = 6.5$ earthquake that occurred on February 22, 2005 at 02:25:20 am in Zarand at (30.76N, 56.74E, depth 14) in central Iran. The data were collected at Kerman station (near Kerman, Iran) at a distance of 86 km from the epicenter. All the statements made above regarding the Saravan earthquake are equally true about the Zarand earthquake. Similar to Saravan earthquake, T_1 and t_M provided a five hour alarm for the earthquake. The thresholds used are, $t_{Mc} = 2$ and $T_{1c} = 4$.

Figure 7 presents $T_1(t)$ and $t_M(t)$ for the same Zarand earthquake, but based on the data from another station, collected at Ashtian station (near Ashtian, Iran) at a distance of ∼150 km from the epicenter. Once again, there was no foreshock for several days before the main event. Once again, T_1 and t_M provided a five hour alarm for the earthquake. The thresholds used are, $t_{Mc} = 12$ and $T_{1c} = 8$.

To estimate the alert times t_a, which are on the order of hours, we carried out an analysis of online data for 14 stations in Iran's broad-band network (the sensors are Guralp CMG-3T broad-band), analyzing the vertical ground velocity data. Our analysis indicates that t_a depends on M, being small for low M, but quite large for large M. Using extensive data for the Iranian earthquakes with $M \geq 4.5$ and $d \leq 150$ km, we have obtained an approximate relation for the broad-band stations, shown in Fig. 8 and represented by

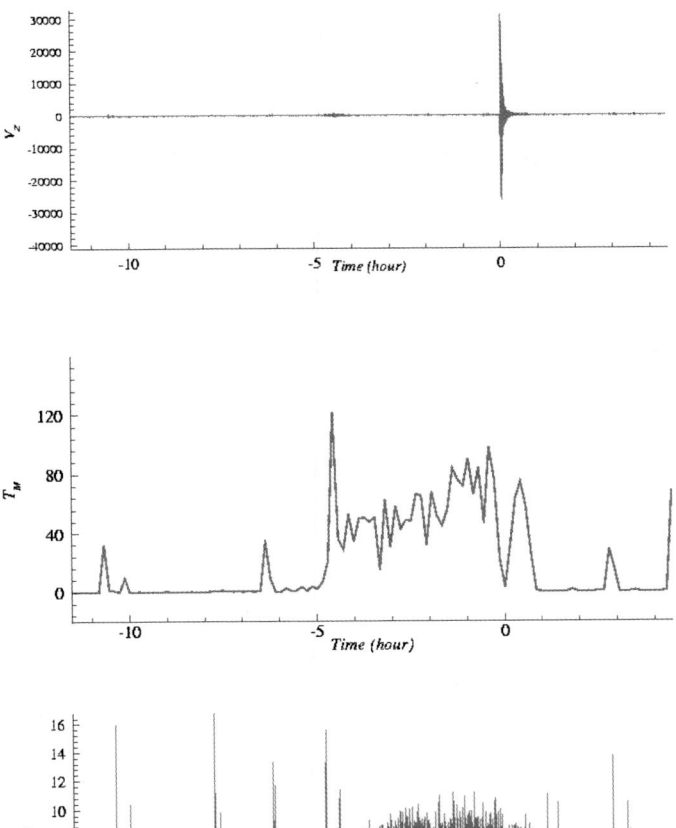

Fig. 5. $T_1(t)$ and $t_M(t)$ for a $M = 5.7$ earthquake that occurred on March 13, 2005, at 03:31:21 am in Saravan at (27.37N, 62.11E, depth 33) in southern Iran. The data were collected at Zahedan station (near Zahedan, Iran) at a distance of ∼150 km from the epicenter. The earthquake catalogue on the internet address given in the text indicates that, for several days before the main event, there was no foreshock in that region. T_1 and t_M provided a five hour alarm for the earthquake. Since the data used for computing t_M and T_1 were, respectively, in strings of 1000 and 200 points, there is no effect of the events *before* they were collected and, hence, the patterns in the figure reflect the events taking place in the time period in which the data were collected. t_M is in number of data points (the frequency at the station is 40 Hz), T_1 is dimensionless, while $V_z(t)$ is in "counts" which, when multiplied by a factor 1.1382×10^{-3}, is converted to μm/sec. The sensors were (broad-band) Guralp CMG-3T that collect data in the east-west, north-south, and vertical directions. The thresholds are, $t_{Mc} = 10$ and $T_{1c} = 4$

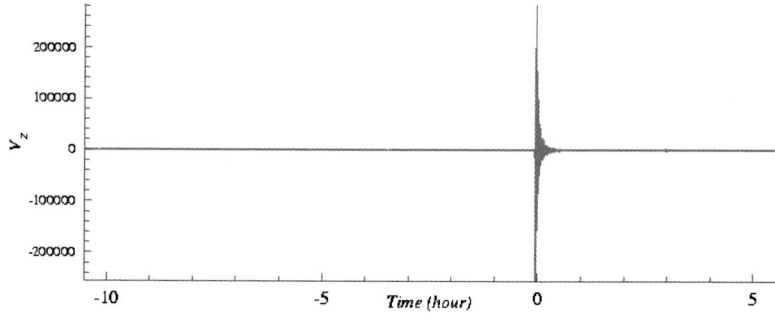

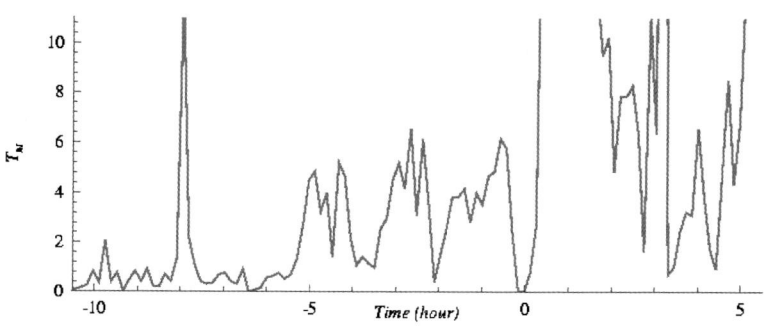

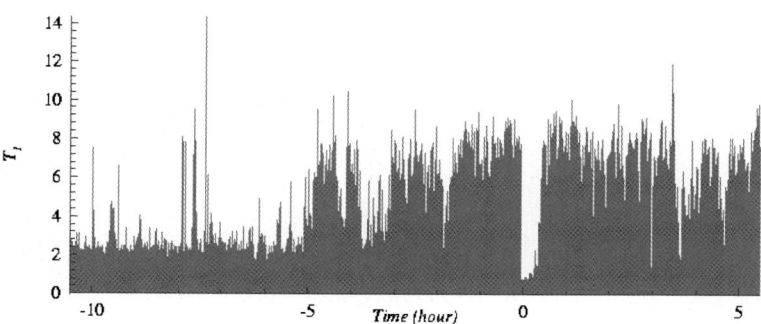

Fig. 6. Same as in Fig. 5, but for a $M = 6.5$ earthquake that occurred on February 22, 2005, at 02:25:20 am in Zarand at (30.76N, 56.74E, depth 14) in central Iran. The data were collected at Kerman station (near Kerman, Iran) at a distance of 86 km from the epicenter. The thresholds are, $t_{Mc} = 2$ and $T_{1c} = 4$

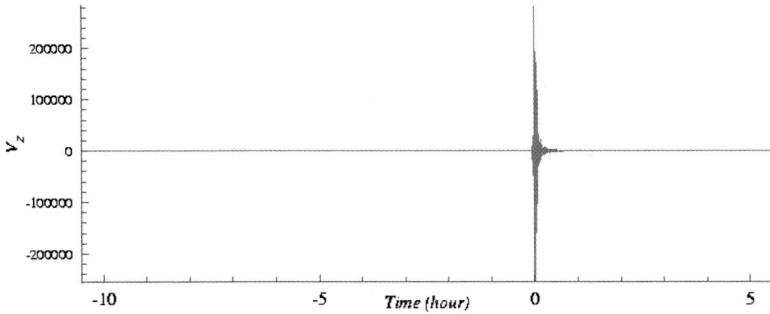

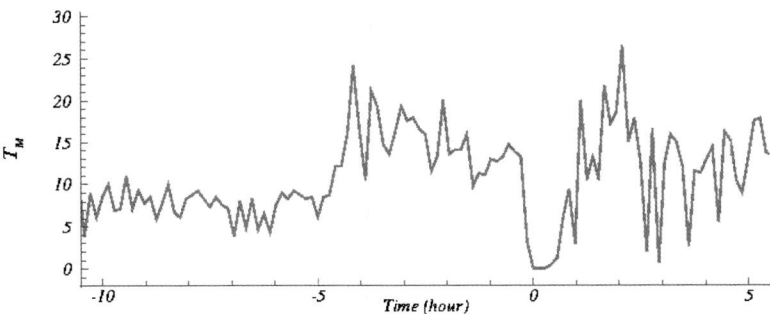

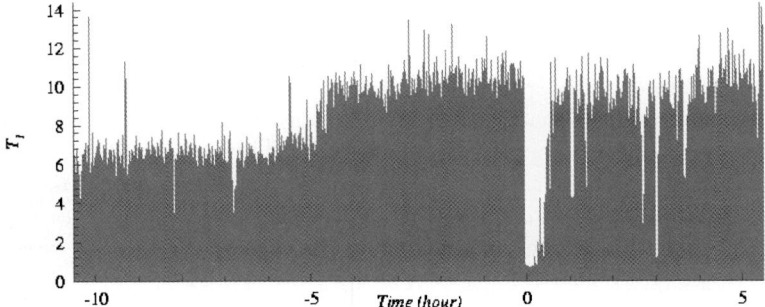

Fig. 7. Same as in Fig. 5, but based on the data collected at Ashtian station (near Ashtian, Iran) at a distance of ∼150 km from the epicenter. Thus, T_1 and t_M provided a about five hour alarm for the earthquake. The thresholds are, $t_{Mc} = 12$ and $T_{1c} = 8$

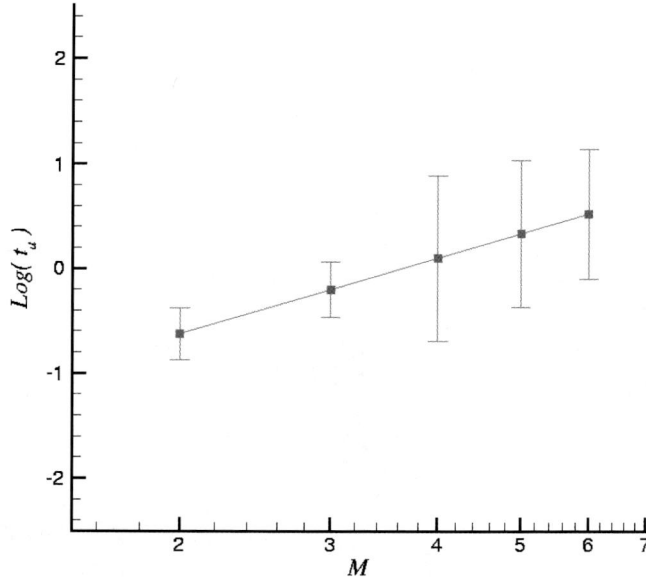

Fig. 8. The dependence of alert time t_a (in hours) on the magnitude M of the earthquakes, obtained based on the data from *broad-band* stations by analyzing 173 earthquakes with magnitudes, $3.2 \leq M \leq 6.3$, that occurred in Iran between 28°N and 40°N latitude, and 47°E and 62.5°E longitude, between January 3 and July 26, 2004

$$\log t_a = -1.35 + 2.4 \log M \ , \tag{13}$$

where t_a is in hours. The numerical coefficients of (13) for each area should be estimated from the data collected for that area. The above analysis can clearly be extended to all the stations around the world. This is currently underway for Iran's network of stations. For an earthquake of magnitude $M = 4.5$, (13) predicts an alert time of about 2 hours. Thus, if, for example, three hours after the alarm is turned on, the earthquake has still not happened, we know that the magnitude of the impending earthquake is $M \geq 5.7$.

5 Summary

In summary, we have proposed a new method for analyzing seismic data and making predictions for when an earthquake may occur with a magnitude $M \geq M_c$ at a distance $d \leq d_c$ from a station that collects seismic data. The method is based on computing the Markov time scale t_M, and a quantity T_1 calculated based on the concept of extended self-similarity of the data, and monitoring them *online* as they evolve with the time. If the two quantities exceed their respective critical thresholds t_{cM} and T_{c1}, estimated based on analyzing the data for the previously-occurred earthquakes, an alarm is turned on.

We are currently utilizing this method for Iran's stations. To do so, we calibrate the method with the data for the stations in one region (i.e., estimate t_{cM} and T_{c1} for distances $d < d_c$). If in a given region there is a single station, then once the online-computed t_M and T_1 exceed their critical values, the alarm is turned on. If there are several stations, then once they declare that their t_M and T_1 have exceeded their thresholds, the alarm is turned on. If after about 2 hours, no earthquake has occurred yet, then we know that the magnitude of the incoming earthquake will be greater $M_c = 4.5$ at a distance $d < d_c$.

In fact, over the past two years, the method has been utilized in the Iranian stations. Our analysis indicates that the method's failure rate decreases to essentially *zero* when t_M and T_1 provide *simultaneous* alarms. That is, practically every earthquake that we have considered, including those that have been occurring while we have been performing online analysis of their incoming data and providing alarms for them (with $M > M_c$), was preceded by an alarm. Of all the earthquakes that we have analyzed so far, the method has failed in only *two* cases. In our experience, if after 10 hours [see (13)] no earthquake occurs, we count that as a failed case. However, as mentioned, we have so far had only two of such cases.

So far, in order to locate the forthcoming earthquake, we have been using the clustering method which means that, for the area in which three stations are in the alert situation, we can determine the approximate location. However, this can be done using the localization property of seismic waves [49, 50]. We will report the method for the location and precise estimation of the magnitude of forthcoming earthquakes hours before their occurence elsewhere.

Finally, it must be pointed out that the most accurate alarms are obtained from stations that receive data from depths of >50 m, and are perpendicular to the active faults that cause the earthquake, since they receive much more correlated data for thedevelopment of the cracks than any other station.

Acknowledgments

We are particulary grateful to K.R. Sreenivasan, R. Mansouri, S. Sohrabpour and W. Nahm for useful discussions, comments, and encouragement. We would also like to thank M. Akhavan, F. Ardalan, H. Arfaei, J. Davoudi, R. Friedrich, M. Ghafori-Ashtiany, S. Ghasemi, M.R. Ghaytanchi, N. Hamadani, K. Hesami, N. Kamalian, V. Karimipour, A. Mahdavi, S. Moghimi Araghi, N. Nafari, A.F. Pacheco, S. Rahvar, M. Rezapour, A. Sadid Khoy, J. Samimi, F. Shahbazi, J. Samimi, H.R. Siahkoohi, N. Taghavinia, and M. Tatar for useful comments.

References

1. C.H. Scholz, *The Mechanics of Eathquakes and Faulting*, Cambridge University Press, Cambridge (1990).
2. V.I. Keilis-Borok and A.A. Soloviev, *Nonlinear Dynamics of the Lithosphere and Earthquake Prediction*, Springer, Heidelberg (2002).
3. D. Sornette, *Critical Phenomena in Natural Sciences*, 2nd ed., Springer, Berlin (2004).
4. L.M. Jones and P. Molnar, J. Geophys. Res. **84**, 3596 (1979).
5. V.I. Keilis-Borok and L.N. Malinovskaya, J. Geophys. Res. **69**, 3019 (1964).
6. G.A. Sobolev and Y.S. Tyupkin, Phys. Solid Earth **36**, 2, 138 (2000).
7. L. Knopoff, et al., J. Geophys. Res. **101**, 5779 (1996).
8. D. Vere-Jones, Math. Geol. **9**, 407 (1977).
9. C.J. Allègre, J.L. Le Mouel and A. Provost, Nature **297**, 47 (1982);
10. D. Stauffer and A. Aharony, *Introduction to Percolation Theory*, 2nd ed., Taylor and Francis, London (1994); M. Sahimi, *Applications of Percolation Theory*, Taylor and Francis, London (1994).
11. M. Sahimi, M.C. Robertson, and C.G. Sammis, Phys. Rev. Lett. **70**, 2186 (1993); H. Nakanishi, M. Sahimi, et al., J. Phys. I. France **3**, 733 (1992); M.C. Robertson, C.G. Sammis, M. Sahimi, and A.J. Martin, J. Geophys. Res. B **100**, 609 (1995).
12. P.J. Reynolds, W. Klein, and H.E. Stanley, Phys. Rev. B (1980).
13. R.F. Smalley, D.L. Turcotte, and S.A. Sola, J. Geophys. Res. **90**, 1884 (1985).
14. M. Sahimi, M.C. Robertson, and C.G. Sammis, Phys. Rev. Lett. **70**, 2186 (1993).
15. H. Nakanishi, M. Sahimi, M.C. Robertson, C.G. Sammis, and D. Rintoul, J. Phys. I. France **3**, 733 (1992).
16. M.C. Robertson, C.G. Sammis, M. Sahimi and A.J. Martin, J. Geophys. Res. B **100**, 609 (1995).
17. A. Sornette and D. Sornette, Tectonophys. **179**, 327 (1990).
18. B. Voight, Nature **332**, 125 (1988); *Science* **243**, 200 (1989).
19. L.R. Sykes and S. Jaumé, Nature **348**, 595 (1990).
20. C.G. Bufe and D.J. Varnes, J. Geophys. Res. **98**, 9871 (1993).
21. D. Sornette and C.G. Sammis, J. Phys. I. France **5**, 607 (1995).
22. W.I. Newman, D.L. Turcotte, and A.M. Gabrielov, Phys. Rev. E **52**, 4827 (1995);
23. H. Saleur, C.G. Sammis, and D. Sornette, J. Geophys. Res. **101**, 17661 (1996).
24. H. Saleur, C.G. Sammis, and D. Sornette, Nonlinear Proc. Geophys. **3**, 102 (1996).
25. A. Johansen, et al., J. Phys. I France **6**, 1391 (1996).
26. A. Johansen, H. Saleur, and D. Sornette, Eur. Phys. J. B **15**, 551 (2000).
27. M. Sahimi and S. Arbabi, Phys. Rev. Lett. **77**, 3689 (1996).
28. Y. Huang, H. Saleur, C.G. Sammis, and D. Sornette, Europhys. Lett. **41**, 43 (1998).
29. D.D. Bowman, G. Ouillon, C.G. Sammis, A. Sornette, and D. Sornette, J. Geophys. Res. **103**, 2435 (1998).
30. D.J. Brehm and L.W. Braile, Bull. Seism. Soc. Am. **88**, 564 (1998); **89**, 275 (1999).
31. G. Ouillon and D. Sornette, Geophys. J. Int. **143**, 454 (2000);
32. A. Johansen and D. Sornette, Eur. Phys. J. B **18**, 163 (2000).

33. P. Mora, et al., in *Geocomplexity and the Physics of Earthquakes*, Ed., J.B. Rundle, D.L. Turcotte, and W. Klein, American Geophysical Union, Washington (2000).
34. P. Mora and D. Place, Pure Appl. Geophys. (2002).
35. D.D. Bowman and G.C.P. King, Geophys. Res. Lett. **28**, 4039 (2001);
36. G. Zoller and S. Hainzl, Geophys. Res. Lett. **29**, 101029/2002GL014856 (2002).
37. G. Zoller and S. Hainzl, Natural Hazards and Earth System Sciences **1**, 93 (2001).
38. G. Zoller and S. Hainzl, Geophys. Res. Lett. **29**, 101029/2002GL014856 (2002).
39. R. Friedrich and J. Peinke, Phys. Rev. Lett. **78**, 863 (1997).
40. R. Friedrich, J. Peinke, and C. Renner, Phys. Rev. Lett **84**, 5224 (2000).
41. M. Siefert, A. Kittel, R. Friedrich, and J. Peinke, Europhys. Lett. **61**, 466 (2003).
42. M. Davoudi and M.R. Rahimi Tabar, Phys. Rev. Lett. **82**, 1680 (1999).
43. G.R. Jafari, S.M. Fazlei, F. Ghasemi, S.M. Vaez Allaei, M.R. Rahimi Tabar, A. Iraji Zad, and G. Kavei, Phys. Rev. Lett. **91**, 226101 (2003).
44. F. Ghasemi, J. Peinke, M. Sahimi, and M.R. Rahimi Tabar, Eur. Phys. J. B **47**, 411 (2005).
45. M.R. Rahimi Tabar, F. Ghasemi, J. Peinke, R. Friedrich, K. Kaviani, F. Taghavi, S. Sadeghi, G. Bijani, and M. Sahimi, Comput. Sci. Eng., in press (2006).
46. H. Risken, *The Fokker-Planck Equation*, Springer, Berlin (1984).
47. R. Benzi, et al., *Physica D* **96**, 162 (1996).
48. A. Bershadskii and K.R. Sreenivasan, Phys. Lett. A **319**, 21 (2003).
49. F. Shahbazi, A. Bahraminasab, S.M. Vaez Allaei, M. Sahimi, and M. R. Rahimi Tabar, Phys. Rev. Lett. **94**, 165505 (2005).
50. A. Bahraminasab, S.M. Vaez Allaei, F. Shahbazi, M. Sahimi, and M.R. Rahimi Tabar, Phys. Rev. B (to be published).

Why Does Theoretical Physics Fail to Explain and Predict Earthquake Occurrence?

Y.Y. Kagan

Department of Earth and Space Sciences, University of California, Los Angeles, California, USA
kagan@equake.ess.ucla.edu

Several reasons for the failure can be proposed:

1. The multidimensional character of seismicity: time, space, and earthquake focal mechanism need to be modeled. The latter is a symmetric second-rank tensor of a special kind.
2. The intrinsic randomness of earthquake occurrence, necessitating the use of stochastic point processes and appropriate complex statistical techniques.
3. The scale-invariant or fractal properties of earthquake processes; the theory of random stable or heavy-tailed variables is significantly more difficult than that of Gaussian variables and is only now being developed. Earthquake process theory should be capable of being renormalized.
4. 4. Statistical distributions of earthquake sizes, earthquake temporal interactions, spatial patterns and focal mechanisms are largely universal. The values of major parameters are similar for earthquakes in various tectonic zones. The universality of these distributions will enable a better foundation for earthquake process theory.
5. The quality of current earthquake data statistical analysis is low. Since little or no study of random and systematic errors is performed, most published statistical results are artifacts.
6. During earthquake rupture propagation, focal mechanisms sometimes undergo large 3-D rotations. These rotations require non-commutative algebra (e.g., quaternions and gauge theory) for accurate models of earthquake occurrence.
7. These phenomenological and theoretical difficulties are not limited to earthquakes: any fracture of brittle materials, tensile or shear, would encounter similar problems.

1 Introduction

The difficulties of seismic analysis are obvious. Earthquake processes are inherently multidimensional [76, 101]: in addition to the origin time, 3-D locations, and measures of size for each earthquake, the orientation of the rupture surface and its displacement requires for its representation either second-rank tensors or quaternions (see more below). Earthquake occurrence is characterized by extreme randomness; the stochastic nature of seismicity is not reducible by more numerous or more accurate measurements. Even a cursory inspection of seismological datasets suggests that earthquake occurrence as well as earthquake fault geometry are scale-invariant or fractal ([76, 101, 114, 143, 149] see also http://www.esi-topics.com/earthquakes/interviews /YanYKagan.html.)

Adequate mathematical and statistical techniques have only recently become available for analyzing fractal temporal, spatial, and tensor patterns of point process data generally and earthquake data in particular. Such methods are still in the development stage. Moreover, it is only in the past 25–30 years that the quality, precision and completeness of earthquake datasets and the processing power of modern computers have become sufficient to allow detailed, full-scale investigation of earthquake occurrence patterns.

After looking at recent publications on earthquake physics (for example, [103, 110, 136]), one gets the impression that knowledge of earthquake process is still at a rudimentary level. Why has progress in understanding earthquakes been so slow? Kagan [72] compared the seismicity description to another problem in physics: turbulence of fluids. Both phenomena are characterized by multidimensionality and stochasticity. Their major statistical ingredients are scale-invariant, and both have hierarchically organized structures. Moreover, the scale of self-similar structures in seismicity and turbulence extends over many orders of magnitude. The size of major structures which control deformation patterns in turbulence and brittle fracture is comparable to the maximum size of the region (see more in [76]).

Yaglom [159, p. 4] commented that turbulence status differs from many other complex problems which twentieth century physics has solved or has considered.

> " '[These problems] deal with some very special and complicated objects and processes relating to some extreme conditions which are very far from realities of the ordinary life... However, turbulence theory deals with the most ordinary and simple realities of the everyday life such as, e.g., the jet of water spurting from the kitchen tap. Therefore, the turbulence is well-deservedly often called 'the last great unsolved problem of the classical physics.' "

Although solving the Navier-Stokes equations, describing turbulent motion in fluids is one of the seven mathematical millennium problems for the 21st century (see http://www.claymath.org/millennium/), the turbulence problem is not among the ten millennium problems in physics presented by the University of Michigan, Ann Arbor (see http://feynman.physics.lsa.umich.edu/-strings2000/millennium.html) or among the 11 problems by the National Research Council's board on physics and astronomy [49]. In his extensive and wide-ranging review of current theoretical physics, [128] does not include the turbulence or Navier-Stokes equations in the book index.

Like fluid turbulence, the brittle fracture of solids is commonly encountered in everyday life, but so far there is no real theory explaining its properties or predicting outcomes of the simplest occurrences, such as a glass breaking. Although computer simulations of brittle fracture (for example, see [121]) are becoming more realistic, they cannot yet provide a scientifically faithful representation. Brittle fracture is a more difficult scientific problem than turbulence, and while the latter attracted first-class mathematicians and physicists, no such interest has been shown in the mathematical theory of fracture and large-scale deformation of solids.

One sees the degree of difficulty in assessing the effects of brittle fracture by looking at the investigation results of the space shuttle Columbia disaster [24, p. 83]. The only way to test the possible cause of the accident – a breach on the edge of the shuttle wing – was to conduct a full-scale experiment. No realistic computation of the breach was possible.

Below we first review the seismological background information necessary for further discussion as well as basic models of earthquake occurrence (Sect. 2). Short Sect. 3 describes the available earthquake catalogs. In Sects. 4–7 evidence for the scale-invariance of earthquake process is presented, in particular, marginal distributions for the multidimensional earthquake process. Fractal distributions of earthquake size, time intervals, spatial patterns, focal mechanism, and stress are discussed. Section 8 describes several multidimensional stochastic models used to approximate earthquake occurrence. They are all based on the theory of branching processes; in this case the multidimensional structure of earthquake occurrence is modeled. In Sect. 8.3 we discuss the branching model of earthquake rupture: a physical multidimensional model based on random stress interactions. The model uses very few free parameters and appears to reproduce all the fundamental statistical properties of earthquake occurrence. Section 8.4 briefly describes the application of statistical models to forecast an earthquake occurrence. The final discussion (Sect. 9) summarizes the results obtained thus far and discusses problems and challenges still facing seismologists.

2 Seismological Background

2.1 Earthquakes

Since this paper is intended for seismologists, physicists, and mathematicians, we briefly describe earthquakes and earthquake catalogs as primary objects of the statistical study. A more complete discussion can be found in Kagan [76, pp. 162–165], Bolt [15], Lee et al. [110], Scholz [135], Kanamori and Brodsky [103]. As a first approximation, an earthquake may be represented by a sudden shear failure – the appearance of a large quasi-planar dislocation loop [1] in rock material.

Figure 1a shows a fault-plane trace on the surface of the Earth. Earthquake rupture starts on the fault-plane at a point called the hypocenter (the "epicenter" is a projection of the hypocenter on the Earth's surface), and propagates with a velocity close to that of shear waves (2.5–3.5 km/s). The "centroid" is in the center of the ruptured area. Its position is determined by a seismic moment tensor inversion (Ekström et al., [33], and references therein). As a result of the rupture, two sides of the fault surface are displaced by a vector along the fault-plane. For large earthquakes, such displacement is on the order of a few meters.

The earthquake rupture excites seismic waves which are registered by seismographic stations. The seismograms are processed by computer programs to obtain a summary of the earthquake's properties. Routinely, these seismogram inversions characterize earthquakes by their origin times, hypocenter (centroid) positions, and second-rank symmetric seismic moment tensors.

Figure 1c represents ("beachball") the quadrupolar radiation patterns of earthquakes. The focal plots involve painting on a sphere the sense of the first motion of the primary, P-waves: solid for compressional motion and open for dilatational. Two orthogonal planes separating these areas are the fault and the auxiliary planes. During the routine determination of focal mechanisms, it is impossible to distinguish these planes. Their intersection is the null-axis (N-axis), the P-axis is in the middle of the open lune, and the T-axis in the middle of the closed lune. These three axes are called the "principal axes of an earthquake focal mechanism," and their orientation defines the mechanism.

In the system of coordinates of TPN axes, shown in Fig. 1c, the seismic moment tensor matrix is

$$\mathbf{M} = M \times \mathrm{diag}[1, -1, 0] \,, \tag{1}$$

where M is a scalar seismic moment of an earthquake, measured in Newton-m (Nm). In an arbitrary system of coordinates all entries in 3×3 matrix (1) are non-zero. However, the tensor is always traceless, with a zero determinant. Hence it has only four degrees of freedom: one for the norm of the tensor (proportional to the scalar seismic moment) and three for orientation (they define the focal mechanism of an earthquake). Another equivalent representation of the earthquake focus is a quadrupole source of a particular type (Fig. 1b) known in seismology as a "double-couple" [18, 1, 85]. The three representations of focal mechanism shown in Fig. 1 as well as in (1) are mathematically equivalent; [85] discusses interrelations between these parameterizations.

2.2 Description of Earthquake Catalogs

Modern earthquake catalogs are collections of estimated earthquake origin times, hypocenter or centroid locations, measures of earthquake size (scalar

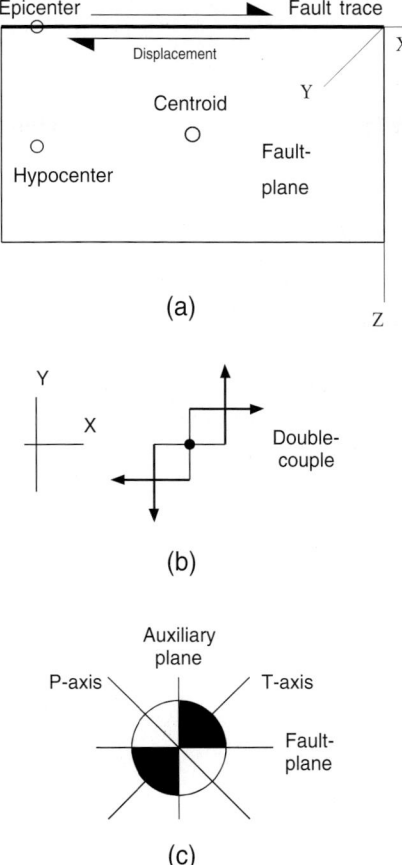

Fig. 1. Schematic diagrams of earthquake focal mechanism. (**a**) Fault-plane trace on the surface of the Earth. Earthquake rupture starts at the hypocenter (epicenter is the projection of a hypocenter on the Earth's surface), and propagates with velocity close to that of shear waves (2.5–3.5 km/s). (**b**) Double-couple source, equivalent forces yield the same displacement as the extended fault (a) rupture in a far-field. (**c**) Equal-area projection (Aki and Richards, 2002, p. 110) of quadrupole radiation patterns. The null (N) axis is orthogonal to the T- and P-axes, or it is located on the intersection of fault and auxiliary planes, i.e., perpendicular to the paper sheet in this display

seismic moment or appropriate magnitude), and earthquake focal mechanisms or seismic moment tensors [1]. Such datasets in a certain sense fully describe each earthquake; for instance one can compute far-field, low-frequency seismic radiation using the above information. However, detailed studies of earthquake occurrences show that this description is far from complete, since each earthquake represents a process with moment tensor or focal mechanism varying

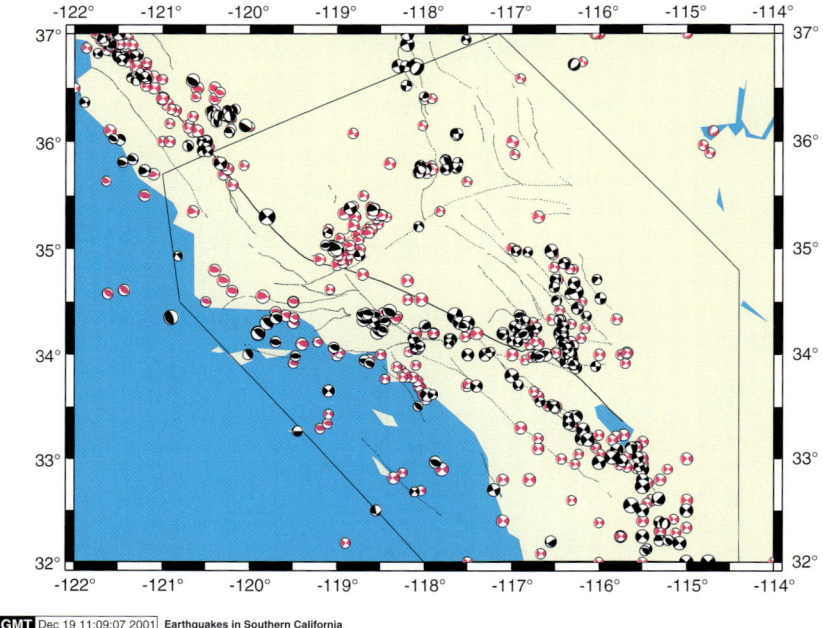

Fig. 2. The southern California catalog and polygon: a region wherein the catalog is believed to be accurate and complete. Time period 1800–2002. Black beachballs – known solutions; orange beachballs – imputed solutions, obtained through interpolation from known focal mechanisms. Thin *curved lines* are active earthquake faults, including the San Andreas fault on which many earthquakes are concentrated

in extended time-space. Moreover, because earthquakes have fractal features, even defining an "individual" earthquake is problematic: earthquake catalog records are the result of a complex interaction of fault ruptures, seismographic recordings, and their interpretations (see Sect. 5). Figure 2 displays a map of the local catalog for southern California [91]. Earthquake focal mechanisms are shown by a stereographic projection [1]. The focal mechanisms can be characterized by a 3-D rotation from a fixed position; an alternative, more compact representation of each mechanism is a normalized quaternion [71, 85].

In Fig. 3 we display a map of earthquake centroids in the global Harvard CMT catalog ([33], and references therein). Earthquakes are mostly concentrated at tectonic plate boundaries. Each earthquake in this catalog is characterized by a centroid moment tensor solution.

There are many other datasets which characterize earthquake processes, such as detailed investigations of earthquake rupture for particular events, or earthquake fault maps and descriptions of certain faults. The unique advantages of an earthquake catalog include relative completeness, uniformity of coverage, and quantitative estimates of errors. These properties make catalogs

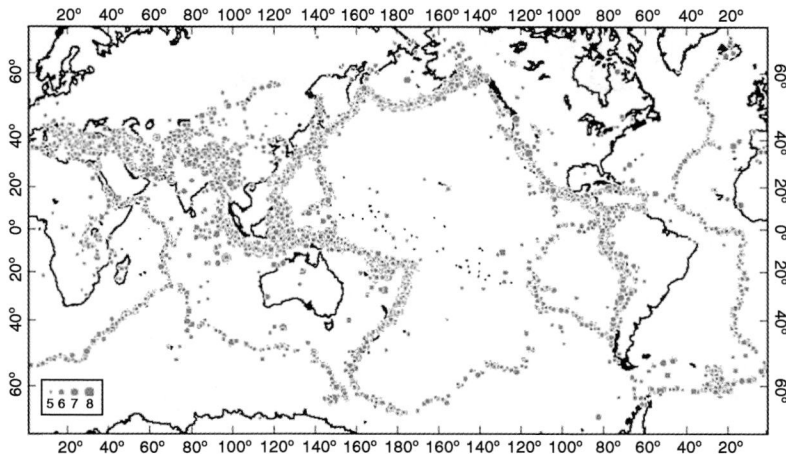

Fig. 3. Location of shallow (depth 0–70 km) earthquakes in the Harvard CMT catalog, 1977/1/1–2002/9/30. Size of a symbol is proportional to earthquake magnitude

especially suitable for statistical analysis and modeling. The catalogs can be roughly subdivided into two categories: global or regional catalogs covering large areas (continents or their large parts), and local catalogs for particular areas such as southern California (Fig. 2) and still smaller areas. As we discuss below, each type has its own properties and problems [86].

A collection of earthquake occurrences can be represented as a multidimensional stochastic process [76]: $T \times \mathbf{R}^3 (= \mathbf{R}^2 \times H) \times M \times \mathbf{SO}(3) (= \Psi \times S^2)$ (time-space-size-orientation), where $\mathbf{R}^3$ or $\mathbf{R}^2$ is the Euclidian space, H is the depth dimension, M is the scalar seismic moment, and $\mathbf{SO}(3)$ is the 3-D special orthogonal (rotation) group. The latter may be represented as a rotation by the angle Ψ around a rotation pole distributed over the 2-D sphere S^2 [71]. Multiple studies summarized in [76] and [101] indicate that marginal earthquake distributions are scale-invariant for all the above variables. The fractal pattern breaks down for large size, distance or time intervals. We discuss this in other sections of the paper.

An important feature of the available earthquake catalogs is the range of the above variables related to the average error in estimating them. The ratio of the range to an error describes roughly the information one can obtain from a catalog. These ratios are only approximate to one order of magnitude (see more in [82]): a summary is shown in Table 1.

From Table 1 we see that the temporal structure of earthquake occurrences can be detailed with great precision. The locations of earthquake foci are estimated relatively accurately in the horizontal plane, but vertical errors are

Table 1. Information available in earthquake catalogs

#	Variable	Accuracy (A)	Range (R)	R/A
1	Origin time T	0.01–1 s	5–25 y	10^9–10^{11}
2	Horiz. space R^2	3–10 km	3000 km	10^3
2'	Horiz. space R^2	0.5 km	200 km	$10^{2.5}$
2''	Horiz. space R^2	0.02 km	2–20 km	10^2–10^3
3	Vert. space R	5–15 km	50 km	10
3''	Vert. space R	0.1 km	10 km	10^2
4	Moment magn. m	0.07	6.0	10^2
5	Rot. angle Ψ	10°	120°	10^3
6	Rot. pole S^2	10°	360°	10^3

2, 3 – global catalogs [82];
2' – local catalogs;
2'', 3'' – wave correlation catalogs (e.g., [50, 140]).

often significantly larger. This effectively reduces available spatial information. The influence of location errors and other nuisance variables often extends well above a catalog's reported accuracy values [69, 82, 94]. Similarly, boundary effects can be observed at distances substantially smaller than a region's total size. Therefore, the scale-invariant range of the spatial distribution is likely to be smaller than the 10^2–10^3 shown in Table 1. Focal mechanisms, which have been reliably obtained in only the last 25 years, have large uncertainties also [78, 82].

Catalogs are a major source of information on earthquake occurrence. Since the late nineteenth century certain statistical features were established: Omori [125] studied temporal distribution; Gutenberg and Richter [46] investigated size distribution; quantitative investigations of spatial patterns started late [94].

Kostrov [107] proposed that earthquake displacement can be described by a second-rank symmetric tensor. Gilbert and Dziewonski [42] were the first to obtain a tensor solution from seismograms. However, statistical investigations remained largely restricted to time-size-space regularities. Tensor or focal mechanism analysis is difficult because we lack appropriate statistical tools to analyze either second-rank tensors or quaternions which properly represent earthquake focal mechanisms [71, 85]. For example, in two recent special issues of the geophysical journals dedicated to statistical analysis of seismicity (*Pure Appl. Geoph.*, **162**(6–7), 2005; *Tectonophysics*, **413**(1–2), 2006) no paper analyzed the statistics of the seismic moment tensor or earthquake focal mechanism. Kagan and Knopoff [97, 98] and Kagan [73, 74, 78, 85] were the first to investigate the statistical properties of the seismic moment tensor (Sect. 7).

2.3 Earthquake Temporal Occurrence: Quasi-Periodic, Poisson, or Clustered?

The periodic or quasi-periodic hypothesis of large earthquake occurrence has long been held by geoscientists. Similar hypotheses are called "seismic gap" or "seismic cycle" models. A seismic gap, according to such a model, is a fault or a plate segment for which the time since the previous large earthquake is long enough that stress builds up. Since earthquake occurrence is multidimensional and periodicity is a property of a one-dimensional process, the seismic record needs to be converted into a temporal series.

The characteristic hypothesis [137] implies a sequence of recognizably similar events and provides the logical basis for discussing recurrence or quasi-periodicity. Recurrence intervals and their statistics are meaningless without a clear definition of the characteristic earthquake [60]. A characteristic earthquake is assumed to release most of the tectonic deformation on a segment. Other earthquakes are significantly smaller than the characteristic one and hence can be ignored when moment release is calculated.

McCann et al. [116] adopted the gap model and produced a colored map of "earthquake potential" for close to a hundred circum-Pacific zones. They assumed that the seismic potential increases with the absolute time since the last large earthquake. Nishenko [120] refined the seismic gap model so that it could be more rigorously tested. He specified the geographical boundaries, characteristic magnitudes, and recurrence times for each segment. He used a quasi-periodic, characteristic recurrence model to estimate conditional earthquake probabilities for 125 plate boundary segments around the Pacific Rim.

Kagan and Jackson [88] compared the model of McCann et al. [116] against later earthquakes. They found that large earthquakes occurred more frequently in the very zones where McCann et al. had estimated low seismic potential. In other words, they found that large earthquakes are rather clustered in time. Kagan and Jackson [88] also found that earthquakes after 1989 did not support Nishenko's [120] gap model. Rong et al. [133] concurred: both predictions were inconsistent with the later earthquake record.

Bakun and Lindh [7] proposed that a magnitude 6 earthquake would occur at the Parkfield, California segment of the San Andreas fault with a 95% probability in the time window 1985–1993. The prediction model was based largely on the characteristic, quasi-periodic earthquake hypothesis. This was the only prediction reviewed and approved by the U.S. government. However, no such earthquake occurred till 28 September 2004, when an earthquake of magnitude 6.0 struck near Parkfield [8]. Meanwhile, a complicated form of the seismic gap model was applied to estimate earthquakes probabilities in the San Francisco Bay region [156]. The Working Group concluded that "there is a 0.62 [0.38–0.85] probability of a major, damaging [$M \geq 6.7$] earthquake striking the greater San Francisco Bay Region over the next 30 years (2002–2031)."

Stark and Freedman [144] argue that the probabilities defined in such a prediction are meaningless because they cannot be validated. They point out that in weather predictions, 50% probability of rain can be tested by counting the ratio of the number of rainy days to the total having this predictive value. No such possibility exists for the predictions concerning San Francisco Bay [156] or Parkfield. Stark and Freedman [144] finish their review of the San Francisco Bay earthquake prediction with the advice that readers "should largely ignore the USGS probability forecast."

This lack of falsifiability and the inability to construct an improved hypothesis both contradict the fundamental requirements of modern scientific method [108, 131]. After the gap model was formulated in its present form 20 or 30 years ago [116, 120, 137] the model proponents did not attempt to verify its fundamental assumptions with a critical test. The apparent failure of the predictions (see above) was not extensively analyzed and explained (see, for example, debate, in the [119] or [8]). Jackson and Kagan [60] review the implications of the Parkfield 2004 event for earthquake prediction, the characteristic earthquake hypothesis, and the earthquake occurrence in general. They argue that a simpler *null* hypothesis based on the Gutenberg–Richter law (see Sect. 4) and Poisson time behavior better explains the Parkfield event sequence. Despite this breakdown of scientific methodology, the potentially incorrect model continued to be in use for seismic hazard assessment in the U.S. and many other countries [60].

How could this happen? Geosciences are largely observational and descriptive disciplines. Earth scientists are not trained to formulate falsifiable hypotheses, critically test them, systematically review possible sources of error, thoroughly rule out alternative explanations, and dismiss or improve the models thereafter. Oreskes [126] discusses how American earth scientists summarily rejected the theory of continental drift for decades before the 1960s, though extensive evidence existed in support of it. Suppe [147] analyzes logical structure of the arguments by one of the most influential papers supporting plate tectonics [118] and considers critical methods to validate hypotheses. Kagan [77] argues that the major challenge facing earthquake seismology is that new methods for hypothesis verification need to be developed. These methods should yield reproducible, objective results, and be as effective, for instance, as double-blind testing in medical research.

2.4 Earthquake Faults: One Fault, Several Faults, or an Infinite Number of Faults?

In Fig. 4a we display the most commonly used geometry of an earthquake fault: a planar boundary between two rigid blocks. Other block boundaries are usually considered to be free. When Burridge and Knopoff proposed this model in 1967, it was the first mathematical treatment of earthquake rupture and a very important development. Since then, hundreds of papers have been published using this model or its variants. We show below why seismology needs a much more complicated geometrical model to represent brittle shear earthquake fracture:

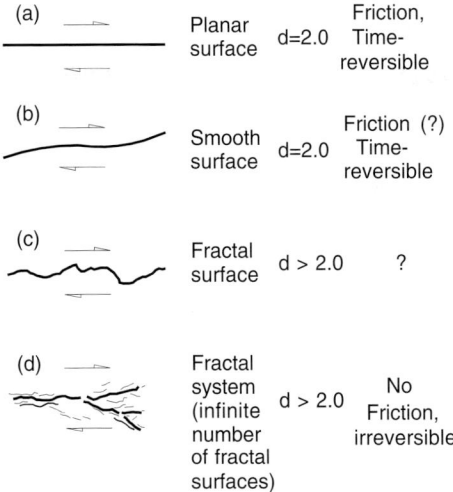

Fig. 4. Earthquake fault models

- The old model (Fig. 4a) is a closed, isolated system, whereas tectonic earthquakes occur in an open environment. This model justifies spurious quasi-periodicity, seismic gaps, and seismic cycle models (Sect. 2.3). No rigorous observational evidence exists for the presence of these features in earthquake records (see above).
- An earthquake fault in the model (Fig. 4a) is a well-defined simple geometrical object – a planar surface with the dimension 2.0. In nature, an earthquake fault *system* is a fractal set. This set is not a surface; its dimension is about 2.2 (Sect. 6.2).
- Two distinct scales are present in the diagram: irregularities of the planar surface and block size. Resistance to block motion due to breakage of microscopical surface inhomogeneities is described as a friction force. The friction is an appropriate model for man-made surfaces, where the scale of inhomogeneities is limited. In contrast, earthquakes are scale-invariant. The geometry and mechanical properties of an earthquake fault zone are the result of self-organization. They are fractal [76, 86].
- A displacement incompatibility problem is wrongly circumvented because of the flat plate boundaries. Real earthquake faults always contain triple junctions (see, for example, Figs. 2 and 3); further deformation is impossible without creating new fractures and rotational defects (disclinations).
- Because the block boundary is planar, stress concentrations are practically absent after a major earthquake. Hence these models have few or no aftershocks.

- All earthquakes in the model have the same focal mechanism. Any variations in mechanisms obvious during even a cursory inspection of maps (as in Fig. 2) are not taken into account.

King [105], Turcotte [148] and Andrews [4] suggested that due to kinematic effects at fault junctions, the fault geometry of earthquakes may be represented as a scale-invariant matrix of faults. Gabrielov et al. [40] developed a mathematical framework for calculating the kinematic and geometric incompatibility in a tectonic block system, both rigid and deformable. They concluded that due to geometric incompatibilities at fault junctions, new ruptures must be created to accommodate large plate tectonic deformations. Indeed, plate tectonic observations indicate that hundreds of km of deformation occur over the several million years of plate boundary existence (e.g. the San Andreas fault system).

Figure 4b,c display a few alternative models of earthquake faults: a smooth surface boundary and a fractal surface. Unless the smooth boundary is a surface of rotation, no large scale block displacement is possible. Similarly, to move blocks along the fractal boundary, one needs to break the surface inhomogeneities. In contrast to the model of Fig. 4a, the largest inhomogeneities are comparable to the block size.

Obviously, if major faults comprising a plate boundary are not strictly parallel, fault junctions are unavoidable. The question is whether large deformations can be accommodated by a few faults delineating a few tectonic blocks (see, for example, [10]), or whether an infinite number of faults must exist to account for such deformations.

The above considerations suggest again that the conventional models of tectonic block deformation need complete revision. If the number of faults and fault junctions is infinite, these junctions, as Gabrielov et al. [41] suggest, constitute "asperities" and "barriers" for fault rupture propagation. These geometric complexities, not friction, should control the developing fault system and the stop-and-go feature of the earthquake rupture propagation. Kagan [67] shows that when the earthquake rupture plane rotates, as in triple junctions, the third-rank seismic moment tensor, which can be identified with asperities or disclinations, becomes non-zero.

In Fig. 4d we show a picture of a fractal boundary zone between two rigid blocks. In this case, a complex fault pattern cannot be characterized as a surface: it is a fractal set of dislocations. In Sects. 6 and 7 we attempt to characterize this pattern quantitatively.

2.5 Statistical and Physical Models of Seismicity

As we mentioned above, hundreds of papers are scattered in geophysical and physical journals on the statistical properties of seismicity. They propose physical or phenomenological models of earthquake occurrence. We will describe the papers and their results briefly.

Several authors, starting with Bak et al. [6], attempt to collapse the time-distance-size earthquake distribution into one plot (see also [9, 23]). Such plots prominently demonstrate the scale-invariant structure of seismicity known previously from marginal distributions. However, as mentioned earlier, although temporal, spatial, and magnitude distributions are scale-invariant for small values of pertinent variables, for larger values the scale-invariant pattern is replaced by finite-scale effects (Sects. 4–6). Moreover, even for small variable values, the distributions are influenced by various random and systematic effects. The study of such errors is difficult in a collapsed multidimensional plot.

There are several groups of physical seismicity models. Most of them employ the geometrical and mechanical scheme illustrated in Fig. 4a as their major paradigm: two blocks separated by a planar surface ([28, 109, 132] see also [103]). Our earlier discussion of this model is also valid for these attempts: they ignore the spatial and mechanical complexity of the earthquake fault zone. Consequently, the deficiencies listed in the previous section are present in these models as well. Moreover, since these paradigms describe only one boundary between blocks, they do not account for a complex interaction between other block boundaries and, in particular, triple junctions. Seismic maps (Figs. 2 and 3) convincingly demonstrate that earthquakes occur mostly at boundaries of relatively rigid blocks. This is a major idea of plate tectonics [118, 126]. However, if blocks are rigid, stress concentrations at other block boundaries and a block's triple junctions should influence earthquake patterns at any boundary. Thus, even after a large earthquake, the stress on a particular boundary can be restored almost immediately due to the influence of the block's other boundaries and its junctions.

Lyakhovsky et al. [113, 112] base their seismicity model on the damage rheology theory. In this case, where the mechanical properties of the rock medium are modeled, even elementary geometrical properties of a fault system are not considered. As a result, the fault geometry and earthquake focal mechanism distribution fall outside their work.

As we mentioned in Sect. 2.3, theoretical developments need to be critically tested against observational evidence. Otherwise, they remain in the realm of speculation. At the present time, numerical earthquake models have shown no predictive capability exceeding or comparable to empirical predictions based on earthquake statistics. Even if a theoretical or physical model exhibits some predictive skill, we should always question whether the predictive power comes from a deeper theoretical understanding, or from the earthquake statistics results imbedded in the model.

The models described above have a large number of adjustable parameters, both obvious and hidden, to simulate a complicated pattern of seismic activity. Dyson [30] says that Enrico Fermi advised him

> ... My friend Johnny von Neumann used to say, with four parameters I can fit an elephant ...

The observational evidence in support of these models generally consists of particular earthquake in specific regions. In a random process there is always the possibility of using a large corpus of data to select a particular series of events which seem to agree with theoretical predictions. For the model confirmation to be credible, the criteria for data selection must be prescribed in advance [60].

Another possibility for perceived confirmation of theoretical models is publication bias. This bias is caused by the fact that research with currently fashionable results is potentially more likely to be submitted and published than work with less appealing outcomes [29].

Therefore, physical models may not have a theoretical predictive capability. As Sect. 8.4 reports, although several phenomenological models issue a quantitative prediction of future seismicity, no physical model has yet attempted to compete in such tests.

2.6 Laboratory and Theoretical Studies of Fracture

In engineering science extensive work has been performed on the conditions of tensile crack initiation and propagation (e.g., [2]). However, these efforts are concentrated on the problem of a single crack: the most important problem for engineers.

The problem of crack propagation and branching, far more relevant to earthquakes, has been recently addressed in several papers. In laboratory experiments, a crack develops instabilities which make its path and propagation velocity highly chaotic and unpredictable [16, 115, 138]. These instabilities and a sensitive dependence on the initial conditions are due to crack propagation, especially at a speed close to the elastic wave velocity. Stress and fracture conditions in laboratory specimens differ significantly from those in earthquake fault zones: in the laboratory the boundary effects are controlled by the researcher. Therefore, fractures can self-organize only at spatial scales much smaller than those of the specimen. In fault zones, the stress, rock mechanical properties, and fault geometry are self-organized as large-scale self-similar patterns develop.

The calculations of molecular dynamics [17, 115, 138] demonstrate that basic properties of tensile fracture can be effectively derived from simple laws. Similarly, precise laboratory measurements of fault propagation demonstrate multiple branching of fault surfaces. These simulations reproduce the fractal character of a fracture. Moreover, calculating the total energy balance in laboratory fracture experiments [138, 139] demonstrates that almost all elastic energy goes into creating new surface. Although the conditions during tensile fracture differ from those of the shear failure in earthquakes, the above result may be significant for the problem of the heat paradox for earthquake faults [135].

3 Modern Earthquake Catalogs

Detailed modern earthquake catalogs with estimates of focal mechanism and/or seismic moment tensor were compiled beginning in the 1970s. Several extensive catalog datasets are available at present. Frohlich and Davis [39] and Kagan [82] discuss the properties of global catalogs and their accuracy.

The global catalog of the centroid moment tensors (CMT) is compiled by the Harvard group [33]. The catalog contains 22,476 solutions over a period from 1976/1/1 to 2004/12/31, see Fig. 3. The Harvard catalog includes seismic moment centroid times and locations as well as estimates of the seismic moment tensor components. Each tensor is constrained to have a zero trace (first invariant): no isotropic component. Double-couple (DC) solutions, or solutions with the tensor determinant equal to zero, are supplied as well. Almost all earthquake parameters are accompanied by internal estimates of error.

The PDE worldwide traditional catalog (Preliminary Determination of Epicenters, 1999, and references therein) is published by the USGS. The catalog measures earthquake size, using several magnitude scales. Body-wave (m_b) and surface-wave (M_S) magnitudes are provided for most moderate and large events since 1965 and 1968, respectively. The catalog contains more than 50,000 shallow earthquakes with $m_b \geq 5$ since 1965.

The problem for almost all catalogs, especially local and regional, is their inhomogeneity: since any local seismographic network is bounded, the accuracy and catalog completeness vary considerably within the catalog area. This inhomogeneity is especially strong for the seismographic networks concentrated on island chains. Where stations represent a quasi-linear array, the location accuracy differs strongly along the direction of the array as compared to the orthogonal direction.

4 Earthquake Size Distribution

The distribution of earthquake sizes is usually invoked as a first confirmation for virtually any model of seismicity. Moreover, this distribution is by far the most studied feature of statistical seismology. Starting with its first discussion by Ishimoto and Iida [59] and then [46], it has been established that earthquakes increase in number as a power-law as their sizes decrease. This dependence is usually referred to as the magnitude-frequency or the Gutenberg–Richter (G-R) relation, and its parameter (see (4) below) is commonly known as the "b-value". A very large body of literature exists concerning the size distribution, its interpretation and possible correlation with geotectonics, stress, rock properties, etc. For example, a search of the ISI Web of Knowledge database for keywords like **"earthquake* and b-value"** yields about 110 publications in the last four years. However, that proliferation has not led to a deeper understanding of earthquake generation.

4.1 Magnitude Versus Seismic Moment

Magnitude is an empirical measure of earthquake size and many different magnitude scales are currently used (see [21]). Several types of errors need to be investigated in earthquake size measurement. Some of them are known to be connected with earthquake magnitude determination: saturation of all magnitude scales [102], which is explained by the finite seismogram frequency for a seismographic network. Other types of errors are common to both magnitude and seismic moment determination [82].

Relatively high-frequency seismic waves are used to determine magnitude, the effects of scattering, multipathing, focussing and unfocussing are stronger as the wave periods decrease. These effects cause great variations of wave amplitude which lead to larger uncertainties and biases in magnitude measurements (cf. [32]).

Seismographic networks are limited in detecting weak earthquakes and their essential parameters such as hypocenter location, origin time and magnitude. This results in another limitation of magnitude distributions: at the lower magnitude end, progressively larger number of events are missing from catalogs. Unfortunately, this lower magnitude cutoff is neither sharp nor uniform over time and space.

In this paper M denotes the scalar seismic moment, and m denotes the magnitude of an earthquake, b is the parameter for magnitude distribution and β is the corresponding parameter for seismic moment distribution. Earthquake moment magnitude m_w is related to the scalar seismic moment M via [47, 102]

$$m_w = \frac{2}{3} \log_{10} M - C , \qquad (2)$$

where seismic moment M is measured in Newton-m, and C is usually taken to be between 6.0 and 6.1. Below we use $C = 6.0$. Equation (2) allows us to use the moment magnitude as a proxy for a physical quantity: seismic moment.

Seismic moment is proportional to the amplitude of seismic waves at zero or close to zero frequency; hence its accuracy is higher than that using magnitudes. Kagan [82] estimates that uncertainty in moment magnitude is on the order of 0.1–0.15 and is by a factor of 2 to 3 smaller than regular magnitude uncertainties.

Because of saturation and other systematic effects, each magnitude can only be evaluated over a limited range of earthquake size. Different magnitude scales are then compared by using a regression relation

$$m_1 = C_1 + C_2 m_2 , \qquad (3)$$

where m_i are magnitudes and C_j are coefficients for a linear regression. Although both magnitudes in (3) usually have errors of similar size, regular, not orthogonal regression, is commonly used [21]. This should cause a significant bias in converting one magnitude into another. Most earthquake catalogs initially use several magnitudes. To obtain a common magnitude value, catalog compilers transform various magnitudes, using variants of (3). Errors and systematic effects of such calculations should significantly shape the estimates of earthquake size.

Inspecting the value of the C_2 coefficient in (3) in various publications, one can see the degree of the problem in determining magnitudes. C_2 should be close to 1.0 at the range of earthquake size where both magnitudes are well-defined. This is rarely the case: the C_2-value often reaches 0.7 or 1.3 (Kagan, 2003, his Fig. 14). These fluctuations of the conversion coefficient may cause spurious variations of the b-value. In contrast, when the moment magnitude in different catalogs is compared, the C_2 coefficient is close to 1.0 (Kagan, 2003, his Fig. 12).

4.2 Seismic Moment Distribution

Gutenberg and Richter's [46] magnitude-frequency relation is usually written as

$$\lg N(m) = a - b\,m \,, \tag{4}$$

where $N(m)$ is the number of earthquakes with magnitude $\geq m$, and a and b are parameters: a characterizes seismic activity or earthquake productivity of a region and b parameter describes the relation between small and large earthquake numbers, $b \approx 1$. The expression (4) has been proposed in the above functional form by Vilfredo Pareto ([127], p. 305, his (1)) for the financial income distribution.

The original G-R distribution (4) can be transformed into the Pareto distribution for the scalar seismic moment M

$$\phi(M) = \beta\, M_t^\beta\, M^{-1-\beta} \quad \text{for} \quad M_t \leq M \,, \tag{5}$$

where β is the index parameter of the distribution, $\beta = \frac{2}{3} b$ (see 2), and M_t is the observational threshold.

Simple consideration of the finiteness of seismic moment flux or the deformational energy, available for earthquake generation, requires that the Pareto relation (5) be modified at the large size end of the moment scale. The distribution density tail must have a decay stronger than $M^{-1-\beta}$ with $\beta > 1$. This problem is generally solved by introducing into the distribution an additional parameter called the *maximum* or *corner* moment (M_x or M_c).

The tapered G-R relation has an exponential taper applied to the cumulative number of events with the seismic moment larger than M [79, 154]

$$\Phi(M) = (M_t/M)^\beta \, \exp\!\left(\frac{M_t - M}{M_c}\right)$$
$$\text{for} \quad M_t \leq M < \infty \,, \tag{6}$$

here M_c is the parameter that controls the distribution in the upper ranges of M ("the corner moment"). Figure 5 displays the seismic moment distribution

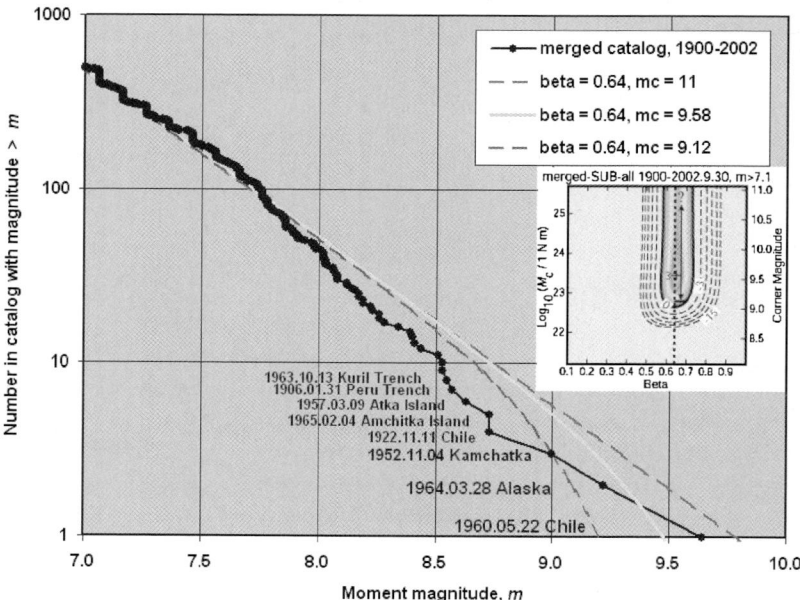

Fig. 5. The size distribution for subduction earthquakes [13]. The inset shows the likelihood map which is used to estimate both the β and the corner moment (M_c) values

and its approximation by (6) with a choice of several corner magnitudes (m_c) for the subduction zones [13].

Many other displays of the moment-frequency relation for earthquakes before 2003 can be found at the power-point file http://moho.ess.ucla.edu/~kagan/Tokyo_Univ.ppt. Updates of the moment distribution parameters after the 2004 great Sumatra earthquake are presented at http://element.ess.ucla.edu/publications/2004_global_coupling-/2004_global_coupling.htm. The corresponding probability density function is

$$\phi(M) = \left[\frac{\beta}{M} + \frac{1}{M_c}\right] (M_t/M)^\beta \exp\left(\frac{M_t - M}{M_c}\right). \quad (7)$$

The above distribution in both expressions (6–7) was proposed by Pareto ([127], pp. 305-306, his (2), 2bis, and 5).

In Fig. 6 we show the result of the maximum likelihood determination of the β-values for eight tectonic provinces [13]. All 95% confidence limits include $\beta \approx 2/3$ value. This can be considered a universal parameter of earthquake size distribution.

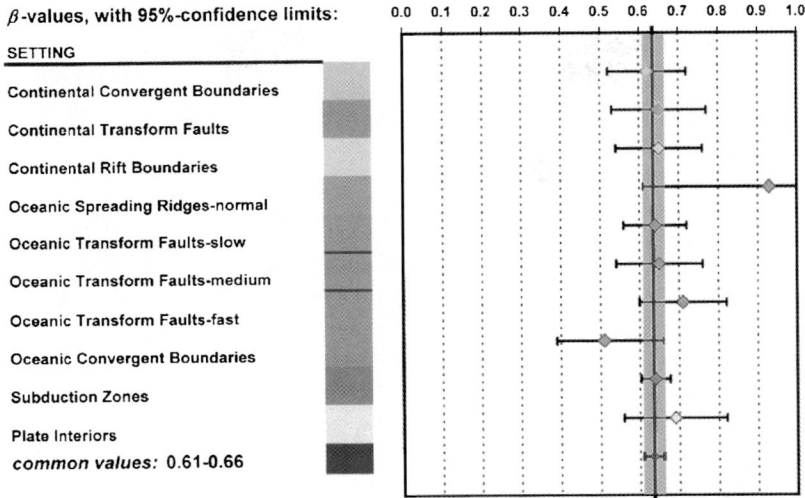

Fig. 6. The β-value global distribution. The β-values ($\beta = \frac{2}{3}b$), determined by the maximum likelihood method for eight tectonic provinces [13], are shown with their 95% confidence limits

The next diagram (Fig. 7) displays the corner moment values evaluated for the same eight provinces. For convenience they are shown on the map of central America, where all the provinces are represented. In contrast to the β-value result, [13] find that at least four distinct values of the corner magnitude seem to be required, based on the 95% confidence limits. These values include Oceanic Spreading Ridge (normal faulting, corner magnitude range, $m = 5.7 - 6.0$); Oceanic Transform Faults (medium and fast velocities, range, 6.4–7.0); all the Continental zones, Oceanic Transform Faults and slow velocity/Oceanic Convergent Boundary (range, 7.4–8.7); and Subduction zone boundaries (range, 9.1–∞).

Using the earthquake size distribution (6, 7), we can calculate the seismic moment rate [81]

$$\dot{M}_s = \frac{\alpha_0 M_0^\beta}{1 - \beta} M_c^{1-\beta} \Gamma(2 - \beta), \qquad (8)$$

where α_0 is the annual rate of earthquakes with moment M_0 or greater and Γ is the gamma function. Subsequently we can compare it to the tectonic rate evaluated by plate motion or by geodesy [13, 81]. Thus, plate tectonic predictions can be quantitatively related to seismic activity. Below we show that because of the power-law property of earthquake size distribution, any naive comparison of the cumulative seismic moment and tectonic deformation yields unreliable results in most cases.

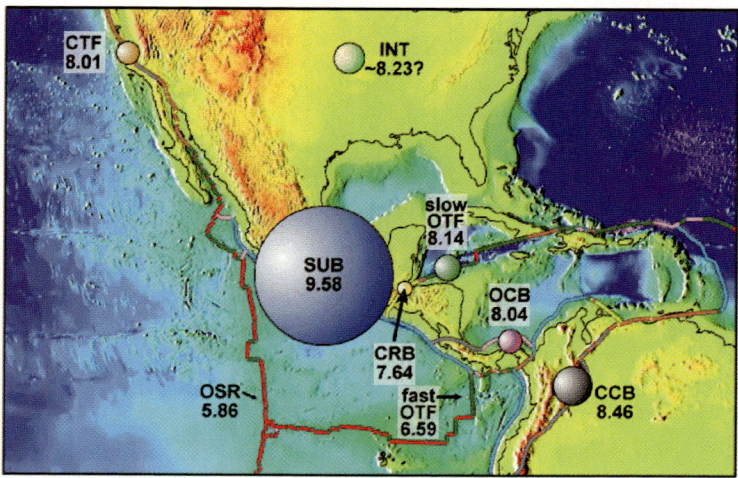

Fig. 7. Corner moment (M_c) distribution for eight tectonic provinces [13]. In this figure the moment magnitude is calculated using $C = 6.05$ in (2). For plate boundary names and abbreviations see Fig. 6

4.3 Seismic Moment Sum Distribution

The global distribution of the seismic moment is well approximated by a power-law (Pareto) distribution with index $\beta \approx 2/3$ (5). This is a heavy-tailed distribution: it has the infinite mean and the standard deviation. Thus, if one uses the pure Pareto model, the Central Limit Theorem does not describe the distribution of the sum of seismic moments. The tapered (6–7) or truncated Pareto distribution appears to eliminate all summation problems. However, a detailed analysis shows that the Gaussian limit is reached for only a large number of observations. For a realistic number of events, the tapered or truncated Pareto still exhibits all the properties of a heavy-tailed distribution.

The cumulative seismic moment released in a region can be used as a proxy for total regional deformation of the Earth surface due to earthquakes. Formally, the strain rate for the volume of the deforming crust is proportional to the sum of the *tensor* moments of individual earthquakes [107]. Thus, evaluating observed seismic moment rates is an important problem connected to the regional earthquake hazard assessment.

Let X_i, $i = 1, \ldots, n$ be independent identically distributed random variables with a common Pareto distribution (5), and let S_n denote their sum

$$S_n = \sum_{i=1}^{n} X_i \ . \tag{9}$$

If the exponent β of the power-law distributed variable is less than 2.0, the sum converges to a stable distribution [134, 150] with the probability density function

$$\phi(X, \beta, \gamma, \mu, \sigma) \ , \tag{10}$$

where γ is a symmetry parameter (for positive variables $\gamma = 1$, i.e., the sum is maximally-skewed), and μ, σ are shift and width parameters. For the Gaussian distribution (see (16) below) only the two last parameters are valid.

An arbitrary quantile z_q of the sum S_n can be approximated as [160]

$$z_q \approx z_q^{(1)} \equiv n^{1/\beta} x_q C_\beta + b_n \ , \tag{11}$$

where x_q solves the equation for the cumulative distribution F_β of the sum

$$F_\beta(x_q) = q \ , \tag{12}$$

and $z_q^{(1)}$ is a quantile for a maximally asymmetrical (maximally-skewed) stable distribution. For $\beta < 1$, $b_n = 0$ and

$$C_n = [\Gamma(1 - \beta) \cos(\pi\beta/2)]^{1/\beta} \ . \tag{13}$$

Figure 8 displays an example of simulated sums (S_n) for the Pareto distribution truncated at $y = M_x/M_t = 3.4 \times 10^4$ compared to the stable distribution quantiles. In this example we take the threshold moment $M_t = 10^{17}$ Nm or $m_t = 5.33$, the threshold of the recent Harvard catalog [82] and the maximum magnitude $m_x = 8.35$ (Sect. 4.2).

According to (11), quantiles of the stable distribution increase as $n^{1/0.66}$, thus, for example, the median of the sum of 40 variables μ_{40} compared to μ_{20} is equal to

$$\mu_{40} \approx 2.86 \times \mu_{20} \quad \text{or} \quad \mu_{40} > \mu_{20} + \mu_{20} \ . \tag{14}$$

This behavior of the stable distribution sums may seem counter-intuitive, as is that of their other properties.

If the exponent β is less than 1.0, the sum of power-law distributed variables is comparable to the value of the largest observation M_n

$$E(S_n) = M_n/(1 - \beta) \ . \tag{15}$$

where E is a mathematical expectation sign [36].

Therefore, (14) means that in a sample of 40 earthquakes, there is a higher chance of encountering a large event which would significantly increase their sum than in a sample of only 20 earthquakes. Pisarenko [129] as well as Huillet and Raynaud [58] also note that for the heavy-tailed distributions, sum quantiles increase non-linearly with n.

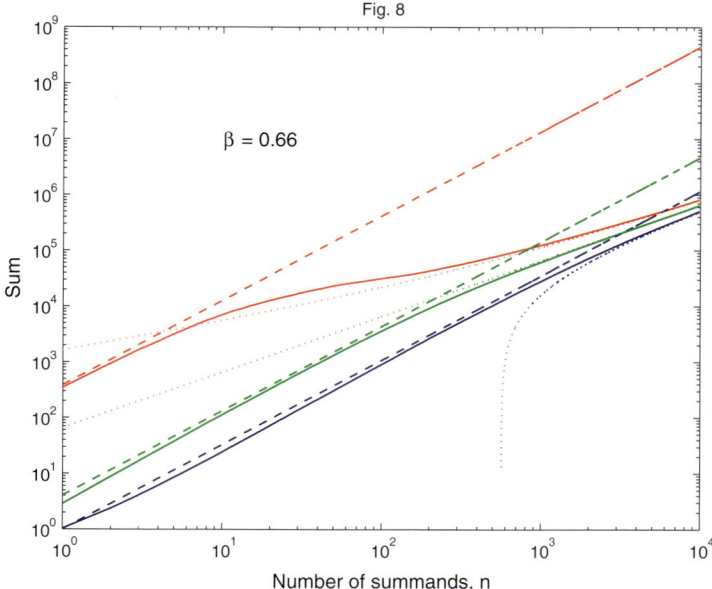

Fig. 8. Quantiles for the sum S_n of truncated Pareto variables (upper limit $y = 3.4 \times 10^4$, (7)) and their approximations as functions of the number of summands, n. Two approximations are considered: via the stable distribution, (11) (*dashed lines*) and Gaussian, (16) (*dotted lines*). *Solid lines* represent quantiles of simulated Pareto sums. Three *thick upper curves* are for the 0.98 quantile, three medium thickness *middle curves* are for the median, and three *thin lower curves* are for the 0.02 quantile

The upper quantiles of the Pareto sum generally approach the stable distribution limit faster than do the lower quantiles [160]. However, in Fig. 8, the upper quantiles depart from the theoretical curve for the stable distribution starting with $n = 2$ because of the upper limit truncation. The behavior of the lower quantile is essentially unaffected by the truncation until n exceeds 10^3.

When the number of summands is large, the truncation point y dominates the behavior of the quantiles. The sum is then distributed asymptotically according to the Gaussian law:

$$\lim_{n \to \infty} F_{S_n}(x) = \Phi\left(\frac{x - n\mu_y}{\sigma_y \sqrt{n}}; 0, 1\right), \tag{16}$$

where Φ is the normal cumulative distribution (17), and the parameters μ_y and σ_y are given by (18), (19).

The standard Gaussian (normal) cdf with expectation μ and standard deviation σ is given by

$$\Phi\left(x;\mu,\sigma^2\right) = \frac{1}{\sigma\sqrt{2\pi}} \int_{-\infty}^{x} \exp\left(-\frac{(y-\mu)^2}{2\sigma^2}\right) dy, \tag{17}$$

whereas

$$\mu_y = \frac{\beta}{1-\beta}\left(y^{1-\beta}-1\right)/\left(1-y^{-\beta}\right), \quad \beta \neq 1, \tag{18}$$

and

$$\sigma_y^2 = \frac{\beta}{2-\beta}\left(y^{2-\beta}-1\right)/\left(1-y^{-\beta}\right) - \mu_y^2, \quad \beta \neq 2, \tag{19}$$

are the conditional mean and variance of each summand [79, 160], given the restriction on the maximum ($X < y$).

From the beginning of the plate tectonics hypothesis, it was assumed that earthquakes are due to plate boundary deformation. Calculations for global tectonics and large seismic regions justified such an approach. However, applying this assumption to smaller regions has usually been inconclusive, given the high variability of seismic moment sums.

Holt et al. [57] compared the observed seismic moment release with the tectonic release inferred from the joint inversion of the GPS and the Quaternary rates of strain for south-east Asia. They also compared strain release with the earthquake record from 1965–1998. Figure 9 shows the seismic coupling χ (the ratio between observed and predicted seismic moment) in 4 large regions and 42 subregions. The coupling is calculated as

$$\chi = \frac{1}{T}\sum_{i=1}^{n} M_i/\dot{M}_{\text{tec}}, \tag{20}$$

where T is the total catalog time and $\dot{M}_{\text{tec}}$ is the rate of tectonic deformation.

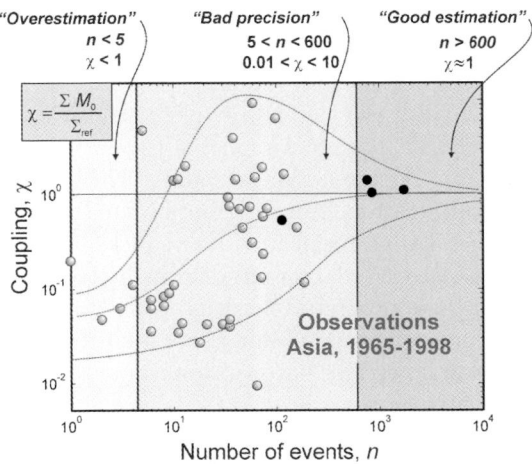

Fig. 9. Coupling of tectonic deformation and seismic moment release for 42 subregions (*grey disks*) of southeast Asia [57]. Black disks are 4 large regions

Three regimes are clearly seen, depending on the number of earthquakes in a region. These regimes are perfectly reproduced by the truncated Pareto model. Figure 9 approximately displays the quantiles (upper, middle, lower) for the ratio between the sum of n random variables and the corresponding mean (which is finite in the truncated model). Notice that we see the conventional Gaussian picture (the sum is proportional to the mean) only with a very large number of events exceeding m_t: $n > 1000$. Thus, the truncated Pareto model explains the non-linear behavior of the cumulative moment release for small to intermediate numbers of earthquakes as a transition between pure power and pure Gaussian approximations.

5 Temporal Earthquake Distribution

Omori [125] showed that the aftershock rate decays approximately as

$$n(t) = \frac{K}{t+c}, \tag{21}$$

where K and c are coefficients, t is the time since the mainshock origin, and $n(t)$ is the aftershock frequency measured over a certain interval of time. Presently a more complicated equation is used to approximate the aftershock rate

$$n(t) = \frac{K}{(t+c)^p}. \tag{22}$$

This expression with the additional exponent parameter p is called the "modified Omori formula" [151].

Kagan [83] and Kagan and Houston [87] argue that the observed saturation in the aftershock numbers described by the "time offset" parameter c in Omori's law is likely an artifact due to the under-reporting of small aftershocks. This under-reporting comes from the difficulty of detecting large numbers of small aftershocks in the mainshock coda, as well as other factors [83]. For even smaller time intervals, close to the rupture time of the mainshock and aftershocks, the point model of the earthquake process breaks down, so that (21) and (22) are no longer valid.

Figure 10 displays the aftershock distribution for the 2004 great Sumatra earthquake. The general time-magnitude aftershock pattern is seen in many other aftershock sequences [83]: larger aftershocks begin early in the sequence, and the occurrence rate is progressively delayed for weaker events. After the aftershocks start in any magnitude band, they seem to be almost uniformly distributed over the log time. This pattern would correspond to the aftershock rate's decay according to Omori's law (21).

Two displays in Fig. 10 exhibit an important property of earthquake catalogs: in the PDE catalog aftershocks start at about 10^{-2} days after the mainshock, whereas in the CMT catalog they start at about 10^{-1} days. The

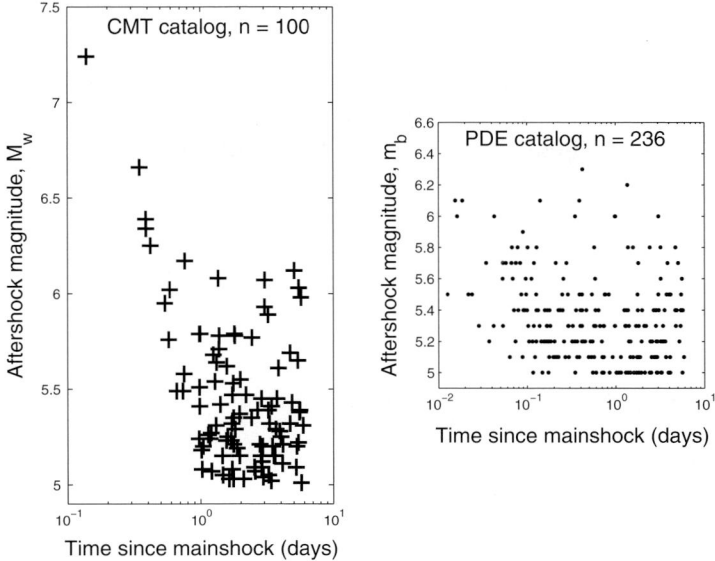

Fig. 10. Time-magnitude distribution of the 2004/12/26 $M = 9.1$ Sumatra aftershocks, n is the aftershock number. The Harvard and the PDE earthquake catalogs are used. Events in the six days following the mainshock and between latitude $0°$N and $15.0°$N and longitude $90.0°$E and $100.0°$E were selected

total number of aftershocks as well as their magnitude range also significantly differ in these two diagrams. The main reason is the frequency range of the seismograms used in compiling both catalogs: in the PDE catalog the aftershocks are determined using waves with 1 s period, whereas the CMT catalog uses low frequency (period 50 s and greater) waves. The magnitude estimates in the PDE catalog saturate at about $m_b = 6.0 - 6.5$ (Sect. 4.1): therefore, we see no large magnitude aftershocks in its display. On the other hand, long-period coda waves of the mainshock and large aftershocks in the CMT catalog extend over a longer time. They make it difficult to discern smaller events in the seismograms. Thus, Fig. 10 as well as the arguments in Kagan [83] and Kagan and Houston [87] demonstrate that the c-value depends on methods of seismogram interpretation. It is probably not a physical parameter.

Therefore, depending on the frequency characteristics of a seismographic network, the number of stations, and the seismogram processing technique, the same earthquake sequence could be variously identified. In one catalog it could be identified as one complex earthquake with some subevents, but in another as a foreshock/mainshock/aftershock sequence with many "individual" events [82, 83]. Thus "an individual earthquake" results from interpretations and selections made by catalog compilers. It is not in itself a physical entity, as tacitly assumed in most statistical analyses of seismicity. It is a naming artifact.

Omori's law has been incorporated in many phenomenological and physical models of earthquake occurrence. Like aftershocks, foreshocks also follow the power-law rate increase before a mainshock [70, 124].

6 Earthquake Location Distribution

6.1 Multipoint Spatial Moments

In the early 1980s, Kagan and Knopoff [94] and Kagan [64, 65] investigated the spatial moment structure of earthquake hypo- and epicenters for global (PDE and other) catalogs and several local earthquake catalogs.

The two-, three-, and four-point moment functions were obtained and analyzed in these studies. The quality and quantity of the earthquake data were relatively poor at that time, and computer capacity was sufficient to study only a small subset of data at a time, especially when computing the higher moments. However, it became apparent that the spatial distribution of earthquakes has a scale-invariant pattern.

Spatial moment functions are basic to the investigation of hypocentral patterns. The principal quantities of the study are the proportions of k-tuples ($k = 2, 3, 4$) points from the catalog (Fig. 11) with the property that the maximum distance between any two points in the k-tuple does not exceed r, as a function of r, and the joint density function of the distances between the points forming such a k-tuple.

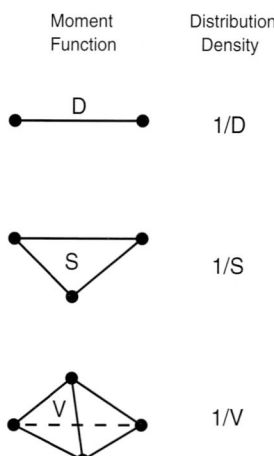

Fig. 11. Schematic representation of 2-, 3-, and 4-point spatial moment functions and their suggested approximate densities

We write
$$q_k(r) = N_k(r)/N_k \;, \tag{23}$$
where $N_k(r)$ is the number of k-tuples with the stated property, and N_k is the total number of k-tuples from the catalog. The quantities $q_k(r)$ are computed first for the epicenters, as points in $\mathbf{R}^2$, and then for the hypocenters, as points in $\mathbf{R}^3$. This function can be interpreted as the average number of k-tuples within a distance r of an "average" point of the catalog.

To overcome the biases in such estimates, which arise from boundary effects, the ratios $q_k(r)$ may be compared to the corresponding values for simulated Poisson catalogs. The simulated catalogs are of the same size and extent as the original catalog, but their epicentral coordinates are uniformly distributed over the region, and the depth distribution is matched to that of the actual catalog.

This results in the ratios
$$Q_k(r) = q_k(r)/\tilde{q}_k(r) \;, \tag{24}$$
where the tilde refers to the simulated catalog. Values of $Q_k(r)$ have been tabulated and graphed in various ways.

The graphs of the ratios $Q_k(r)$ against r typically display three ranges: the initial range, the middle range over which the $1/r$ behavior is observed, and the final range, in which the ratio approaches 1.0, as r approaches the diameter of the observed region. We interpret the first range as dominated by measurement errors. The second range illustrates self-similar behavior, and the third range is dominated by boundary effects. Kagan [86] provides extensive analysis of various errors and biases in the 2-point moment evaluation.

Thus, our key results can be summarized as follows. The growth rates of the moment functions are consistent with a dimensional deficit of approximately 1.0. Within an order of magnitude over different radial and angular combinations:

(i) the distribution of pairs of points selected at random from the catalog is consistent with the density inversely proportional to the distance $1/D$ [69, 86, 92, 94]
(ii) the distribution of triplets of points selected at random from the catalog is consistent with the density inversely proportional to the area, $1/S$ [64];
(iii) the distribution of quadruplets of points selected at random from the catalog is (for the hypocenters only) consistent with the density inversely proportional to the volume, $1/V$ [65].

6.2 Correlation Dimension

Kagan [69, 86, 94] revisited the two-point moment problem and was able to more accurately estimate the correlation dimension for shallow (0–70 km depth), intermediate (70–300 km), and deep (300–700 km) earthquakes. In these papers the dependence of the moment on the time interval between earthquakes was also investigated.

To demonstrate the influence of time limits on the correlation dimension, Fig. 12 shows the distribution of distances between accurately located hypocenters in southern California. The probability density function of distances $N_3(R)$ between these hypocenters, irrespective of the inter-earthquake time interval, is close to a power-law $N_3(R) \propto R^\delta$ in the range $0.1 \leq R \leq 5$ km, where $N_3(R)$ is the number of pairs in the 3-D space. The correlation fractal dimension (measured by the least-square linear regression of $\log(R)$ and $\log[N_3(R)]$ for $0.1 \leq R \leq 5$ km) is $\hat{\delta} \approx 1.5$ (black lines in Fig. 12, see also [54]). The faster decay for $R < 0.1$ km is due to location errors, and the roll-off for distances $R > 5$ km is due to the finite thickness of the seismogenic crust. For larger distances ($R > 50$ km), the $\hat{\delta}$ decrease is caused by catalog boundaries [86].

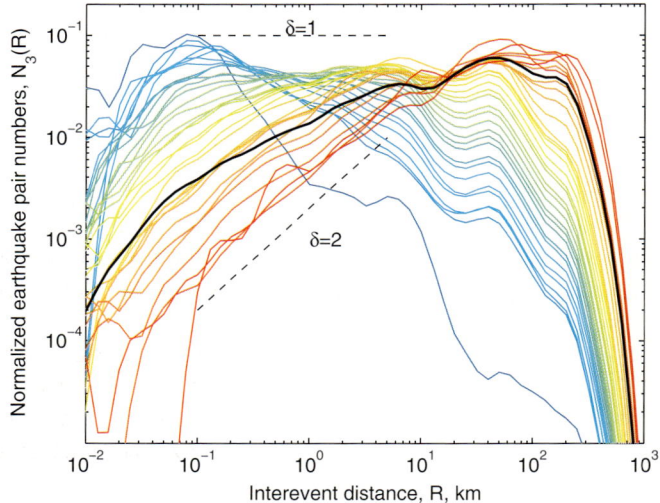

Fig. 12. Distribution of distances between hypocenters $N_3(R, t)$ for the Hauksson and Shearer [50] catalog, using only earthquake pairs with inter-event times in the range $[t, 1.25t]$. Time interval t increases from 1.4 minutes (*blue curve*) to 2500 days (*red curve*). We divide the earthquake pair number by R so that the *horizontal line* would correspond to $\delta = 1$. The *black line* is the function $N_3(R)$ measured for all earthquake pairs; it has a fractal dimension $\hat{\delta} \approx 1.5$ for $0.1 \leq R \leq 5$ km

For $N_3(R, t)$, the correlation dimension $\hat{\delta}$ increases between $\hat{\delta} \approx 0$ at times $t = 5$ minutes up to $\hat{\delta} \to 2$ for $t = 2500$ days. This maximum inter-event time of 2500 days is long enough that earthquake interactions are relatively small compared to the tectonic loading. This value $\hat{\delta} = 2$, measured for $t = 2500$ days, can thus be interpreted as approaching the fractal dimension of the active fault network.

In Fig. 13 we display epicentral and hypocentral moments for earthquakes in the PDE catalog at three depth intervals. We include all the earthquake pairs without taking the inter-earthquake time into account. The curves are normalized, so that the horizontal line corresponds to a self-similar distribution with $\delta = 2.0$. The curves below the horizontal line have $\delta \geq 2.0$ (the fractal dimension is equal to the tangent of the slope angle of the curve plus 2.0). Since the epicentral moments are defined in 2-D, the horizontal line corresponds to $\delta = 1.0$. To show their differences, we combine two types of curves in one plot: epicentral and hypocentral. As in Kagan and Knopoff [94], epicentral moments yield a higher value of the exponent for distance ranges less than or comparable to the thickness of the appropriate layer. From simple geometrical arguments, the hypocentral curves are the preferred data input to calculate the fractal dimension [86].

Comparing Figs. 12 and 13, we conclude that self-similarity of earthquake geometry is established up to the scale length of 0.1 km and less. Since the equations of elasticity lack any intrinsic scale, we expect that the property of

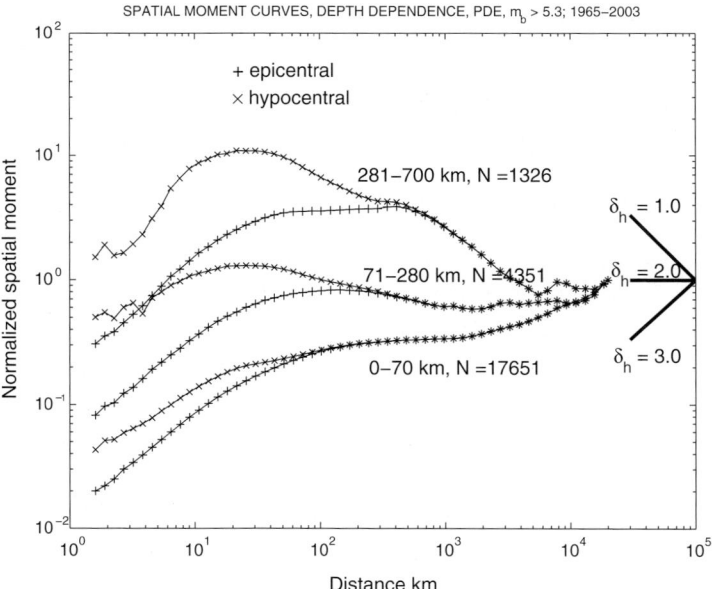

Fig. 13. Hypocentral and epicentral spatial moment curves for various depth intervals. A complete PDE catalog (1965–2003) with $m_b \geq 5.3$ is used. In each of the *coupled curves* the *upper curve* is for the hypocentral moment and the lower curve for the epicentral. The two upper curves are for the depth interval of 281–700 km, the middle ones are for the depth interval of 71–280 km, and the lower ones are for the depth interval of 0–70 km. *Solid lines* at the right show a slope of the curves corresponding to the integer values of the hypocentral correlation dimension of δ_h. For the epicentral moment $\delta_s = \delta_h - 1$

self-similarity can be extended for the brittle fracture of disordered materials (rocks) up to the scale of a few millimeters: the size of rock grains. The upper cutoff for scale-invariance (2000 km) is connected to the size of major tectonic plates. The δ values in Fig. 13 demonstrate that the dimension decreases as the depth increases.

Kagan [86] indicates that evaluating the fractal dimension for earthquake spatial patterns is difficult and prone to many errors and biases. This may explain the contrast with two other classical scale-invariant, universal exponents of earthquake distribution: unlike the G-R relation [13] and Omori's law [70, 87], here the properties and value of the correlation dimension are not yet firmly established.

In most studies of earthquake spatial distribution, errors of location and other errors have not been properly considered [86]. This might explain the high values of fractal dimensions reported in many publications and the great variability of these values. Such findings may reflect not physical and geometrical properties of earthquake fracture, but rather various location and projection errors peculiar to the catalogs studied.

6.3 Spatial Scaling

Kagan [80] investigated the distribution of aftershock zones for large earthquakes (scalar seismic moment $M \geq 10^{19.5}$ Nm, moment magnitude, $m \geq 7$) in global catalogs (CMT and PDE). The dependence of the aftershock zone length, l on the earthquake size was studied for three representative focal mechanisms: thrust, normal, and strike-slip.

The regression curves in Fig. 14 show that $M \propto l^3$ dependence continues up to $m = 9$ earthquakes. Estimated regression parameters for strike-slip and normal earthquakes are similar to those of thrust events, supporting the conjecture that the scaling relation is identical for earthquakes of various focal mechanisms. No observable scaling break or saturation occurs for the largest earthquakes ($M \geq 10^{21}$ Nm, $m \geq 8$). It is natural to assume that the aftershock zone length l is equal or proportional to the rupture length L. Thus, earthquake geometrical focal zone parameters are self-similar.

Using the derived scaling law and moment-frequency relation, we can derive the distribution of earthquake slip not only for a region, but also for a specific place on a fault [84]. This distribution depends on the linear size of earthquake rupture. For example, if the rupture is relatively short, a particular spot on a fault would be ruptured less frequently but would have a larger slip.

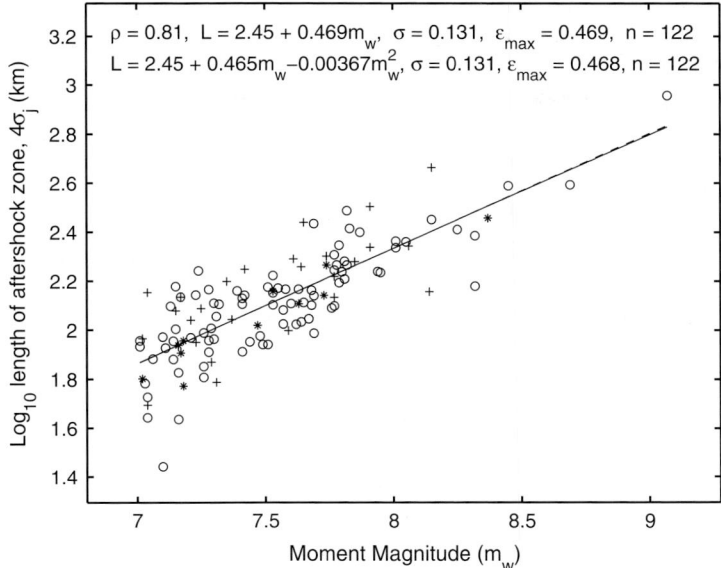

Fig. 14. Plot of the log aftershock zone length (L) against moment magnitude (m). Magnitude values are shifted in formulas shown in the plot ($m_r = m - 8.25$). Rupture length is determined using a 1-day aftershock pattern. The diagram shows the values of the correlation coefficient (ρ), coefficients for linear (*dashed line*) and quadratic (*solid line*) regression, standard (σ) and maximum ($\epsilon_{\max}$) errors, and the total number (n) of aftershock sequences. The *dashed line* is the linear regression, the solid line is a quadratic approximation. Circle – thrust mainshocks; Star – normal mainshocks; Plus – strike-slip mainshocks

7 Focal Mechanism Orientation and Stress Distribution

7.1 Focal Mechanism Distribution

It is difficult to measure the stress tensor itself in the deep interior of the Earth, but rotations of earthquake focal mechanisms may indicate the stress redistribution. Kagan [66] introduced the rotational Cauchy distribution to represent rotations of focal mechanisms of micro-dislocations which comprise the focal zone of an earthquake. The rotational Cauchy distribution can be written as [68]

$$F(\Psi) = \frac{2}{\pi}\left[\arctan(A/\kappa) - \frac{A \times \kappa}{A^2 + \kappa^2}\right], \qquad (25)$$

where $A = \tan(\Psi/2)$ and Ψ is the rotation angle. The scale parameter κ of the Cauchy distribution represents the degree of *incoherence* or *complexity* of an earthquake fault.

An additional complication in studying the 3-D rotation of earthquake focal mechanisms is the symmetry of the source: the double-couple earthquake source has the rotational symmetry of a rectangular box with unequal sides. Due to this symmetry, the maximum rotation angle for the earthquake source cannot exceed 120° [68, 71].

Using the correspondence between the group $\mathbf{SO}(3)$ and the group of normalized quaternions, we solved an inverse problem of a 3-D rotation of double-couple earthquake sources. For each pair of focal mechanisms, we find a minimum 3-D rotation which transform one mechanism into another [71].

In Fig. 15 we display, as an example, the distributions of rotation angle Ψ for shallow earthquake pairs which are separated by a distance of less than 50 km and in a distance range of 400–500 km. We study whether the rotation of focal mechanisms depends on the location of the second earthquake in the pair with regard to the first event. Thus, we measure the rotation angle for hypocenters located in 30° cones around each principal axis (curves marked the T-, P-, and N-axes) of the first event (see Fig. 1). The curves in Fig. 15 for small distances are narrowly clustered, and are clearly well approximated by the rotational Cauchy distribution.

For large distances the curve corresponding to fault-planes (the N-axis) is clearly separated from the curves connected with the T- and P-axes. Although the rotation near the fault-plane is relatively small ($\kappa \approx 0.2$), the earthquakes situated in cones around the T- and P-axes have focal mechanisms essentially uncorrelated with the primary event: the curves are close to the curve corresponding to a completely random rotation of a double-couple (see formulas in [85]).

Figure 16 displays a smoothed map of the average Ψ dependence on time and distance intervals for well-constrained earthquakes in the Harvard catalog [39, 78]. The angle increases with distance between events. The increase with time interval (ΔT) is much less pronounced. For earthquake sequences clustered in time and space, the Ψ difference between focal mechanisms is small, on the order 10–15°. These Ψ-values are close to the minimum uncertainty in Ψ evaluation [82].

7.2 Random Stress Tensor

The aim of earthquake seismology is to rigorously describe the tensor stress field which triggers earthquakes. Until now, extensive attempts to study stress fields have been concentrated on stress tensor properties at particular points, especially at hypocentral locations of potential future earthquakes (e.g., [48, 75, 145, 146]). However, if the earthquake spatial distribution is indeed fractal, the stress field must also be scale-invariant, representing an extremely complicated matrix with critical conditions for earthquake rupture satisfied in an infinite number of points. This would correspond to an infinite number of micro-earthquake occurrences, if one extrapolates the G-R law for earthquake size distribution (Sect. 4.2) toward earthquakes of size zero.

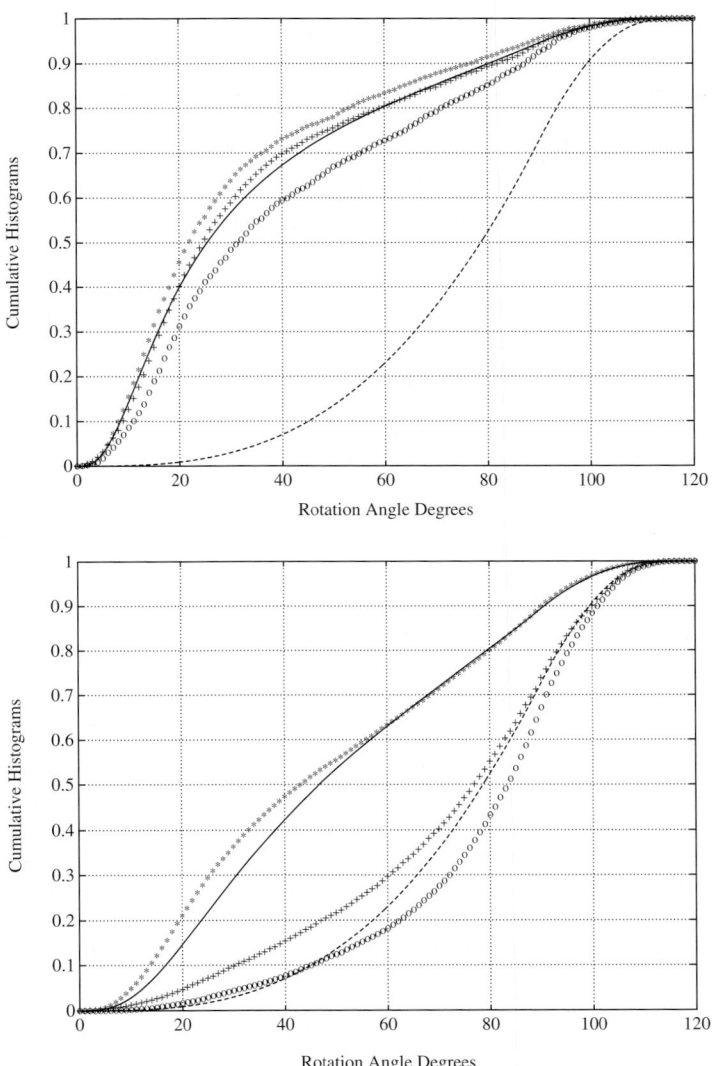

Fig. 15. Distributions of rotation angles for pairs of focal mechanisms of shallow earthquakes from the Harvard catalog. Hypocenters are separated by distances: the upper diagram shows 0–50 km; the lower plots 400–500 km; *circles* indicate hypocenters in 30° cones around the T-axis; plusses hypocenters in 30° cones around the P-axis; stars hypocenters in 30° cones around the N-axis. *Solid line* is for the Cauchy rotation (25) with $\kappa = 0.1$ (*upper plot*); and $\kappa = 0.2$ (*lower plot*). The *dashed line* represents the random rotation

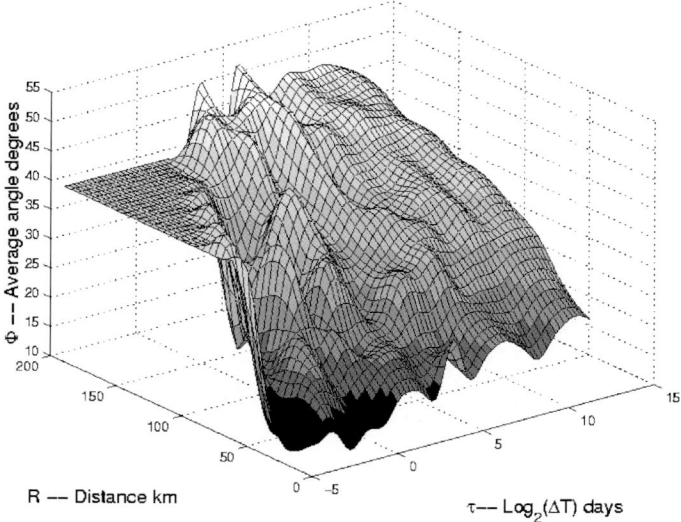

Fig. 16. How the rotation angle Ψ depends on time difference and distance between two earthquakes for shallow well-constrained earthquakes with magnitude $m \geq 5.5$ registered in the time period 1977/1/1–1999/3/31 in the Harvard CMT catalog

While it is apparent that earthquakes are triggered everywhere in seismic regions, the question remains unsolved why small earthquake ruptures develop into giant events which can cause massive destruction. Answering this question adequately will require a detailed description of the 3-D stress field geometry, including its singularities, limit cycles, and possible bifurcations [40]. This is an extremely difficult and open problem: Gabrielov and Keilis-Borok ([40], p. 489) comment that "The [mathematical] problem of the complete description [of the topology of the field...] has not as yet been solved."

The Cauchy distribution is especially important for representing earthquake geometry. It can be shown by theoretical arguments ([162], pp. 45–46; [68]) and by simulations [68] that the stress tensor in the medium with defects follows this distribution. Kagan [68] argues that the Cauchy distribution of the stress should produce the rotational Cauchy distribution of earthquake focal mechanisms.

For any point in an elastic medium which is surrounded by defects, the characteristic function for the random stress distribution can be written as

$$\log \phi(\zeta, \alpha) = \int_0^\infty [\exp(i\zeta\sigma r^{-3}) - 1]\nu(r)r^2 dr , \qquad (26)$$

where $\nu(r)$ is the density of defects which might depend on r, the distance of the defect from the reference (measurement) point, and σ is the normalized (for $r = 1$) stress Green function of an earthquake; stress decays with distance as r^{-3}. For the uniform 3-D distribution of defects, $\nu = \nu_0$. In this case (26) yields the Cauchy stress distribution.

Earthquake spatial distribution, as described in Sect. 6.2, is fractal. In (26) we should substitute the fractal distribution of sources $\nu = \nu_0 r^{\delta-D}$, where $D = 3$ is the Euclidean dimension of the space, and δ is a fractal correlation dimension of earthquake hypocenters. Then (cf. [162])

$$\log \phi(\zeta, \alpha) = \nu_0 \int_0^\infty [\exp(i\zeta\sigma u) - 1] u^{(\delta/3)-1} du$$
$$= \nu_0 \Gamma(-\alpha)|\zeta|^\alpha \, , \qquad (27)$$

with $\alpha = \delta/3$. The above formula means that if $\delta = 3$, the resulting distribution is the Cauchy law [162, 68], whereas for a fractal spatial distribution of earthquakes, $\alpha < 1$.

Analyzing seismic moment and stress tensors has been basic to earthquake seismology. Although tensors are fundamentally important, they have not been sufficiently investigated or interpreted from a statistical point of view in the earth sciences, with few exceptions. A linear error propagation was first independently proposed to derive the error estimate of the principal stresses and their orientations by Angelier et al. [5] and Soler and van Gelder [142]. The correlation study on the invariant quantities of seismic moment tensors was investigated by Kagan and Knopoff [97, 98]. Kagan [78, 80, 81] further extended the correlation results on invariant quantities to analyze earthquake catalogs and interpret faulting geometry.

The study of random tensors has its root in nuclear physics (see e.g., [43, 117]) and multivariate statistics (see e.g., [3]). For nuclear physics, a simple rotation-invariant distribution has been widely investigated. But in multivariate statistics, only a handful of large sample or asymptotic distribution results involving such distributions are available. These results, despite their significance, can not be applied directly to the Earth sciences, because the number of tensors derived from the same original source is generally small. In particular, the ratio of signal to noise is not large enough to neglect the effect of nonlinearity. More importantly, efforts in nuclear physics and statistical mathematics have largely been focused on the principal invariants, namely, the principal eigenvalues. Very little attention has been paid to random eigendirections, which are equally important in the Earth sciences. Moreover, the nonlinearity of the mapping onto the eigendirections and eigenvalues has been insufficiently studied. This nonlinearity could strongly affect the estimated eigenvalues and directions if the noise level is high. Xu [158, 157] and Cai et al. ([20], and references therein) have attempted to develop a probabilistic approach in dealing with random/stochastic tensors in geoscience. The main new results from such studies include exact distributions for the random eigenvalues and eigendirections. They also include accuracy estimates of a higher order and bias computations.

8 Stochastic Processes and Earthquake Occurrence

Most distributions considered so far have been one-dimensional marginal distributions of the earthquake point process. Two enhancements of this picture need to be presented: multidimensional distributions are to be constructed and the point structure of the process needs revision. Figure 10, for example, shows that earthquake rupture duration needs to be taken into account when very small time intervals are considered. In Fig. 12 we show the influence of inter-earthquake time intervals on the spatial structure of earthquake distribution. In Sect. 6.3 we show that the focal zone of an earthquake, especially a large one, cannot be regarded as a point.

A more basic way to study the multidimensional structure of earthquake process is to apply the theory of stochastic point processes [26], not ordinary statistical methods. The first applications of this theory to earthquake occurrence were made by Vere-Jones [152], Kagan [62, 63], and Ogata [123]. Many researchers ([25, 53], and others) have recently applied the theory of stochastic point processes to analyze earthquake occurrence and clustering. The major impetus for these investigations is application of statistical methods for earthquake forecasting, both long- and short-term. Below we briefly review the available methods for earthquake occurrence analysis and their application for earthquake forecasting. We then discuss how these methods can be improved.

8.1 Earthquake Clustering

Almost any earthquake forecast requires proper accounting for earthquake clustering, mainly for aftershocks. If present, foreshocks may be used to calculate a mainshock probability. Even if we are mainly interested in a long-term earthquake forecast, the influence of short-term earthquake clustering on our results should be estimated. Moreover, a faithful modeling of the earthquake clustering is needed for any short-term forecast.

Clustering presents a special challenge since modern local catalogs have a magnitude range extending over several units: in California and Japan, the lower magnitude threshold is close to 1.0, whereas the largest earthquake may exceed 8.0. In such catalogs one should expect the aftershock numbers approaching or even exceeding millions after a very strong event. Handling these earthquakes and accounting for various systematic and random effects both present serious challenges.

Figure 17 displays a sketch of earthquake catalog data in the magnitude-time format. The left part of all the diagrams is the past for which no information is available, and similarly for the right or future part. Some earthquakes are detected below the magnitude threshold, shown as a dashed line.

Aftershock sequences have traditionally been taken into account by catalog declustering. Declustering can be used only as a preliminary step in seismicity analysis: it is subjective; and many different techniques are available but they

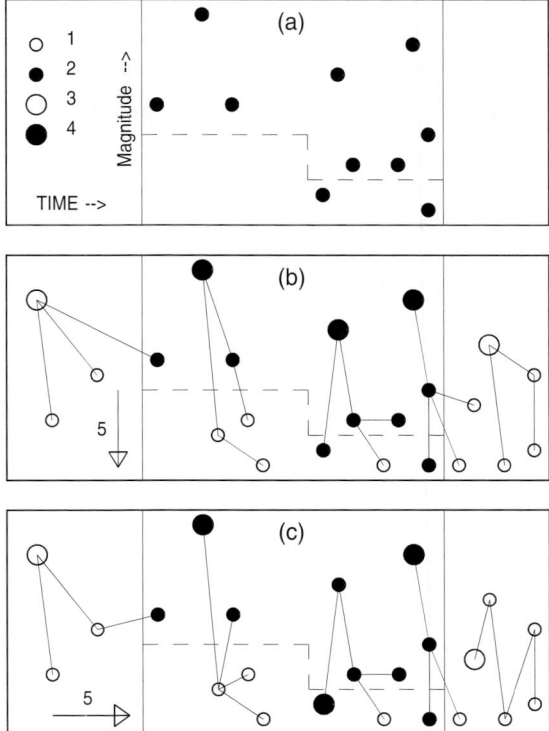

Fig. 17. Earthquake branching models: *Filled circles* (2) indicate observed earthquakes; *open circles* (1) unobserved, modeled events. The *dashed line* represents observational magnitude threshold; the earthquake record above the threshold is complete. Many small events are not registered below this threshold. Large circles (3, 4) denote the initial (main) event of a cluster. *Arrows* (5) indicate direction of branching process: down magnitude axis in (b) and down time axis in (c). (**a**) Observational data; (**b**) Branching-in-moment (magnitude) model; (**c**) Branching-in-time model

are not optimized and have not been rigorously tested. We must use quantitative statistical methods to rigorously describe earthquake clustering. Only an application of stochastic point process theory can provide a robust solution to the problem.

However, the multidimensional nature of earthquake occurrence, fractal or power-law properties of earthquake statistical distributions, and inhomogeneities of earthquake distributions all make it difficult to create and statistically analyze stochastic models. Over the years several such models of earthquake occurrence have been proposed and all are based on the theory of branching processes. Branching is expected to model the well-known property of primary and secondary clustering for aftershock sequences: a strong aftershock (or foreshock) tends to have its own sequence of dependent events. These multidimensional models are:

(A) Point process branching along the magnitude axis, introduced by Kagan (1973a;b) and shown in Fig. 17b.
(B) Point process branching along the time axis ([51, 52, 100, 123] – called Hawkes self-exciting process, see Fig. 17c). Hawkes and Tukey (see discussion section in [63]) debate the difference between branching in earthquake size and in time.
(C) Continuum-state critical branching process develops along the time axis ([95, 66]; see Sect. 8.3).

The first two models (A-B) use the Poisson cluster process to approximate the earthquake occurrence. In these models, earthquake clusters are assumed to follow the Poisson temporal occurrence. Earthquakes within a cluster are modeled by a multidimensional branching process which reproduces a temporal-spatial pattern of dependent events (mostly aftershocks) around the initial event of a sequence [62, 63, 100, 123, 124].

These models employ in one form or another the classical statistical properties of earthquake occurrence: the G-R relation and Omori's law. Model (A) reproduces the G-R relation as the result of branching along the magnitude axis and uses Omori's law to describe earthquake clustering in time. Model (B) combines the G-R relation and Omori's law in a fairly empirical fashion to approximate seismicity. Physical model (C) yields the G-R law as the consequence of critical branching [153]. It applies a version of Omori's law to the temporal distribution of micro-dislocations and simulates the position and orientation of dislocations to reproduce the entire earthquake process (Sect. 8.3). As we discuss below, other models may have certain advantages in earthquake forecasting and the representation of seismicity. But phenomenological model (B) is now almost exclusively used to statistically analyze and simulate earthquake occurrence [89, 100, 124].

Models (A) and (B) can be parameterized to analyze earthquake catalogs. The optimal parameter values can then be found by the maximum likelihood method [70, 123, 124]. To account for earthquake clustering, one can put the obtained parameter values back into the model and find the probabilities for each event to be foreshock/mainshock/aftershock [92, 161]. If these probabilities are known, a catalog can be either declustered in an objective manner, or dependent events can be taken into account.

Most of the statistical models for earthquake occurrence [70, 123] treat earthquake catalogs as a population set, with earthquakes considered as individual entities. As we discuss in Sect. 5, "an individual earthquake" is not a physical entity. Instead it is the result of interpretation and selection by catalog compilers. Thus, extrapolations of observed features toward smaller inter-earthquake time intervals, smaller size earthquakes, etc., may see a model breakdown. Such approximation deterioration is caused not by physical properties of earthquake occurrence, but by peculiarities of earthquake identification technique and catalogs. Why is this?

8.2 Several Problems and Challenges

1. Earthquake spatial distribution is very complex: the depth inhomogeneity, the fractal character of the spatial pattern, and various hypocenter location errors all make model parameterization difficult and create various biases in estimating parameters. Recent applications of stochastic point processes for seismicity analysis often yield results which are incompatible or unstable: slight variations in the data, assumptions, or processing techniques yield significantly different parameter values [70]. It is difficult to see whether these contradictions are caused by biases of analysis, data defects, or differences in parametrization.
2. A critical and careful analysis of errors in the earthquake catalogs needs to be performed before each statistical analysis. Otherwise, unless the effect being studied is very strong, the results are almost surely artifacts. The problem is that most errors in the earthquake data are caused by systematic effects, so they are more difficult to identify and to correct [82].
3. There is no effective statistical tool to select proper models and check whether they fit the data. Likelihood methods and the "Akaike Information Criterion" (AIC) dependent on them (see [27, 124]) apparently work only for regular processes: quasi-Gaussian in a continuous case and quasi-Poisson for discrete (point) processes. However, an earthquake occurrence is controlled by scale-invariant, fractal distributions, diverging to infinity. Although these infinities can be regularized by using renormalization procedures similar to techniques used in model (C), statistical tests applicable to such distributions have not been developed yet. Calculating the likelihood function for aftershock sequences illustrates this point: the rate of aftershock occurrence after a strong earthquake increases by a factor of thousands. $\mathrm{Log}(1000) = 6.9$; hence, one close aftershock yields a contribution to the likelihood function analogous to about 7 free parameters.
4. What can be done in the present situation to obtain reliable statistical results? The model's number of degrees of freedom should be kept as small as possible: the new adjustable parameters are to be introduced only if they are critically tested against the data in various catalogs and against different tectonic environments.
5. Earthquake catalogs are incomplete in a wake of strong events (Sect. 5). They are also incomplete generally for small earthquakes (Sect. 4.2). Both of these effects need to be carefully accounted for [83].
6. Until now, only worldwide seismicity or seismicity in certain seismic zones has been analyzed. Several tectonic provinces have not been investigated sufficiently: deep earthquakes, oceanic earthquakes, earthquakes in stable continental areas, and volcanic earthquakes. The dependence of earthquake clustering on the rate of tectonic deformation should also be investigated:

for example, in continental areas (and specifically in California) aftershock sequences occur in zones of fast and slow deformation rate. Are the clustering properties of earthquakes the same in these conditions? A study of earthquake occurrence in these tectonic environments should yield important information on general properties of seismicity.

7. Apparently all the statistical models based on Omori's law fail to capture the properties of long-term earthquake clustering. Kagan and Jackson [88] argued that, in addition to short-term clustering which manifests in foreshock/mainshock/aftershock shallow event sequences, long-term clustering also occurs. The latter phenomenon is common both to shallow and deep earthquakes. Kagan and Jackson [88] conjectured that short-term clustering results from stress redistribution in a brittle crust; long-term clustering is most likely due to mantle convection.

8. Earthquake probabilities calculated using model (B) have a serious defect: if a strong event is preceded by a foreshock or a number of foreshocks, this large quake is considered dependent. Model (A) does not present this difficulty; the largest event in a cluster is always the mainshock.

9. Point models by definition provide only a point forecast. Each future earthquake is characterized by its location, magnitude, time, and possibly its focal mechanism. In reality, earthquakes are spatially extended and they are not instantaneous. This is especially important for large events. Therefore, to compute seismic hazard, a point forecast needs to be supplemented by an extended source model. In contrast with models (A) and (B), model (C) is in principle a continuum which can simulate realistic, complex rupture process extended in time, space, and fault orientation.

8.3 Critical Continuum-State Branching Model of Earthquake Rupture

Kagan [66] proposed a model of earthquake rupture which incorporated results of 2-, 3-, and 4-point moment studies [64, 65, 92, 94] and tried to reproduce the inferred geometrical properties of hypocenter distributions. The model was based on the propagation (governed by a critical branching process) of infinitesimal dislocation loops.

The simulation proceeds in three stages. In the first stage the branching family trees are started from a number of initial ancestors as in Fig. 18. The second stage of simulation involves adding time delays between the appearance of the parent and the offspring. The delay is power-law distributed (Fig. 19a)

$$X(t) \propto t^{-1-u} . \tag{28}$$

For shallow earthquakes Kagan and Knopoff [95] find that $u \approx 1/2$.

Kagan and Knopoff [99] show that the distribution (28) may have a simple explanation: stresses at the end of an earthquake rupture are below the critical value and thereafter change randomly according to a one-dimensional

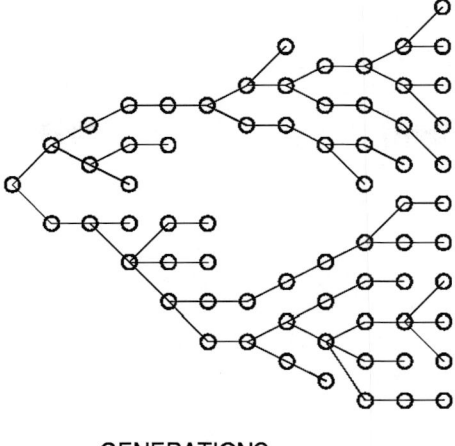

GENERATIONS -->

Fig. 18. An example of a "genealogical" tree of a critical branching process. The process starts with one "particle" of zero generation. Each of the particles produces the Poissonian number of descendants with the mean equal to one. The development of any particle is independent of all other particles in this or previous generations. Time, position, and orientation of descendant offspring are shown in Fig. 19

Brownian motion. A new rupture starts when stress reaches a critical level. The level-set of this motion is a fractal set with a dimension $u = 0.5$ [114]. The distribution of time intervals is Lévy type which has density [160]

$$f_{1/2}(x) = \frac{1}{x\sqrt{2\pi x}} \exp\left(-\frac{1}{2x}\right), \tag{29}$$

and cdf

$$F_{1/2}(x) = 2\left[1 - \Phi\left(\frac{1}{\sqrt{x}}; 0, 1\right)\right], \tag{30}$$

where Φ is the Gaussian distribution (17).

With this information available, a cumulative plot of the number of elementary events against time can be obtained. (In seismological terms, each elementary event is supposed to contribute a fixed amount to a scalar moment release, so that cumulative plots can be interpreted as analogues to the cumulative moment-release plots used in discussing real earthquake catalogues). The intense clustering of the near critical process results in this cumulative plot taking on a self-similar, step-function appearance. By convoluting the derivative of this cumulative function with a suitably shaped Green's function, a record can be obtained which may be compared with the trace of a seismograph or its envelope in reality (Fig. 20). By applying similar criteria to those used to identify real, particular events, Kagan and Knopoff [95] were

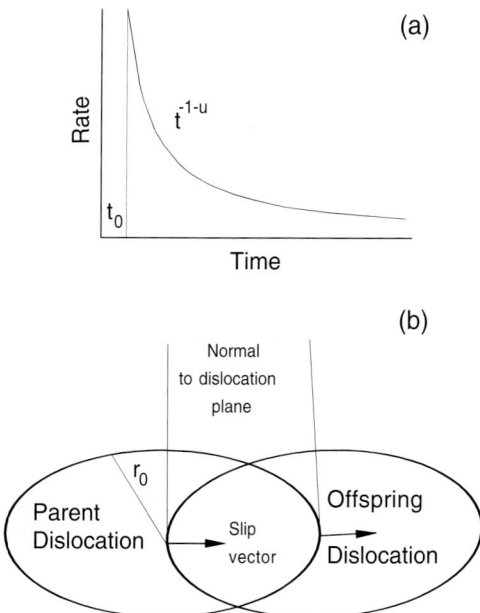

Fig. 19. Schematic diagram of fault propagation. (**a**) Temporal rate of occurrence of dependent shocks. For shallow earthquakes $u \approx 0.5$, t_0 corresponds to rupture time for a dislocation disk in (b). (**b**) Spatial propagation of a synthetic fault. The initial infinitesimal circular dislocation of radius r_0 gives rise to a secondary event. The center of this dislocation is situated on the boundary of the initial dislocation loop. *Solid lines* indicate the vector that is normal to the fault plane in both dislocations, arrows show slip vectors. The fault-plane and slip vector of the secondary dislocation rotate following the Cauchy distribution (25). The secondary dislocations can produce new events according to the same law

able from the time series record to list simulated "events," each with its own "magnitude".

In the third stage of modeling, the spatial coordinates (location of disc center, orientation, and direction) are simulated according to Fig. 19b. Although in principle dislocations are infinitesimal, in practical simulations the dislocation loops are finite with a disc radius r_0. However, this radius can be taken to be as small as possible. In such a case the critical branching process converts into a continuum-state process [61].

The rotation of the focal mechanisms follows the 3-D rotational Cauchy distribution (25). Most rotations are infinitesimal, though in rare cases large rotations give rise to fault branching [66]. As we explained above, the 3-D rotations are described by the group **SO**(3); hence, our model is a random branching walk on non-commutative groups. From the results of this stage, it

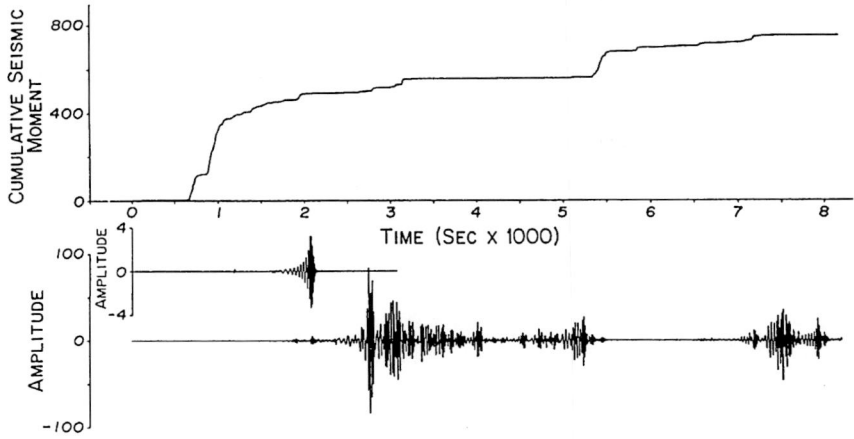

Fig. 20. Cumulative event curve (interpreted here as "cumulative seismic moment" for a realisation of the branching process model with (*below*) an illustration of the filtered signal (using the theoretical seismogram in the *middle*) from which events and their "seismic moments" can be determined

is possible to obtain a visual picture of the resulting "fractures" by plotting the intersection of the elementary discs with a fixed plane (see Fig. 21).

It is partly from such pictures that the angular Cauchy distribution, rather than some other angular analogue of the stable distributions, has been chosen. As can be seen from Sect. 7.2, this distribution also has a simple, physical explanation. The obtained distribution of fault traces looks like actual earthquake fault maps. The spatial moment functions are also qualitatively similar to those in Fig. 11: the 1/D and 1/V behaviors have been reproduced, although no rigorous tests have been attempted [101].

By averaging the locations of the elementary dislocations resulting in such an "event", an approximate location for the "hypocenter" of the event can be determined. The centroid, representing roughly the center of gravity of the locations of the elementary events contributing to the cluster (Fig. 1), can be determined as well. In this way a synthetic seismic catalogue can be produced in which events are listed in time sequence and associated with a hypocenter, magnitude and a focal mechanism. Processing the synthetic catalog through a maximum likelihood procedure similar to that used for real catalogs (Sect. 8.1) yields similar values of basic parameters describing an earthquake occurrence.

For extended rupture, like that shown in Fig. 21, we can calculate the seismic moment tensor of an earthquake or earthquake sequence

$$\mathbf{M} = \sum_{i=0}^{N} \mu \left(\prod_{\xi_i} q_i \times \cdots \times q_j \times \cdots \times q_0 \right) , \qquad (31)$$

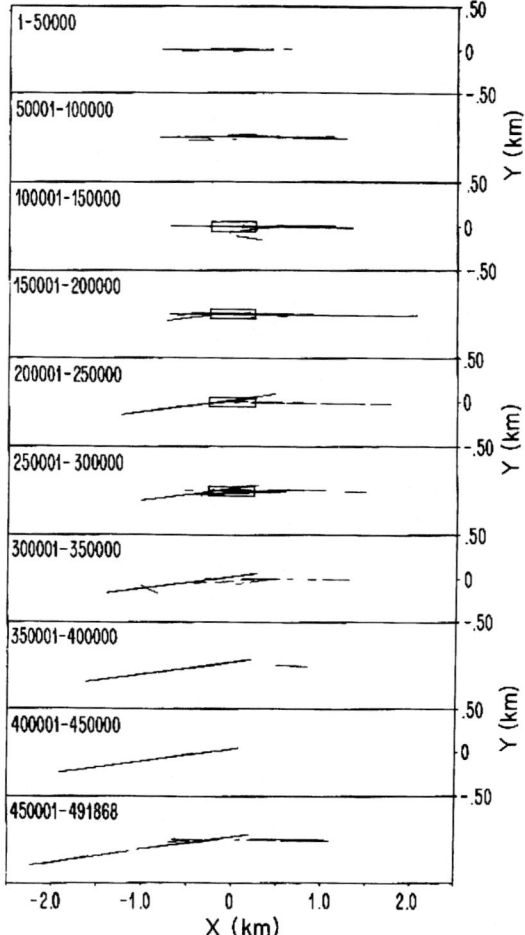

Fig. 21. Stages in the evolution of an episode in the branching simulation model: intersection with a fixed plane of the dislocation discs

where q_0 is a quaternion corresponding to the initial dislocation in Fig. 18. The quaternion product, ending with q_i, describes a combination of 3-D rotations at the path ξ_i in a branching process leading to the i-the dislocation. In a branching process such a path is unique. Each of the quaternion product q_j components follows the rotational Cauchy distribution (25). Thus, the quaternion product in the formula represents the orientation of the i-th dislocation. The operator $\mu(.)$ converts the orientation (quaternion) into the seismic moment tensor (Kagan and Jackson, 1994, their Appendix).

The quaternion multiplication is non-commutative, i.e., in general

$$q_1 \times q_2 \neq q_2 \times q_1 \,. \tag{32}$$

Therefore, the resulting probability structure should be studied by a non-commutative probability theory (e.g., [155]). Moreover, the moment tensor in (31) should have its smallest eigenvalue as non-zero. Hence the combined source would not be a double-couple (1), although it will likely only insignificantly differ from a double-couple [97]. In this case normalized quaternions are insufficient to characterize complex moment tensors as in (31). These tensors, even after normalization, require four degrees of freedom for representation. Higher-order seismic moment tensors [67] can be used to characterize the complex geometry of a fault rupture. The gauge theory of dislocations and disclinations in solids [31], most likely needs to be employed to describe complex earthquake rupture fully.

There has been renewed interest in the branching earthquake fault model. In a limited test, Libicki and Ben-Zion [111] used a simplified procedure to reproduce some properties of the Kagan [66] model.

8.4 Earthquake Forecasting Attempts

The fractal dimension of the earthquake process is lower than the embedding dimension: in time (1-D) the fractal dimension is 0.5 (28); the correlation dimension in 3-D space is 2.2 (Sect. 6.2). This allows us to forecast the spatial and temporal probability of earthquake occurrence.

Phenomenological Branching Models and Earthquake Hazard Estimation

For the phenomenological forecast of seismic activity, we employed both methods shown in Figs. 17b;c: branching-in-magnitude and branching-in-time (see [93, 100] respectively). It is not currently clear which technique would be more appropriate for earthquake forecasting. The advantages or drawbacks may depend on catalog properties or on goals of forecasting. With the first method (A), it is easier to calculate the earthquake rate at the detection threshold and extend the forecasts below the threshold. Forecasting in forward time would involve simulating various cluster probabilities [93].

The second technique (B) is convenient for calculating the earthquake rate at the forward time boundary of the available catalog. This method is currently widely used in earthquake prediction efforts (see citations in the beginning of Sect. 8). However, to extend the forecast horizon into future, simulation is needed [53, 89], since we need to consider earthquakes that occur between the end of catalog data and the prediction time. Similarly, if we want to take into account past seismicity (as shown in Figs. 17b;c, then simulation is needed in both models (A-B).

Since 1999 we have been running experimental short- and long-term forecasts of the west Pacific seismic activity [89]. In Fig. 22 we display long-term forecast maps computed in 2000 for the north-west Pacific region.

We have tested the long-term forecast by a Monte-Carlo simulation (Kagan and Jackson 1994; 2000), see also http://scec.ess.ucla.edu/~ykagan/tests_index.html. The test involves comparing the forward prediction issued before the test period with a retrospective prediction optimized after 2002, when the earthquakes which occurred in 2000-2002 were known. If these two forecasts differ within the 95% confidence limit estimated by a simulation procedure, we consider the forward prediction successful. In effect, instead of competing against a null hypothesis which cannot be effectively defined for the inhomogeneous spatial distribution of seismicity, we test our results against the "perfect", ideal model, specified on the basis of retroactively adjusting the model parameters. Using a similar technique, we produce a daily short- and long-term earthquake forecast for southern California [55].

Kagan and Knopoff [100] tested the short-term forecast by the maximum likelihood technique for a retrospective earthquake forecast at the San Andreas fault. Kossobokov [106] tested our short-term western Pacific forecast and

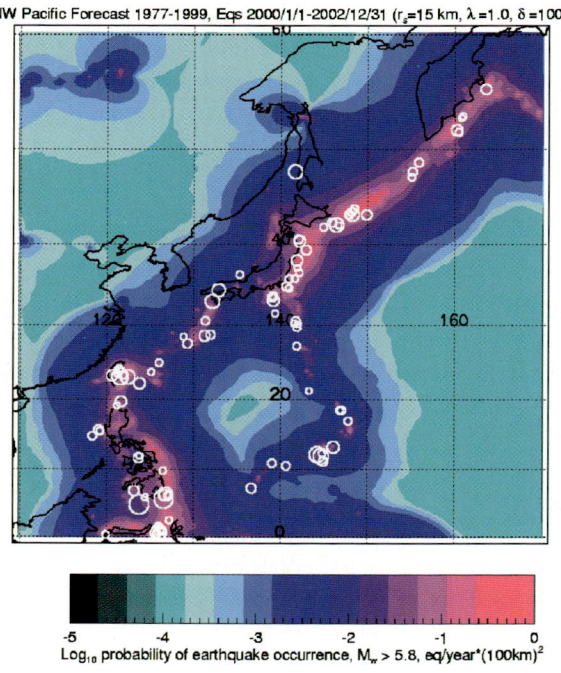

Fig. 22. Northwest Pacific long-term seismicity forecast: latitude limits from 0.25°S to 60.25°N, longitude limits from 109.75°E to 170.25°E; Color scale tones show the probability of earthquake occurrence calculated using the Harvard 1977–1999 catalogue; earthquakes from 2000/1/1 to 2002/12/31 are shown in white. We demonstrate forecast effectiveness: displayed earthquakes occurred after a smoothed seismicity forecast was calculated

found that "... the achieved statistics are much better than random guessing." (However, see also our comment on his paper – Kagan and Jackson, [90].)

Earthquake Fault Propagation Modeling and Earthquake Hazard Estimation

One way of using the continuum-state branching model of fault propagation (Sect. 8.3) is to apply it to maps of geologic faults and past earthquakes to predict the propagation of future earthquake ruptures. As we discussed earlier (Sect. 2.4), geometric compatibility conditions imply that faults must rupture virgin rock. The model which reproduces branching properties of real earthquake faults can be applied for extrapolation to fault data. Kagan and Knopoff [96] made an early attempt to see how such a forecast could be formulated. They extrapolated the fault traces as the result of model simulation. In principle, if appropriate Green's functions are available, this model can generate a set of seismograms for each synthetic sequence.

9 Discussion

As we show in this paper, the earthquake process is controlled by scale-invariant, fractal distributions. Thus, mathematical "monsters" that were produced more than a hundred years ago [35, 114] are directly related to earthquake science and representations of seismicity.

Now we understand that we see and experience these monsters when, for example, we pour milk into tea or step over cracked pavement. Moreover, Sect. 8.3 proposes that these scale-invariant distributions can be at least generally explained by using simple assumptions on random stress behavior: (i) earthquake time behavior by the Brownian motion-like history of stress change and (ii) fault geometry by the Cauchy distribution of stress tensors due to randomly distributed defects in rock medium. This Cauchy distribution induces the Cauchy 3-D rotation of focal mechanisms (25). Such a physical and mathematical explanation is a relatively rare case in the study of fractal distributions [114].

However, it has not yet been explored whether simulated earthquakes (Sect. 8.3) are faithful representation of seismicity. Several reasons complicate the comparison: our observational data are not sufficiently detailed, especially with regard to spatial and angular resolution (see Table 1). The mathematical and logical structure of the stochastic model needs to be explored to see if it is consistent and can be extended to the continuum limit.

As we mentioned in several parts of this review paper, many of the mathematical techniques necessary to describe earthquake geometry and its occurrence are still being developed: (i) the theory of stable distributions and their statistics; (ii) statistics of 3-D rotations; (iii) random branching walk on non-commutative groups; (iv) the gauge theory of deformation in solids, etc.

But developing a comprehensible theory of earthquake rupture may encounter serious mathematical difficulties. Earthquake faults, as shown in Fig. 21, are stochastic fractal objects. The stress at the fractal boundary should be nowhere a differentiable function. Thus, it is possible that calculating earthquake rupture criteria for points close to a "fault-tip" cannot be carried out effectively.

For example, for the deterministic Mandelbrot set ([114], pp. 188–189) it has been shown ([14], p. 55) that even if we use *real-number* arithmetic operations, no algorithm can decide in a finite number of steps whether an arbitrary point in a complex plane is in the set. The reason for the "undecidability" of the Mandelbrot set and many similar complex mathematical objects is that their boundary has a fractal Hausdorff dimension. Thus, it is possible that we cannot effectively calculate the boundary of earthquake rupture faults [77].

Even if the above difficulties are resolved, more "menacing" monsters are on the horizon: the Banach-Tarski theorem [35, 38] states that in a space of three and more dimensions a ball can be divided into several pieces and the pieces rearranged into two balls of the same size. This paradoxical result may mean that new ideas in the mathematics foundations may be needed to solve our problem.

In the introduction we mentioned the similarity of two major problems in classical physics: turbulence of fluids and fracture of solids [72] and an apparent lack of interest by theoretical physicists in solving the former problem. This, most likely, has a simple explanation. (Goldstein [45], p. 23) remarks:

> "It was at a meeting of the British Association in London in 1932 that I remember that [Horace] Lamb remarked 'I am an old man now, and when I die and go to Heaven there are two matters on which I hope for enlightenment. One is quantum electrodynamics, and the other is the turbulent motion of fluids. And about the former I am really rather optimistic.'"

In other, apocryphal, more recent versions of the story Lamb is replaced by Einstein, von Neumann, Heisenberg, Feynman, and others ([44, 159], pp. 121, 329). Does it reflect a general feeling among mathematical physicists that the turbulence problem may be unsolvable? The more complex problem of fracture in solids, including earthquake rupture process, may not be solved either.

This opinion starkly contradicts the optimism expressed by David Hilbert [56] who said that any mathematical problem could be solved (see also comments by Feferman, [34], p. 14):

> "This conviction of the solvability of every mathematical problem is a powerful incentive to the worker. We hear within us the perpetual call: There is the problem. Seek its solution. You can find it by pure reason, for in mathematics there is no **ignorabimus** [we shall not know]."

If we revisit our question in the title of this paper, the simple answer is that the theoretical explanation of earthquake occurrence is very difficult. It requires applying mathematical methods that are unfamiliar to geophysicists and physicists. For example, many papers and a few monographs (e.g., [37]) consider vector and axial statistics in 2-D and 3-D. However, there are almost no publications, except for those cited in Sect. 7, dealing with the statistics of 3-D rotations. Perhaps, a recent development of statistical theory for topological manifolds [104, 141] could be adapted for describing the complex geometry of earthquake faulting, including 3-D rotations of focal mechanisms.

What can be done? Clearly the level of mathematics employed in earthquake physics is inadequate. Presently, the mathematical tools used in seismological research go back to the 18th or to the first half of 19th century. As we explained above, the level of mathematics needs to be raised by the order of a magnitude. Results in the forefront of modern mathematical research should be employed to describe earthquake occurrence and the geometry of earthquake faults in particular. Mathematical disciplines, such as tensor analysis, matrix theory, group theory, topology, and theory of stochastic processes must be involved in the solution. Otherwise no significant progress is possible.

Although applied and pure mathematicians work in other geoscience disciplines like atmospheric and plasma physics or geodynamo theory, until now only statisticians have been studying earthquake occurrence problems. But if we look at the development of earthquake science in the U.S., no professional statisticians have been involved full time in the research. This situation contrasts with earthquake investigations in other countries: in Japan, Russia, and New Zealand statisticians have been involved. It is not surprising that earthquake prediction efforts in the U.S. have been particularly unsuccessful (see Sect. 2.3).

To summarize our discussion, we see that there are major, perhaps fundamental difficulties in creating a comprehensive physical/mathematical theory of brittle fracture and earthquake rupture process. On the other hand, developing quantitative models of earthquake occurrence needed to evaluate probabilistic seismic hazard is within our reach. It will require a combined effort of earth scientists, physicists, statisticians, and pure and applied mathematicians.

Acknowledgements

I appreciate partial support from the National Science Foundation through grants EAR 00-01128 and DMS-0306526, as well as from the Southern California Earthquake Center (SCEC). SCEC is funded by NSF Cooperative Agreement EAR-0106924 and USGS Cooperative Agreement 02HQAG0008. The author thanks D. D. Jackson, P. Bird, F. Schoenberg and I. V. Zaliapin of UCLA, D. Vere-Jones of Wellington University and P. L. Xu of Kyoto University for very useful discussions. Invitation to the Workshop by

Bikas K. Chakrabarti that led to this paper and his help with various problems in producing this paper is greatly appreciated. I am also very grateful to Kathleen Jackson who edited the final manuscript version.

[Publication 968, SCEC. Figures with higher resolution and color may be obtained from the author by sending request to kagan@equake.ess.ucla.edu, or from the webpage: "http://scec.ess.ucla.edu/ykagan.html".]

References

1. K. Aki and P. Richards, *Quantitative Seismology*, 2nd ed., Sausalito, Calif., University Science Books (2002)
2. T.L. Anderson, *Fracture Mechanics: Fundamentals and Applications*, 3rd ed., Boca Raton, Taylor and Francis (2005)
3. T.W. Anderson, *An Introduction to Multivariate Statistical Analysis*, John Wiley and Sons, New York (1958)
4. D.J. Andrews, Mechanics of fault junctions, J. Geophys. Res., **94**, 9389–9397 (1989)
5. J. Angelier, A. Tarantola, B. Valette and S. Manoussis, Inversion of field data in fault tectonics to obtain regional stress - I. Single phase fault populations: a new method of computing stress tensor, Geophys. J.R. astr. Soc., **69**, 607–621 (1982)
6. P. Bak, K. Christensen, L. Danon and T. Scanlon, Unified scaling law for earthquakes, Phys. Rev. Lett., **88**, 178501, pp. 1–4 (2002)
7. W.H. Bakun and A.G. Lindh, The Parkfield, California, earthquake prediction experiment, Science, **229**, 619–624 (1985)
8. W.H. Bakun et al., Implications for prediction and hazard assessment from the 2004 Parkfield earthquake, Nature, **437**, 969–974 (2005)
9. M. Baiesi and M. Paczuski, Complex networks of earthquakes and aftershocks, Nonlinear Processes Geophys., **12**(1), 1–11 (2005)
10. Y. Ben-Zion and C.G. Sammis, Characterization of Fault Zones, Pure Appl. Geophys., **160**, 677–715 (2003)
11. P. Bhattacharyya, Of overlapping Cantor sets and earthquakes: analysis of the discrete Chakrabarti-Stinchcombe model, Physica A, **348**, 199–215 (2005)
12. P. Bhattacharyya, A. Chatterjee and B.K. Chakrabarti, A common mode of origin of power laws in models of market and earthquake, http://arxiv.org/abs/physics/0510038 (2005)
13. P. Bird and Y.Y. Kagan, Plate-tectonic analysis of shallow seismicity: apparent boundary width, beta, corner magnitude, coupled lithosphere thickness, and coupling in seven tectonic settings, Bull. Seismol. Soc. Amer., **94**(6), 2380–2399 (2004)
14. L. Blum, F. Cucker, M. Shub and S. Smale, *Complexity and Real Computation*, New York, Springer (1998)
15. B.A. Bolt, *Earthquakes*, 5th ed., New York, W.H. Freeman (2003)
16. E. Bouchbinder, I. Procaccia and S. Sela, Disentangling scaling properties in anisotropic fracture, Phys. Rev. Lett., **95**(25), 255503 (2005)
17. M.J. Buehler and H.J. Gao, Dynamical fracture instabilities due to local hyperelasticity at crack tips, Nature, **439**(7074), 307–310 (2006)

18. R. Burridge and L. Knopoff, Body force equivalents for seismic dislocations, Bull. Seismol. Soc. Amer., **54**, 1875–1888 (1964)
19. R. Burridge and L. Knopoff, Model and theoretical seismicity, Bull. Seismol. Soc. Amer., **57**, 341–371 (1967)
20. J.Q. Cai, E.W. Grafarend and B. Schaffrin, Statistical inference of the eigenspace components of a two-dimensional, symmetric rank-two random tensor, J. Geodesy, **78**(7–8), 425–436 (2005)
21. S. Castellaro, F. Mulargia and Y.Y. Kagan, Regression problems for magnitudes, Geophys. J. Int., accepted (2006) http://scec.ess.ucla.edu/~ykagan/silvia_index.html
22. B.K. Chakrabarti and R.B. Stinchcombe, Stick-slip statistics for two fractal surfaces: a model for earthquakes, Physica A, **270**(1–2), 27–34 (1999)
23. A. Corral, Renormalization-group transformations and correlations of seismicity, Phys. Rev. Lett., **95**(2), 028501 (2005)
24. *Columbia Accident Investigation Board, Report*, Vol. I, NASA, U.S. government, Washington DC (2003)
25. R. Console, M. Murru and A.M. Lombardi, Refining earthquake clustering models, J. Geophys. Res., **108**(B10), Art. No. 2468 (2003)
26. D.J. Daley and D. Vere-Jones, *An Introduction to the Theory of Point Processes*, Springer-Verlag, New York, 2–nd ed., Vol. 1 (2003)
27. D.J. Daley and D. Vere-Jones, Scoring probability forecasts for point processes: The entropy score and information gain, J. Applied Probability, **41A**, 297–312, (Sp. Iss.) (2004)
28. J. Dieterich, A constitutive law for rate of earthquake production and its application to earthquake clustering, J. Geophys. Res., **99**, 2601–2618 (1994)
29. S. Duval and R. Tweedie, Trim and fill: a simple funnel-plot-based method of testing and adjusting for publication bias in meta-analysis, Biometrics, **56**, 455–463 (2000)
30. F. Dyson, A meeting with Enrico Fermi – How one intuitive physicist rescued a team from fruitless research, Nature, **427**(6972), 297 (2004)
31. D.G.B. Edelen and D.C. Lagoudas, *Gauge Theory and Defects in Solids*, Amsterdam, North-Holland (1988)
32. G. Ekström and A.M. Dziewonski, Evidence of bias in estimation of earthquake size, Nature, **332**, 319–323 (1988)
33. G. Ekström, A.M. Dziewonski, N.N. Maternovskaya and M. Nettles, Global seismicity of 2003: Centroid-moment-tensor solutions for 1087 earthquakes, Phys. Earth Planet Inter., **148**(2–4), 327–351 (2005)
34. S. Feferman, Deciding the undecidable: Wrestling with Hilbert's problems, math.stanford.edu/~feferman/papers/deciding.pdf (1994)
35. S. Feferman, Mathematical intuition vs. mathematical monsters, Synthese, **125**(3), 317–332 (2000)
36. W. Feller, *An Introduction to Probability Theory and its Applications*, **2**, 2nd ed., J. Wiley, New York (1971)
37. N.I. Fisher, T. Lewis, B.J.J. Embleton, *Statistical Analysis of Spherical Data*, Cambridge, Cambridge University Press (1987)
38. R.M. French, The Banach-Tarski theorem, Math. Intelligencer, **10**(4), 21–28 (1988)
39. C. Frohlich and S.D. Davis, How well constrained are well-constrained T, B, and P axes in moment tensor catalogs?, J. Geophys. Res., **104**, 4901–4910 (1999)

40. A.M. Gabrielov and V.I. Keilis-Borok, Patterns of stress corrosion: Geometry of the principal stresses, Pure Appl. Geophys., **121**, 477–494 (1983)
41. A. Gabrielov, V. Keilis-Borok and D.D. Jackson, Geometric incompatibility in a fault system, P. Natl. Acad. Sci. USA **93**, 3838–3842 (1996)
42. F. Gilbert and A.M. Dziewonski, An application of normal mode theory to the retrieval of structural parameters and source mechanisms from seismic spectra, Phil. Trans. R. Soc. Lond. A, **278**, 187–269 (1975)
43. V.L. Girko, *Theory of Random Determinants*, Boston, Kluwer Academic Publishers (1990)
44. J. Gleick, *Chaos, Making a New Science*, Viking, New York (1987)
45. S. Goldstein, Fluid mechanics in the first half of this century, Annual Rev. Fluid Mech., **1**, 1–29 (1969)
46. B. Gutenberg and C.F. Richter, Frequency of earthquakes in California, Bull. Seism. Soc. Am., **34**, 185–188 (1944)
47. T.C. Hanks and H. Kanamori, A moment magnitude scale, J. Geophys. Res., **84**, 2348–2350 (1979)
48. R.A. Harris, Introduction to special section: Stress triggers, stress shadows, and implications for on seismic hazard, J. Geophys. Res., **103**, 24,347–24,358 (1998)
49. E. Haseltine, The 11 greatest unanswered questions of physics, Discover, **23**(2) (2002)
50. E. Hauksson and P. Shearer, Southern California hypocenter relocation with waveform cross-correlation, Part 1: Results using the double-difference method, Bull. seism. Soc. Am., **95**(3), 896–903 (2005)
51. A.G. Hawkes and L. Adamopoulos, Cluster models for earthquakes - Regional comparisons, Bull. Int. Statist. Inst., **45**(3), 454–461 (1973)
52. A.G. Hawkes and D. Oakes, A cluster process representation of a self-exciting process, J. Appl. Prob., **11**, 493–503 (1974)
53. A. Helmstetter and D. Sornette, Predictability in the epidemic-type aftershock sequence model of interacting triggered seismicity, J. Geophys. Res., **108**(B10), Art. No. 2482 (2004)
54. A. Helmstetter, Y.Y. Kagan and D.D. Jackson, Importance of small earthquakes for stress transfers and earthquake triggering, J. Geophys. Res., **110**(5), B05S08 (2005)
55. A. Helmstetter, Y.Y. Kagan and D.D. Jackson, Comparison of short-term and time-independent earthquake forecast models for Southern California, Bull. Seismol. Soc. Amer., **96**(1), 90–106 (2006)
56. D. Hilbert, Mathematical Problems, Lecture at the International Congress of Mathematicians at Paris, English translation by M.W. Newson, Bull. Amer. Math. Soc., **8** (1902), 437–479 (1900)
57. W.E. Holt, N. Chamot-Rooke, X. Le Pichon, A.J. Haines, B. Shen-Tu and J. Ren, Velocity field in Asia inferred from Quaternary fault slip rates and Global Positioning System observations, J. Geophys. Res., **105**, 19,185–19,209 (2000)
58. T. Huillet and H.F. Raynaud, On rare and extreme events, Chaos, Solitons and Fractals, **12**, 823–844 (2001)
59. M. Ishimoto and K. Iida, Observations sur les seismes enregistres par le microsismographe construit dernierement (1), *Bull. Earthquake Res. Inst. Tokyo Univ.*, **17**, 443–478, (in Japanese) (1939)

60. D.D. Jackson and Y.Y. Kagan, The 2004 Parkfield earthquake, the 1985 prediction, and characteristic earthquakes: Lessons for the future, Bull. Seismol. Soc. Amer., submitted (2006) http://scec.ess.ucla.edu/~ykagan/parkf2004_index.html
61. M. Jirina, Stochastic branching processes with continuous state space, Czech. Math. J., **8**, 292–313 (1958)
62. Y.Y. Kagan, A probabilistic description of the seismic regime, Izv. Acad. Sci. USSR, Phys. Solid Earth, 213–219, (English translation) (1973a)
63. Y.Y. Kagan, Statistical methods in the study of the seismic process (with discussion: Comments by M.S. Bartlett, A.G. Hawkes and J.W. Tukey), Bull. Int. Statist. Inst., **45**(3), 437–453 (1973b)
64. Y.Y. Kagan, Spatial distribution of earthquakes: The three-point moment function, Geophys. J. Roy. Astr. Soc., **67**, 697–717 (1981a)
65. Y.Y. Kagan, Spatial distribution of earthquakes: The four-point moment function, Geophys. J. Roy. Astr. Soc., **67**, 719–733 (1981b)
66. Y.Y. Kagan, Stochastic model of earthquake fault geometry, Geophys. J. Roy. Astr. Soc., **71**, 659–691 (1982)
67. Y.Y. Kagan, Point sources of elastic deformation: Elementary sources, static displacements, Geophys. J. Roy. Astr. Soc., **90**, 1–34 (1987) (Errata, Geophys. J. R. Astron. Soc., **93**, 591, 1988.)
68. Y.Y. Kagan, Random stress and earthquake statistics: Spatial dependence, Geophys. J. Int., **102**, 573–583 (1990)
69. Y.Y. Kagan, Fractal dimension of brittle fracture, J. Nonlinear Sci., **1**, 1–16 (1991a)
70. Y.Y. Kagan, Likelihood analysis of earthquake catalogues, Geophys. J. Int., **106**, 135–148 (1991b)
71. Y.Y. Kagan, 3-D rotation of double-couple earthquake sources, Geophys. J. Int., **106**, 709–716 (1991c)
72. Y.Y. Kagan, Seismicity: Turbulence of solids, Nonlinear Sci. Today, **2**, 1–13 (1992a)
73. Y.Y. Kagan, On the geometry of an earthquake fault system, Phys. Earth Planet Inter., **71**, 15–35 (1992b)
74. Y.Y. Kagan, Correlations of earthquake focal mechanisms, Geophys. J. Int., **110**, 305–320 (1992c)
75. Y.Y. Kagan, Incremental stress and earthquakes, Geophys. J. Int., **117**, 345–364 (1994a)
76. Y.Y. Kagan, Observational evidence for earthquakes as a nonlinear dynamic process, Physica D, **77**, 160–192 (1994b)
77. Y.Y. Kagan, Is earthquake seismology a hard, quantitative science?, Pure Appl. Geoph., **155**, 233–258 (1999)
78. Y.Y. Kagan, Temporal correlations of earthquake focal mechanisms, Geophys. J. Int., **143**, 881–897 (2000)
79. Y.Y. Kagan, Seismic moment distribution revisited: I. Statistical results, Geophys. J. Int., **148**, 520–541 (2002a)
80. Y.Y. Kagan, Aftershock zone scaling, Bull. Seismol. Soc. Amer., **92**(2), 641–655 (2002b)
81. Y.Y. Kagan, Seismic moment distribution revisited: II. Moment conservation principle, Geophys. J. Int., **149**, 731–754 (2002c)
82. Y.Y. Kagan, Accuracy of modern global earthquake catalogs, Phys. Earth Planet. Inter., **135**(2–3), 173–209 (2003)

83. Y. Y. Kagan, Short-term properties of earthquake catalogs and models of earthquake source, Bull. Seismol. Soc. Amer., **94**(4), 1207–1228 (2004)
84. Y.Y. Kagan, Earthquake slip distribution: A statistical model, J. Geophys. Res., **110**(5), B05S11 (2005a)
85. Y.Y. Kagan, Double-couple earthquake focal mechanism: Random rotation and display, Geophys. J. Int., **163**(3), 1065–1072 (2005b)
86. Y.Y. Kagan, Earthquake spatial distribution: The correlation dimension, Geophys. J. Int., submitted (2006)
   http://scec.ess.ucla.edu/∼ykagan/p2rev_index.html.
87. Y.Y. Kagan and H. Houston, Relation between mainshock rupture process and Omori's law for aftershock moment release rate, Geophys. J. Int., **163**(3), 1039–1048 (2005)
88. Y.Y. Kagan and D. D. Jackson, Long-term earthquake clustering, Geophys. J. Int., **104**, 117–133 (1991)
89. Y.Y. Kagan and D.D. Jackson, Probabilistic forecasting of earthquakes, Geophys. J. Int., **143**, 438–453 (2000)
90. Y.Y. Kagan and D.D. Jackson, Comment on "Testing earthquake prediction methods: 'The West Pacific short-term forecast of earthquakes with magnitude MwHRV $\geq$ 5.8'" by V.G. Kossobokov, Tectonophysics, **413**(1–2), 33–38 (2006)
91. Y.Y. Kagan, D.D. Jackson and Y.F. Rong, A new catalog of southern California earthquakes, 1800–2005, Seism. Res. Lett., **77**(1), 30–38 (2006)
92. Y. Kagan and L. Knopoff, Statistical search for non-random features of the seismicity of strong earthquakes, Phys. Earth Planet. Inter., **12**(4), 291–318 (1976)
93. Y. Kagan and L. Knopoff, Earthquake risk prediction as a stochastic process, Phys. Earth Planet. Inter., **14**(2), 97–108 (1977)
94. Y.Y. Kagan and L. Knopoff, Spatial distribution of earthquakes: the two-point correlation function, Geophys. J. Roy. Astr. Soc., **62**, 303–320 (1980)
95. Y.Y. Kagan and L. Knopoff, Stochastic synthesis of earthquake catalogs, J. Geophys. Res., **86**, 2853–2862 (1981)
96. Y.Y. Kagan and L. Knopoff, A stochastic model of earthquake occurrence, Proc. 8-th Int. Conf. Earthq. Eng., San Francisco, Calif., **1**, 295–302 (1984)
97. Y.Y. Kagan and L. Knopoff, The first-order statistical moment of the seismic moment tensor, Geophys. J. Roy. Astr. Soc., **81**, 429–444 (1985a)
98. Y.Y. Kagan and L. Knopoff, The two-point correlation function of the seismic moment tensor, Geophys. J. Roy. Astr. Soc., **83**, 637–656 (1985b)
99. Y.Y. Kagan and L. Knopoff, Random stress and earthquake statistics: Time dependence, Geophys. J. Roy. Astr. Soc., **88**, 723–731 (1987a)
100. Y.Y. Kagan and L. Knopoff, Statistical short-term earthquake prediction, Science, **236**, 1563–1567 (1987b)
101. Y.Y. Kagan and D. Vere-Jones, Problems in the modelling and statistical analysis of earthquakes in: *Lecture Notes in Statistics (Athens Conference on Applied Probability and Time Series Analysis)*, **114**, C.C. Heyde, Yu. V. Prohorov, R. Pyke, and S.T. Rachev, eds., New York, Springer, pp. 398–425 (1996)
102. H. Kanamori, The energy release in great earthquakes, J. Geophys. Res., **82**, 2981–2987 (1977)
103. H. Kanamori and E.E. Brodsky, The physics of earthquakes, Rep. Prog. Phys., **67**, 1429–1496 (2004)

104. D.G. Kendall, D. Barden, T.K. Carne and H. Le, *Shape and Shape Theory*, New York, Wiley (1999)
105. G. King, The accommodation of large strains in the upper lithosphere of the Earth and other solids by self-similar fault systems: The geometrical origin of b-value, Pure Appl. Geophys., **121**, 761–815 (1983)
106. V.G. Kossobokov, Testing earthquake prediction methods: "The West Pacific short-term forecast of earthquakes with magnitude MwHRV $\geq$ 5.8", Tectonophysics, **413**(1–2), 25–31 (2006)
107. B.V. Kostrov, Seismic moment and energy of earthquakes, and seismic flow of rock, Izv. Acad. Sci. USSR, Phys. Solid Earth, January, 13–21 (1974)
108. T.S. Kuhn, Logic of discovery or psychology of research?, In *Criticism and the Growth of Knowledge*, eds., I. Lakatos and A. Musgrave, pp. 1–23, Cambr. Univ. Press, Cambrigde (1965)
109. J.S. Langer, J.M. Carlson, C.R. Myers and B.E. Shaw, Slip complexity in dynamic models of earthquake faults, Proc. Nat. Acad. Sci. USA, **93**, 3825–3829 (1996)
110. W.H.K. Lee, H. Kanamori, P.C. Jennings and C. Kisslinger, Eds., *IASPEI Handbook of Earthquake and Engineering Seismology, Part A*, Boston, Academic Press (2002)
111. E. Libicki and Y. Ben-Zion, Stochastic branching models of fault surfaces and estimated fractal dimension, Pure Appl. Geophys., **162**(6–7), 1077–1111 (2005)
112. V. Lyakhovsky, Y. Ben-Zion and A. Agnon, A viscoelastic damage rheology and rate- and state-dependent friction, Geophys. J. Int., **161**(1), 179–190; Correction, Geophys. J. Int., **161**(2), 420 (2005)
113. V. Lyakhovsky, Y. Ben-Zion and A. Agnon, Distributed damage, faulting and friction, J. Geophys. Res., **102**, 27,635–27,649 (1997)
114. B.B. Mandelbrot, *The Fractal Geometry of Nature*, W.H. Freeman, San Francisco, Calif., 2nd ed. (1983)
115. M. Marder, Computational science – Unlocking dislocation secrets, Nature, **391**, 637–638 (1998)
116. W. R. McCann, S. P. Nishenko, L. R. Sykes and J. Krause, Seismic gaps and plate tectonics: Seismic potential for major boundaries, Pure Appl. Geophys., **117**, 1082–1147 (1979)
117. M.L. Mehta, *Random Matrices*, 2nd ed., Boston, Academic Press (1991)
118. W.J. Morgan, Rises, trenches, great faults and crustal blocks, J. Geophys. Res., **73**(6), 1959–1982 (1968)
119. Nature magazine, February-April of 1999. Debate on earthquake prediction, http://www.nature.com/nature/-debates/earthquake/equake_frameset.html
120. S.P. Nishenko, Circum-Pacific seismic potential – 1989-1999, Pure Appl. Geophys., **135**, 169–259 (1991)
121. J.F. O'Brien and J.K. Hodgins, Graphical modeling and animation of brittle fracture, Proceedings of Assoc. Computing Machinery (ACM) SIGGRAPH 99, 137–146 (1999)
122. A. Okabe, B. Boots, K. Sugihara and S. Chiu, *Spatial Tessellations*, 2nd ed., Wiley, Chichester (2000)
123. Y. Ogata, Statistical models for earthquake occurrence and residual analysis for point processes, J. Amer. Statist. Assoc., **83**, 9–27 (1988)
124. Y. Ogata, Space-time model for regional seismicity and detection of crustal stress changes, J. Geophys. Res., **109**(B3), Art. No. B03308. Correction J. Geophys. Res., **109**(B6), Art. No. B06308 (2004)

125. F. Omori, On the after-shocks of earthquakes, J. College Sci., Imp. Univ. Tokyo, **7**, 111–200 (with Plates IV-XIX) (1894)
126. N. Oreskes, *The Rejection of Continental Drift: Theory and Method in American Earth Science*, Oxford University Press, USA (1999)
127. V. Pareto, *Cours d'Économie Politique*, Tome Second, Lausanne, F. Rouge, see also V. Pareto, 1964. *Œuvres Complètes*, Publ. by de Giovanni Busino, Genève, Droz, Vol. II (1897)
128. R. Penrose, *The Road to Reality: A Complete Guide to the Laws of the Universe*, New York, Knopf (2005)
129. V.F. Pisarenko, Non-linear growth of cumulative flood losses with time, Hydrological Processes, **12**, 461–470 (1998)
130. *Preliminary Determination of Epicenters, (PDE), Monthly Listings*, U.S. Dept. Interior/Geol. Survey, Nat. Earthquake Inform. Center, January, pp. 47 (1999) http://www.neic.cr.-usgs.gov/neis/data_services/ftp_files.html.
131. K.R. Popper, *Logic of Scientific Discovery*, 2nd ed., London, Hutchinson (1980)
132. J.R. Rice and Y. Ben-Zion, Slip complexity in earthquake fault models, Proc. Nat. Acad. Sci. USA, **93**, 3811–3818 (1996)
133. Y.-F. Rong, D.D. Jackson and Y.Y. Kagan, Seismic gaps and earthquakes, J. Geophys. Res., **108**(B10), 2471, **ESE-6** (2003)
134. G. Samorodnitsky and M.S. Taqqu, *Stable non-Gaussian Random Processes: Stochastic Models with Infinite Variance*, New York, Chapman and Hall (1994)
135. C.H. Scholz, Faults without friction?, Nature, **381**, 556–557 (1996)
136. C.H. Scholz, *The Mechanics of Earthquakes and Faulting*, Cambr. Univ. Press, Cambridge, 2nd ed., (2002)
137. D.P. Schwartz and K.J. Coppersmith, Fault behavior and characteristic earthquakes: Examples from Wasatch and San Andreas fault zones, J. Geophys. Res., **89**, 5681–5698 (1984)
138. E. Sharon and J. Fineberg, Confirming the continuum theory of dynamic brittle fracture for fast cracks, Nature, **397**, 333–335 (1999)
139. E. Sharon and J. Fineberg, Microbranching instability and the dynamic fracture of brittle materials, Physical Review B, **54**, 7128–7139 (1996)
140. P. Shearer, E. Hauksson and G.Q. Lin, Southern California hypocenter relocation with waveform cross-correlation, Part 2: Results using source-specific station terms and cluster analysis, Bull. seism. Soc. Am., **95**(3), 904–915 (2005)
141. C.G. Small, *The Statistical Theory of Shape*, New York, Springer (1996)
142. T. Soler and B.H. W. van Gelder, On covariances of eigenvalues and eigenvectors of second-rank symmetric tensors, Geophys. J. Int., **105**, 537–546 (1991)
143. D. Sornette, *Critical Phenomena in Natural Sciences (Chaos, Fractals, Self-organization, and Disorder: Concepts and Tools)*, New York, Springer, 2nd ed. (2003)
144. P. B. Stark and D.A. Freedman, What is the chance of an earthquake?, Chapter 5.3 in *"EARTHQUAKE SCIENCE AND SEISMIC RISK REDUCTION"*, eds. F. Mulargia and R.J. Geller, pp. 201–213, Kluwer, Dordrecht (2003) preliminary draft of the document is available at http://oz.berkeley.edu/~stark/Preprints/611.pdf.
145. S. Steacy, J. Gomberg and M. Cocco, Introduction to special section: Stress transfer, earthquake triggering, and time-dependent seismic hazard, J. Geophys. Res., **110**(B5), B05S01 (2005)
146. R.S. Stein, The role of stress transfer in earthquake occurrence, Nature, **402**, 605–609 (1999)

147. F. Suppe, The structure of a scientific paper, Phil. Sci., **65**(3), 381–405; also see pp. 417–424 (1998)
148. D.L. Turcotte, A fractal model for crustal deformation, Tectonophysics, **132**, 261–269 (1986)
149. D.L. Turcotte, *Fractals and Chaos in Geology and Geophysics*, 2nd ed., Cambridge Univ. Press, Cambridge (1997)
150. V.V. Uchaikin and V.M. Zolotarev, *Chance and Stability: Stable Distributions and Their Applications*, Utrecht: VSP International Science Publishers (1999)
151. T. Utsu, Y. Ogata and R.S. Matsu'ura, The centenary of the Omori formula for a decay law of aftershock activity, J. Phys. Earth, **43**, 1–33 (1995)
152. D. Vere-Jones, Stochastic models for earthquake occurrence (with discussion), J. Roy. Stat. Soc., **B32**, 1–62 (1970)
153. D. Vere-Jones, A branching model for crack propagation, Pure Appl. Geophys., **114**, 711–725 (1976)
154. D. Vere-Jones, R. Robinson and W.Z. Yang, Remarks on the accelerated moment release model: Problems of model formulation, simulation and estimation, Geophys. J. Int., **144**, 517–531 (2001)
155. D. Voiculescu, Lectures on free probability theory, in: *Lectures Notes in Mathematics, 1738*, pp. 283–349, ed., P. Bernard, Springer, Berlin (2000)
156. Working Group on California Earthquake Probabilities (WG02), *Earthquakes probabilities in the San Francisco Bay region: 2002 to 2031*, USGS, Open-file Rept., 03-214; (2003) http://pubs.usgs.gov/of/2003/of03-214.
157. P.L. Xu, Isotropic probabilistic models for directions, planes, and referential systems, Proc. R. Soc., London, **458A**(2024), 2017–2038 (2002)
158. P.L. Xu, Spectral theory of constrained second-rank symmetric random tensors, Geophys. J. Int., **138**, 1–24 (1999)
159. A. Yaglom, The century of turbulence theory: The main achievements and unsolved problems, In: *New Trends in Turbulence*, eds., M. Lesieur, A. Yaglom, and F. David, pp. 1–52, NATO ASI, Session LXXIV, Springer, Berlin (2001)
160. I.V. Zaliapin, Y.Y. Kagan and F. Schoenberg, Approximating the distribution of Pareto sums, Pure Appl. Geoph., **162**(6–7), 1187–1228 (2005)
161. J.C. Zhuang, Y. Ogata and D. Vere-Jones, Analyzing earthquake clustering features by using stochastic reconstruction, J. Geophys. Res., **109**(B5), Art. No. B05301 (2004)
162. V.M. Zolotarev, *One-Dimensional Stable Distributions*, Amer. Math. Soc., Providence, R.I. (1986)

Part III

Modelling Related Phenomena

Aeolian Transport and Dune Formation

H.J. Herrmann

Departamento de Fisica, Univesridade Federal do Ceara, Campus do Pici,
60451-970, CE, Brazil[**]
hans@ica1.uni-stuttgart.de

1 Introduction

Granular materials like sand typically exist inside of a fluid like air or water. If this fluid is in motion, it will exert forces on the grains and in that way create a particle flux. This transport of granular material is responsible for the formation of dunes and beaches. The first to systematically study airborne sand transport was the British brigadier R. Bagnold who, during the time of World War II did experiments in wind channels and field measurements in the Sahara. He presented the first expression for the sand flux as function of the wind velocity. Since then more refined expressions have been proposed. Bagnold also described for the first time the two basic mechanisms of sand transport: saltation and creep, and wrote the classic book on the subject which still is consulted very much [1].

If the ground is covered by sand and has no vegetation the sand flux on the surface modifies the shape of the landscape and spontaneously creates patterns on different scales: ripples in the range of ten to twenty centimeters and dunes in the range of two to two hundred meters. One example are the Barchan dunes shown in Fig. 1 which arise when the wind always blows from the same direction and not much sand is available. The change of the topography can be described by a set of coupled equations of motion which contain as variable fields the shear stress of the wind and the sand flux. These equations allow to explain among others the different dune morphologies, their velocity and their formation.

In this article we will first introduce the properties of the turbulent wind field, then present the mechanisms of sand transport and then we will discuss dune formation.

[**] On sabbatical leave from ICP, University of Stuttgart.

Fig. 1. Field of Barchan dunes near Laâyoune, Morocco

2 The Aeolian Field

The air is a Newtonian fluid of density $\rho = 1,225\,\mathrm{kgm^{-3}}$ and has a dynamic viscosity $\mu = 1.78 \times 10^{-5}\,\mathrm{kgm^{-1}s^{-1}}$ which is defined as

$$\tau = \mu \frac{dv}{dz}$$

where τ is a small applied shear stress and $\frac{dv}{dz}$ the resulting velocity gradient. Its state is fully described by the velocity field $\mathbf{v}(\mathbf{r})$ and the pressure field $p(\mathbf{r})$ when we assume constant temperature and density. Its time evolution is given by the Navier-Stokes equations and the incompressibility condition. The solution of this equation is mainly characterized by the dimensionless Reynolds number defined through

$$Re = \frac{Lv}{\nu}$$

where $\nu = \frac{\mu}{\rho}$ is the kinematic viscosity. L and v are a characteristic length and a characteristic velocity of the problem as it could be given by the boundary conditions. Re represents the ratio of inertial forces to viscous forces. For low Reynolds numbers the flow is laminar which means that it attains a time independent final state which minimizes all the spatial gradients. We encounter this situation at small wind velocities, in narrow channels (for instance between two grains) and at a close distance to a surface. For high Reynolds numbers the flow is turbulent which means that there are strong spatial and temporal fluctuations on different scales all the time. This situation is typical outdoor even at moderate wind velocities due to the enormous size of the atmosphere. This complex behaviour arises from the fact that for large Re the Navier-Stokes equation is dominated by the non-linear inertia term.

Since the turbulent fluctuations in the velocity and pressure fields are random, one can only make predictions about temporal averages. The critical Reynolds number at which the atmospheric boundary layer becomes turbulent is in the order of 6000 [2]. Hence, even small wind speeds create turbulent flows. The velocity field at which aeolian sand transport occurs is always turbulent. The wind can locally exert higher shear stresses due to turbulent fluctuations and eddies, as compared to the laminar case. To model this effect, a turbulent viscosity η or a turbulent shear stress τ_T can be introduced,

$$\tau = \tau_l + \tau_T = (\mu + \eta)\frac{dv}{dz} . \tag{1}$$

where we assume that the flow is in the x-direction. The turbulent shear stress τ_T is equal to the transfer of momentum per unit time and unit area and can be given by [3, 4, 5],

$$\tau_T = -\overline{\rho u'w'} = \eta \frac{dv}{dz} . \tag{2}$$

The bar denotes the time average, u' and w' velocity fluctuations in the x and z-direction, respectively. The velocity fluctuations are defined according to $u' = u - \bar{u}$ and $w' = w - \bar{w}$, where u, w denote the velocities and $\bar{u}, \bar{w}$ the time–averaged velocities. Using the mixing length theory [4] for the absorption of vertical momentum, the turbulent shear stress can be expressed by,

$$\tau_T = \rho l^2 \left(\frac{dv}{dz}\right)^2 , \tag{3}$$

here l is the vertical mixing length. Prandtl [4] assumed that the mixing length $l = \kappa z$ increases linearly with the distance from the surface, where $\kappa \approx 0.4$ is the von Kármán universal constant for turbulent flow. In a fully turbulent flow, where fluctuations of all length scales have been developed the dynamic viscosity μ is much smaller than the turbulent viscosity η and can be neglected. Therefore, the turbulent shear stress τ_T is identified with the overall shear stress τ. By integrating (3) from z_0 to z, one obtains the well known logarithmic profile of the atmospheric boundary layer illustrated in Fig. 2.

$$v(z) = \frac{u_*}{\kappa} \ln \frac{z}{z_0} , \tag{4}$$

where z_0 denotes the roughness length of the surface and $u_* = \sqrt{\tau/\rho}$ the shear velocity. The shear velocity u_* characterizes the flow and has the dimensions of a velocity, although it is actually a measure of the shear stress. The roughness length z_0 is either defined by the thickness of the laminar sublayer for aerodynamically smooth surfaces or by the size of surface perturbations for aerodynamically rough surfaces, as is illustrated in Fig. 3.

The spatial and temporal fluctuations can be of small scale and high frequency and therefore it is generally too expensive to simulate them directly in practical applications. Instead, the Navier Stokes equations can be time–averaged or ensemble–averaged, or otherwise manipulated to remove the small scale dynamics, which results in a modified set of equations that are computationally more accessible.

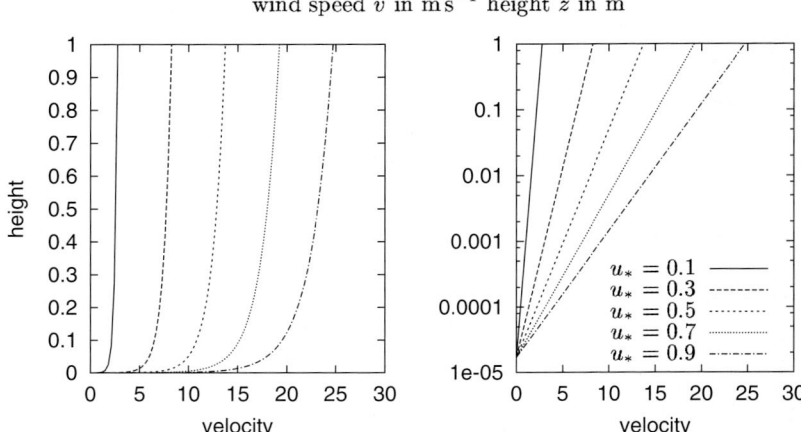

Fig. 2. Velocity profile of the atmospheric boundary layer above a surface with a roughness length $z_0 = 1.7\ 10^{-5}$ m; *left*: linear scale, *right*: semi-log plot

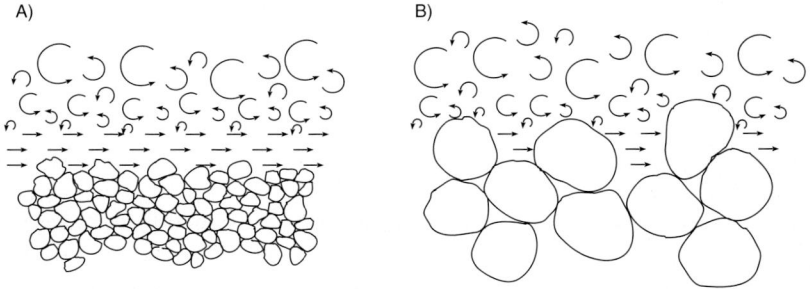

Fig. 3. (a) Aerodynamically smooth surface due to small grains forms a laminar sublayer. (b) Grains larger than the laminar sublayer create an aerodynamically rough surface

One of these approaches for turbulence are models in which the solution of two separate transport equations allows the turbulent velocity and length scales to be independently determined. The semi-empirical standard k-ϵ model [6] is based on transport equations for the turbulent kinetic energy k and its dissipation rate ϵ. In the derivation of the k-ϵ model one assumes that the flow is fully turbulent, and the effects of molecular viscosity are negligible. The standard k-ϵ model is therefore only valid for fully developed homogeneous turbulence.

There are many programs, packages, and libraries available that have been developed to solve the turbulent Navier Stokes equation with different boundary conditions using for instance the k-ϵ model. Nevertheless, three-dimensional turbulent flow on large scales is still a challenge and limited by

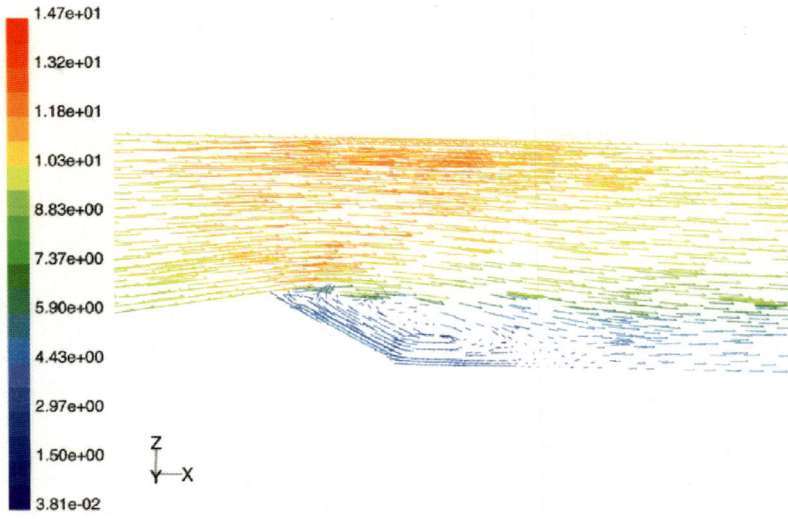

Fig. 4. Velocity field in a longitudinal cut along the central slice of a Barchan dune. One clearly sees the flow separation and a large eddy that forms in the wake of the dune

the performance of processors and memory. We have chosen here the commercial code FLUENT V5.0 [7].

We show in Figs. 4 and 5 the velocity field of the wind over a crescent-shaped obstacle which is in fact the topography of a real Barchan dune (see Fig. 1) measured in Marocco. We see from the cut in wind direction Fig. 4 that behind the dune an eddy of relatively low velocity is formed while the strong wind seems to follow above an imaginary continuation of the initial hill following the line $s(x)$ that delimits the eddy (separation line). The projection of Fig. 5 shows that in fact there are two such eddies.

The three dimensional calculations using FLUENT are very time consuming from a computational point of view. It is not possible to use it in an iterative calculation as needed to follow the evolution of a dune where the surface and thus the boundary evolves in time. Furthermore, the theoretical understanding is limited by using such a "black–box" model.

A dune or a smooth hill can be considered as a weak modification of the surface that causes a perturbation of the air flow. An analytical calculation of the shear stress perturbation due to a two dimensional hill has been performed first by Jackson and Hunt [8]. Later, the work has been extended to three dimensional hills and further refined [9, 10, 11, 12]. The following discussion is mainly based on the work of [13].

After a rather lengthy calculation they obtain for the Fourier transformation of the shear stress perturbation $\hat{\tau}_x$ in wind direction,

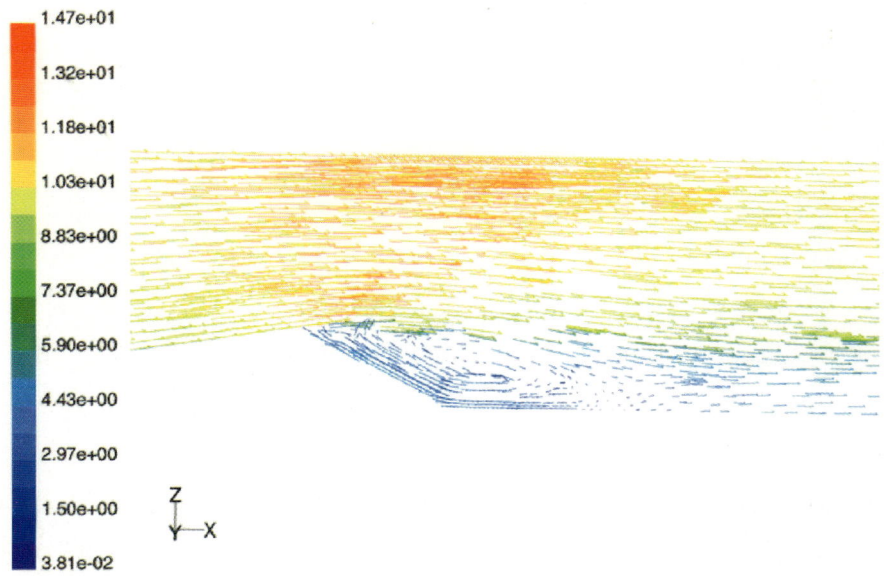

Fig. 5. Velocity field on the ground of a Barchan dune. The vectors show the existence of two neighboring wake eddies

$$\hat{\tau}_x(k_x, k_y) = \frac{h(k_x, k_y)k_x^2}{|k|} \frac{2}{v^2(l)} \left(1 + \frac{2\ln L|k_x| + 4\gamma + 1 + i\,\mathrm{sign}(k_x)\pi}{\ln l/z_0}\right) , \quad (5)$$

and the shear stress perturbation $\hat{\tau}_y$ in lateral direction,

$$\hat{\tau}_y(k_x, k_y) = \frac{h(k_x, k_y)k_x k_y}{|k|} \frac{2}{v^2(l)} , \quad (6)$$

where $h(k_x, k_y|k| = \sqrt{k_x^2 + k_y^2}$ and $\gamma = 0.577216$ (Euler's constant). Furthermore, $v(l)$ denotes the dimensionless velocity at the height l normalized by the velocity v_0,

$$v(l) = \frac{u_*}{v_0 \kappa} \ln \frac{l}{z_0} , \quad (7)$$

where $\kappa \approx 0.4$ is the von Kármán's constant and v_0 the velocity of the undisturbed upwind profile at the intermediate height. We further simplify (5) by approximating the logarithmic term $\ln L|k|$ by the constant value that corresponds to the wave length $4L$ of the hill and obtain,

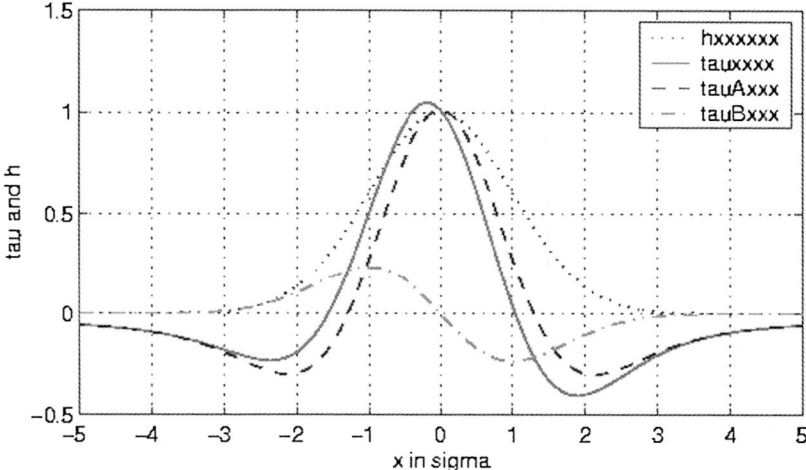

Fig. 6. Shear stress at the suface calculated using (8). The first part τ_A is responsible for the depressions before and behind the hill. The second part τ_B leads to an upwind shift of the maximum of the shear stress with respect to the maximum of the profile $h(x)$

$$\hat{\tau}_x(k_x, k_y) = A \frac{h(k_x, k_y) k_x}{\sqrt{k_x^2 + k_y^2}} (k_x + iB|k_x|) \,, \qquad (8)$$

where A and B depend logarithmically on $\ln L/z_0$.

The non-local convolution integral term is a direct consequence of the pressure perturbation over the hill. The second local term is a correction that comes from the non-linearity of the Navier Stokes equation and represents the effect of inertia. Both terms are depicted in Fig. 6 for a Gaussian hill. The first term mainly determines the speed–up and is symmetric whereas the second term is asymmetric. The sum of both terms leads to an upwind shift of the maximum of the shear stress perturbation with respect to the top of the hill.

The calculation of the shear stress of the air onto a smooth surface using (8) is computationally very efficient. The limitation of this analytical formula is that it can only be used for surfaces with slopes having less than 30 degrees. However, for a dune with a slip face the flow separation occurs at the brink which is a sharp edge as seen in Fig. 4 and where consequently the surface is not smooth.

One way to treat this problem of flow separation is to divide the flow into two parts by the separating streamline $s(x)$ that reaches from the brink at which one has the flow separation to the ground. The area enclosed by the separating streamline and the surface, called the *separation bubble*, a re-circulating flow develops,whereas the (averaged) flow outside is laminar as shown in Fig. 4. The general idea, suggested by [10], is that the air shear

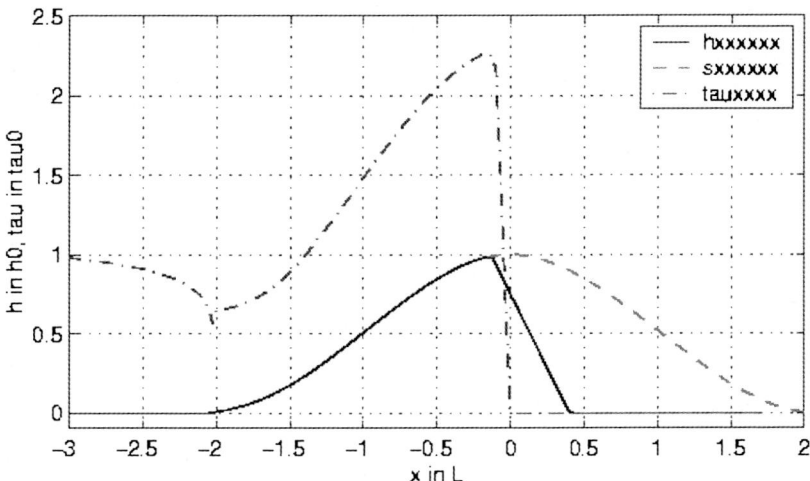

Fig. 7. The envelope $\tilde{h}(x)$ of the windward profile of a dune $h(x)$ and the separating streamline $s(x)$ form together a smooth line on which one can calculate the air shear stress $\tau(x)$ using (8). Inside the region of re-circulation the air shear stress τ is set to zero

stress $\tau(x)$ on the windward side can be calculated using the envelope that comprises the dune and the separation bubble Fig. 7. The envelope $\tilde{h}(x)$ is defined as,

$$\tilde{h}(x) = \max\left(h(x), s(x)\right) . \qquad (9)$$

The minimal Ansatz for a "smooth" separating streamline $s(x)$ is a third order polynomial as was used in [35, 36, 37]. Recent, more systematic studies using FLUENT were able to give more detailed and complex picture [14].

In Fig. 8 we see the streamline calculated using FLUENT for a test dune that is modelled by a circle segment and a brink position ten meters before the maximum of this circle segment. The dotted line is a fit using an ellipse segment.

3 Aeolian Transport of Sand

Sand consists of grains with diameters d which range from $d \approx 2\,\text{mm}$ for very coarse sand to $d \approx 0.05\,\text{mm}$ for very fine sand. The sand itself is mostly composed of quartz (SiO_2) which has a density ρ_{quartz} of $2650\,\text{kg}\,\text{m}^{-3}$ being more than 2000 times larger than the density of air. Dune sand has a quite sharply peaked distribution of diameters because the transport produces a natural mechanism of size segregation. One can also distinguish sand grains with respect to their shape [15]. An moving fluid such as air exerts two types of forces on grains when blowing over a bed of sand. The first is called *drag*

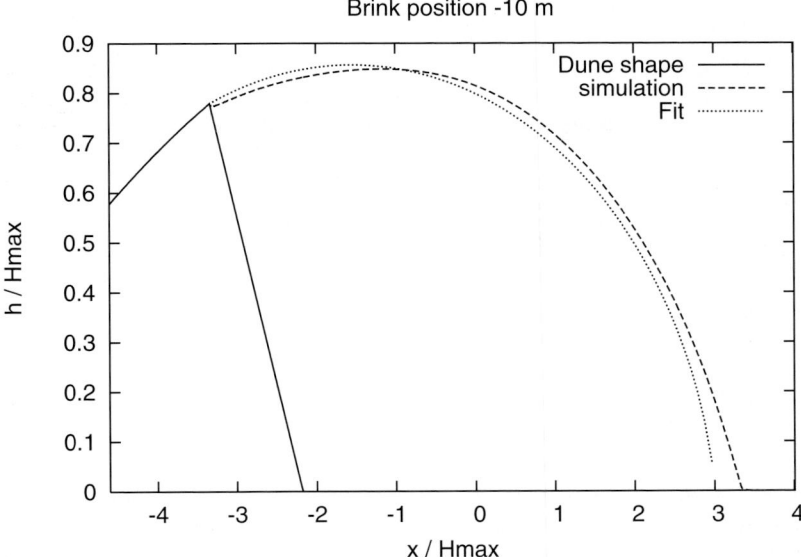

Fig. 8. Separation streamline as calculated with FLUENT (*dashed line*) and elliptic fit (*dotted line*) for a dune with a brink position ten meters before the maximum (from [14])

force F_d and acts in the direction of the flow. For turbulent flow it scales quadratically with the velocity due to Newton's drag law:

$$F_d = \beta \rho u_*^2 \frac{\pi d^2}{4} \;, \tag{10}$$

where β is a phenomenological parameter. The second force is called *lift force* F_l and arises from the static pressure difference Δp between the bottom and the top of a grain, caused by the strong velocity gradient of the air near the ground,

$$F_l = \Delta p \frac{\pi d^2}{4} = C_L \rho v^2 \frac{\pi d^2}{2} \;, \tag{11}$$

where $C_L = 0.0624$ [16]. v denotes the air velocity at a height of $0.35d$ with respect to the zero level at z_0. In a turbulent flow, where high velocity and pressure fluctuations occur, the lift forces can be sufficiently large to eject grains. Chepil [16] showed further that the ratio $c = 0.85$ between the drag force and the lift force is constant within the relevant range of Reynolds numbers,

$$F_l = c\,F_d \;. \tag{12}$$

Gravity and inertia oppose the aerodynamic forces. In particular does the grains' weight F_g directly counteract the lift force,

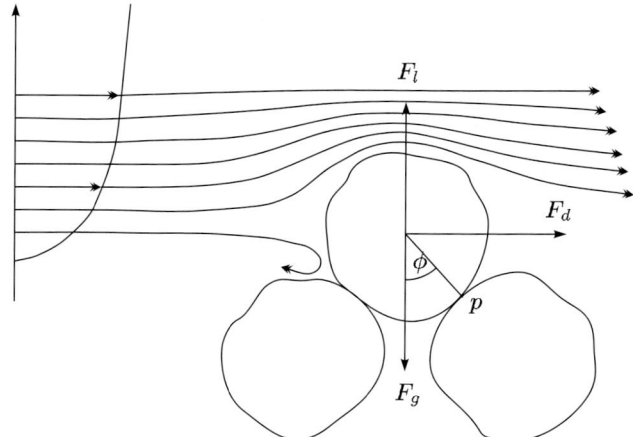

Fig. 9. The grain starts to move when the drag and lift forces exceed the gravitational force. This can be expressed by the momentum balance with respect to the pivot point p

$$F_g = \rho' g \frac{\pi d^3}{6} \qquad (13)$$

$\rho' = \rho_{\text{quartz}} - \rho_{\text{air}}$ being the reduced density of the sand grains in the air. In addition, cohesive and adhesive forces would have to be taken into account for small grains and wet sand beds, but are neglected here. The uppermost layer of grains in the bed is free to move when the aerodynamic forces overcome the gravitational force. The balance of momentum for a grain that rotates around its pivot point as depicted in Fig. 9 yields

$$F_d \frac{d}{2} \cos\phi = (F_g - F_l) \frac{d}{2} \sin\phi \qquad (14)$$

Inserting (10), (12), and (13), (14) defines the minimal shear stress required to move a grain called *aerodynamic entrainment threshold* $\tau_{ta} = \rho_{\text{air}} u_{*ta}^2$,

$$\frac{\tau_{ta}}{\rho' g d} = \frac{2}{3\beta} \left(\frac{\sin\phi}{\cos\phi + c \sin\phi} \right) . \qquad (15)$$

The aerodynamic entrainment fluid threshold shear stress τ_{ta} on a flat surface is directly proportional to the reduced density ρ' and the diameter d of the grains. The packing of the grains is reflected by the angle ϕ which can be understood in a similar way as the angle of internal friction of a sand pile, however, its value may be different. The shape of the grains is taken into account by the constant β.

Shields [17] introduced a dimensionless coefficient Θ that expresses the ratio of the applied tangential force to the inertial force of the grain,

$$\Theta(Re_*) \equiv \frac{\tau_{ta}}{\rho' g d}. \tag{16}$$

This Shields parameter Θ depends on the Reynolds friction number $Re_* = u_* d\nu^{-1}$. For $Re_* > 1$, Θ is constant with a value that ranges from 0.01 to 0.014. [1] used the dimensionless Shields parameter Θ to define the fluid threshold shear velocity u_{*ta},

$$u_{*ta} = \sqrt{\Theta \frac{\rho' g d}{\rho_{\text{air}}}}. \tag{17}$$

This expression is only valid as long as cohesive and adhesive forces can be neglected and thus for grain diameters larger than 0.2 mm. The typical value for the fluid threshold shear velocity $u_{*ta} = 0.25 \, \text{m s}^{-1}$ is obtained for $d = 250 \, \mu\text{m}$ using $\Theta = 0.012$.

During sediment transport when sand grains are flying in the air, they impact onto the bed. The momentum transfer from an impacting grain to grains resting on the ground lowers the threshold for entrainment. This has already been observed by Bagnold [18] who called this lowered threshold *impact threshold* u_{*t}. The impact threshold shear velocity u_{*t} can be calculated in an analogous way and expressed by (17) replacing the Shields parameter by an effective value $\Theta_{eff} = 0.0064$.

Gravity strongly determines the value of the threshold velocity so that it is for instance very different on Mars and also different on inclined chutes. The definition of a single threshold becomes more and more inacurate if sediments are poorly sorted or if the effect of moisture and cementing agents becomes important. A detailed discussion of these effects can be found in the book of Pye and Tsoar [15].

Different mechanisms of aeolian sand transport such as suspension and bed–load can be distinguished according to the degree of detachment of the grains from the ground. Bed-load transport can further be divided into saltation, reptation, and creep. However, the different subclassifications of bed–load do vary from author to author [15].

Small grains are suspended in air and can travel long distances on irregular trajectories before reaching again the ground. A turbulent air flow can keep grains in suspension when the vertical component of the turbulent fluctuations w', (2), exceeds the settling velocity w_f of the grains.

During typical sand storms, when shear velocities are in the range of 0.18 to $0.6 \, \text{m s}^{-1}$ [15], particles with a maximum diameter of 0.04–0.06 mm can be transported in suspension, as seen in Fig. 10. The grains of typical dune sand have a diameter in the order of 0.25 mm and are transported via bed-load and mainly by saltation. For this reason, suspension will be neglected in the following discussion and we will focus on the saltation process in the following.

Saltation is the most relevant bed-load mechanism transport mechanism. To initiate saltation some grains have to be entrained directly by the air. This is called *direct aerodynamic entrainment*. However, if there is already a sufficiently large amount of grains in the air, the direct aerodynamicentrainment is

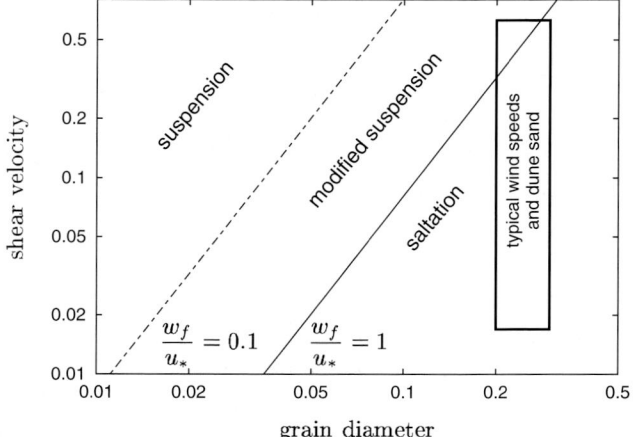

Fig. 10. The transport mechanism as function of the wind shear velocity and the grain diameter. The area enclosed by the rectangle shows the region in which aeolian sand transport on Earth occurs by saltation for typical dune sand (0.2 mm < d < 0.3 mm) and typical wind velocities (0.2 m s^{-1} < u_* < 0.6 m s^{-1})

negligible and grains are mainly ejected by collisions of impacting grains. The entrained grains are accelerated by the wind along their trajectory mainly by the drag force before they impact onto the bed again. The interaction between an impacting grain and the bed is called *splash process* and can produce a jet of grains that are ejected into the air. It is currently the subject of theoretical and experimental investigations [19, 20]. Finally, the momentum transferred from the air to the grains gives rise to a deceleration of the air. Due to this negative feedback mechanism saltation reaches a constant transport rate after some transient time.

From (15) we see that direct aerodynamic entrainment occurs if the air shear stress τ exceeds τ_{ta}. A simple way to model the rate of aerodynamically entrained grains has been proposed by Anderson [21]. He claims that the number of entrained grains is proportional to the excess shear stress,

$$N_a = \zeta(\tau - \tau_{ta}) , \qquad (18)$$

N_a being the number of entrained grains per time and ζ a proportionality constant, chosen by [21] to be of the order of 10^5 grains N^{-1} s^{-1}.

When already flying in the air, aerodynamic forces (lift and drag) and gravity act on the grain and determine its trajectory [1, 22, 23, 24]. In a saltation trajectory, the vertical motion is mainly determined by the gravitational force $\mathbf{F}_g$,

$$\mathbf{F}_g = m\mathbf{g} , \qquad (19)$$

m denoting the mass of the grain and $\mathbf{g}$ the gravitational acceleration. The acceleration in the horizontal direction is due to the drag force $\mathbf{F}_d$,

$$\mathbf{F}_d = \frac{1}{2}\rho_{\text{air}} C_d \frac{\pi d^2}{4}\left(\mathbf{v}(z) - \mathbf{u}\right)|\mathbf{v}(z) - \mathbf{u}|\,, \qquad (20)$$

d being the grain diameter, $\mathbf{v}(z)$ the velocity of the air, $\mathbf{u}$ the velocity of the grain, and C_d the drag coefficient that depends on the local Reynolds number $Re = |\mathbf{v} - \mathbf{u}|d/\nu$. Hence, the trajectory is close to that of a simple ballistic trajectory, of height h and flight time T,

$$T = \frac{2u_{z0}}{g}\,; \quad h = \frac{u_{z0}^2}{2g}\,, \qquad (21)$$

u_{z0} being the vertical component of the initial velocity. More elaborate calculations [23, 26] have shown that the simple approximation using the ballistic formula gives values which are overstimated by 10–20%. Wind tunnel measurements [19] have lead to: $T(u_*) \approx 1.7\,u_*/g$, $h(u_*) \approx 1.5\,u_*/g$, and the saltation length $l(u_*) \approx 18\,u_*^2/g$. For $u_* = 0.5\,\text{m s}^{-1}$ a flight time $T \approx 0.08\,\text{s}$, hop height $h \approx 3.8\,\text{cm}$ and hop length $l \approx 45\,\text{cm}$.

The transfer of momentum from the air to the grains lowers the wind speed near the ground. This is called the *negative feedback* mechanism and it finally limits the number of grains in the air and drives the system into a steady state. A possibility to include this effect is to add to the right of the Navier Stokes equation a body force $\mathbf{f}$ that models the average momentum transfer from the air to the grains.

$$\rho_{\text{air}}\partial_t \mathbf{v} + \rho_{\text{air}}(\mathbf{v}\nabla)\mathbf{v} = -\nabla p + \nabla\tau + \mathbf{f} \qquad (22)$$

Anderson [21] used an averaged drag force $\mathbf{f}$ obtained from microscopic trajectory calculations to calculate the velocity profile according to momentum conservation:

$$\nabla\tau + \mathbf{f} = 0\,. \qquad (23)$$

Using again FLUENT it is possible to calculate the saltation layer on the grain level and obtain the loss of velocity of the wind due to the negative feedback for different heights as done in [27]. An example is given in Fig. 11 were we see that the velocity loss occurs mostly around a specific height, namely the one which is typical for the grain trajectories. One also notices that the loss is proportional to the amount of transported grains given through the particle flux q as illustrated in the inset.

The inclusion of saltation on the grain level as discussed before is too detailed for many macroscopic applications concerning sand filling, desertification or dune formation. They have to be linked to macroscopic variables such as the sand flux q per unit width and time. The sand flux q depends on the shear velocity u_*, the threshold u_{*t}, the grain diameter d, and many other properties. It may even depend on the history if non-equilibrium conditions like transients are important. However, most of the known sand flux

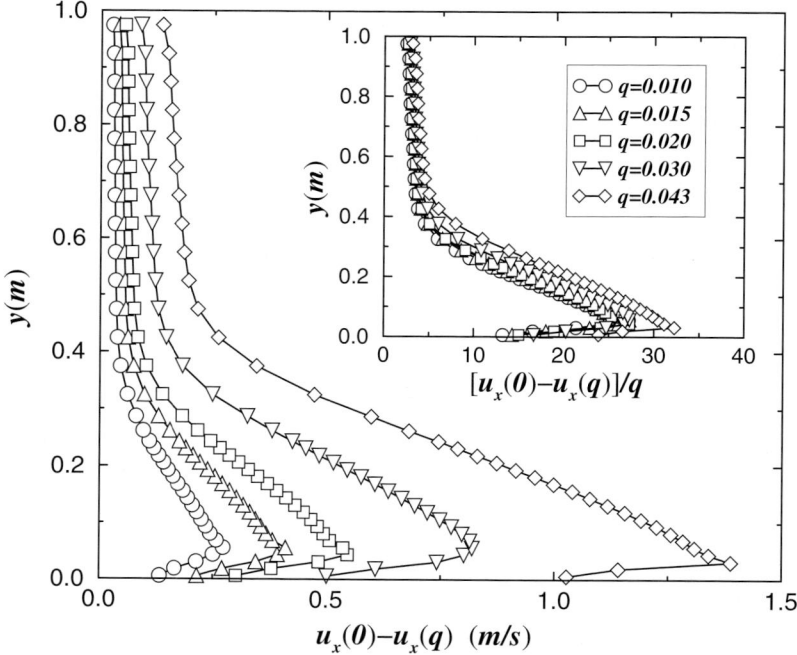

Fig. 11. Profile of the difference of the wind velocity without and with grain transport as function of the height for different fluxes q. The inset shows that this velocity loss scales with q (from [27])

relations are restricted to the saturated case where the sand flux is a function $q(u_*, u_{*t}, d, ...)$ depending in a unique way of its parameters.

Measurements performed in wind tunnels [28, 29] show that sand flux starts at a threshold u_{*t} and scales with the cube of the shear velocity ($q \propto u_*^3$) for high shear velocities ($u_* \gg u_{*t}$). In the vicinity of the threshold the functional dependence is not well understood and empirical and theoretical flux relations differ considerably. In fact recent calculations using FLUENT yield a quadratic dependence of the form $q \propto (u_* - u_{*t})^2$ [27]. The simplest flux relation that predicts a cubic relation between sand flux q and shear velocity u_* was proposed by Bagnold [1],

$$q_B = C_B \frac{\rho_{\text{air}}}{g} \sqrt{\frac{d}{D}} u_*^3 , \qquad (24)$$

being d the real grain diameter and $D = 250\,\mu\text{m}$ a reference grain diameter. This simple relation is still of theoretical interest at high wind speeds, far from the threshold, where most of the sand is transported. Later the threshold u_{*t} was incorporated into the sand flux relations in order to account for the fact that sand transport cannot be maintained below a certain shear velocity.

Many phenomenological sand flux relations have been proposed and have been summarized for instance in [15]. A sand flux relation that is widely used is the one by Lettau and Lettau [30],

$$q_L = C_L \frac{\rho_{\text{air}}}{g} u_*^2 (u_* - u_{*t}) \tag{25}$$

C_L being a fit parameter. Analytical calculations that predict the sand flux by averaging over the microscopic processes have deepened very much the understanding of aeolian sediment transport [22, 31, 32, 33]. A result of such a theoretical calculation was obtained by Sørensen [23],

$$q_S = C_S \frac{\rho_{\text{air}}}{g} u_* (u_* - u_{*t})(u_* + 7.6 * u_{*t} + 2.05 \, \text{m s}^{-1}) , \tag{26}$$

C_S being a parameter that has been determined analytically. The fit of (26) to wind tunnel data revealed that the analytical obtained value C_S is about four times too small. The functional structure of the relation reproduces the data, however, quite well [29].

The relations of the form $q(u_*,\ldots)$ discused up to now assume that the sediment transport is in steady state, i.e. the sand flux is saturated. In order to overcome this limitation and to get information about the dynamics of the aeolian sand transport, numerical simulations based on the grain scale have been performed [21, 24, 34]. They showed that on a flat surface the typical time to reach the equilibrium state in saltation is approximately two seconds, which was later confirmed by wind tunnel measurements [28].

Assuming that each splash event produces on average the same number of ejected new particles the number of saltating grains would increase exponentially in time. Each accelerated grain, however, removes momentum from the wind field. Therefore after a saturation time T_s the flux must saturate to a value q_s. From this microscopic picture Sauermann et al. [35, 36, 37] have derived an equation describing this evolution of the flux towards saturation

$$\frac{\partial q}{\partial x} = \frac{1}{l_s} q \left(1 - \frac{q}{q_s}\right) , \tag{27}$$

l_s being the "saturation length". In Fig. 12 we see solutions of (27).

Let us emphasize that $T_s(\tau, u)$ and $l_s(\tau, u) = T_s u$ are not constant, but depend on the external shear stress τ of the wind and on the mean grain velocity u. We can relate the characteristic time T_s and length l_s of the saturation transients to the saltation time T and the saltation length l of the average trajectory of a saltating grain,

$$T_s = T \frac{\tau_t}{\gamma(\tau - \tau_t)} , \qquad l_s = l \frac{\tau_t}{\gamma(\tau - \tau_t)} \tag{28}$$

τ_t being the entrainment threshold shear stress and γ a constant. For typical wind speeds, the time to reach saturation is in the order of 2 s [21, 24, 34].

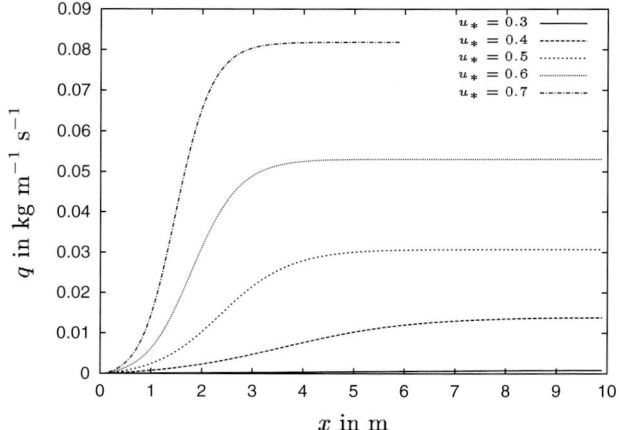

Fig. 12. Numerical solution of the sand flux equation (27) for different shear velocities u_*. The model parameter $\gamma = 0.2$ of (28) of that defines the length and time of the saturation transients was chosen here so that saturation is reached between 1 s and 2 s

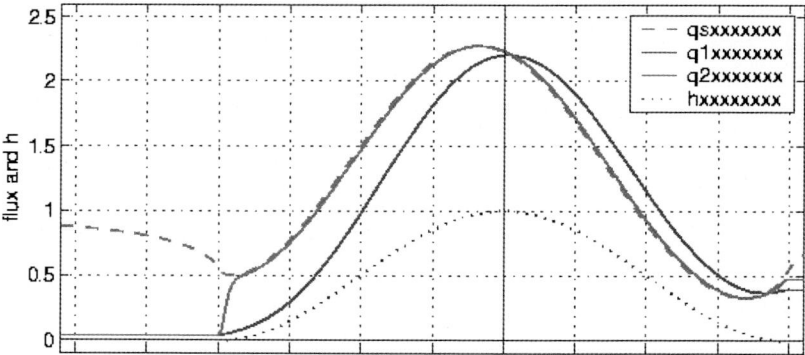

Fig. 13. The figure shows the saturated flux q_s, (26), and the flux q, (27), including saturation transients, calculated for two cosine shaped hills with an aspect ratio of $H/L = 1/8$. The hills have a height of $H = 1\,\text{m}$ and $H = 10\,\text{m}$, the saturation length is $l_s = 0.8\,\text{m}$. The saturated flux q_s is scale invariant and thus identical for both hills, but the flux q gets completely different in the case of the small hill, $L \approx 10 l_s$

Assuming a grain velocity of 3–5 m s^{-1} [25] we obtain a length scale of the order of 10 m for saturation. This length scale is large enough to play an important role in the formation of dunes.

We depict in Fig. 13 the solution of (27) together with the saturated sand flux q_s for two cosine shaped heaps with the same aspect ratio. We see that the solution of the large hill is close to the saturated flux except in the region near the foot of the heap. The situation at the foot is different, there is no deposition and the surface velocity decreases, which predicts a steepening in the foot area. For small hills the shear stress gradient, $\partial \tau / \partial x \propto 1/L$, gets large and the real sand flux lags a certain distance behind the saturated one. Let us point out that up to a certain degree this lag compensates the upwind shift of the shear stress, or may even overcompensate it, as it is the case for the small hill in Fig. 13. The effect of this lag can be seen in the surface velocity of the windward side (left side). The large hill shows a tendency to steepen due to a decreasing surface velocity whereas the small hill will flatten due to an increasing surface velocity. Finally, a further consequence of the saturation transients is the existence of a minimal size for a dune to be stable.

4 Inclusion of the Slip Face

We discussed in the previous section the lee side of a hill has the tendency to steepen. If the wind blows long enough from the same direction, the lee side will reach the angle of repose $\Theta \approx 34°$, which is the steepest stable angle of a free sand surface. If this angle is exceeded, avalanches start to slide down the hill until the surface has relaxed to a slope equal or below the angle of repose. In that case the responsible for sand flux is not the wind but gravity.

Without having to take into account the individual avalanches this effect can be implemented by redistributing the sand in such a way that the slip face is always a straight line with a slope corresponding to the angle of repose Θ. In two dimensions this is easy. In three dimensions this process, however is not straightforward. Bouchaud et al. (BCRE) [38] proposed a set of equations to describe avalanches which allows implementing locally and iteratively even in three dimensions the formation of surfaces having the angle of repose as their steepest inclination. Therefore these BCRE equations seem adequate to describe the dynamics of the slip face.

The complete model is defined by the three variable fields $h(x,y)$, $q(x,y)$ and $\tau(x,y)$. τ is calculated from h through the Fourier transformation of (8). Then q is obtained from τ through (27) using q_s from (25) and l_s from (27). The new topography h is then obtained from q using mass conservation:

$$\frac{\partial h}{\partial t} = \frac{1}{\rho_{\text{sand}}} \nabla_s q \qquad (29)$$

In regions where $\nabla_s h > \tan \Theta$ slip occurs and the just mentioned BCRE equations are applied. ∇_s denotes the spatial derivative in direction of the strongest gradient. Once $h(x,y)$ is obtained one goes back to calculate again $\tau(x,y)$ etc. In this way one iteratively obtains the time evolution of the three fields.

The above system of iteratively solved coupled equations describes fully the motion of the free granular surface under the action of wind and gravity and can be used to calculate formation, evolution and shape of dunes. A natural consequence of the two different driving mechanisms, wind and gravity, is that

Fig. 14. Typical desert landscape showing the characteristic sharp edges separating the slip faces from wind driven regions

the solution will separate in two regions: those for which the slope was larger than Θ and where therefore the BCRE equation was applied, ie the slip faces, and those where this was not the case. Theses two regions are separated by characteristic sharp edges which are the typical feature of sandy landscapes as seen in Fig. 14.

5 Dunes

Dunes are land formations of sand of heights, ranging typically from 1 to 500 meters which have been shaped by the wind. These topographical structures are found typically where large masses of sand have accumulated, which can be in the desert or along the beach. Correspondingly one distinguishes desert dunes and coastal dunes. Dunes can be mobile or fixed. Fixed dunes are older and are either "fossilized" which means transformed into a cohesive material, precurser to sand stone, or fixed because the average wind at their site over some period is zero. Otherwise the sand moves if the winds are strong enough, that means typically stronger than 4 meters a second.

As we all know, the beautiful landscapes (Fig. 14) formed by dunes are characterized by very gentle hills interruped by sharp edges called brink lines, delimiting regions of steeper slope, called slip-faces, lying in the wind shadow.

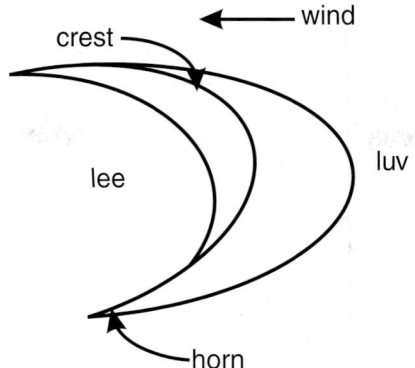

Fig. 15. Schematic diagram of a Barchan dune

Depending on the amount of available sand and the variation of the wind direction, one distinguishes different typical dune morphologies that have been classified by geographers into over 100 categories. The most well-known are longitudinal, transverse and Barchan dunes; other common dunes are star dunes, ergs, parabolic dunes and draas.

If the wind always comes from the same direction, one obtains transverse dunes if there is much sand and Barchan dunes (from an Arabic word) if little sand is available. They exist in large fields in Marocco, Peru, Namibia etc. typically parallel to the coast (see Fig. 1). As shown schematically in Fig. 15, Barchans are crescent-shaped dunes. Their velocity ranges from 5 to 50 m per year and is inversely proportional to their height. It can be calculated in the following way.

If a dune moves shape invariantly with velocity $\mathbf{v}_d$ one has

$$\frac{\partial h}{\partial t} = \mathbf{v}_d \nabla_s h \tag{30}$$

which inserted in (29) gives

$$\mathbf{v}_d = \frac{1}{\rho}\frac{dq}{dh} \tag{31}$$

or if q_b is the flux over the maximum (brink) of the dune of height H

$$\mathbf{v}_d = \frac{q_b}{\rho H} \tag{32}$$

This result that dune velocity is inversely proportional to its height is also known since Bagnold [1].

Another interesting question about Barchans is their shape. In Fig. 16 we see a longitudinal cut through the highest point of Barchans of different size normalized in such a way that they all have the same maximum [39]. On the

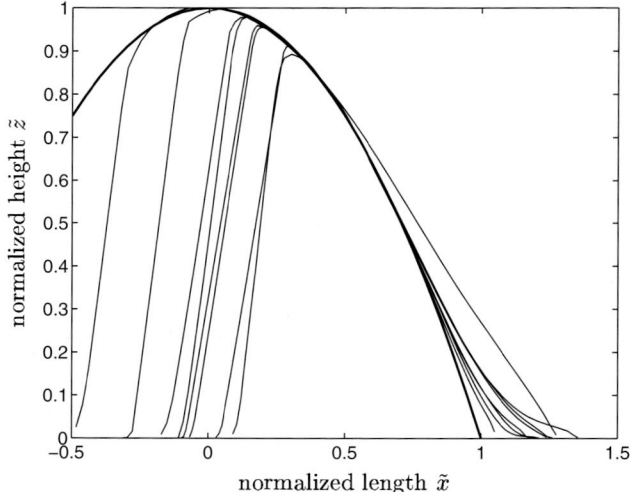

Fig. 16. Longitudinal profiles of eight dunes along their symmetry plane normalizing the length scales such that the shapes collapse on top of each other (from [39])

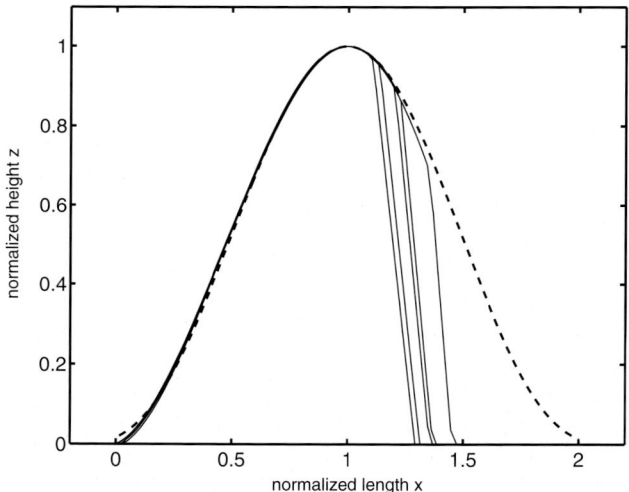

Fig. 17. Numerical calculation of the normalized longitudinal profiles of the dunes. A fit with $\cosh^4(x)$ (*dashed line*) reproduces quite well the windward side. The shear velocity is $u_* = 0.5$ ms^{-1} (from [40])

windward side all curves fall on top of each other, while the crest lies more inwards for increasing dune height. In Fig. 17 we show the corresponding results from the numerical solution of the equations of the last section. Only now the wind comes from the opposite side. We see the same behaviour as function of dune height as in In Fig. 16. The dashed curve is a fit to the windward side. We see that the theoretical equations do reproduce very well quantitatively the measurements, but on top they do not have the uncontrollable fluctuations that come always from field data. Similarly also the transverse cuts scale with height and the numerical calculation also agrees with the observation.

One consequence of the above similarity relations is a well-known linear dependence between dune height, length and width as has been already reported by Bagnold [1]. Viewing the dune from the top, the brink has the shape of a parabola [39]. Due to the competition between the saturation length and the size of the separation bubble behind the dune one can calculate for the minimal height of a stable dune to be about 1,5 meters. The shear stress of the wind and the sand flux on aBarchan dune have also been measured and very favourably been compared to the numerical results of the [41].

With these computer dunes it has lately been shown [42] that when a small Barchan bumps into a larger one it can either be swallowed (if it is too small), or it can coalesce but produce at each horn a new baby Barchan (breeding), or it can, if the two initial dunes are of similar size separate again after some exchange of sand (solitary behaviour). In this last case, it looks as if the dunes do cross each other unaltered except for an eventual change in their size.

The system of equations of motion for dunes has also been used to calculate entire systems of dunes and virtual landscapes. An example is shown in Fig. 18

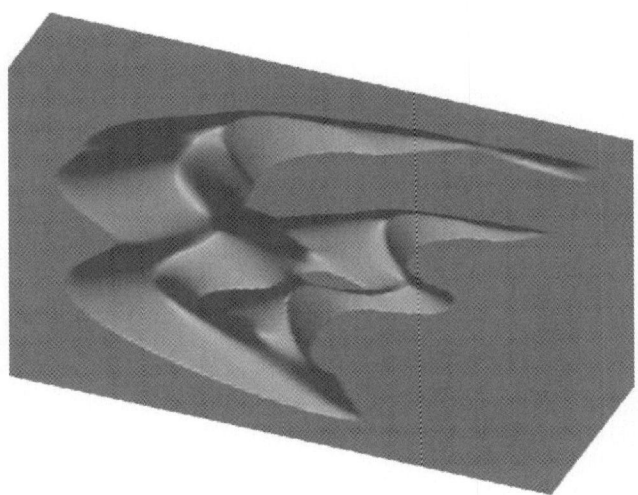

Fig. 18. Complex dune pattern, calculated with the full three dimensional model. Wind is blowing from the *left* to the *right*. When Barchan dunes are too close they interact, get eventually connected, and form complex dune structures. The large dunes are shielding the small dunes from the arriving sand flux which then constantly loose volume

where a constant influx of sand is used as boundary condition on the left while on the right one has a free boundary.

6 Conclusion

In this review we have shown that using known expressions for turbulent flow and using the transport mechanism of saltation, it is possible to formulate up a set equations of motion for a wind driven free granular surface. These three coupled equations containing as variable fields the shear stress of the wind, the sand flux at the surface and the profile of the landscape must be complemented by the BCRE equations in regions where the slope exceeds the angle of repose in order to correctly describe the slip faces. The resulting system of equations can be solved iteratively using appropiate boundary conditions and initializations. The solutions produce patterns that not only ressemble those observed in nature but also agree very well quantitatively with field measurements of shapes, sand fluxes and dune velocities.

The simulation of dune motion on the computer allows make predictions over long time scales since in the real world dune motion is very slow. One can also predict the effect of protective measures like the BOFIX-technique of Meunier [43] and even calculate the dunes on Mars [44].

References

1. R.A. Bagnold, *The physics of blown sand and desert dunes*. London: Methuen (1941).
2. J.T. Houghton, *The physics of atmospheres*, Volume 2nd edn. Cambridge: Cambridge Univ. Press (1986).
3. T. Kármán, Some aspects of the turbulence problem. Proc. 4th Int. Congr. Appl. Mech., Cambridge, 54–91 (1935).
4. L. Prandtl, The mechanics of viscous fluids. In W.F. Durand (Ed.), *Aerodynamic theory*, Volume Vol. III, pp. 34–208. Berlin: Springer (1935).
5. O.G. Sutton, *Micrometeorology*. New York: McGraw–Hill (1953).
6. D.B. Launder, B.E. and Spalding, *Lectures in Mathematical Models of Turbulence*. London, England: Academic Press (1972).
7. Fluent Inc., Fluent 5, Finite Volume Solver (1999).
8. P.S. Jackson and J.C.R. Hunt, Turbulent wind flow over a low hill. Q.J.R. Meteorol. Soc. **101**, 929 (1975).
9. R.I. Sykes, An asymptotic theory of incompressible turbulent boundary layer flow over a small hump. J. Fluid Mech. **101**, 647–670 (1980).
10. O. Zeman and N.O. Jensen, Progress report on modeling permanent form sand dunes. *Risø National Laboratory* **M-2738** (1988).
11. D.J. Carruthers and J.C.R. Hunt, *Atmospheric Processes over Complex Terrain*, Volume 23, Chapter Fluid Mechanics of Airflow over Hills: Turbulence, Fluxes, and Waves in the Boundary Layer. Am. Meteorological. Soc (1990).

12. W.S. Weng, J.C.R. Hunt, D.J. Carruthers, A. Warren, G.F.S. Wiggs, I. Livingstone and I. Castro, Air flow and sand transport over sand–dunes. Acta Mechanica (Suppl.) **2**, 1–22 (1991).
13. J.C.R. Hunt, S. Leibovich and K.J. Richards,. Turbulent wind flow over smooth hills. Q.J.R. Meteorol. Soc. **114**, 1435–1470 (1998).
14. V. Schatz and H.J. Herrmann, Numerical investigation of flow separation in the lee side of transverse dunes. preprint for Geomorphology (2005).
15. K. Pye and H. Tsoar, *Aeolian sand and sand dunes*. London: Unwin Hyman (1990).
16. W.S. Chepil, The use of evenly spaced hemispheres to evaluate aerodynamic forces on a soil surface. Trans. Am. Geophys. Union **39**, 397–403 (1958).
17. A. Shields, Applications of similarity principles and turbulence research to bed–load movement. Technical Report Publ. No. 167, California Inst. Technol. Hydrodynamics Lab. Translation of: Mitteilungen der preussischen Versuchsanstalt für Wasserbau und Schiffsbau. W.P. Ott and J.C. van Wehelen (translators) (1936).
18. R.A. Bagnold, The size–grading of sand by wind. Proc. R. Soc. London **163**(Ser. A), 250–264 (1937).
19. P. Nalpanis, J.C.R. Hunt and C.F. Barrett, Saltating particles over flat beds. J. Fluid Mech. **251**, 661–685 (1993).
20. F. Rioual, A. Valance and C. Bideau, C. Experimental study of the collision process of a grain on a two–dimensional granular bed. Phys. Rev. E **62**, 2450–2459 (2000).
21. R.S. Anderson, Wind modification and bed response during saltation of sand in air. Acta Mechanica (Suppl.) **1**, 21–51 (1991).
22. R.S. Owen, Saltation of uniformed sand grains in air. J. Fluid. Mech. **20**, 225–242 (1964).
23. M. Sørensen, An analytic model of wind-blown sand transport. Acta Mechanica (Suppl.) **1**, 67–81 (1991).
24. I.K. McEwan, I. and B.B. Willetts, Numerical model of the saltation cloud. Acta Mechanica (Suppl.) **1**, 53–66 (1991).
25. B.B. Willetts and M.A. Rice, Inter-saltation collisions. In O.E. Barndorff-Nielsen (Ed.), *Proceedings of International Workshop on Physics of Blown Sand*, Volume 8, pp. 83–100. Memoirs (1985).
26. R.S. Anderson and B. Hallet, Sediment transport by wind: toward a general model. Geol. Soc. Am. Bull. **97**, 523–535 (1986).
27. M.P. Almeida, J.S. Andrade Jr and H.J. Herrmann, Aeolian transport layer. preprint, cond-mat/0505626 (2005).
28. G.R. Butterfield, Sand transport response to fluctuating wind velocity. In N.J. Clifford, J.R. French, and J. Hardisty (Eds.), *Turbulence: Perspectives on Flow and Sediment Transport*, Chap. 13, pp. 305–335. John Wiley & Sons Ltd (1993).
29. K.R. Rasmussen and H.E. Mikkelsen, Wind tunnel observations of aeolian transport rates. Acta Mechanica *Suppl* **1**, 135–144 (1991).
30. K. Lettau and H. Lettau, Experimental and micrometeorological field studies of dune migration. In H.H. Lettau and K. Lettau (Eds.), *Exploring the world's driest climate*. Center for Climatic Research, Univ. Wisconsin: Madison (1978).
31. J.E. Ungar and P.K. Haff, Steady state saltation in air. Sedimentology **34**, 289–299 (1987).

32. M. Sørensen, Estimation of some eolian saltation transport parameters from transport rate profiles. In O.E.B.-N. et al. (Ed.), *Proc. Int. Wkshp. Physics of Blown Sand.*, Volume 1, Denmark, pp. 141–190. University of Aarhus (1985).
33. B.T. Werner, A steady-state model of wind blown sand transport. J. Geol. **98**, 1–17 (1990).
34. R.S. Anderson and P.K. Haff, Simulation of eolian saltation. Science **241**, 820 (1988).
35. G. Sauermann, K. Kroy and H. Herrmann (2001), A continuum saltation model for sand dunes. Phys. Rev. E **64**, 31305 (2001).
36. K. Kroy, G. Sauermann and H.J. Herrmann, A minimal model for sand dunes. Phys. Rev. Lett. **88**, 054301 (2002).
37. K. Kroy, G. Sauermann and H.J. Herrmann, Minimal model for aeolian sand dunes Phys. Rev. E **66**, 31302 (2002).
38. J.P. Bouchaud, M.E. Cates, J. Ravi Prakash and S.F. Edwards, Hysteresis and metastability in a continuum sandpile model. em J. Phys. France I **4**, 1383 (1994).
39. G. Sauermann, A. Poliakov, P. Rognon and H.J. Herrmann, The shape of theBarchan dunes of southern Marocco, Geomorphology **36**, 47–62 (2000).
40. V. Schwämmle and H.J. Herrmann, A model ofBarchan dunes including lateral shear stress, submitted to Euro. Phys. J.E, cond-mat/0304695 (2003).
41. G. Sauermann, J.S. Andrade, L.P. Maia, U.M.S. Costa, A.D. Araújo and H.J. Herrmann, Wind velocity and sand transport on a Barchan dune, Geomorphology **1325**, 1–11 (2003).
42. V. Schwämmle and H.J. Herrmann, Budding and solitary wave behaviour of dunes submitted to Nature (2003).
43. J. Meunier and P. Rognon P. (2000), Une méthode écologique pour détruire les dunes mobiles, Secheresse **11**, 309–316 (2000).
44. E.J. Ribeiro Parteli, V. Schatz V. and H.J. Herrmann, (2005), Barchan dunes on Mars and on Earth, *Powders and Grains 2005*, eds. R. Garcia-Rojo, H.J. Herrmann and S. McNamara (Balkema, Leiden), pp. 959–962 (2005).

Avalanches and Ripples in Sandpiles

A. Mehta

SN Bose National Centre for Basic Sciences, Block JD, Sector III Salt Lake, Calcutta 700 098, India
anita@bose.res.in

In this paper we unify several approaches taken to model sandpile dynamics, with a focus on avalanches and ripples. Our approaches include a coupled-map lattice model of sand in a rotating cylinder as well as noisy coupled nonlinear equations to model sandpile dynamics and ripple formation.

1 Introduction

Avalanches have been of great significance in the development of sandpile modeling because experimental results are at variance with the original model of Bak et al. [1]; while the latter predicts a simple power law for the avalanche size distribution function, indicating scale-invariant dynamics, the former show that in practice, larger avalanches occur more frequently than predicted by the power law describing the distribution of the smaller events. These experimental results demonstrate that one particular range of avalanche sizes is preferred, and have led to a large body of theoretical and numerical work which aims to identify the underlying physical mechanisms for such behavior.

Ripples on a sand dune also originate as avalanches on the dune; they can be viewed as their propagating extension once the limit of the dune has been reached. They are interesting examples of pattern formation in and of themselves, and their study involves a combination of hydrodynamics as well as granular dynamics on a sandpile.

In this review paper, we first focus on the avalanche size distribution in a rotating cylinder [2]. Next we move on to the modelling of generic avalanches on sandpile surfaces [3, 4], using coupled noisy nonlinear equations. Finally, we use this formulation to study ripple formation [5], one which has subsequently been generalised to the study of sand dunes, to be presented in another chapter in these proceedings. The interested reader is referred to a monograph on granular materials [6] which will have many more details on these and other subjects of relevance in this field.

2 Avalanches in a Rotating Cylinder

We model [2] an experimental situation which forms the basis of many traditional as well as modern experiments; a sandpile in a rotating cylinder. We consider the dynamics of sand in a half-cylinder that is rotating slowly around its axis, and we suppose that the sand is uniformly distributed in the direction of the axis. Our model is therefore essentially one-dimensional. The driving force arising from rotation continually affects the stability of the sand at all positions in the pile and is therefore distinct from random deposition. We include both surface flow and internal restructuring as mechanisms of sandpile relaxation, and focus on a situation where reorganization within the pile dominates the flow. This situation, of a sandpile subjected to slow rotation or tilt, has been formulated elsewhere [3, 4] in terms of continuum equations. Finally, we look at the effect of random driving forces in our model and compare the results with those from other models.

2.1 The Model

Since the effect of grain reorganization driven by slow tilt is easiest to visualize from a continuum viewpoint, our model incorporates grains which form part of a continuum; the column heights, h_i, are real variables, while the column identities, $1 < i < L$, are discrete as usual. We consider granular driving forces, f_i, that include, in addition to a term that drives the normal surface flow, a contribution that is proportional to the deviation of the column height from an "ideal" height; this ideal height is a simple representation of a natural random packing of the grains in a column, so that columns which are taller (shorter) than ideal would be relatively loosely (closely) packed, and driven to consolidate (dilate) when the sandpile is perturbed externally.

We generalize this here to include driving in the form of rotation and deposition. Thus

$$f_i = k_1(h_i - iaS_0) + k_2(h_i - h_{i-1} - aS_0), i \neq 1 \quad (1)$$

where h_i are the column heights, k_1 and k_2 are constants, a is the lattice spacing, and iaS_0 is the ideal height of column i. We note that

- the first term, which depends on the absolute height of the sandpile, corresponds to a force that drives column compression or expansion towards the ideal height. Since we normally deal with columns which are more dilated than their normal height, we will henceforth talk principally about column compression;
- the second term is the usual term driving surface flow, which depends on local slope, or height differences; the offset of S_0 is the ideal slope from which differences are measured.

Equation (1) suggests the redefinition of heights $az_i \equiv h_i - iaS_0$ which leads to the dimensionless representation:

$$fi = (k_1/k_2)z_i + (z_i - z_{i-1}), i \neq 1 \qquad (2)$$

When column i is subject to a force greater than or equal to the threshold force f_{th}, the height changes are as follows:

$$z_i \to z_i - \delta z,$$
$$z_{i-1} \to z_{i-1} + \delta z', i \neq 1. \qquad (3)$$

The column-height changes that correspond to a typical relaxation event are illustrated in Fig. 1. Thus, the height δz removed from column i due to a local driving force that exceeds the threshold force, leads to a flow of grains with total height increment $\delta z'$ from column i onto column $i-1$, and a consolidation of the grains in column i which reduces the column height by $(\delta z - \delta z')$. This reorganization clearly expresses the action of two relaxation mechanisms. The decomposition of the relaxation, that is, a particular choice for δz and $\delta z'$, is discussed below; such a coupling between the column heights may lead to the propagation of instabilities along the sandpile and hence to avalanches. Avalanches have also been discussed widely in the context of earthquakes [7], and one aim of this work is to draw analogies between sandpiles and earthquakes. We choose a discrete model of earthquakes, put forward by Nakanishi [8], to highlight those features which are common to our sandpile model and earthquake models. We choose the force relaxation function to be [8]:

$$f_i - f'_i = f_i - f_{th}(((2 - \delta f)^2/\alpha)/((f_i - f_{th})/f_{th} + (2 - \delta f)/\alpha) - 1) \qquad (4)$$

where f_i and f'_i are the granular driving forces on column i before and after a relaxation event. This function has a minimum value $(= \delta f f_{th})$ when $f_i = f_{th}$,

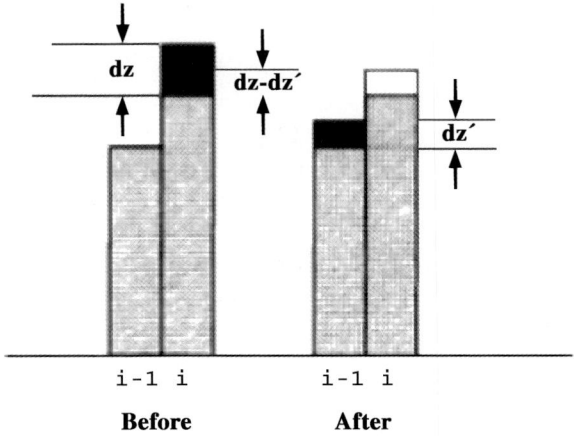

Fig. 1. A schematic diagram showing the column height changes that describe a single relaxation event in the CML sandpile model

and increases monotonically with increasing f_i; this form models the stick-slip friction associated with sandpiles and earthquakes. For driving forces f_i below the threshold force f_{th}, nothing happens but for forces that exceed this threshold, the size of relaxation events increases in proportion to the excess force. Accordingly, the minimum value of the function (1) is known as the minimum event size and its initial rate of increase, $\alpha = d(f_i - f_i')/d(f_i - f_{th})$ at $f_i = f_{th}$, is called the *amplification* [8]. In our sandpile model, amplification refers to the phenomenon whereby grains collide with each other during an avalanche so that their inertial motion contributes to its buildup; thus α is an expression of grain inertia. Using (2) we can rewrite the map in terms of the driving forces as [2]:

$$f_i - f_i' = 2\delta z/\Delta ,$$
$$\delta z' = (\delta z)/(1 + k_1/k_2) \quad (5)$$

$$f_{i-1}' - f_i = f_{i+1}' - f_{i+1} = -\Delta(f_i' - f_i)/2, i \neq 1 \text{ or } L . \quad (6)$$

In both sandpile and earthquake models the amount of redistributed force at a relaxation event is governed by the parameter $\Delta = 2(1 + k_1/k_2)/[1 + (1 + k_1/k_2)^2]$; since the undistributed force is "dissipated", $(1 - \Delta)$ becomes the dissipation coefficient [2]. Note, however, that in our sandpile model, this dissipation is linked to nonconservation of the sandpile volume arising from the compression of columns towards their ideal heights; it is therefore linked to the phenomenon of granular *consolidation*.

We use boundary conditions appropriate to a sandpile in a rotating cylinder; open at $i = 0$ and closed at $i = L$. Equations (4) and (6) give a prescription for the evolution of forces $f_i, i = 1, L$, so that any forces in excess of the threshold force are relaxed according to (6) and redistributed according to (4). Alternatively this sequence of events can be followed in terms of the redistribution of column heights according to (3) and (5). The prescription (4)–(6), classifies the model as a local and unlimited sandpile in the framework given by Kadanoff et al. [9]. We will show below that for all $\Delta \neq 1$, the largest part of the volume change during relaxation occurs as a result of consolidation; the quantity of interest is thus the difference between the old and new configurations, rather than the mass exiting the sandpile [10]. A measure of this change is the quantity $\ln M = \ln \Sigma_i(f_i - f_i') = \ln[\Sigma_i((k_1/k_2)(z_i - z_i') + z_L - z_L']$ where z_i' is the height of column i immediately after a relaxation event; this quantity is the analogue of the event magnitude in earthquake models [7, 8]. We will discuss the variation of this quantity as a function of model parameters in the next subsection; in particular, we will compare the response of our rotated sandpile model to that of the same model subjected to random deposition.

2.2 Results

Rotated Sandpile

For a sandpile in a rotating cylinder, the tilting of the sandpile results in changes of slope over the complete surface in a continuous manner (in contrast to the case of random deposition where the slopes change locally and discontinuously at a deposition event [10]). Our coupled map lattice (CML) model is driven continuously; from a configuration in which all forces f_i are less than the threshold force, elements of height, z_i^+, are added onto each column with

$$z_i^+ = i(f_{th} - f_j)/(1 + jk_1/k_2), i = 1, L ,\qquad(7)$$

where $f_j = max(f_i)$. This transformation describes the effect of rotating the base of the sandpile with a constant angular speed until a threshold force arises at column j. (We note that this is distinct from the external driving force in the earthquake model [8], which would correspond, in a sandpile model, to the uniform addition of height elements across the surface.) The response to the tilting is, as described above, a flow of particles down the slope as well as reorganisation of particles within the sandpile. The predominant effect of our model is to cause volume changes by consolidation, rather than to generate surface flow. Using the relation between force and column height (2), and integrating from the left, we can construct the shape of a critical sandpile which has driving forces equal to the threshold force on all of its columns; in terms of the variable $\zeta \equiv (1 + \sqrt{1 - \Delta^2})/\Delta$, the critical sandpile has column heights z_i^c given by

$$z_i^c = f_{th}[1 - \zeta^{-1} \exp{(1 - \zeta)(i - 1)a}]/(\zeta - 1), \Delta < 1 . \qquad(8)$$

This shows that for all $\Delta < 1$, the critical sandpile starts at $i = 1$ with a slope greater than S_0, and subsequently the slope decreases until it becomes steady at S_0 for $i \gg 1$, where the constant deviation of the column heights from their ideal values is given by $f_{th}\Delta/(1 - \Delta + \sqrt{1 - \Delta^2})$ (Fig. 2).

We have verified by simulation that the corresponding state is an attractor. We emphasize that this sandpile shape is quite distinct from that generated by standard lattice sandpiles, and is close to the S-shaped sandpile observed in rotating cylinder experiments [11]. This description of the critical sandpile leads to the assertion that our model is one in which reorganization of grains predominates over surface flow, as follows: From (1), it is clear that any value of steady slope which differs from S_0 would lead to a linear growth in the first term – this is therefore unstable. Thus stability enforces solutions where the average slope, for $i \gg 1$, is S_0. For a truly critical pile the second term in (1) is identically zero for $i \gg 1$ so that, except in the small i region, the threshold force that drives relaxation arises solely from the compressive component. The same reasoning is true on average for model sandpiles near criticality and makes for threshold forces which are predominantly compressive (although

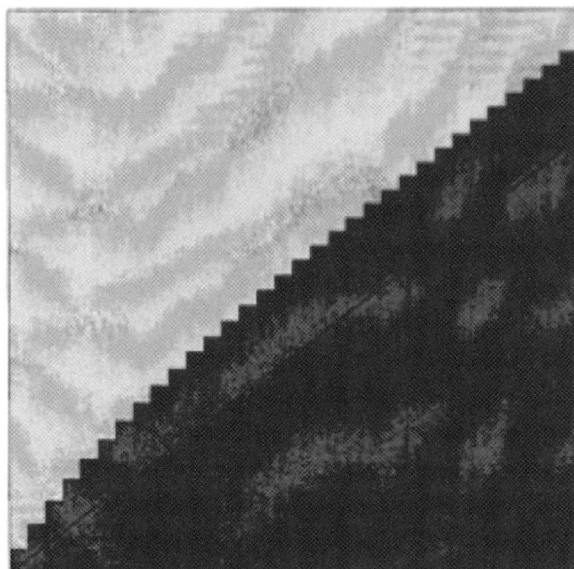

Fig. 2. The shape of a critical CML model sandpile with $L = 32$ and $\Delta = 0.95$. The line indicates the "ideal" column heights

surface flow events arising from local slope inhomogeneities also exist, particularly near the bottom of the sandpile). This predominance of the compressive term then leads to column height changes that are typically $\sim f_{th}\Delta\delta f$ and, in the parameter range under consideration, are small compared to the "column grain size" $S_0 a$ (the average step size in a lattice slope with gradient S_0 and column width a). In other words, *typical events are likely to be due to internal rearrangements generating volume changes that are small fractions of "grain sizes"*, and they can be visualized as the slow rearrangement of grains within their clusters; this is in contrast to the surface flow events in standard lattice models [1] where entire grains flow down the surface independently of their clusters [12]. This preponderance of reorganizational events over large surface avalanches is consistent with the dynamics of sand in a slowly rotating cylinder [11, 12].

The steady state response of the driven sandpile may be represented as a sequence of events, each of which corresponds to a set of column height changes. Each avalanche is considered to be instantaneous, so that the temporal separation of consecutive events is defined by the driving force (7). We choose a timescale in which the first column has unit growth rate, and begin each simulation at $t = 0$ with a sandpile containing columns which have small and random deviations from their natural heights; also, we set $a = S_0 = 1$ to fix the arbitrary horizontal and vertical length scales, and we fix $f_{th} = 1$ to

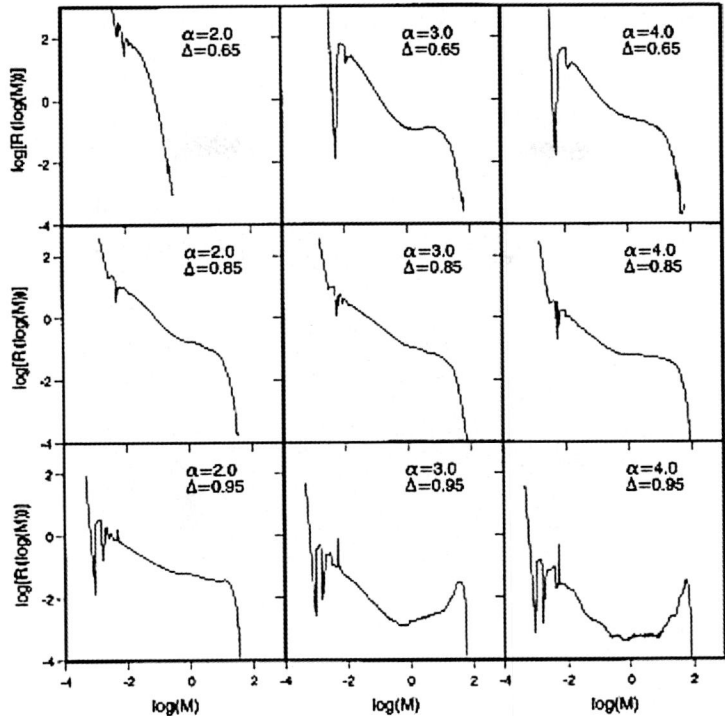

Fig. 3. A logarithmic plot of the distribution function of event sizes, $R\mathrm{Log}\,(M)$, for 10^7 consecutive events in a CML model sandpile with $L = 512$ and parameter values $\delta f = 0.01, \alpha = 2, 3, 4$, and $\Delta = 0.6, 0.85, 0.95$

define units of "force". The dynamics of events do not depend explicitly on these choices.

In Fig. 3, we plot the distribution function per unit time and length $R\log(M)$ against $\log(M)$, for sandpiles with size $L = 512$ and parameter values $\delta f = 0.01, \alpha = 2, 3, 4$, and $\Delta = 0.6, 0.85, 0.95$. We note in particular the small value of δf, and mention that our results are qualitatively unaffected by choosing δf in the range $0.001 < \delta f < 0.1$; given its interpretation in terms of the smallest event size, this reflects our choice of the quasistatic regime, where small cooperative internal rearrangements predominate over large single-particle motions. The distribution functions in Fig. 3 indicate a scaling behavior in the region of small magnitude events and, for larger magnitudes, frequencies that are larger than would be expected from an extension of the same power law. The phase diagram in the $\Delta - \alpha$ plane indicates qualitatively distinct behavior for low-inertia, strongly consolidating (low α and Δ) systems where the magnitude distribution function has a single peak, and high-inertia, weakly consolidating (high α and Δ) systems for which the magnitude distribution has a clearly distinct second peak.

These results are in accord with the Nakanishi earthquake automaton [8]; however, their interpretation in the context of our sandpile model is quite novel and distinct (see below). In Fig. 4, we show the relative column heights z_i plotted against the distance of the column from the axis of rotation. The solid line denotes the configuration before, and the dotted line that after, a large avalanche; we note that a section of the sandpile has "slipped" quite considerably during the event.

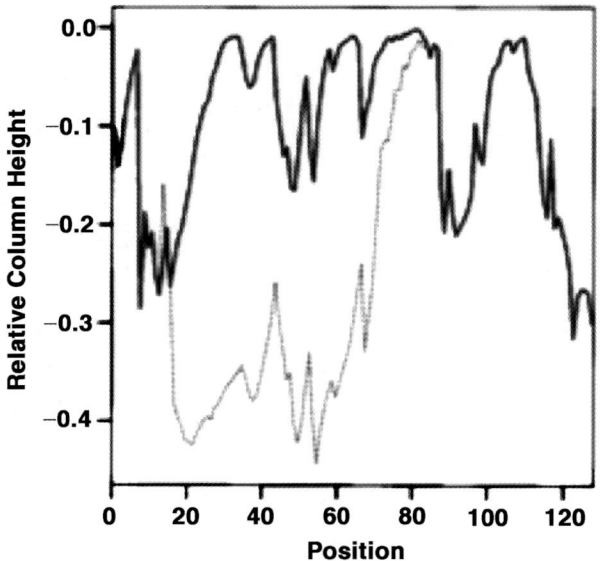

Fig. 4. A plot of column heights, relative to their critical heights, for a CML model sandpile with $L = 128$ and parameter values $\delta f = 0.01$, $\alpha = 3$, and $\Delta = 0.85$. The full (*dotted*) line shows the configuration before (after) a large event

The rotation of the sandpile causes a uniform increase of the local slopes and a preferential increase of absolute column heights in the upper region of the sandpile. The sandpile is thus driven towards its critical shape where relaxation events are triggered locally. These events will be localized ("small") or cooperative ("large") depending on α and Δ. For strongly consolidating systems with small amplification α, a great deal of excess volume is lost via consolidation, and the effect of surface granular flow is small; in these circumstances, the propagation and buildup of an instability is unlikely, so that events are in general localized, uncorrelated and hence small. This leads to the appearance of the single peak in the distributions in the upper left corner of Fig. 3. Alternatively, for weakly consolidating systems with large amplification, surface flow is large, dilatancy predominates, and there are many space-wasting configurations; this situation favours (see Fig. 11) the appearance of large avalanches, which are manifested by the appearance of a second peak in the distributions at the lower right corner of Fig. 3. In principle, these large avalanches would be halted by strong configurational inhomogeneities such as a "dip" on the surface, where the local driving force (1) is far below threshold; our simulations show that such configurations are rare in sandpiles that are close to criticality, and this leads to the appearance of the special scale for large avalanches.

Figure 5 shows a time series of avalanche locations that occur for a model sandpile in the two-peak region. The large events are almost periodic and each one is preceded by many small precursor events; this is in accord with previous work [10, 13, 14] on sandpiles as well as earthquakes [8]. In addition it is apparent that large avalanches tend to occur repeatedly at or around the same regions of the sandpile, whose location changes only very slowly compared to the interval between the large avalanches; these correlations in both the positions and the times of large events are often referred to as "memory"

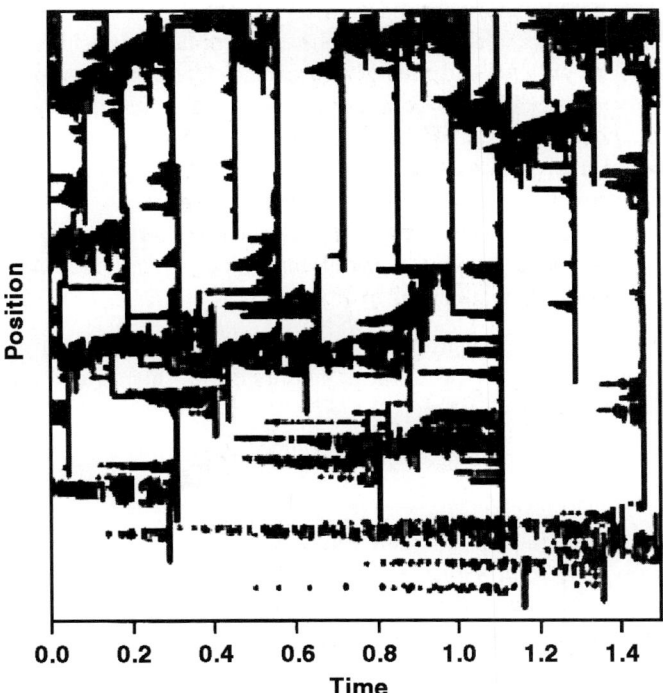

Fig. 5. A plot showing the locations of relaxation events (changes in column heights), that occur during an interval of length 1.5 which begins at $t = 10^4$, for a CML model sandpile with $L = 256$ and parameter values $\delta f = 0.01$, $\alpha = 3$, and $\Delta = 0.85$

[11, 12, 15]. We have previously given a qualitative explanation for memory in sandpiles [13, 14] which, in the context of our present model, becomes more quantitative along the following lines:

Since the relaxation function (4) is a smooth function of the excess force $f_i - f_{th}$ (which depends on the configuration of the pile), the propagation of a large event across the sandpile causes a smoothing of small configurational inhomogeneities. In turn, this reduces the probability that a new event will be initiated in the same region of the sandpile, until the whole region is again driven towards its critical configuration. In contrast, large configurational inhomogeneities such as large dips or surface voids (which are able to halt the progress of large avalanches) remain as significant features (often slightly weakened and displaced) in the sandpile configuration following a large event; these can then have an effect on the spatial extent of subsequent events. Thus, for those regions of phase space (large α and Δ) where granular inertia plays a large role in amplifying avalanches, and where consolidation is not effective, our model shows the existence of quasiperiodic large events which repeatedly disrupt the same regions of the system, thus manifesting configurational memory. On the other hand, when the inertial effect is weak and when the void space is honed down via consolidation (small α and Δ), the predominant effect is that of small uncorrelated events which do not leave a persistent mark on the sandpile configuration, so that no memory effects are observed. For moderate values of α and Δ, both small and large events will be seen (Fig. 5); note also that for the specific case of the rotated sandpile, these large events occur predominantly towards the top of the pile, where structural reorganization brought on by rotation is most effective.

The shape of the critical sandpile, which we discussed earlier, leads to another interesting feature, namely, an intrinsic size dependence. As mentioned before, the shape is characterized by

- the length of the decreasing slope region and
- the constant deviation of column heights from their ideal values in the steady slope region which follows (Fig. 2).

The length Λ of the increasing slope region at small i has a finite extent given by

$$\Lambda \sim (1 + \sqrt{1 - \Delta^2}/(1 - \Delta + \sqrt{1 - \Delta^2}) \tag{9}$$

and this can be made an arbitrary fraction of the sandpile by an appropriate choice of system size. An intrinsic size dependence resulting from the physics of competition between the surface and bulk relaxation processes is of interest because it has been observed in sandpile experiments [16, 17]; however, we will defer the full effects of this to future work, and for the present limit our results to sandpiles whose length $L \gg \Lambda$.

As mentioned in the introductory section, we would expect few events to result in mass exiting the pile, as our model is one in which internal volume reorganizations dominate surface flow. Thus, mass will exit a pile either via

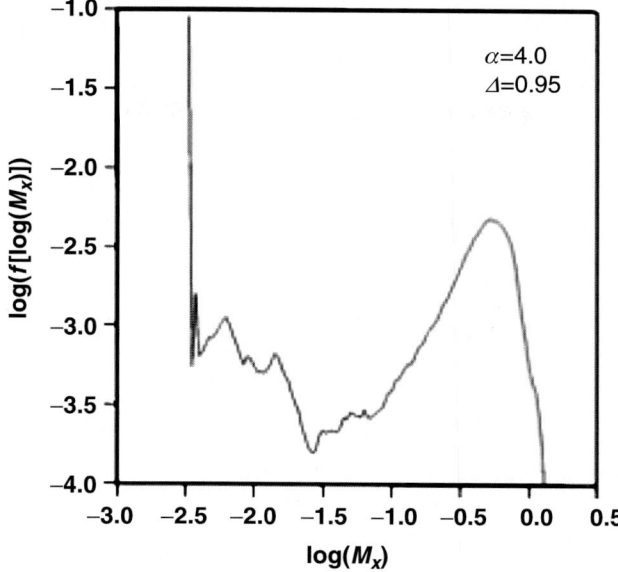

Fig. 6. A logarithmic plot of the distribution function of exit mass sizes, $f\log(M_x)$, for a sandpile of size $L = 128$ with $\delta f = 0.01, \alpha = 4$ and $\Delta = 0.95$ for 10^7 consecutive events. The exit mass M_x is the sum of height increments $\delta z'$ that topple from the first column during an event

the propagation of large events (which occur for α and Δ large) or if surface flow is significant (typically events initiated in the increasing slope region Λ).

Figure 6 shows the logarithm of the exit mass size distribution function, $f\log(M_x)$, for a sandpile of size $L = 128$ with $\delta f = 0.01, \alpha = 4$, and $\Delta = 0.95$. While the absolute magnitudes of the event sizes are suppressed in comparison to Fig. 3, we see the two-peak behavior consistent with the corresponding event size distribution function. The large second peak indicates that a significant proportion of the exit mass is due to large events referred to above; also, we have checked that this is the only part of the distribution that survives for larger system sizes, in agreement with the length dependence above. Finally, we mention that the two-peak behavior obtained for the exit mass distribution is in accord with behavior we have observed previously in a cellular automaton model of a reorganizing sandpile [10], thus confirming our conclusion [12, 15] that the presence of reorganization as a "second" mechanism of relaxation causes the breakdown of scale invariance observed in simpler models [1].

Sandpile Driven by Random Deposition

The perturbation more usually encountered in sandpiles is random deposition. We may replace the organized addition (7) with the random sequential addition of height elements, $z_i^+ = z_g$, onto columns $i = 1 \ldots L$. When the added elements are small compared to the minimum event size, so that $z_g \ll f_{th}\delta f$, random addition is statistically equivalent to uniform addition, which was the case considered in the context of earthquakes [8].

The distribution of event sizes, shown by the full lines (corresponding to $z_g = 0.01$) in Fig. 7, is then not markedly distinct from that shown in Fig. 3 for the rotational driving force, and both are similar to the distributions presented in [8]. In most of the parameter ranges we consider, the event size distribution functions are size-independent, indicating that intrinsic properties of the sandpile are responsible for their dominant features, which include a second peak representing a preferred scale for large avalanches.

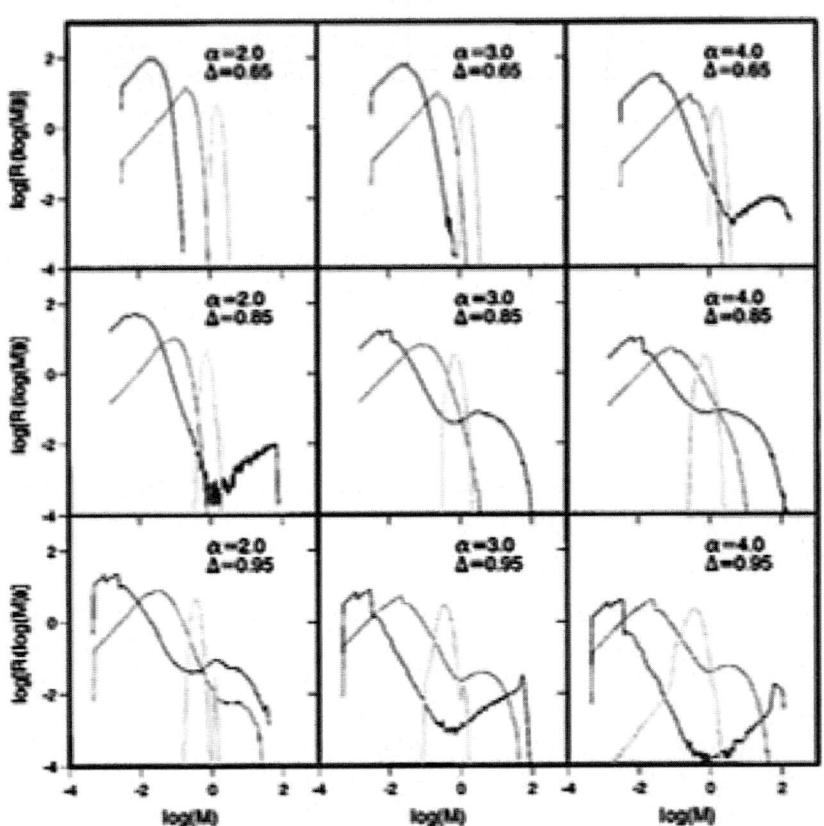

Fig. 7. A logarithmic plot of the distribution function of event sizes, $RLogM$ (*full lines*) for 10^7 consecutive events in a randomly driven CML model sandpile with $L = 512$ and parameter values $\delta f = 0.01, \alpha = 2, 3, 4, \Delta = 0.6, 0.85, 0.95$ and $z_g = 0.01$. Faint lines show the corresponding distribution functions for $zg = 0.1$ and 1.0

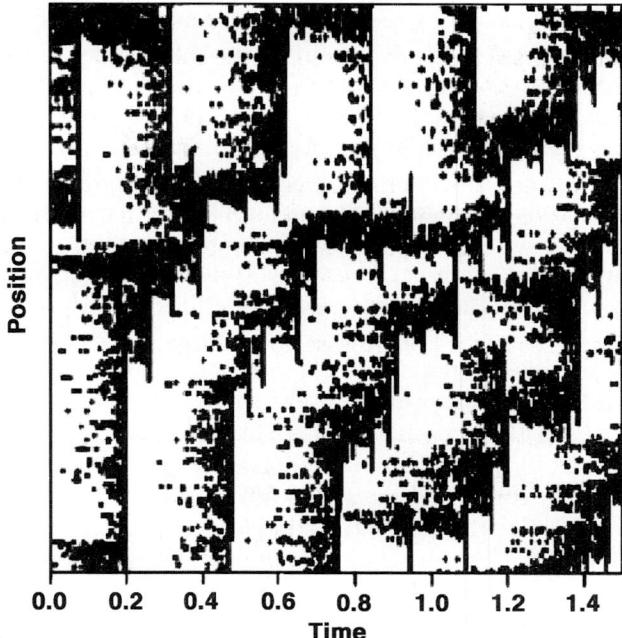

Fig. 8. A plot showing the locations of relaxation events (changes in column heights), that occur during an interval of length 1.5 which begins at $t - 10^3$, for a randomly driven CML model sandpile with $L = 256$ and parameter values $\delta f = 0.01, \alpha = 3, \Delta = 0.85$, and $z_g = 0.01$.

Figure 8 shows the corresponding time series of event locations – note that random driving leads to events which are relatively evenly spread over the sandpile and to repeated large events, each preceded by their precursor small events. Note also that after the passage of a large event and/or catastrophe over a region, there is an interval before events are generated in response to the deposition; this underlines our picture referred to earlier, whereby large events leave their signatures on the landscape in the form of dips, for instance. These configurational sinks are associated in our model with forces well below threshold, so that grains deposited on them will, for a while, not cause any relaxation events until the appropriate thresholds are reached. If we now increase the size of the incoming height elements so that they are larger than the minimum event size but are still small compared to the column grain size (i.e. $f_{th}\delta f < z_g < aS_0$), there are two direct consequences. First, the driving force leads to local column height fluctuations $\sim z_g$, so that the surface is no longer smooth; these height fluctuations play the role of additional random barriers which impede the growth of avalanches, thus reducing the probability for large, extended events. Second, given that the added height elements are much larger than the minimum event size, their ability to generate small events is also reduced; the number of small events therefore also decreases. The size distributions for this case consequently have a domed shape with apparently two scaling regions. This case is illustrated by one of the fainter lines in Fig. 7 (corresponding to $z_g = 0.1$). Note that for large α and Δ, the large events, being more persistent, are able to overcome the configurational barriers ($\sim z_g$ fluctuations in column heights) referred to above and we still see a second peak indicating the continued presence of a preferred avalanche size.

As the perturbation strength becomes even stronger, $(z_g > aS_0)$ so that column height fluctuations are comparable with the column grain size, there are frequent dips on the landscape, which can act as configurational traps for large events. All correlations between events begin to be destroyed and relaxation takes place locally, giving a narrow range of event sizes determined only by the size of the deposited grains. This situation is illustrated by the second faint line in Fig. 7, which corresponds to $z_g = 1.0$.

Finally, and for completeness, we link up with the familiar scaling behavior of lattice sandpiles [1]; as mentioned before, scale invariance is recovered when the driving force is proportional to slope differences alone and no longer contains the second mechanism of compression and/or reorganization, so that $k_1 = 0$ and $\Delta = 1$. In order, more specifically, to match up with the local and limited $n_f = 2$ model of Kadanoff et al. [9], we start with the randomly driven model and

- set $k_1 = 0$ and $\Delta = 1$ in 1,
- set $z_g = 1$,
- choose a relaxation function $f_i - f_i' = f_{th} \delta f$; this is a constant independent of f_i for $f_i > f_{th}$, and in particular contains no amplification;
- set $\delta f = 4.0$, so that each threshold force causes a minimum of two "grains" to fall onto the next column at every event, so that $n_f = 2$

This special case of our CML model is then identical with the scale-invariant model of Kadanoff et al. [9].

Figure 9 shows the smooth (scaling) exit mass size distribution in the limit of no dissipation, and the corresponding spatial distribution of scaling events is shown in Fig. 10. Uncorrelated events are observed over many sizes indicating a return to scale invariance.

2.3 Discussion

Before discussing our main results, we review the motivation behind this work. The earliest cellular automaton models of sandpiles [1, 9], which were put forward to illustrate self-organized criticality, relied on simple pictures of grains flowing down sloping surfaces. While they were entirely adequate in their aim of acting as paradigms of SOC, their predictions were at variance with experiments on real sandpiles [16, 17]; this anomaly stimulated a great deal of work on the dynamics of *real* granular systems [11, 15]. The importance of a

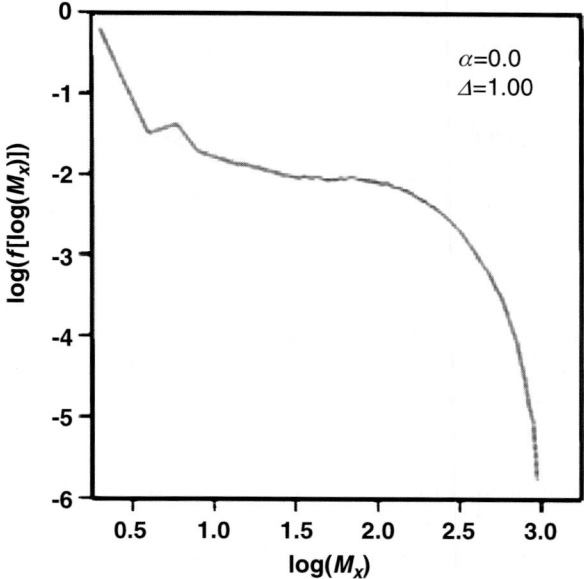

Fig. 9. A logarithmic plot of the distribution function of exit mass sizes, $f \operatorname{Log}(M_x)$, for 10^7 consecutive events in a randomly driven CML model sandpile with $L = 512$ and parameter values $\delta f = 4, \Delta = 1$, and $z_g = 1$

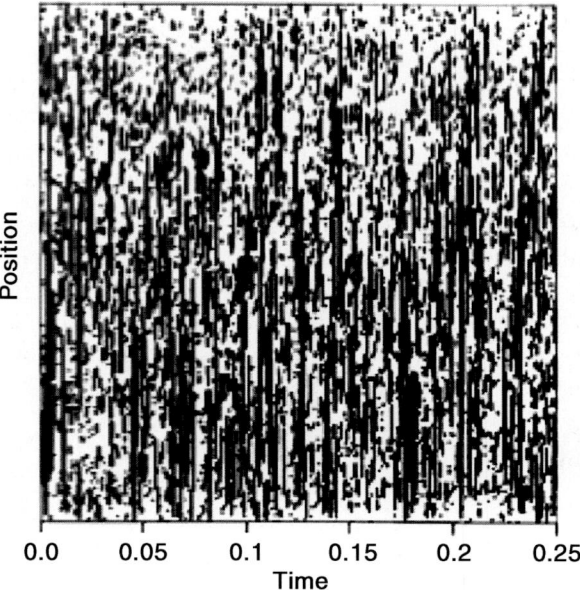

Fig. 10. A plot showing the locations of relaxation events (changes in column heights) that occur during an interval of length 0.25 which begins at $t = 10^3$, for a randomly driven CML model sandpile with $L = 256$ and parameter values $\delta f = 4, \Delta = 1$, and $z_g = 1$

coupling between freely flowing surface grains and relatively immobile clusters in the bulk material soon became apparent; this idea was put forward in a model [18] encapsulating the competition between independent-particle and collective dynamics in a sandpile subjected to external perturbation, where mobile grains and clusters were respectively responsible for the two dynamical modes. A body of work proceeded to examine the relative effects of the two relaxation mechanisms on the material properties of the system [3, 4, 10, 12, 19] – this yielded results in agreement with experiment, thus validating the idea of grain-cluster coupling on which they were based.

Here we have provided a decorated lattice model to represent grain and cluster couplings in a sandpile. The coupled map lattice model has two important parameters; α (amplification) which determines the strength of surface flow or grain inertia effects, and Δ, which is related to internal reorganization such that $1 - \Delta$ represents consolidation.

Our main result is that for large α and Δ, there is a preferred size for large avalanches, which is manifested as a second peak in the distribution of event sizes (Fig. 3). In terms of a simple picture this is because, for large α and Δ, grains have enough inertia to speed past available traps, and there is consequently a large amount of *dilatancy* – i.e. unrelaxed excess volume – on the surface. This excess volume can be visualized, for instance, as a precariously balanced cluster (Fig. 11); the oncoming (dark) grain will knock off the shaded grains when it hits them, unleashing a large avalanche. For small α and Δ we see, by contrast, mainly small events leading to a single peak in Fig. 3; we visualize this by imagining slowly moving grains (low inertia) drifting down the surface, locking into voids and dissipating excess volume efficiently. This qualitative picture also indicates that initiated avalanches will be terminated relatively rapidly, leading to many small events.

The large avalanches mentioned above are quasiperiodic and tend to occur repeatedly around the same regions of the surface (Fig. 5), providing an important representation of configurational memory. In terms of the simple

Fig. 11. A schematic diagram illustrating the mechanism for large-avalanche formation. When Δ is large, there is a great deal of undissipated volume in the cluster, resulting in the upper (*shaded*) grains being unstable to small perturbations. When α is large, the black grain hitting the cluster has large inertia so that a large avalanche results when it dislodges the shaded grains

picture above, this is because regions of the sandpile which look like Fig. 11 are wiped clean by the effect of the large avalanche, so that further deposition or rotation has no effect for a while. However, the effect of large α and Δ mean that once again, excess void space will be created around the same region after high-inertia grains flow down the surface; this will be the case after a number of small events have occurred (Fig. 5). These spatial and temporal correlations result in a quasiperiodic repetition of large avalanches around the same regions of the sandpile, resulting in configurational memory [13, 14].

We have examined the response of the CML model to random deposition, with particular reference to the size z_g of the deposited grains. Three distinct regimes are observed:

- When z_g is of the order of a "minimum event", i.e., it is comparable to the smallest fractional change in volume caused by a reorganizing grain, the response is similar to that of rotation (Fig. 7).
- When z_g is intermediate between the minimum event size and the typical column grain size of the sandpile, reorganizations of grains corresponding to the smallest volume changes are ruled out; on the other hand, there are moderately sized barriers ($\sim z_g$) across the landscape impeding the progress of large events. The appropriate size distribution in Fig. 7 has, consequently, a shape which lacks the extreme small and large events of the previous case.
- When z_g is larger than the column grain size of the sandpile, large configurational barriers are generated by deposition, and these act as traps for large events. Correlations between events are destroyed, leading to a narrow distribution of event sizes corresponding to local responses to deposition.

The central analogy between our work and work on spring-block models of earthquakes [7, 8] is that small events build up configurational stress (in the sandpile context this is via landscapes which look like Fig. 11) which then leads, quasiperiodically, to the large events able to release the stress. While we have drawn this analogy in previous work [12, 13, 14], the work presented here [2] is a quantification of this analogy with an explicit model. Thus, while our model is necessarily similar to earthquake models in some respects, such as the choice of Δ governing the force redistribution, our interpretation in terms of fluctuating column heights (1, 3) is actually quite distinct. The relaxation process (6) is symmetric and is, in that sense, similar to that presented in [8]. However the CML sandpile model has a preferred direction of flow (down the slope) created by the boundary conditions so that, on average, $z_i > z_{i-1}$; this predominantly leads to a situation for which $f_{i-1} > f_{i+1}$. Thus, even though the relaxation process in (6) is symmetric, the propagation of relaxation events in our sandpile model [2] is biased in the direction down the slope. The effect of this asymmetry has far-reaching consequences[1], one of which is

---

[1] The effect of symmetry involving the difference between the presence and absence of a preferred direction of flow, has also been explored in the analytical version

that it is possible to solve a boundary value problem in our sandpile model [2] which cannot be solved in the spring-block model [8]; this leads to the explicit shape of the critical pile [17]. Another crucially important consequence of this difference is that in the sandpile model, most of the flow comes from the compression term, whereas this is not the case for the earthquake model [8]. This feature, with a change of driving force, enables us to model the experimentally important situation of a sandpile in a slowly rotating cylinder.

We note that the limit $\Delta = 1$ is a special case; this corresponds to the situation with no reorganization ($k_1 = 0$) and describes a sandpile which is constantly at an ideal density. The granular driving force no longer has a compressive component and, as for standard sandpile models [1], depends only on height differences. The approach to this limit is also of interest, involving a discontinuous transition to a regime in which the critical sandpile has a constant slope ($S_0 + f_{th}$). The neighborhood of the limit $\Delta \sim 1$ is a region of very weak dissipation, and, as has been seen in other deterministic nonlinear dynamical systems [20, 21], could well be characterized by complex periodic motion at large times; this has been argued to be especially relevant to models with periodic boundaries [22]. We have investigated this issue and find that, for the regions of parameter space explored here, we do not see periodic features in sequences containing up to 5×10^7 events.

To conclude, we have designed a coupled-map lattice model of a sandpile which includes surface flow and internal reorganization as the two principal mechanisms of relaxation, focusing on the situation where reorganization dominates flow. We have, by an appropriate choice of driving forces, examined its response to rotation, as well as to deposition. In the former case, we have presented the event size distribution corresponding to different regions of parameter space, and explained our results in terms of the inertia of the flowing grains and the reorganization of clusters. In the latter case, we have analyzed the event size distribution function in terms of the size of the deposited grains. We hope that these results will stimulate new investigations, both experimental and theoretical, which include an explicit examination of the couplings between internal and surface degrees of freedom for flowing sandpiles.

3 Coupled Continuum Equations: The Dynamics of Sandpile Surfaces

The dynamics of sandpiles have intrigued researchers in physics over recent years [12, 15] with a great deal of effort being devoted to the development of techniques involving for instance cellular automata [1, 23], continuum equations [3, 24] and Monte Carlo schemes [25] to investigate this very complex subject. However, what have often been lost sight of in all this complexity

---

[3] of this problem; the two different symmetries are characterized by different critical exponents, underlining the importance of symmetry in this problem.

are some of the extremely simple phenomena that are exhibited by granular media which still remain unexplained.

One such phenomenon is that of the smoothing of a sandpile surface after the propagation of an avalanche [4, 19]. It is clear what happens physically: an avalanche provides a means of shaving off roughness from the surface of a sandpile by transferring grains from bumps to available voids [12, 23], and thus leaves in its wake a smoother surface. In particular what has not attracted enough attention in the literature is the qualitative difference between the situations which obtain when sandpiles exhibit intermittent and continuous avalanches [26]. In this section, we examine both the latter situations, via distinct models of sandpile surfaces.

A particular experimental paradigm that we choose, for reasons of continuity with the first section, is that of sand in rotating cylinders [17, 27]. In the case when sand is rotated slowly in a cylinder, intermittent avalanching is observed; thus, sand accumulates in part of the cylinder to beyond its angle of repose [28], and is then released via an avalanche process across the slope. This happens intermittently, since the rotation speed is less than the characteristic time between avalanches. By contrast, when the rotation speed exceeds the time between avalanches, we see continuous avalanching on the sandpile surface. Though this phenomenon has been observed [29] and analysed physically [26] in terms of avalanche statistics, we are not aware of measurements which measure the characteristics of the resulting surface in terms of its smoothness or otherwise.

What we focus on here is precisely this aspect, and make predictions which we hope will be tested experimentally. In order to discuss this, we review the notion that granular dynamics is well described by the competition between the dynamics of grains moving independently of each other and that of their collective motion within clusters [12]. A convenient way of representing this is via coupled continuum equations with a specific coupling between mobile grains ρ and clusters h on the surface of a sandpile [3]. In the regime of intermittent avalanching, we expect that the interface will be the one defined by the "bare" surface, i.e. the one defined by the relatively immobile clusters across which grains flow intermittently. This then implies that the roughening characteristics of the h profile should be examined. The simplest of the three models we discuss in this paper (an exactly solvable model referred to hereafter as Case A) as well as the most complex one (referred to hereafter as Case C) treat this situation, where we obtain in both cases an asymptotic smoothing behaviour in h. When on the other hand, there is continuous avalanching, the flowing grains provide an effective film across the bare surface and it is therefore the species ρ which should be analysed for spatial and temporal roughening. In the model hereafter referred to as Case B we look at this situation, and obtain the surprising result of a gradual crossover between purely diffusive behaviour and hypersmooth behaviour. In general, the complexity of sandpile dynamics leads us to equations which are coupled, nonlinear and noisy: these equations present challenges to the theoretical physicist in more ways than the obvious ones to do with their detailed analysis and/or their numerical solutions. In particular, our analysis of Case C reveals the presence of hidden length scales whose existence was suspected analytically, but not demonstrated numerically in earlier work [3].

The normal procedure for probing temporal and spatial roughening in interface problems is to determine the asymptotic behaviour of the interfacial width with respect to time and space, via the single Fourier transform. Here only one of the variables, (x,t) is integrated over in Fourier space, and appropriate scaling relations are invoked to determine the critical exponents which govern this behaviour. However, it turns out that this leads to ambiguities for those classes of problems where there is an absence of simple scaling, or to be more specific, where multiple length scales exist – this is further discussed in [30]. In such cases it is the double Fourier transform (where *both* time and space are integrated over) that yields the correct results. This point is illustrated by Case A, an exactly solvable model that we introduce.

In order to make some of these ideas more concrete, we now review some general facts about rough interfaces [31]. Three critical exponents, α, β, and z, characterise the spatial and temporal scaling behaviour of a rough interface. They are conveniently defined by considering the (connected) two-point correlation function of the heights

$$S(x-x',t-t') = \langle h(x,t)h(x',t')\rangle - \langle h(x,t)\rangle\langle h(x',t')\rangle . \qquad (10)$$

We have

$$S(x,0) \sim |x|^{2\alpha} \quad (|x|\to\infty) \quad \text{and} \quad S(0,t) \sim |t|^{2\beta} \quad (|t|\to\infty) ,$$

and more generally

$$S(x,t) \approx |x|^{2\alpha} F\bigl(|t|/|x|^z\bigr)$$

in the whole long-distance scaling regime (x and t large). The scaling function F is universal in the usual sense; α and $z = \alpha/\beta$ are respectively referred to as the roughness exponent and the dynamical exponent of the problem. In addition, we have for the full structure factor which is the double Fourier transform $S(k,\omega)$

$$S(k,\omega) \sim \omega^{-1} k^{-1-2\alpha} \Phi(\omega/k^z)$$

which gives in the limit of small k and ω,

$$S(k,\omega=0) \sim k^{-1-2\alpha-z} \quad (k\to 0) \quad \text{and} \quad S(k=0,\omega) \sim \omega^{-1-2\beta-1/z} \quad (\omega\to 0) \qquad (11)$$

The scaling relations for the corresponding single Fourier transforms are

$$S(k,t=0) \sim k^{-1-2\alpha} \quad (k\to 0) \quad \text{and} \quad S(x=0,\omega) \sim \omega^{-1-2\beta} \quad (\omega\to 0) \qquad (12)$$

In particular we note that the scaling relations for $S(k,\omega)$ (11) always involve the simultaneous presence of α and β, whereas those corresponding to $S(x,\omega)$ and $S(k,t)$ involve these exponents *individually*. Thus, in order to evaluate the double Fourier transforms, we need in each case information from the growing as well as the saturated interface (the former being necessary for β and the latter for α) whereas for the single Fourier transforms, we need only information from the saturated interface for $S(k,t=0)$ and information from the growing interface for $S(x=0,\omega)$. On the other hand, the information that we will get out of the double Fourier transform will provide a more unambiguous picture in the case where multiple length scales are present, something which cannot easily be obtained in every case with the single Fourier transform.

In the following subsections, we present, analyse and discuss the results of Cases A, B and C respectively. Finally, we reflect on the unifying features of these models, and make some educated guesses on the dynamical behaviour of real sandpile surfaces.

3.1 Case A: The Edwards-Wilkinson Equation with Flow

Our first model involves a pair of linear coupled equations, where the equation governing the evolution of clusters ("stuck" grains) h is closely related to the very well-known Edwards-Wilkinson (EW) model [32]. The equations are:

$$\frac{\partial h(x,t)}{\partial t} = D_h \nabla^2 h(x,t) + c\nabla h(x,t) + \eta(x,t) \qquad (13)$$

$$\frac{\partial \rho(x,t)}{\partial t} = D_\rho \nabla^2 \rho(x,t) - c\nabla h(x,t) \qquad (14)$$

where the first of the equations describes the height $h(x,t)$ of the sandpile surface at (x,t) measured from some mean $\langle h \rangle$, and is precisely the EW equation in the presence of the flow term $c\nabla h$. The second equation describes the evolution of flowing grains, where $\rho(x,t)$ is the local density of such grains at any point (x,t). As usual, the noise $\eta(x,t)$ is taken to be Gaussian so that:

$$\langle \eta(x,t)\eta(x',t') \rangle = \Delta^2 \delta(x-x')\delta(t-t').$$

with Δ the strength of the noise. Here, $\langle \cdots \rangle$ refers to an average over space as well as over noise.

For the purposes of analysis, we focus on the first of the two coupled equations (13) presented above,

$$\frac{\partial h}{\partial t} = D_h \nabla^2 h + c\nabla h + \eta(x,t)$$

noting that this equation is essentially decoupled from the second[2]. We note that this is entirely equivalent to the Edwards-Wilkinson equation [32] in a frame moving with velocity c

---

[2] This statement is, however, not true in reverse, which has implications discussed in [4].

$$x' = x + ct, \quad t' = t$$

and would on these grounds expect to find only the well-known EW exponents $\alpha = 0.5$ and $\beta = 0.25$ [32]. This would be verified by naive single Fourier transform analysis of (13) which yields these exponents via (12).

Equation (13) can be solved exactly as follows. The propagator $G(k,\omega)$ is

$$G_h(k,\omega) = (-i\omega + D_h k^2 + ikc)^{-1}$$

This can be used to evaluate the structure factor

$$S_h(k,\omega) = \frac{\langle h(k,\omega) h(k',\omega') \rangle}{\delta(k+k') \delta(\omega+\omega')}$$

which is the Fourier transform of the full correlation function $S_h(x-x', t-t')$ defined by (10). The solution for $S_h(k,\omega)$ so obtained is:

$$S_h(k,\omega) = \frac{\Delta^2}{(\omega - ck)^2 + D_h^2 k^4} \tag{15}$$

It is obvious from (15) that $S_h(k,\omega)$ does not show simple scaling. More explicitly, if we write

$$S_h^{-1}(k, \omega = 0) = \frac{\omega_0^2}{\Delta^2} \left(\frac{k}{k_0}\right)^2 \left[1 + \left(\frac{k}{k_0}\right)^2\right]$$

with $k_0 = c/D_h$, and $\omega_0 = c^2/D_h$, we see that there are two limiting cases:

- for $k \gg k_0$, $S_h^{-1}(k, \omega = 0) \sim k^4$; using again $S_h^{-1}(k=0,\omega) \sim \omega^2$, we obtain $\alpha_h = 1/2$ and $\beta_h = 1/4$, $z_h = 2$ via (11).
- for $k \ll k_0$, $S_h^{-1}(k, \omega = 0) \sim k^2$; using the fact that the limit $S_h^{-1}(k=0,\omega)$ is always ω^2, this is consistent with the set of exponents $\alpha_h = 0$, $\beta_h = 0$ and $z_h = 1$ via (11).

The first of these contains no surprises, being the normal EW fixed point [32], while the second represents a new, "smoothing" fixed point.

We now explain this smoothing fixed point via a simple physical picture. The competition between the two terms in (13) determines the nature of the fixed point observed: when the diffusive term dominates the flow term, the canonical EW fixed point is obtained, in the limit of large wavevectors k. On the contrary, when the flow term predominates, the effect of diffusion is suppressed by that of a travelling wave whose net result is to penalise large slopes; this leads to the smoothing fixed point obtained in the case of small wavevectors k. This combination of drift and diffusion is a classic ingredient of what we have elsewhere termed *anomalous ageing* [30].

3.2 Case B: When Moving Grains Abound

Our model equations, first presented in [3] involve a simple coupling between the species h and ρ, where the transfer between the species occurs only in the presence of the flowing grains and is therefore relevant to the regime of continuous avalanching when the duration of the avalanches is *large* compared to the time between them. The equations are:

$$\frac{\partial h(x,t)}{\partial t} = D_h \nabla^2 h(x,t) - T(h,\rho) + \eta_h(x,t) \qquad (16)$$

$$\frac{\partial \rho(x,t)}{\partial t} = D_\rho \nabla^2 \rho(x,t) + T(h,\rho) + \eta_\rho(x,t) \qquad (17)$$

$$T(h,\rho) = -\mu\rho(\nabla h) \qquad (18)$$

where the terms $\eta_h(x,t)$ and $\eta_\rho(x,t)$ represent Gaussian white noise as usual:

$$\langle \eta_h(x,t)\eta_h(x',t') \rangle = \Delta_h^2 \delta(x-x')\delta(t-t')$$
$$\langle \eta_\rho(x,t)\eta_\rho(x',t') \rangle = \Delta_\rho^2 \delta(x-x')\delta(t-t')$$

and the $\langle \cdots \rangle$ stands for average over space as well as noise.

A simple physical picture of the coupling or 'transfer' term $T(h,\rho)$ between h and ρ is the following: flowing grains are added in proportion to their local density to regions of the interface which are at less than the critical slope, and vice versa, *provided that the local density of flowing grains is always non-zero*. This form of interaction becomes zero in the absence of a finite density of flowing grains ρ (when the equations become decoupled) and is thus the simplest form appropriate to the situation of continuous avalanching in sandpiles. We analyse in the following the profiles of h and ρ consequent on this form.

It turns out that a singularity discovered by Edwards [33] several decades ago in the context of fluid turbulence is present in models with a particular form of the transfer term T; the above is one example, while another example is the model due to Bouchaud et al. (BCRE) [24] where

$$T = -\nu \nabla h - \mu\rho(\nabla h)$$

and the noise is present only in the equation of motion for h. This singularity, the so-called infrared divergence, largely controls the dynamics and produces unexpected exponents [4].

We focus now on our numerical results for Case B. The exponents α and β have been calculated for both stuck and mobile species: the details are presented elsewhere [4], It turns out, remarkably, that smoothing occurs only in the mobile species ρ: the single Fourier transform $S_\rho(k, t=0)$ (Fig. 12) shows a crossover behaviour from

$$S_\rho(k, t=0) \sim k^{-2.12 \pm 0.017}$$

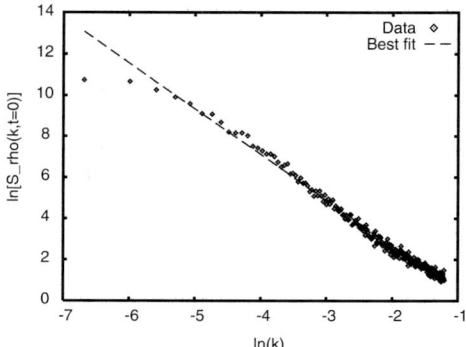

Fig. 12. Log-log plot of the single Fourier transform $S_\rho(k, t = 0)$ vs k (Case B) showing a crossover from a slope of $-1 - 2\alpha_\rho = 0$ at small k to -2.12 ± 0.017 at large k

for large wavevectors to

$$S_\rho(k, t = 0) \sim \text{constant}$$

as $k \to 0$. While the range of wavevectors over which crossover in $S_\rho(k, t = 0)$ is observed was restricted by our computational constraints [4], the form of the crossover appears conclusive. Checks (with fewer averages) over larger system sizes revealed the same trend.

Finally, we focus on the underlying physics of the equations whose results we have just presented. In the regime of continuous avalanching in sandpiles, the major dynamical mechanism is that of mobile grains ρ present in avalanches flowing into voids in the h landscape as well as the converse process of unstable clusters (a surfeit of ∇h above some critical value) becoming destabilised and adding to the avalanches. Our results for the critical exponents in h, presented elsewhere [4], indicate no further spatial smoothing beyond the diffusive; however, those in the species ρ indicate a crossover from purely diffusive to an asymptotic hypersmooth behaviour. Our claim for continuous avalanching is as follows: *the flowing grains play the major dynamical role as all exchange between h and ρ takes place only in the presence of ρ. These flowing grains therefore distribute themselves over the surface, filling in voids in proportion both to their local density as well as to the depth of the local voids; it is this distribution process that leads in the end to a strongly smoothed profile in ρ*. Additionally, since in the regime of continuous avalanches, the effective interface is defined by the profile of the *flowing* grains, it is this profile that will be measured experimentally for, say, a rotating cylinder with high velocity of rotation. It is thus all the more remarkable that our model yields *smoothing in the appropriate species* for this case, as well as for the one below.

3.3 Case C: Tilt Combined with Flowing Grains

The last case we discuss in this section involves a more complex coupling between the the stuck grains h and the flowing grains ρ as follows [4]:

$$\frac{\partial h(x,t)}{\partial t} = D_h \nabla^2 h(x,t) - T + \eta(x,t) \tag{19}$$

$$\frac{\partial \rho(x,t)}{\partial t} = D_\rho \nabla^2 \rho(x,t) + T \tag{20}$$

$$T(h,\rho) = -\nu(\nabla h)_- - \lambda \rho (\nabla h)_+ \tag{21}$$

with $\eta(x,t)$ representing white noise as usual. Here,

$$\begin{aligned} z_+ &= z \quad \text{for} \quad z > 0 \\ &= 0 \quad \text{otherwise} \end{aligned} \tag{22}$$

$$\begin{aligned} z_- &= z \quad \text{for} \quad z < 0 \\ &= 0 \quad \text{otherwise} \end{aligned} \tag{23}$$

This equation was also presented in earlier work [3] in the context of the surface dynamics of an evolving sandpile. The two terms in the transfer term T represent two different physical effects which we will discuss in turn. The first term represents the effect of tilt, in that it models the transfer of particles from the boundary layer at the 'stuck' interface to the flowing species whenever the local slope is steeper than some threshold (which is zero in this case, leading to negative slopes being penalised). The second term is restorative in its effect, in that in the presence of 'dips' in the interface (regions where the slope is shallower, i.e. more positive than the zero threshold used in these equations), the flowing grains have a chance to resettle on the surface and replenish the boundary layer [12]. We notice that because one of the terms in T is independent of ρ, we are no longer restricted to a coupling which exists only in the presence of flowing grains: i.e., this model is applicable to intermittent avalanches when ρ may or may not always exist on the surface. In the following we examine the effect of this interaction on the profiles of h and ρ respectively.

The complexity of the transfer term with its discontinuous functions has only allowed numerical solutions to date [4]. The structure factor $S_h(k, \omega = 0)$ signals a dramatic behaviour of the roughening exponent α_h, which crosses over from

- A value of 1.3 indicating anomalously large roughening at intermediate wavevectors, to
- A value of about -1 for small wavevectors indicating asymptotic hyper-smoothing.

The anomalous roughening $\alpha_h = 1$ seen here is consistent with that observed via the single Fourier transform. We suggest [4] that the present model

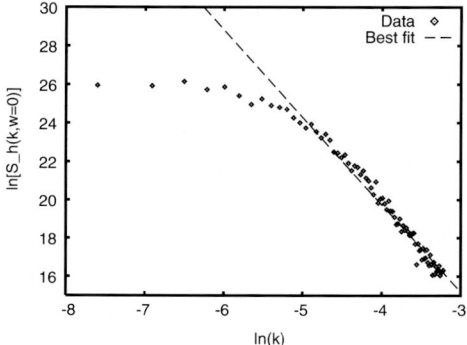

Fig. 13. Log-log plot of the double Fourier transform $S_h(k, \omega = 0)$ vs k obtained for Case C. The best fit for high wavevector has a slope of $-(1 + 2\alpha_h + z_h) = -4.54 \pm 0.081$. As $k \to 0$ we observe a crossover to slope of zero

is an integrated version of the earlier two, reducing to their behaviour in different wavevector regimes; we speculate therefore that there could *two* dynamical exponents ($z_h = 1$ and $z_h = 2$) in the problem.

3.4 Discussion of Coupled Equations

We have presented in the above a discussion of three models of sandpiles, all of which manifest asymptotic smoothing: Cases A and C manifest this in the species h of stuck grains, while Case B manifests this in the species ρ of flowing grains. We reiterate that the fundamental physical reason for this is the following: Cases A and C both contain couplings which are independent of the density ρ of flowing grains, and are thus applicable for instance to the dynamical regime of intermittent avalanching in sandpiles, when grains occasionally but not always flow across the "bare" surface. In Case B, by contrast, the equations are coupled only when there is continuous avalanching, i.e. in the presence of a finite density ρ of flowing grains.

The analysis of Case A is straightforward, and was undertaken really only to explain features of the more complex Case C; that of Case B shows satisfactory agreement between perturbative analysis [4] and simulations. Anomalies persist however when such a comparison is made in Case C, because the discontinuous nature of the transfer term makes it analytically intractable. These are removed when the analysis includes a mean-field solution which is able to reproduce the asymptotic smoothing observed [4].

We suggest therefore an experiment where the critical roughening exponents of a sandpile surface are measured in

1. a rapidly rotated cylinder, in which the time between avalanches is much less than the avalanche duration. Our results predict that for small system

sizes we will see only diffusive smoothing, but that for large enough systems, we will see extremely smooth surfaces.
2. a slowly rotated cylinder where the time between avalanches is much more than the avalanche duration. In this regime, the results of Case C make a fascinating prediction: anomalously large spatial roughening for moderate system sizes crossing over to an anomalously large spatial smoothing for large systems.

Finally we make some speculations in this context concerning natural phenomena. The qualitative behaviour of blown sand dunes [34, 35, 36] is in accord with the results of Case B, because sand moves swiftly and virtually continuously across their surface in the presence of wind. By contrast, on the surface of a glacier, we might expect the sluggish motion of boulders to result in intermittent flow across the surface, making the results of Case C more applicable to this situation. It would be interesting to see if the predictions of anomalous roughening at moderate, and anomalous smoothing at large, length scales is applicable here. Could this model be the answer to why mountains look smooth from a distance?

3.5 Application to Ripple Formation

Here we use the formalism of the previous subsections to investigate ripple formation [5]. Aeolian sand ripples are formed by the action of the wind on the sand bed in the desert and at the seashore. They have also recently been observed on Mars [37]. Aeolian ripples are a few centimetres in wavelength and their crests lie perpendicular to the prevailing wind direction. Bagnold made an influential early study [36] – he identified the importance of saltation, where sand grains are entrained by the wind, and whipped along the sand bed, colliding with it at high speed (of order ≤ 1m/s [38]), and causing other grains to jump out of and along the bed, thus sculpting ripples. The impact angle remains roughly constant at about $10°$–$16°$ despite the gusting of the wind. The stoss slope of the ripple lies in the range $8°$–$10°$ [39]. The lee slope is composed of a short straight section near the crest at an angle of about $30°$–$34°$ to the horizontal [36, 39, 40], followed by a longer and shallower concave section. The deposition of grains on the lee slope of sand dunes leads to oversteepening and avalanching [41, 42], which maintains the lee slope at an angle of around $30°$–$34°$ near the crest. A similar, if less dramatic, mechanism is likely to hold for ripples.

Numerical simulations of ripples and dunes based on tracking individual sand grains or on cellular automata have in the past few years yielded good qualitative agreement with observations [43]. However, these methods are computationally expensive. Continuum models provide a complementary approach allowing faster calculation of ripple evolution, and the possibility of obtaining certain information, such as the preferred ripple wavelength, analytically. Anderson [44] produced an analytical model of the initial generation of ripples from a flat bed. A one-species analytical continuum model was formulated by Hoyle and Woods [45]. In this section, we present work [5] which extends the one-species model [45] to include relaxation effects, inspired by models of sandpiles [3, 4], in order to obtain more realistic ripple profiles and to predict the ripple wavelength and speed.

We consider two-dimensional sand ripples: following earlier work on sandpiles [3, 4, 12, 18] we assume that the effective surface of the ripple comprises a "bare" surface defined by the local heights of clusters $h(x,t)$ sheathed by a thin layer of flowing mobile grains whose local density is $\rho(x,t)$, with x the horizontal coordinate, and t time. The ripple evolves under the influence of two distinct types of mechanism. Firstly, the impact of a constant flux of saltating grains, knocks grains out of the "bare" surface, causing them to hop along the ripple surface and land in the layer of flowing grains. This is the underlying cause of ripple formation. Secondly, the ripples are subject to intracluster and intercluster granular relaxation mechanisms which result in a smoothing of the surface.

Following [45] we consider grains to be bounced out of the "bare" surface by a constant incoming saltation flux, which impacts the sand bed at an angle β to the horizontal. These hopping or "reptating" [47] grains subsequently land in the flowing layer. The saltating grains are highly energetic and continue in saltation upon rebounding from the sand bed. We assume that the number $N(x,t)$ of sand grains ejected per unit time, per unit length of the sand bed, is proportional to the component of the saltation flux perpendicular to the sand surface, giving

$$N(x,t) = J\sin(\alpha+\beta) = \frac{J(\sin\beta + h_x \cos\beta)}{(1+h_x^2)^{1/2}}, \qquad (24)$$

where $\alpha = \tan^{-1}(h_x)$, and J is a positive constant of proportionality. We assume that each sand grain ejected from the surface hops a horizontal distance a, with probability $p(a)$, and then lands in the flowing layer. We consider the flight of each sand grain to take place instantaneously, since ripples evolve on a much slower timescale than that of a hop. The hop length distribution $p(a)$ can be measured experimentally [47, 48], so we consider it to be given empirically. It is possible that $p(a)$ and hence the mean and variance of hop lengths, could depend upon factors such as wind speed. The number $\delta n_o(x,t)$ of sand grains leaving the surface between positions x and $x+\delta x$ in time δt, where δx and δt are infinitesimal is given by $\delta n_o(x,t) = N(x,t)\delta x \delta t$. The change δh in the surface height satisfies $\delta x \delta h(x,t) = -a_p \delta n_o(x,t)$, where a_p is the average cross-sectional area of a sand grain. In the limit $\delta t \to 0$ we find that the contribution to the evolution equation for $h(x,t)$ from hopping alone is

$$h_t = -a_p N(x,t) = -a_p J \frac{\sin\beta + h_x \cos\beta}{(1+h_x^2)^{1/2}}. \qquad (25)$$

There may be regions on the sand bed which are shielded from the incoming saltation flux by higher relief upwind. In these regions there will be no grains bounced out of the surface, and there will be no contribution to the h_t equation from hopping. The number $\delta n_i(x,t)$ of sand grains arriving on the layer of flowing grains between positions x and $x + \delta x$ in time δt is given by

$$\delta n_i(x,t) = \int_{-\infty}^{+\infty} p(a) N(x-a,t) da \delta x \delta t \,. \tag{26}$$

The change in depth of the flowing layer satisfies $\delta x \delta \rho(x,t) = a_p \delta n_i(x,t)$, and hence the contribution to the evolution equation for the flowing layer depth from hopping alone is

$$\rho_t = a_p J \int_{-\infty}^{+\infty} p(a) \frac{\sin\beta + h_x(x-a,t)\cos\beta}{(1+h_x^2(x-a,t))^{1/2}} da \,. \tag{27}$$

We incorporate diffusive motion [32] as well as processes governing the transfer between flowing grains and clusters, following [3, 4], leading to the equations

$$h_t = D_h h_{xx} - T(x,t) - a_p J \frac{\sin\beta + h_x \cos\beta}{(1+h_x^2)^{1/2}}, \tag{28}$$

$$\rho_t = D_\rho \rho_{xx} + \chi(\rho h_x)_x + T(x,t)$$
$$+ a_p J \int_{-\infty}^{+\infty} p(a) \frac{\sin\beta + h_x(x-a,t)\cos\beta}{(1+h_x^2(x-a,t))^{1/2}} da \,, \tag{29}$$

where D_h, D_ρ and χ are positive constants and where $T(x,t)$, which represents the transfer terms, is given by

$$T(x,t) = -\kappa \rho h_{xx} + \lambda \rho(|h_x| - \tan\alpha) \,, \tag{30}$$

for $0 \leq |h_x| \leq \tan\alpha$ and by

$$T(x,t) = -\kappa \rho h_{xx} + \frac{\nu(|h_x| - \tan\alpha)}{(\tan^2\gamma - h_x^2)^{1/2}} \,, \tag{31}$$

for $\tan\alpha \leq |h_x| < \tan\gamma$, with κ, λ and ν also positive constants. The term $D_h h_{xx}$ represents the diffusive rearrangement of clusters while the term $D_\rho \rho_{xx}$ represents the diffusion of the flowing grains. The flux-divergence term $\chi(\rho h_x)_x$ models the flow of surface grains under gravity. The current of grains is assumed proportional to the number of flowing grains and to their velocity, which in turn is proportional to the local slope to leading order [45]. The $-\kappa \rho h_{xx}$ term represents the inertial filling in of dips and knocking out of bumps on the 'bare' surface caused by rolling grains flowing over the top. The $\lambda \rho(|h_x| - \tan\alpha)$ term represents the tendency of flowing grains to stick onto the ripple surface at small slopes; it is meant to model the accumulation of slowly flowing grains at an obstacle. Clearly for this to happen the obstacle must be stable, or else it would be knocked off by the oncoming grains, so that this term only comes into play for slopes less than $\tan \alpha$, where α is the *angle of repose*. The term $\nu(|h_x| - \tan \alpha)(\tan^2 \gamma - h_x^2)^{-1/2}$ represents tilt and avalanching; it comes into play only for slopes greater than $\tan \alpha$ and models the tendency of erstwhile stable clusters to shed grains into the flowing layer when tilted. This shedding of grains starts when the surface slope exceeds the angle of repose. For slopes approaching the angle γ, which is the *maximum angle of stability*, the rate of tilting out of grains becomes very large: an avalanche occurs. This term, among other things, is a novel representation of the well-known phenomena of *bistability* and *avalanching* at the angle of repose [11, 15, 46].

We renormalise the model equations setting $x \to x_0 \tilde{x}$, $t \to t_0 \tilde{t}$, $a \to x_0 \tilde{a}$, $\rho \to \rho_0 \tilde{\rho}$, $h \to h_0 \tilde{h}$, where $x_0 = D_h/a_p J \cos \beta$, $t_0 = D_h/(a_p J \cos \beta)^2$, $h_0 = D_h \tan \gamma / a_p J \cos \beta$, $\rho_0 = a_p J \sin \beta / \lambda \tan \alpha$, which gives for $0 \le h_x \le \tan \alpha / \tan \gamma$

$$h_t = (1 + \hat{\kappa}\rho)h_{xx} - \rho\frac{\tan \beta}{\tan \alpha}\left(|h_x| - \frac{\tan \alpha}{\tan \gamma}\right) - f(x), \tag{32}$$

$$\rho_t = \frac{h_0}{\rho_0}\left\{-\hat{\kappa}\rho h_{xx} + \rho\frac{\tan \beta}{\tan \alpha}\left(|h_x| - \frac{\tan \alpha}{\tan \gamma}\right)\right\}$$
$$+ \frac{h_0}{\rho_0}\int_{-\infty}^{+\infty} p(a)f(x-a)da + \frac{D_\rho}{D_h}\rho_{xx} + \hat{\chi}(\rho h_x)_x, \tag{33}$$

and for $\tan \alpha / \tan \gamma \le h_x < 1$

$$h_t = (1 + \hat{\kappa}\rho)h_{xx} - f(x) - \frac{\hat{\nu}(|h_x| - \tan \alpha/\tan \gamma)}{(1 - h_x^2)^{1/2}}, \tag{34}$$

$$\rho_t = \frac{h_0}{\rho_0}\left\{-\hat{\kappa}\rho h_{xx} + \frac{\hat{\nu}(|h_x| - \tan \alpha/\tan \gamma)}{(1 - h_x^2)^{1/2}}\right\}$$
$$+ \frac{h_0}{\rho_0}\int_{-\infty}^{+\infty} p(a)f(x-a)da \frac{D_\rho}{D_h}\rho_{xx} + \hat{\chi}(\rho h_x)_x, \tag{35}$$

where the tildes have been dropped and where

$$f(x) = \left(h_x + \frac{\tan \beta}{\tan \gamma}\right)(1 + h_x^2 \tan^2 \gamma)^{-1/2}, \tag{36}$$

and $\hat{\kappa} = \kappa \rho_0/D_h$, $\hat{\nu} = \nu t_0/h_0$, $\hat{\chi} = \chi h_0/D_h$. Wherever the sand bed is shielded from the saltation flux, the hopping term must be suppressed by removing the term $-f(x)$ in the h_t equation.

Close to onset of the instability that gives rise to sand ripples, the slopes of the sand bed will be small, since surface roughness is of small amplitude; hence the regime $0 \le \tan \alpha / \tan \gamma$ is relevant. There are no shielded regions at early times, since the slope of the bed does not exceed $\tan \beta$. Note that $h_x = 0$, $\rho = 1$ is a stationary solution of (32) and (33). Setting $h = \hat{h}e^{\sigma t + ikx}$ and $\rho = 1 + \hat{\rho}e^{\sigma t + ikx}$, where $\hat{h} \ll 1$ and $\hat{\rho} \ll 1$ are constants, linearising, and

Taylor-expanding the integrand gives a dispersion relation for σ in terms of k. The presence of the $|h_x|$ term means that strictly we are considering different solutions for sections of the ripple where $h_x > 0$ and sections where $h_x < 0$, but in fact the effect of the $|h_x|$ terms appears only as a contribution $\pm \tan\beta/\tan\alpha$ to the bracket $(1 \pm \tan\beta/\tan\alpha)$, where the $h_x > 0$ case takes the $+$ sign and $h_x < 0$ case takes the $-$ sign. Since typically we have $\tan\beta/\tan\alpha \ll 1$, the two solutions will not be very different. One growth rate eigenvalue is given by $\sigma = -h_0 \tan\beta/\rho_0 \tan\gamma + O(k)$ and is the rate of relaxation of ρ to its equilibrium value of 1. To $O(k^4)$ the other eigenvalue is

$$\sigma = (\bar{a} - 1 - \hat{\chi}\rho_0/h_0)k^2 + iAk^3 + Bk^4 \tag{37}$$

where

$$A = -\frac{1}{2}\overline{a^2} + \frac{\rho_0 \tan\gamma}{h_0 \tan\beta}\left(1 + \hat{\chi}\frac{\rho_0}{h_0} - \bar{a} - \frac{D_\rho}{D_h}\right)\left(1 \pm \frac{\tan\beta}{\tan\alpha}\right), \tag{38}$$

$$B = -\frac{1}{6}\overline{a^3} - \frac{1}{2}\overline{a^2}\frac{\rho_0 \tan\gamma}{h_0 \tan\beta}\left(1 \pm \frac{\tan\beta}{\tan\alpha}\right)$$
$$- \frac{\rho_0 \tan\gamma}{h_0 \tan\beta}\left(\bar{a} - 1 - \hat{\chi}\frac{\rho_0}{h_0} + \frac{D_\rho}{D_h}\right)\left\{\bar{a} + \hat{\kappa} + \frac{\rho_0 \tan\gamma}{h_0 \tan\beta}\left(1 \pm \frac{\tan\beta}{\tan\alpha}\right)^2\right\}, \tag{39}$$

and where $\overline{(.)}$ denotes $\int_{-\infty}^{+\infty}(.)p(a)da$. We have neglected higher order terms as we are looking for long wave modes where $|k|$ is small, since short waves are damped by the diffusion terms. Sand ripples grow if $\bar{a} > 1 + \hat{\chi}\rho_0/h_0$, which is equivalent to requiring that $\bar{a}a_p J \cos\beta > D_h + \chi a_p J \sin\beta/\lambda \tan\alpha$ holds in physical variables, giving a threshold saltation flux intensity for ripple growth. This is in agreement with the threshold found in [45]. Since B is negative ($\beta < \alpha$), the fastest growing mode has wavenumber $k^2 = -(\bar{a}-1-\hat{\chi}\rho_0/h_0)/2B$ with growth rate $\sigma = -(\bar{a} - 1 - \hat{\chi}\rho_0/h_0)^2/4B$. The allowed band of wavenumbers for growing modes is $0 < k^2 < -(\bar{a}-1-\hat{\chi}\rho_0/h_0)/B$. The wave speed is given by $c = -Ak^2 > 0$; it is higher for larger k^2 which implies that shorter waves move faster, as indeed was seen in the numerical simulations described below. The speed is higher for $h_x > 0$ than for $h_x < 0$, leading to wave steepening.

The renormalised model equations (32)–(35) were integrated numerically using compact finite differences [49] with periodic boundary conditions. The $-f(x)$ term in the h_t equation was suppressed in shielded regions. We used a normal distribution for the hop lengths with mean $\bar{a}$ and variance s^2. In the run illustrated, we chose $\bar{a} = 3.1$, $s = 0.1$, $D_\rho/D_h = 1.0$, $h_0/\rho_0 = 20.0$, $\hat{\chi} = 0.1$, $\hat{\nu} = 1.0$, $\hat{\kappa} = 0.1$, $\beta = 10°$, $\alpha = 30°$ and $\gamma = 35°$. The angles were chosen to agree with observational evidence, the ratio h_0/ρ_0 to ensure a thin layer of flowing grains, and the remaining parameters to allow ripple growth. The output was rescaled back into physical variables using $D_h = 1.0$ and $\lambda = 10.0$. The initial conditions for the dimensionless variables were $h = 1.0 + 0.1\eta_h$, $\rho = 0.95 + 0.1\eta_\rho$, where η_h and η_ρ represent random noise generated by random variables on $[0, 1)$ in order to model surface roughness.

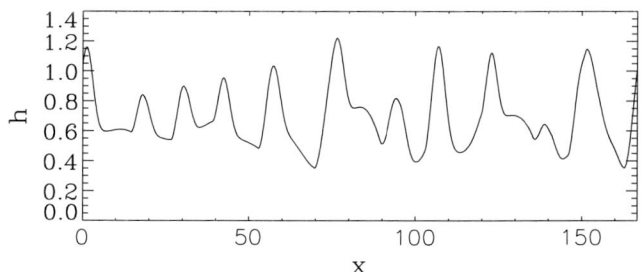

Fig. 14. The surface height h at time $t = 10.0\Delta t$

In this case $B = -8.47$ and $A = -5.26$ (taking the minus sign in the brackets), giving a preferred wavenumber of $k = 0.352$, and a wave speed of $c = 5.26k^2$. The length of the integration domain was chosen to be ten times the linearly preferred wavelength.

Figure 14 shows the surface height at time $t = 10.0\Delta t$, where $\Delta t = 2.78$. Note the emergence of a preferred wavelength, with wavenumber $k \approx 0.457$ lying in the permitted band for growing modes predicted by the linear stability analysis and arrived at by a process of ripple merger. The wave speed close to onset was also measured and found to be $c \approx (4.13\pm 0.39)k^2$, which is reasonably close to the predicted value. Ripple merger typically occurs when a small fast ripple catches up and merges with a larger slower ripple (Fig. 16), the leading ripple transferring sand to its pursuer until only the pursuer remains. Occasionally a small ripple emerges from the front of the new merged ripple and runs off ahead. Figure 15 shows the surface height h at time $t = 89\Delta t$, with one shallow and one fully developed ripple.

Note the long shallow stoss slopes, and the shorter steeper lee slopes with straight sections near the crests and concave tails. The leftmost ripple has a maximum stoss slope angle of 3.3°, and a maximum lee slope of 9.2°, whereas the more fully developed rightmost ripple has a maximum stoss slope angle of 24.8° and a maximum lee slope angle of 33.5°, which lies between the

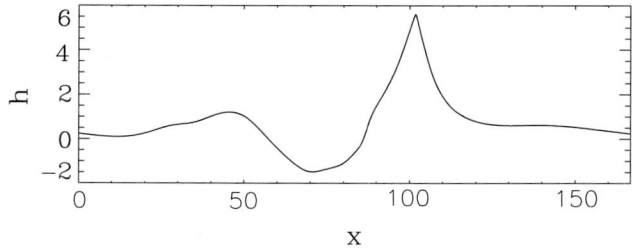

Fig. 15. The surface height h at time $t = 89\Delta t$, showing fully developed ripples. Note the straight segments on the lee slopes close to the crests

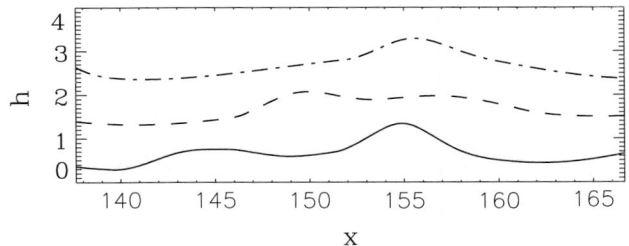

Fig. 16. A sequence of profiles showing a small ripple catching up and merging with a larger ripple. Sand is transferred from the larger to the smaller ripple until only the latter remains. The profiles are shown at times $t = 11.0\Delta t$ (*solid line*), $t = 12.0\Delta t$ (*dashed line*) and $t = 13.0\Delta t$ (*dot-dash line*). The later profiles are each offset by one additional unit in height

angle of repose and the maximum angle of stability. The height to length ratio of the ripples is in the range 1:8–1:22, which is in reasonable agreement with observations [39]. In the long time limit, we would expect sand ripples to grow until the maximum lee slope angle reaches an angle close to $\tan\gamma$. In reality, there is only a relatively shallow layer of loose sand available for incorporation into ripples, and this together with the maximum slope condition will determine the size of the fully-developed ripples.

In summary, we have formulated an analytical continuum model for aeolian sand ripples using a two-species model embodying intracluster and intercluster relaxation, in a description that leads naturally to bistable behaviour at the angle of repose, and its cutoff at the angle of maximal stability [17]. We have predicted analytically the preferred ripple wavelength, the wave speed and the threshold saltation flux required for ripples to form. Our numerical simulations show the development of realistic ripple profiles from initial surface roughness via growth and ripple merger.

References

1. P. Bak, C. Tang and K. Wiesenfeld, Phys. Rev. Lett. **59**, 381 (1987); Phys. Rev. A **38**, 364 (1988).
2. G.C. Barker and A. Mehta, Phys. Rev. E **53**, 5704 (1996).
3. A. Mehta, J.M. Luck and R.J. Needs, Phys. Rev. E **53**, 92 (1996).
4. P. Biswas, A. Majumdar, A. Mehta, and J.K. Bhattacharjee, Physical Review E **58**, 1266 (1998).
5. R.B. Hoyle and A. Mehta Phys. Rev. Lett. **83**, 5170 (1999).
6. A. Mehta, *The Physics of Granular Materials*, Cambridge University Press, Cambridge (2007, to be published).
7. J.M. Carlson and J.S. Langer, Phys. Rev. A **40**, 6470 (1989).
8. H. Nakanishi, Phys. Rev. A **43**, 6613 (1991).
9. L.P. Kadanoff, S.R. Nagel, L. Wu and S. Zhou, Phys. Rev. A **39**, 6524 (1989).

10. A. Mehta and G.C. Barker, Europhys. Lett. **27**, 501 (1994).
11. H.M. Jaeger and S.R. Nagel, Science **255**, 1523 (1992).
12. A. Mehta, Ed., *Granular Matter: an Interdisciplinary Approach,* Springer Verlag, New York (1996).
13. A. Mehta and G C Barker, New Scientist **40**, (1991).
14. G.C. Barker and A. Mehta, Avalanches in real sandpiles – the role of disorder. In *Powders and Grains,* pp 315–319, Ed. C. Thornton, Balkema, Rotterdam (1993).
15. A. Mehta and G.C. Barker, Rep. Prog. Phys. **57**, 383 (1994).
16. G.A. Held, D.H. Solina, D.T. Keane, W.J. Haag, P.M. Horn and G. Grinstein, Phys. Rev. Lett **65**, 1120 (1990).
17. H.M. Jaeger, C. Liu and S.R. Nagel, Phys. Rev. Lett. **62**, 40 (1989).
18. A. Mehta, Physica A **186**, 121 (1992).
19. G.C. Barker and A. Mehta, Phys. Rev. E **61**, 6765–6772 (2000).
20. A. Mehta and J.M. Luck, Phys. Rev. Lett. **65**, 393 (1990).
21. J.M. Luck and A. Mehta, Phys. Rev. E **48**, 3988 (1993).
22. A. Gabrielov, W.I. Newman and L. Knopoff, Phys. Rev. E **50**, 188 (1994); H.J. Xu and L. Knopoff, Phys. Rev. E **50**, 3577 (1994).
23. A. Mehta, G.C. Barker, J.M. Luck and R.J. Needs, Physica A **224**, 48 (1996).
24. J.P. Bouchaud, M.E. Cates, J. Ravi Prakash, and S.F. Edwards, J. de Physique I **4**, 1383 (1994); Phys. Rev. Lett.**74**, 1982 (1995).
25. A. Mehta and G.C. Barker, Phys. Rev. Lett. **67**, 394 (1991).
26. P.G. de Gennes, Rev. Mod. Phys. **71**, S374 (1999).
27. R. Franklin and F. Johanson F, Chem. Eng. Sci. **4**, 119 (1955).
28. J.M. Luck and A. Mehta, JSTAT P10015 (2004).
29. S.R. Nagel, Rev. Mod. Phys. **64**, 321 (1992).
30. J.M. Luck and A. Mehta, Europhysics Letters, **54**, 573, (2001).
31. T. Halpin-Healy T. and Y.C. Zhang: Phys. Rep. **254**, 215 (1995); J. Krug and H. Spohn, in *Solids far from Equilibrium,* Ed. C. Godreche C, Cambridge University Press, Cambridge (1992).
32. S.F. Edwards and D.R. Wilkinson, Proc. Roy. Soc. A **381**, 17 (1982).
33. S.F. Edwards, J. Fluid Mech. **18**, 239 (1964).
34. R.A. Bagnold, Proc. Roy. Soc. London A **225**, 49 (1954).
35. R.A. Bagnold, Proc. Roy. Soc. London A **295**, 219 (1966).
36. R.A. Bagnold, *The physics of blown sand and desert dunes* (Methuen and Co., London 1941).
37. R. Greeley et al, J. Geophys. Res. **104**, 8573 (1999).
38. R.S. Anderson and K.L. Bunas, Nature **365**, 740 (1993).
39. R.P. Sharp, J. Geol. **71**, 617 (1993).
40. K. Pye and H. Tsoar, *Aeolian sand and sand dunes,* Unwin Hyman, London (1990).
41. R.E. Hunter, Sedimentology **32**, 409 (1985).
42. R.S. Anderson, Sedimentology **35**, 175 (1988).
43. R.S. Anderson, Earth-Sci.Rev. **29**, 77 (1990); S.B. Forrest and P.K. Haff, Science **255**, 1240 (1992); W. Landry and B.T. Werner, Physica D **77**, 238 (1994); H. Nishimori and N. Ouchi, Phys. Rev. Lett.**71**, 197 (1993).
44. R.S. Anderson, Sedimentology **34**, 943 (1987).
45. R.B. Hoyle and A.W. Woods, Phys. Rev. E **56**, 6861 (1997).
46. H.M. Jaeger, S.R. Nagel, and R.P. Behringer, Rev. Mod. Phys. **68**, 1259 (1996).

47. S. Mitha, M.Q. Tran, B.T. Werner and P.K. Haff, Acta Mechanica **63**, 267 (1986).
48. J. Ungar and P.K. Haff, Sedimentology **32**, 267 (1987).
49. S.K. Lele, J. Comput. Phys. **103**, 16 (1992).

Dynamics of Stick-Slip: Some Universal and Not So Universal Features

G. Ananthakrishna[1] and R. De[2]

[1] Materials Research Centre and Centre for Condensed Matter Theory, Indian Institute of Science, Bangalore-560012, India
garani@mrc.iisc.ernet.in
[2] Materials Research Centre, Indian Institute of Science, Bangalore-560012, India. Department of Materials and Interfaces. Weizmann Institute of Science, Rehovot 76100, Israel
rumi.de@weizmann.ac.il (present address)

Stick-slip is usually observed in driven dissipative threshold systems. In these set of lectures, we discuss, some generic and system specific features of stick-slip systems by considering a few examples wherein there has been some progress in understanding the associated dynamics. In most stick slip systems, both at low and high drive rates, the system slides smoothly, but within a window of drive rates, the motion becomes intermittent; the system alternately "sticks" till the stress builds up to a threshold value, and then "slips" when the stress is rapidly released. This intermittent motion can be traced to the existence of an unstable branch separating the two resistive branches in the force-drive-rate relation. While the two resistive branches are experimentally measurable, the unstable branch is usually not measurable and is only inferred.

We shall consider a few examples drawn from solid-on-solid, the Portevin Le Chatlier effect and peeling of an adhesive tape to extract common features and system specific features. Even though the negative flow rate characteristic (NFRC) is a common feature and is an input into most models of stick-slip, the origin of this stable and unstable branches has remained largely ill understood except in the case of the Portevin-Le Chatelier effect. We shall deal this case in some detail where it has been convincingly demonstrated that force-pull rate function is a consequence of competing time scales.

Finally, we shall examine a puzzle related to power law distribution of event sizes seen in stick-slip systems. As such systems are scale free, this has been interpreted to mean the impossibility of predicting individual avalanches. We address the issue of predictability of individual avalanches in the context of a modified stick-slip spring-block model for earthquakes by introducing a dissipative term which acts as a source of a precursor. We show that individual events are predictable without violating the scale invariance property.

1 Introduction

A large number of dissipative threshold systems driven at constant drive rates (velocity, strain rate, etc.) often exhibit stick-slip oscillations, for instance, frictional sliding, earthquakes faults (seismic motion), peeling of an adhesive tape, jerky flow of alloys or the Portevin-Le Chatelier effect (a kind of plastic instability), flow of granular materials (avalanches) and martensite transformation to name a few. The nature of motion of such systems can range from very simple to complex dynamics depending on many factors, including the type of metastable states in the system, the time needed to transform between states, and the mechanical device that imposes the drive rate. Both at low and high drive rates, the system slides smoothly and correspondingly, there are two stable resistive branches in the the force-drive-rate relation. However, within a window of drive rates that separates the two stable branches, the motion becomes intermit tent which corresponds to the unstable branch of the force-drive-rate relation. In this regime of intermittent motion, the system alternately "sticks" till the stress builds up to a threshold value, and then "slips" when the stress is rapidly released. Stick-slip instabilities manifest only when systems are driven to a critical threshold value of the respective drive parameters. In experiments, it has been observed that the slopes of the two resistive branches are different. This is due to the fact that the dissipation mechanisms at low and high velocity resistive branches are different. Further, the high velocity branch is usually much steeper compared to the low velocity resistive branch. Indeed, the steeper slope of the high velocity branch is reflected in the characteristic feature of the system spending most of the time in the stuck state and a short time in the slip state. The two resistive branches can be experimentally measured. In contrast, the intermediate branch is not accessible in experiments as it is unstable. Thus, except in the case of solid friction, the unstable branch is only inferred whose assumed form is the one that smoothly interpolates between the two stationary branches. Stick-slip dynamics is usually seen in systems subjected to a constant drive rate where the force developed in the system is measured by dynamically coupling the system to a measuring device. This method of measurement should be contrasted with conventional physics experiments where a force is applied and the response of the system is measured.

As the negative flow rate characteristic (NFRC) is an input into most models of stick-slip, it is necessary to understand the origin of this feature. To begin with, we recall some simple features of stick-slip dynamics and then briefly discuss some experimental and theoretical attempts to understand the underlying physics, both generic as well as system specific features. To illustrate basic ideas of stick-slip, consider a block of mass m attached to a spring k that is being pulled at a constant velocity V. As the mass is pulled, initially the block "sticks" to the surface (AB in Fig. 1a where the velocity is zero). During this period, the spring gets stretched and the spring force F rises linearly with time till it reaches the static friction threshold value F_s between

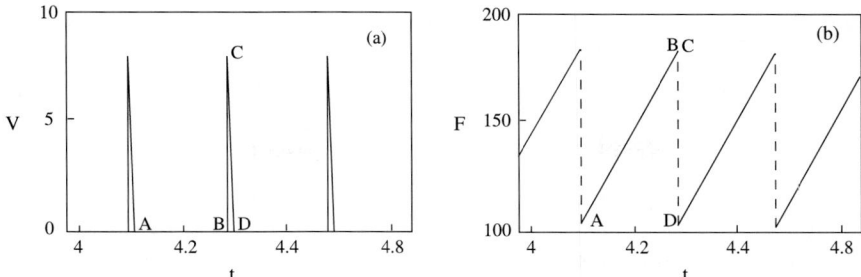

Fig. 1. Stick-slip dynamics: (a) velocity vs time, (b) corresponding force vs time plot

the block and the surface (AB in Fig. 1b). When F exceeds F_s, the block begins to "slip" f orward, concomitantly there is burst in the velocity (BC Fig. 1a). Since the kinetic friction F_d that resists sliding is less than F_s (as we shall see), the block accelerates and F drops rapidly (CD in Fig. 1a). When it falls enough, the surfaces stick once more causing the cycle to repeat. The burst in the velocity represents a slip event. These plots also show that the system spends most of the time in the stuck state and a very short time in the slip state.

As mentioned earlier, in steady state experiments, the force-velocity curve shows two stable resistive branches separated by an unstable branch as shown in Fig. 2a (the dotted curves separated by the unstable dashed curve). In terms of the $F - V$ phase plot, the state of the system can then be seen to move on the friction curve in cyclic way. Noting that during the period when $V(t)$ is nearly zero (AB region in Fig. 1a) where F increases almost linearly, one can see that the state of the system moves up on the AB branch till it reaches the threshold value for the slip (B in Fig. 2a). The abrupt increase in the velocity (from B to C) without any noticeable change in the force (Fig. 1b) translates to a jump from B to C in the phase plot (Fig. 2a). Once the system is on the high velocity resistive branch (CD), F drops rapidly from C to D, i.e., the state of the system moves from C to D on t he phase plot and then on to the left branch. The cycle repeats.

2 Solid Friction

To illustrate a few ideas of stick-slip, we begin with a very old problem relating to dry friction. Basic understanding on dry friction can be attributed to the early studies of Bowden and Tabor [1]. In case of solid-on-solid, the frictional force is due to the shear strengths of junctions of real contact called asperities. These asperities are distributed in a random way with typical dimensions ranging from 1 to 10 μm. A simple physical model for stick slip based on random arrangement of asperities of arbitrary heights was proposed by Rabinowicz [2]. Rapid slips can occur whenever an asperity on one surface goes over the top of an asperity on the other surface. The extent of "slip" will depend on the asperity height and slope, on the speed of sliding, and on the elastic compliance of the surfaces in contact. As in all cases of stick-slip motion, the drive velocity may be constant but the resulting motion at the surfaces will display large variations in slips.

As mentioned above, the frictional force is determined by the strength of the junctions. The shear force is then proportional to $F_s = \sigma_s A_{real}$, where σ_s is the shear strength of the asperities and A_{real} is the real area of contact. The normal load then is given by $Mg = p_m A_{real}$, where p_m is the normal flow stress. Then, $\mu_s = \sigma_s/p_m$ is the static friction. This accounts for the Amonton-Coulomb law which states that the frictional force is independent of the apparent area of contact between the two surfaces. In case of the kinetic friction, μ_d, the frictional force is lower than μ_s i.e., $\mu_d(V) < \mu_s$, where μ_d depends on the pull velocity V.

More recent and carefully controlled experiments of paper-on-paper [3] show a number interesting features that clarify the nature of dry friction as well as that of the kinetic friction. First, while the static friction is considered to be constant, in reality it is time dependent. Indeed, these experiments show that it increases logarithmically with time [3]. Second, the dynamic friction $F/Mg = \mu$, undergoes stick-slip oscillations within a window of velocities as shown in Fig. 2b. As can be seen from the figure, μ increases linearly during a stick time τ_{stick} and then slip appears as a quasi-discontinuity over a very short time τ_{slip} ($\ll \tau_{stick}$). (Note that $\mu = \mu(t)$ is different from the kinetic friction μ_d which refers to the steady state value.) Third, there is no truly stuck state where the system is completely at rest. At low pulling velocities, there is a slow relaxation under load. Thus, instead of truly motionless "stuck state",

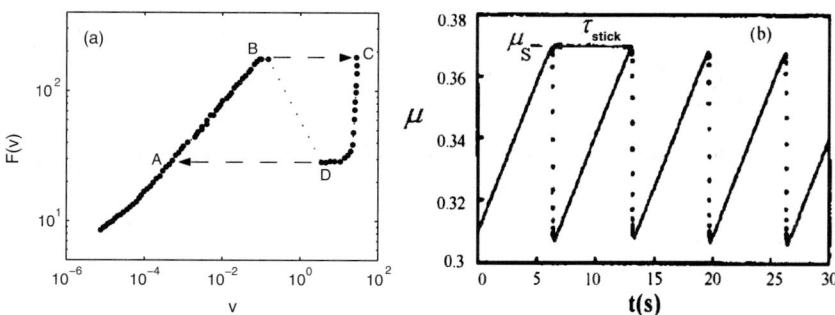

Fig. 2. (a) The cyclic changes in the state of the system along the force-velocity curve (after [12]). (b) Normalized force $\mu = F/Mg$ vs time t recorded in the stick-slip regime for $M = 2.1$ kg, $k = 1.5 \times 10^4$ N m^{-1}, and $V = 10$ μm s^{-1} in an experiment on dry-friction dynamics of a paper-on-paper system (after [3])

it shows a stress-induced creep suggesting a continuous crossover between stick and slip states rather than an abrupt stick-slip jump. Thus, the force function or the friction coefficient $\mu(t)$ does not merely increase linearly; there is a contribution coming from creep of the asperities. Fourth, it has also been observed that the kinetic friction μ_d not only depends on the instantaneous velocity but also on the previous dynamical history of the solid-solid contacts, usually referred to as the "memory effect".

To demonstrate that the low-velocity dynamics is controlled by a creep process, these authors monitor the displacement $x(t)$ of the slider with respect to the track directly [3]. A perfect stuck state corresponds to $x(t) = 0$, whereas a departure from this state has been observed that increases nonlinearly with time and is mostly visible just before the rapid slip event as can be seen from Fig. 3a. This is also in agreement with the results from rock samples [4]. The method also gives a direct measure of "creep length". This discussion also suggests that in the low pull velocity regime, there is enough time for plastic relaxation (creep) to occur. In contrast, for high pull velocities, there is very little time for the plastic relaxation to occur and other dissipative forces come into play.

The crucial physical ingredient responsible for the complex stick-slip behavior is the "velocity-weakening" phenomena which interpolates between the static threshold μ_s and a smaller kinetic value μ_d by defining a region of decreasing friction coefficient. In general, this velocity weakening branch is not accessible. However, Baumberger et al. [5] were able to measure this branch as shown in Fig. 3a. As this branch is unstable, it requires a great care in ensuring that the imposed velocity is fully transferred to the system which is done by using a stiff coupling to the measuring device that transfers the entire pull speed to the system. This implies that a softer spring gives rise to amplification of any deviation and thus a soft spring cannot be used for measuring this unstable branch. Their experiments shows [3] a logarithmic velocity ($\ln V$) weakening dependence of the kinetic friction μ_d. However, due

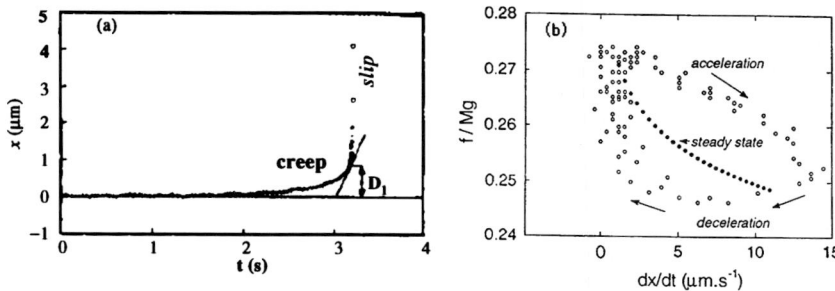

Fig. 3. (a) Direct measurement of the slider displacement x vs time t for $M = 0.32$ kg, $k = 1.5 \times 10^4$ N m^{-1}, $V = 5$ μm s^{-1}, and D_1 is the creep length (after [3]). (b) Stationary $\mu_d(V)$ and amplification of even small perturbations from these values (after [5])

to the unstable nature of this branch, even very small deviations from the steady state values are amplified. This leads to local accelerations as shown in Fig. 3b which in turn emphasizes the difficulty in observing the unstable branch.

The solid friction case is one situation where there is a threshold at near zero velocity followed by a velocity weakening law while most other stick-slip situations exhibit two resistive branches separated by an unstable branch. In spite of growing interest in understanding the origin of the NFRC, there is still little understanding of the dissipation mechanisms leading to the presence of the threshold at zero velocity or $ln\,V$ dependence of the kinetic friction observed in experiments. This can be largely attributed to a lack of understanding of the underlying physical processes.

It is obvious that physical mechanisms leading stick-slip are different in different systems, but the common feature turns out to be the bistable nature. The importance of understanding the origin of these underlying mechanisms becomes clear when one considers the fact that the "negative flow rate characteristic" (NFRC) is an important input into models of stick-slip dynamics seen in different physical situations. For example, in a simple mechanical model of an earthquake fault originally introduced by Burridge and Knopoff [6] and later studied extensively by many others, a velocity weakening friction law has been used as an input [7, 8]. This is also true for most models of the Portevin-Le Chatelier (PLC) effect as well [9, 10, 11]. Even in the context of the dynamics of an adhesive tape [12, 13] which we shall also discuss, this feature is an input.

3 Stick-Slip Instability During Plastic Flow: The Portevin–Le Chatelier (PLC) Effect

In contrast to the solid-on-sold case, the physical mechanisms are well known in the case of the Portevin–Le Chatelier (PLC) effect [10], but a quantitative derivation of each of these branches based on microscopic dissipative mechanisms is still lacking. However, this is one example where NFRC macroscopic law has been derived and a new interpretation provided based on Ananthakrishna's model for the PLC effect [14, 15]. In this sense, this model occupies an important place in stick-slip dynamics. We shall therefore discuss this in some detail.

When metallic alloys are subjected to a tensile test at constant strain rate, one usually finds a single yield drop. However, in a certain range of temperatures and strain rates, dilute substitutional as well as interstitial solid solutions, subjected to tensile tests exhibit repetitive stress drops or serrations in the stress-strain curve. The phenomenon was first studied by Le Chatelier in 1909 on mild steel specimens [10], subsequently in Duralumin by Portevin and Le Chatelier in 1923, hence the name Portevin–Le Chatelier (PLC) effect [10]. The phenomenon is also referred to as jerky flow. Here a uniform

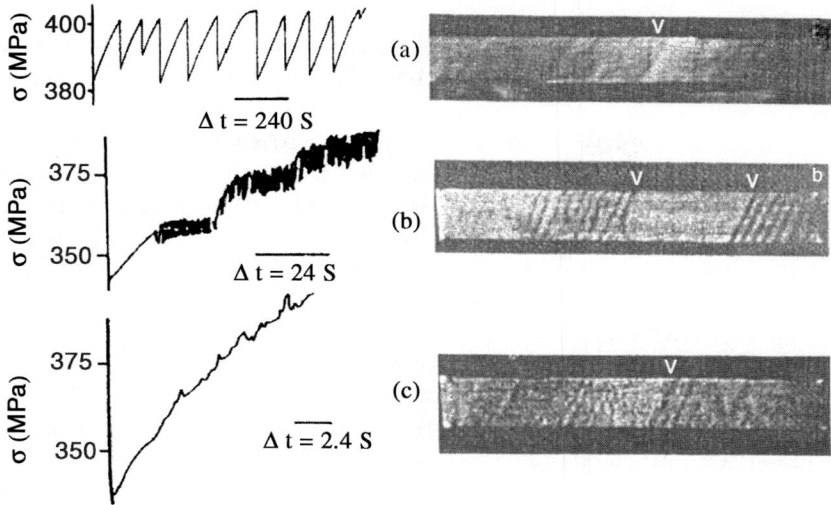

Fig. 4. Stress-time (σ-t) curves for Al-5at%Mg alloy at $T = 300$ K showing change over from type $C \to$ type $B \to$ type A serrations with increasing strain rate (after [21]). **(a)** Type C, $\dot{\epsilon}_a = 5 \times 10^{-6}$ s^{-1}, **(b)** Type B, $\dot{\epsilon}_a = 5 \times 10^{-4}$ s^{-1}, **(c)** Type A, $\dot{\epsilon}_a = 5 \times 10^{-3}$ s^{-1} (after [19])

deformation mode becomes unstable leading to a spatially and temporally inhomogeneous state. Each stress drop is associated with the nucleation and often the propagation of a band of localized plastic deformation. These bands and the associated serrations are classified into three generic types shown in Fig. 4. On increasing the applied strain rate or decreasing the temperature, one first finds the type C band, identified with randomly nucleated static bands with large characteristic stress drops. The serrations are quite regular. Then the type B "hopping" bands are seen with each band forming ahead of the previous one in a spatially correlated way. The serrations are more irregular with amplitudes that are smaller than that for the type C. Finally, one observes the continuously propagating type A bands associated with small stress drops. These different types of PLC bands are believed to represent distinct correlated states of dislocations in the bands.

From a dynamical point of view this jerky or stick-slip kind of behavior is related to the physical mechanism of the discontinuous motion of dislocations, namely, the pinning (stick) and depinning (slip) of dislocations. The well accepted classical explanation of the PLC effect is via the dynamic strain aging (DSA) concept first introduced by Cottrell [16] and later extended by others [9, 11, 17, 18]. In the Cotrell's picture, the dynamic strain aging refers to the interaction of mobile dislocations with the diffusing solute atoms. At low strain rates (or high temperatures) the average velocity of dislocations is low and there is sufficient time for the solute atoms to diffuse to the dislocations and pin them (called as aging). Thus, longer the dislocations are arrested, larger will be the stress required to unpin them. When these dislocations are unpinned, they move at large speeds till they are arrested again. At high strain rates (or low temperatures), the time available for solute atoms to diffuse to the dislocations decreases and hence the stress required to unpin them decreases. Thus, in a range of strain rates and temperatures where these two time scales are of the same order of magnitude, the PLC instability manifests. The competition between the slow rate of pinning and sudden unpinning of the dislocations, at the macroscopic level translates into a negative strain rate sensitivity (SRS) of the flow stress as a function of strain rate. This is the basic instability mechanism used in most phenomenological models [9]. Penning [11], in his landmark paper, was the first to recognize that negative "stress rate sensitivity" (SRS) as a condition for repeated yielding. Even though, negative SRS is a function of macroscopic variables such as stress, strain, and strain rate, these variables are actually treated as local variables.

A schematic diagram of the negative strain rate sensitivity is shown in Fig. 5a. In the physical context of the PLC effect, the branch AB corresponds to the dissipation arising when solute atoms are dragged along by the dislocations. As the branch CD corresponds to the high velocity, dislocations are torn free of the solute atoms and the associated damping mechanisms arise from phonons, electrons and other dissipative sources.

A natural question is: how is this negative SRS measured in experiments on the PLC effect? In this case, there are a few attempts to "measure" the unstable branch as a function of the strain rate [20], even though there is a full recognition of the limitations of such a measurement. Here we describe the procedure adopted by Kubin's group [21]. Following Pennings decomposition (the flow stress σ is assumed to decomposable into a sum of two terms, $\sigma = \frac{\partial \sigma}{\partial \epsilon}\epsilon + F(\dot{\epsilon})$, where $F(\dot{\epsilon})$ is a N-shaped dynamic curve), the strain is fixed at $\epsilon = 8 \times 10^{-2}$ and the mean of the upper stress values of the serrations is taken

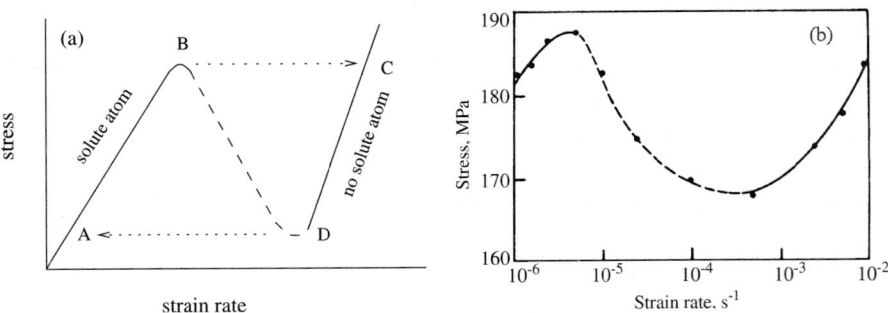

Fig. 5. (a) A schematic plot of the flow stress vs strain rate in case of the PLC. (b) The flow stress at strain $\epsilon = 8 \times 10^{-2}$ for $Al - 5\% Mg$ tested at 300 K under constant strain rate (after [21])

to represent the unstable branch. (See for more details in [21].) A plot of the flow stress as a function of the strain rate is shown in Fig. 5b. The alloy used is $Al - 5\% Mg$ tested at 300 K. Note the logarithmic scale along the x-axis.

3.1 Dynamical Interpretation of Negative Strain Rate Sensitivity

Early theories based on DSA do not deal with the temporal aspect [11, 17, 18] and thus are unsuitable for analyzing the dynamical aspects of the PLC effect. The first dynamical model that uses dislocations and their interactions to explain generic features of the PLC effect was due to Ananthakrishna and his coworkers [14, 15]. The dynamical basis of the model by its very nature allows for explicit inclusion and interplay of different time scales inherent in the dynamics of dislocations. The model consists of rate equations for different types of dislocation densities coupled to the machine equation. Despite the simplicity of the model, many generic features of the PLC effect such as the existence of a window of strain rates and temperatures within which it occurs, etc., were correctly reproduced. More importantly, the *negative SRS was shown to emerge naturally* in the model as a result of nonlinear interaction of the participating defects [14, 15]. One prediction that is specific to the model is the chaotic nature of serrations at low and medium strain rates which has been later confirmed by analysing experimental time series [22, 23, 24].

3.2 The Ananthakrishna's Model

In the dynamical model [14], the well separated time scales mentioned in the DSA are mimicked by three types of dislocations, namely, the fast mobile, the slow immobile and the intermediate "decorated" Cottrell type dislocations. The basic idea of the model is that all the qualitative features of the PLC effect emerge from the nonlinear interaction of these few dislocation populations, assumed to represent the collective degrees of freedom of the system. Following the notation in [15], we shall briefly outline the model in the scaled variables. In our model, a natural basis for including the spatial coupling is through the cross-slip mechanism proposed earlier [9] with an important difference (see below). The model consists of densities of mobile, immobile, and Cottrell's type dislocations denoted by $\rho_m(x,t)$, $\rho_{im}(x,t)$i, and $\rho_c(x,t)$ respectively, in the scaled form. The evolution equations are

$$\frac{\partial \rho_m}{\partial t} = -b_0 \rho_m^2 - \rho_m \rho_{im} + \rho_{im} - a\rho_m + \phi_{eff}^m \rho_m + \frac{D}{\rho_{im}} \frac{\partial^2 (\phi_{eff}^m(x) \rho_m)}{\partial x^2}, \quad (1)$$

$$\frac{\partial \rho_{im}}{\partial t} = b_0 (b_0 \rho_m^2 - \rho_m \rho_{im} - \rho_{im} + a\rho_c), \quad (2)$$

$$\frac{\partial \rho_c}{\partial t} = c(\rho_m - \rho_c). \quad (3)$$

The model includes the following dislocation mechanisms: immobilization of two mobile dislocations due to the formation of locks ($b_0 \rho_m^2$), the annihilation of a mobile dislocation with an immobile one ($\rho_m \rho_{im}$), the remobilization of the immobile dislocation due to stress or thermal activation (ρ_{im}). It also includes the immobilization of mobile dislocations due to solute atoms ($a\rho_m$). Once a mobile dislocation starts acquiring solute atoms we regard it as the Cottrell's type dislocation ρ_c. As they progressively acquire more solute atoms, they eventually stop, then they are considered as immobile dislocations ρ_{im}. Alternately, the process of aggregation of solute atoms can be regarded as the definition of ρ_c, i.e., $\rho_c = \int_{-\infty}^{t} dt' \rho_m(t') K(t-t')$, where $K(t)$ is an appropriate kernel which for the sake of simplicity is modeled using a single time scale, $K(t) = ce^{-ct}$. The convoluted nature of the integral physically implies that the mobile dislocations to which solute atoms aggregate earlier will be aged more than those which acquire solute atoms later (see [15]). The fifth term in Eqn.(1) represents the rate of multiplication of dislocations due to cross-slip. This depends on the velocity of the mobile dislocations taken to be $V_m(\phi) = \phi_{eff}^m$, where $\phi_{eff} = (\phi - h\rho_{im}^{1/2})$ is the scaled effective stress, ϕ the scaled stress, m the velocity exponent and h a work hardening parameter. The last term in (1) corresponds to the spatial coupling. Within the scope of our model, cross-slip is a natural source of spatial coupling as dislocations generated due to cross slip at a point spread over to the neighboring elements. We also note that cross-slip spreads only into regions of minimum back stress which is taken to result from the immobile dislocation density ahead of it. Finally, a, b_0, and c are the scaled rate constants referring, respectively, to the concentration of solute atoms slowing down the mobile dislocations, the thermal and athermal reactivation of immobile dislocations, and the rate at which the solute atoms are gathering around the mobile dislocations. These equations are coupled to the machine equation

$$\frac{d\phi(t)}{dt} = d[\dot{\epsilon} - \frac{1}{l}\int_0^l \rho_m(x,t)\phi_{eff}^m(x,t)dx] , \qquad (4)$$

where $\dot{\epsilon}$ is the scaled applied strain rate, d the scaled effective modulus of the machine and the sample, and l the dimensionless length of the sample. We also note here that there is a feed back mechanism between (4) and (1). It is interesting to point out that the machine equation (4) that determines the stress depends on the difference between the applied strain rate and average plastic strain rate generated in the sample. Thus, the nature of internal relaxation can influence stress generated in the sample which in turn determines the dislocation multiplication in (1).

3.3 Slow Manifold Analysis

Later, Rajesh and Ananthakrishna [15] provided a dynamical interpretation of the negative branch of the stress-strain rate relation in terms of the structure of the slow manifold of the model [25]. The methodology of slow manifold analysis is basically a "dimensional reduction" procedure that provides a smaller dimensional version of the original dynamical system that retains the essential dynamics. This is best suited for analysis of nonlinear slow-fast dynamical systems wherein all trajectories are attracted, in the long time limit, to a subspace of $\mathcal{R}^n$ which forms a topological invariant manifold.

In practice, the slow manifold of the model (without spatial degrees of freedom) is conventionally done by setting the derivative of the fast variable to zero [15, 25]

$$\dot{\rho}_m = g(\rho_m, \phi) = -b_0 \rho_m^2 + \rho_m \delta + \rho_{im} = 0 \,. \tag{5}$$

where $\delta = \phi^m - \rho_{im} - a$. The variable δ has been shown to have all the features of an effective stress and thus plays an important physical role [25], particularly in studying the pinning-unpinning of dislocations. We note that δ is a function of the two slow variables ϕ and ρ_{im}. Thus, δ takes on small positive and negative values. Indeed, it can be easily shown that for $\delta < 0$, the mobile dislocation density is small and this part of the slow manifold is designated by S_2, while for $\delta > 0$, the mobile density is large (see Fig. 6a). Figure 6a shows the two pieces of the slow manifold S_1 and S_2. The slow manifold pieces S_1 and S_2 arise from two different physical processes, i.e., pinning and unpinning of dislocations. On S_1, the dynamics is largely controlled by the fast evolution of mobile dislocation density ρ_m. However, on S_2, ρ_m is small and is nearly constant as can be seen from Fig. 6a, i.e., the state of the system is one in which most dislocations are in the pinned state on S_2. In contrast, on S_1, as the mobile dislocation density is large compared to the immobile de nsity ρ_{im}, it is referred to as the unpinned state of dislocations. Rajesh and Ananthakrishna [15] then use the equations of motion to make a correspondence with

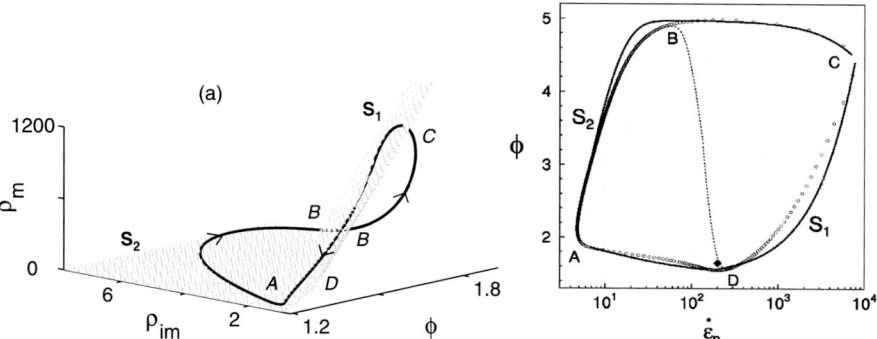

Fig. 6. (a) Evolution of the trajectory along with the bent-slow manifold (S_1 and S_2) structure in $(\rho_m, \rho_{im}, \phi)$ space indicated by the gray plane. Here, ρ_m denotes mobile dislocation density, ρ_{im} immobile dislocation, and ϕ scaled stress [27]. (b) The *empty circles* show the phase space projection of stress ϕ vs. strain rate $\dot{\epsilon}_p$. The *dotted line* represents the negative strain rate sensitivity (SRS) region. The *thick lines* are analytical approximations of the corresponding regions [15]

the negative strain rate sensitivity. This dynamics actually translates into the two stable dissipative branches of the "N" shaped strain rate sensitivity curve shown in Fig. 6b. Note that S_1 and S_2 are stable manifold pieces. The left branch AB of Fig. 6a corresponds to the pinned state of dislocations on S_2, while the right branch CD corresponds to the unpinned state of dislocations on S_1. The region BC (following the direction of the arrow in Fig. 6a) where the density of mobile dislocations rapidly increases denotes the unstable region. Precisely for this reason, this region maps to unstable region of the SRS curve (Fig. 6b). This analysis stresses the importance of using sound dynamical tools such as the slow manifold approach, as a basis for studying complex dynamics rather than using phenomenological concepts such as the negative SRS.

The success of the model can be attributed to the fact that the underlying physical mechanism-the dynamic strain aging-has been adopted in a subtle way. Indeed, the aging of mobile dislocations is represented as a transformation of the mobile density to the Cottrell density. Further, its influence on other dislocation densities is also included. The net result is that the negative strain rate sensitivity comes out naturally. Apart from this, the model predicts several features observed in experiments such as the crossover from a low dimensional chaotic state found at low and medium strain rates [22, 23, 24, 26] to a power law state of stress drops seen at high strain rates [24, 26, 27]. The model also predicts different band types as a function of strain rate [27, 28].

The above discussion on solid-on-solid and the PLC effect shows that in former case, there is a static threshold at near zero velocities followed by a velocity weakening law. The velocity weakening law itself is experimentally measurable quite unlike other situations where this branch has to be deduced. In a sense, the low velocity resistive branch is singular which in principle could correspond to a steep resistive branch. Indeed, the fact that there is creep at low velocities is suggestive of such a steep branch. Further, the velocity weakening law has $\ln V$ dependence. (It is possible that there is a resistive branch at high velocities as well.) In most other cases, one usually measures only the two resistive branches and not the unstable branch. For example, in the case of the PLC effect, only the resistive branches are measurable in the strict sense and the negative branch that is "measured" should be taken as a trend. One aspect that is common between the two examples discussed is the logarithmic dependence of the stress on drive rates in the drive rate weakening regime. Further, it is clear that in both cases, the NFRC arises as a consequence of a competion between the internal relaxation time scales and driving time scale. Finally, the PLC case is one of the few cases where one can build a model that clarifies the concept of negative SRS of the flow stress.

In a general context, however, the nature of stick-slip motion of a physical system is not merely determined by the force-drive-rate function. In fact, other relevant degrees of freedom and the associated time scales do have influence on the dynamics. This will be clearly demonstrated in the case of a model for peeling of an adhesive tape.

In the next section, we shall briefly summarize experimental and theoretical works on the problem of peeling of an adhesive tape. A model that has been introduced a decade and half ago that is relevant to the experimental situation has remained improperly understood as it belongs to a special class of differential equations. We shall also show that these equations are not complete as they ignore one more time scale which lifts the singular nature of the equations.

4 Peeling of Adhesive Tapes: Stick-Slip Instabilities

The phenomena of adhesion has attracted renewed interest due to scientific challenges it poses as well as their industrial importance. The literature on this subject is immensely diverse. A large body of work addresses the interfacial properties, interactions, debonding, and rupture of adhesive bonds. Science of adhesion is truly interdisciplinary involving a great variety of different interrelated physical phenomena like friction, fracture, mechanics of contact, and visco-plastic deformation. The detailed mechanisms of such a complicated mixture of phenomena are not yet well understood.

Tests of adhesion are essentially fracture tests designed to study adherence of solids and generally involve normal pulling off, shearing and peeling. Peeling also provides a rich insight into fracture mechanics as the dynamics is highly nonlinear and shows a variety of complex instabilities. Furthermore, peeling experiments are comparatively easy to set up in laboratory and recorded response helps to extract useful information on the nonlinear features of the system.

There are several attempts to get detailed understanding of the peeling process. Early studies by Bikermann [29] and Kaeble [30] have attempted to explain the results by considering the system as a fully elastic object. This is clearly inadequate as it ignores the viscoelastic nature of the glue at the contact surface and therefore cannot capture many important features of the dynamics. The first detailed experimental study on peeling of an adhesive tape was due to Maugis and Barquins [12]. The peeling experiments carried out at constant pull speed condition show that peeling is jerky within a window of pull speeds accompanied by acoustic emission [31, 32, 33]. More recently, constant load experiments have also been carried out [34, 35]. An important characteristic feature of the peeling process is that the experimental strain energy release rate G shows two stable branches separated by an unstable branch as shown in Fig. 2a. It is well established that there are three different modes of failure during the peeling process. At low applied velocities, the peeling front keeps pace with the pull velocity and the failure mode is cohesive whereas at high pull velocities, the failure is adhesive. If the pull speed is within some intermediate range, one sees oscillations between the two fracture mechanisms that exhibits load values appropriate to both crack initiation and crack arrest. Maugis and Barquins [12] report that the pull force shows a rich variety of behavior ranging from sinusoidal, sawtooth and highly irregular (chaotic as these authors refer to) wave patterns with increasing pull speeds. They also report that the average amplitude of the pull force decreases with increasing pull speeds.

Apart from detailed experimental investigation of the peeling process, Maugis and Barquins [12], have also contributed substantially to the understanding of the dynamics of the peeling process. However, the first dynamical analysis is due to Hong and Yue [13] who use a "N" shaped function to mimic the dependence of the peel force on the rupture speed. They showed that the system of equations exhibits periodic and chaotic stick-slip oscillations. However, the jumps in the rupture speed were introduced *externally* once the rupture velocity exceeded the limit of stability [32, 36]. Thus, the stick-slip oscillations are *not* obtained as a natural consequence of the equations of motion and hence these results [13] are the artifacts of the numerical procedure followed. Ciccotti et al. [32] interpret the stick-slip jumps as catastrophes. Again, the belief that the jumps in the rupture velocity cannot be obtained from the equations of motion appears to be the motivation for introducing the action of discrete operators on the state of the system to interpret the stick-slip jumps [32]. Lastly, there are no reports that explain the decreasing amplitude of the peel force with increasing pull speed as observed in experiments.

In the following, we first derive these equations using an appropriate Lagrangian [37]. We then show that a time scale corresponding to the kinetic energy of the tape is missing in modeling the physical situation. Once this term is included, the singular nature of the solutions are lifted demonstrating the importance of including all time scales [38]. When this time scale is ignored, one obtains the same equations used in [12, 13] which fall into the class of differential-algebraic equations (DAE) requiring an appropriate algorithm for their solutions [39]. (See for details [37, 38].)

4.1 Equation of Motion

We start by considering the geometry of the experimental setup shown schematically in Fig. 7a. An adhesive roll of radius R is mounted on an axis passing through O normal to the paper and is pulled at a constant velocity V by a motor positioned at O' with a force F acting along PO'. Let the distance between O and O' be l, and that between the contact point P and O' be L. The point P moves with a local velocity v which can undergo rapid bursts in the velocity during rupture. The line PO' makes an angle θ with the tangent at the contact point P. The point P subtends an angle α at O with the horizontal line OO'. We denote the elastic constant of the adhesive tape by k, the elastic displacement of the tape by u, the angular velocity by ω and the moment of inertia of the roller tape by I. The angular velocity itself is identified by $\omega = \dot{\alpha} + v/R$. The geometry of the setup gives $L\cos\theta = -l\sin\alpha$ and $L\sin\theta = l\cos\alpha - R$ which further gives, $L^2 = l^2 + R^2 - 2lR\cos\alpha$. The total velocity V at O' is then made up of three contributions [12], given by

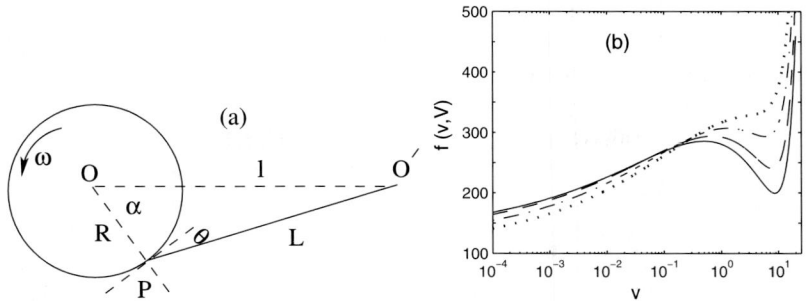

Fig. 7. (a) A schematic diagram of the experimental setup [12]. (b) Plots of $f(v, V)$ as a function of v (x axis in log scale) for $V = 1$ (*solid curve*), $V = 2$ (*dashed curve*), $V = 4$ (*dashed and dotted curve*), $V = 6$ (*dotted curve*) [37]

$V = v + \dot{u} - \dot{L}$, which gives

$$v = V + \dot{L} - \dot{u} = V - R\cos\theta\,\dot{\alpha} - \dot{u}\,. \tag{6}$$

Following standard methods in mechanics, it is straightforward to derive the equations of motion for α and ω by considering $(\alpha, \dot{\alpha}, u, \dot{u})$ as the generalized coordinates. The corresponding Lagrangian of the system can be written as [38]

$$\mathcal{L}(\alpha, \dot{\alpha}, u, \dot{u}) = \frac{I}{2}[\omega(\alpha, \dot{\alpha}, u, \dot{u})]^2 + \frac{1}{2}m\dot{u}^2 - \frac{k}{2}u^2\,. \tag{7}$$

The first term on the right hand side is the kinetic energy of the roller tape, the second term is the kinetic energy of the stretched tape with m referring to the mass of stretched tape, and the last term is the potential energy of the tape. We write the dissipation function as

$$\mathcal{R} = \Phi(v, V) = \int f(v, V)\,dv\,, \tag{8}$$

where $f(v, V)$ physically represents the peel force function which we assume is dependent on rupture speed v as well as the pull speed V. Further we shall assume $f(v, V)$ to be derivable from a potential function $\Phi(v, V)$. The physical origin of the dissipation is the competition between the internal relaxation time scale of the viscoelastic fluid and the time scale determined by the applied velocity [33]. When the applied velocity is low, there is sufficient time for the viscoelastic fluid to relax. As we increase the applied velocity, the relaxation of the fluid gets increasingly difficult and thus behaves much like an elastic solid. The effect of competing time scales is well represented by Deborah number [40] which is the ratio of the time scale for structural relaxation to the characteristic time scale for deformation. Indeed, in the studies on Hele-Shaw cell with mud as the viscous fluid, one observes a transition from viscous fingering to viscoelastic fracturing [41] with increasing rate of invasion of the displacing fluid.

As stated earlier, the decreasing nature of the amplitude of pull force with pull velocity has not been explained so far. A similar feature observed in the PLC serrations has been modeled using a scheme referred to as dynamization of the negative strain rate sensitivity (SRS) of the flow stress $f(\dot{\epsilon}_p)$ [42, 43], where $\dot{\epsilon}_p$ is the plastic strain rate. Kubin et al. modify this function to depend on the applied strain rate, $\dot{\epsilon}_a$ as well, i.e., the negative SRS of the flow stress is taken to be a function of $\dot{\epsilon}_p, \dot{\epsilon}_a$ such that the gap between the maximum and the minimum of the function $f(\dot{\epsilon}_p, \dot{\epsilon}_a)$ decreases with increasing $\dot{\epsilon}_a$. The underlying physics for this modification is the incomplete plastic relaxation occurring at high applied strain rates. In the present case, as explained in the preceding paragraph, incomplete viscoelastic relaxation should be anticipated. Thus, we consider f to depend on V also apart from v, in a way that the gap in f decreases as a function of the pull speed V (see Fig. 7b).

Using the Lagrange equations, we obtain the following equations of motion

$$\dot{\alpha} = -\frac{\dot{v}}{R} + \frac{R}{I} \frac{\cos\theta}{(1-\cos\theta)} f(v, V), \tag{9}$$

$$m\ddot{u} = \frac{1}{(1-\cos\theta)}[f(v, V) - ku(1-\cos\theta)]. \tag{10}$$

Now, consider the case when $m = 0$ which corresponds to ignoring the kinetic energy of the stretched part of the tape. Then, we get [37],

$$\dot{\alpha} = \omega - v/R, \tag{11}$$

$$I\dot{\omega} = RF\cos\theta, \tag{12}$$

$$\dot{F} = k(V-v) - k\cos\theta\,(\omega R - v), \tag{13}$$

$$f(v, V) = F(1-\cos\theta). \tag{14}$$

These equations are the same as derived and used earlier [12, 13, 36] while (11)–(13) are differential equations, (14) is an algebraic constraint necessitating the use of differential-algebraic scheme to obtain the numerical solution of (11–14) [37, 39]. We shall refer to (11–14) as differential-algebraic equations (DAE). We note that even though α, ω, F (or u) evolve with time determined by the above equations, the values should satisfy the algebraic constraint given by (14) at all times. This is the root cause of the problems encountered in obtaining the numerical solutions which have forced artificial approaches in the past [36].

However, when m is finite, however small, (9, 10) can be handled, although not in their present form as they have to satisfy the constraint (6). In the spirit of classical mechanics of systems with constraints (see [44]), we derive the equation for the acceleration variable $\dot{v}$ by differentiating (6) giving the following equations of motion for α, ω, u and v

$$\dot{\alpha} = \omega - v/R, \tag{15}$$

$$\dot{\omega} = \frac{R}{I} \frac{\cos\theta}{(1-\cos\theta)} f(v,V), \tag{16}$$

$$\dot{u} = V - v - R\cos\theta\, \dot{\alpha}, \tag{17}$$

$$\dot{v} = -\ddot{u} + R\sin\theta\, \dot{\theta}\, \dot{\alpha} - R\cos\theta\, \ddot{\alpha}. \tag{18}$$

Using (9), (10) in (18), we get

$$\dot{v} = \frac{1}{(1-\cos\theta)}\left[\frac{ku}{m} - \frac{f(v,V)}{m(1-\cos\theta)} - \frac{(R\cos\theta)^2 f(v,V)}{I(1-\cos\theta)}\right.$$
$$\left. + \frac{R}{L}\dot{\alpha}^2(l\cos\alpha - R(\cos\theta)^2)\right]. \tag{19}$$

We note that (15, 17, 19) constitute ordinary differential equations which pose no difficulty in obtaining solutions unlike (11–13) and (14). The latter however requires an appropriate algorithm which we shall discuss in the following.

It is easy to show that these set of equations are unstable when $f'(v = V,V)$ changes from positive to negative value and the system undergoes a Hopf bifurcation leading to a limit cycle solution. The limit cycle reflects the abrupt jumps between the two positive slope branches of the function $f(v,V)$.

We also note that in most cases, α is small and one can approximate (14) by

$$F(1-\cos\theta) - f(v,V) \simeq F(1+\alpha) - f(v,V) = 0. \tag{20}$$

5 Algorithm

Equations (11–14) can be written as [37]

$$M\dot{\mathbf{X}} = \phi(\mathbf{X}), \tag{21}$$

where $\mathbf{X} = (\alpha, \omega, F, v)$, ϕ is a vector function that governs the evolution of $\mathbf{X}$ and M is a singular "*mass matrix*" [39] given by,

$$M = \begin{pmatrix} 1 & 0 & 0 & 0 \\ 0 & 1 & 0 & 0 \\ 0 & 0 & 1 & 0 \\ 0 & 0 & 0 & 0 \end{pmatrix}.$$

Equation (21) can be solved using singular perturbation technique [39] in which the singular matrix M is perturbed by adding a small constant ϵ such that the singularity is removed. The resulting equations can then be solved numerically and the limit solution obtained as $\epsilon \to 0$. We have checked the numerical solutions for ϵ values ranging from 10^{-7} to 10^{-11} in some cases and the results do not depend on the value of ϵ used as long as it is small. The results presented below, however, are for $\epsilon = 10^{-7}$. We have solved (21) using a standard variable-order solver, MATLAB ODE15S program.

We have parametrized the form of $f(v, V)$ as

$$f(v, V) = 400v^{0.35} + 110v^{0.15} + 130e^{(v/11)} - 2V^{1.5}$$
$$-(415 - 45V^{0.4} - 0.35V^{2.15})v^{0.5}, \qquad (22)$$

to give values of the extremum of the peel velocity that mimic the general form of the experimental curves [12]. The measured strain energy release rate $G(V)$ from stationary state measurements is shown in Fig. 2a. The decreasing nature of the gap between the maximum and minimum of $f(v, V)$ for increasing V is clear from Fig. 7b.

5.1 Results

We have solved the set of ordinary differential equations (15–17) and (19), henceforth referred to as ODE and differential-algebraic equations (11–14), as the DAE. We have studied the dynamics over a wide range of values of I (kg m^2), V (m s^{-1}) for m (kg) ranging from 10^{-4} to 0.1 (keeping $k = 1000$ N m^{-1}, $R = 0.1$ m, and $l = 1$ m). Henceforth, we suppress the units for the sake of brevity. (For detailed analysis see [37, 38].)

A rough idea of the nature of the dynamics can be obtained by comparing the frequency $\Omega_u = (k/m)^{1/2}$ associated with u with $\Omega_\alpha = (Rf/I)^{1/2}$ corresponding to α. Since f (in N) is limited to 180–280, the range of Ω_α (s^{-1}) is 1342–1673 for small I ($\sim 10^{-5}$) decreasing to 42–53 for large I (0.01). In comparison, Ω_u is 3162 for $m = 10^{-4}$ decreasing to 100 for $m = 0.1$. Thus, one expects the nature of solutions to be influenced with increasing m for fixed I.

As stated earlier, the solutions for $m \to 0$ are similar to the DAE solutions, and wherever necessary we shall provide the DAE solutions also. Here, we present results (obtained after discarding the initial transients) for a few representative values of the m when (I, V) are at low and high values. Further we show that mass of the tape has a strong influence on the nature of the dynamics.

It is clear that we should recover the DAE solutions for low mass limit. As an illustration, we first show that the low mass limit of the ODE is essentially the same as the DAE solutions. Figures 8a and 8b show the phase plots in the $v - F$ plane obtained using the DAE algorithm (11–14) and ODE equations [(15–17) and (19)] for $m = 10^{-4}$ respectively keeping $I = 10^{-5}, V = 1$. It is evident that the DAE solution is similar to the ODE solution. From henceforth, for all practical considerations, we can use $m = 10^{-4}$ to represent the DAE solutions. Much more complex dynamics emerges as a result of a competition between this additional time scale and other time scales present in the system. Consider the results for $m = 10^{-4}$ in the low mass limit and 0.1 in the high mass limit, for $I = 10^{-2}$ and $V = 1$. The small m plots

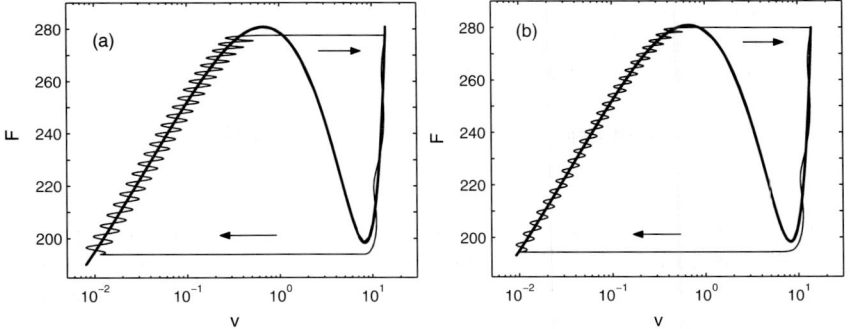

Fig. 8. (a) The $v - F$ phase plots for $I = 10^{-5}, V = 1$ corresponding to the DAE solutions. *Bold line* shows $f(v,1)$. (b) The $v - F$ phase plots for $I = 10^{-5}$ and $V = 1$ corresponding to the ODE solutions for $m = 10^{-4}$. *Bold line* shows $f(v,1)$. (Unit of m is in kg, v in m s^{-1}, and F in N [38])

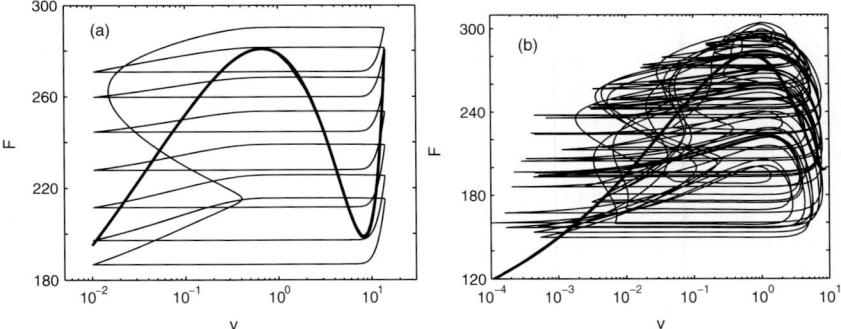

Fig. 9. (a), (b) Phase plots of $v - F$ obtained for $m = 10^{-4}$ and 0.1 for $I = 10^{-2}, V = 1$. *Bold line* shows $f(v,1)$. (Units of v, V are in m s^{-1}, F in N, I in kg m^2, and t in s. See [37, 38])

are provided for the sake of comparison as they essentially correspond to the DAE solution. Consider the phase plots in $v - F$ plane shown in Fig. 9a for $m = 10^{-4}$ and Fig. 9b for $m = 0.1$. It is clear that the influence of increasing m is significant. In particular, note that while the orbit corresponding to the low mass case visits the high velocity branch, the orbits corresponding to the large mass case does not. Indeed, such solutions which have a tendency to stick to the unstable branch are known as canard solutions [45]. The effect of the additional time scale due to finite mass of the tape is even more evident in the plots of $v(t)$ for the low and high mass cases shown in Figs. 10a and 10b respectively. The sharp changes in the velocity corresponding to the small mass case are rendered smooth when the mass of the tape is increased. It may be noted from (10), that the inertial contribution from the stretched tape increases with m. This force has to be balanced by the difference between the

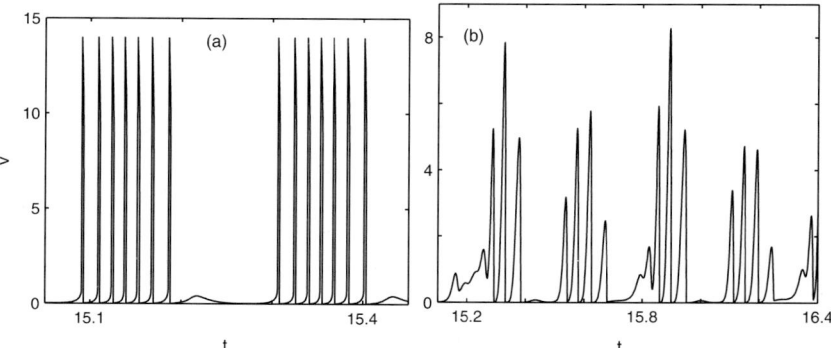

Fig. 10. (a), (b) Plots of $v(t)$ for $m = 10^{-4}$ and 0.1 for $I = 10^{-2}, V = 1$. (Units of v, V are in m s^{-1}, I in kg m^2, and t in s. See [37, 38])

pull force $F = ku$ and peel force $f(v, V)$. Finally, it is clear that the phase plot (Fig. 9b) fills the space and is suggestive of chaotic dynamics.

The chaotic nature can be ascertained by calculating the Lyapunov spectrum. Using the QR decomposition method [46], we have calculated the Lyapunov spectrum and find a large positive exponent with a value $\sim 4.4\,\mathrm{s}^{-1}$.

Increasing m does not always increase the level of complexity of the solutions. As an example, Figs. 11a and 11b show plots of $v(t)$ for $m = 10^{-4}$ and 0.1 respectively for high roller inertia $I = 10^{-2}$ and high pull velocity $V = 4$. While the ODE solution for small m (similar to that of DAE) exhibits several sharp spikes in velocity (Fig. 11a), the solution for large mass ($m = 0.1$) is surprisingly simple and is periodic [Fig. 11b]. Indeed, this is better seen in the phase plots $v - F$ for $m = 10^{-4}$ and 0.1 shown in Figs. 12a and 12b respectively. In contrast to the low mass $v - F$ plot which is chaotic, a simple limit cycle emerges for $m = 0.1$.

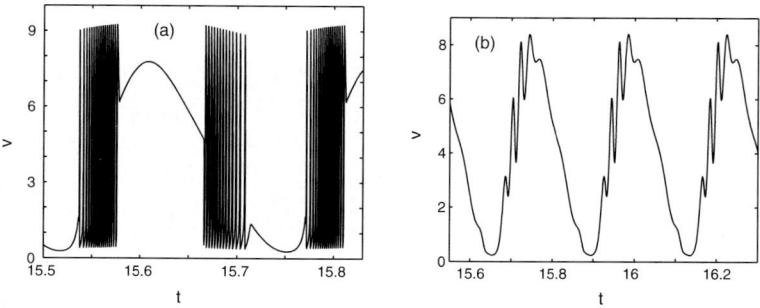

Fig. 11. (a), (b) Plots of $v(t)$ for $m = 10^{-4}$ and 0.1 for $I = 10^{-2}, V = 4$. (Units of v, V are in m s^{-1}, I in kg m^2 and t in s. See [37, 38])

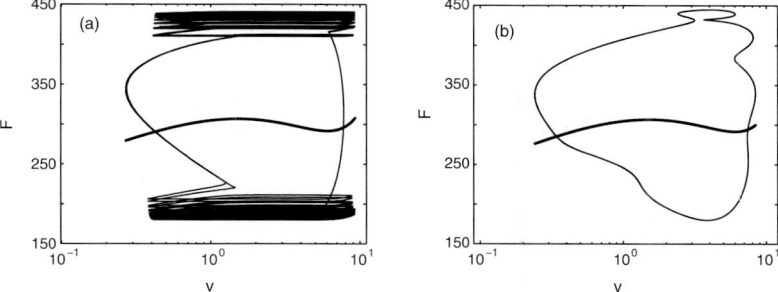

Fig. 12. (a), (b) Phase space trajectories for $m = 10^{-4}$ and 0.1 for $I = 10^{-2}, V = 4$. $f(v, V)$ is shown by a *bold line*. (Units of v, V are in m s^{-1}, F in N, I in kg m^2 and t in s. See [37, 38])

As the nature of the dynamics can vary from a simple limit cycle to a chaotic attractor as the three parameters are varied, these results can be summarized as phase-diagrams in the $I - m$ plane for different values of pull velocities, V, as shown in Fig. 13. Apart from the chaotic state (∗) seen for a few values of the parameters, for most values, the system is periodic (•) and a few other values, the attractor is long periodic (×). We also find a (marginally) chaotic attractor (◦) for $V = 1, m = 10^{-4}, I = 10^{-3}$ for which the positive Lyapunov exponent is ∼0.03 s^{-1} (which is much beyond the error in computation).

In summary, we have demonstrated that the missing time scale arising from the kinetic energy of the stretched part of the tape plays an important role in the peeling dynamics of the adhesive tape. As the inclusion of this term lifts the singularity in the equations of motion hitherto considered, stick-slip jumps across the two resistive branches emerge as a consequence of the inherent dynamics. Further, our study shows that the mass of the tape has a strong influence on the nature of the dynamics. For low pull velocities, and low inertia of the roller tape I, the complexity increases, i.e., trajectories that are not chaotic for low mass become chaotic with increasing m. In contrast, for high V, the trajectories that are chaotic for low m are rendered nonchaotic with increase of m.

It is pertinent to stress here that there are other issues related to the peeling problem. For instance, there are no theoretical models which address the dynamics of the peel front. In an addition, acoustic emission (AE) emitted during the peeling process has also remained ill understood. These problems have been addressed recently by including the spatial degrees of freedom of the peel front to study the contact line dynamics. We include an additional Rayleigh dissipative functional [47] which depends on the strain rate arising due to the rapid movement of the peeling front. The form of this dissipation functional is the same as the energy released $E_{ae}(r)$ in the form of AE

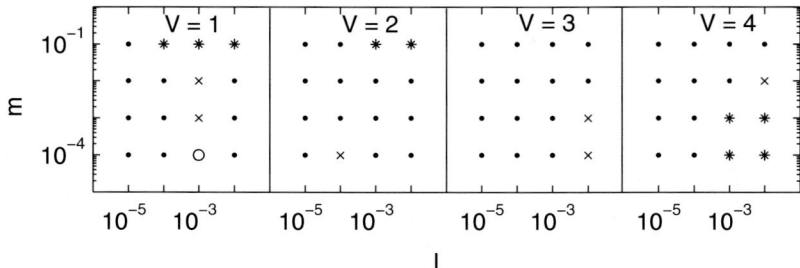

Fig. 13. Phase diagram in the $I - m$ plane for various values of V. Periodic •, long periodic ×, chaotic (marginal) ○ and chaotic ∗. (Unit of I is in kg m² and m in kg. See [38])

signals from abrupt movement of dislocations. This then forms the basis for our studies on the peel front and acoustic emission.

6 Predictability of Slip Events in Power Law States

One of the interesting result of the extended model for the peeling front is that the distribution of the burst of energy dissipated in the form of AE signals obeys a power law, both in experiments and in the model. However, power law distributions are ubiquitous features in slowly driven systems. Indeed, the conceptual framework of self-organized criticality (SOC) was introduced to explain the universality of power laws in varied systems [48, 49]. This brings up a puzzle related to the lack of intrinsic length scales and time scales in power law systems which in turn suggests that *large avalanches are inherently unpredictable* [50, 51]. Clearly, this comment is applicable to model systems as well. This has triggered considerable debate regarding earthquake predictability, as seismically active fault systems are considered to be in a SOC state [51]. However, there has been reports of increased levels of AE signals before failure of rock samples [52, 53, 54], in experiments on laboratory samples [55, 56] and precursory effects in individual earthquakes as well [50, 57, 58]. There has been attempts to predict failure (at the laboratory level as well on geological scale) within the general context of time-to-failure (TTF) models [56, 59]. Some efforts have also been made on the predictability of avalanches in SOC state [60, 61, 62]. The problem of looking for precursory effect and predictability of events has been addressed recently within the scope of modified stick-slip model [63] representing an earthquake fault originally introduced by Burridge and Knopoff [6].

Failure of materials is a common but undesirable property of materials and its prediction is of general interest in science and engineering (electrical breakdown, fracture of laboratory samples to engineering structures). Predicting earthquakes looked at as slip failure is particularly important in seismology due to the enormous damage earthquakes can cause. Whether it is at a laboratory or geological scale, this amounts to identifying useful precursors at a statistically significant level. One important non-destructive tool in fracture studies is the acoustic emission (AE) technique as it is sensitive to the microstructural changes occurring in the sample. Insight into earthquake dynamics has been obtained through fracture studies of rock samples (usually precut samples to mimic slip on preexisting tectonic faults) [64]. Such studies have established that there is a considerable overlap between AE and seismology as both are concerned about the generation and propagation of elastic waves. Quite early, the statistics of the AE signals was shown to exhibit a power law [64, 65, 66] similar to the Gutenberg–Richters law for the magnitudes of earthquakes [67] and Omori's law for aftershocks [50, 65]. These prompted further investigations to look for precursor effects that can be used for earthquake predictability [50, 54, 58, 68].

Interestingly apart from the fact that power law statistics is observed in AE signals during fracture, acoustic activity of unusually large number of situations as varied as volcanic activity [69], micro-fracturing process [70, 71], and collective dislocation motion [23, 24, 27, 72], exhibit power laws. Though the general mechanism attributed to AE is the release of stored strain energy, the details are system specific. Thus, the ubiquity of the power law statistics of AE signals suggests that the details of the underlying processes are irrelevant.

Recently, studies on rock samples report an interesting crossover in the exponent value from small amplitude regime of the AE amplitudes to large [73], a result that is similar to the well noted observation on the change in the power law exponent for small and large magnitude earthquakes [74]. The exponent value is also found to be sensitive to the deformation rate [75]. *There has been no explanation of these observations.* This can partly be traced to the lack of efforts to model AE signals.

One simple model for earthquakes that has attracted considerable attention which mimics stick-slip events on preexisting faults and the Gutenberg–Richter law is the Burridge-Knopoff (BK) model and its variants [6, 7, 8]. We introduce an additional dissipative term into the BK model that helps us to identify to precursory effect and hence to predict a major slip event within the scope of the model. We show that the new dissipative term mimics the AE bursts and captures the essential features stated above.

Deformation and/or breaking of the asperities results in an accelerated motion of the local areas of slip. We consider this accelerated motion of the local slip as responsible for acoustic emission. However, a rapid movement also prevents the system from attaining a quasi-static equilibrium which in turn generates dissipative forces that resist the motion of the slip at asperities. Such dissipative forces are modeled by the Rayleigh dissipation functional which depends on the gradient of the local velocity [47]. Indeed, such a dissipative term has proved useful in explaining the power law statistics of the AE signals during martensitic transformation [76].

The Burridge-Knopoff model for earthquakes, though known for a long time [6], has been recently popularized by Carlson and Langer [7]. Despite its limitation (lack of appropriate continuum limit, absence of long range interaction, etc.) [8, 77], it forms a convenient platform to investigate the question of predictability of large avalanches as the dissipated energy bursts themselves follow a power law and hence the state of the system is scale invariant. The model consists of a chain of blocks of mass m coupled to each other by coil springs of strength k_c and attached to a fixed surface by leaf springs of strength k_p as shown in Fig. 14a. The blocks are in contact with a rough surface moving at constant speed v (mimicking the points of contact between two tectonic plates). A crucial input into the model is the velocity-dependent frictional force between the blocks and the surface (see Fig. 14b).

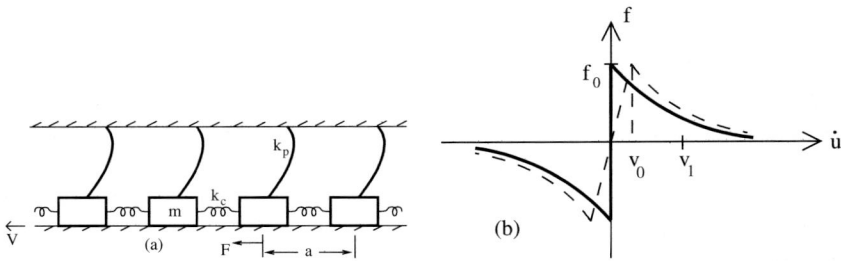

Fig. 14. (a) The Burridge-Knopoff spring block model. (b) The two forms of friction laws used

The additional dissipative force is introduced through the Lagrange's equations of motion given by $\frac{d}{dt}(\frac{\delta L}{\delta \dot{u}(x)}) - \frac{\delta L}{\delta u(x)} = -\frac{\delta F}{\delta \dot{u}(x)}$. The Lagrangian is given by $L = T - P$, where the kinetic energy T and potential energy P are respectively defined by $T = \frac{m}{2} \int (\frac{\partial u(x)}{\partial t})^2 dx$, and $P = \frac{1}{2} \int [k_c(\frac{\partial u(x)}{\partial x})^2 + k_p(u(x))^2] dx$. Here u is the displacement of the blocks measured from the initial equilibrium position. The Rayleigh dissipative functional [47] is given by $R = \frac{\gamma}{2} \int (\frac{\partial \dot{u}(x)}{\partial x})^2 dx$, where γ is a dissipation coefficient. The total dissipation F is the sum of $R(t)$ and frictional dissipation is given by $F_{fr} = \int [f(v + \dot{u}(x))] dx$, where the frictional force is taken to be derivable from a potential like function. Then the equation of motion is given by

$$m\frac{\partial^2 u}{\partial t^2} = k_c \frac{\partial^2 u}{\partial x^2} - k_p u - f(\dot{u} + v) + \gamma \frac{\partial^2 \dot{u}}{\partial x^2}, \qquad (23)$$

where the over dot refers to the time derivative. The discretized version, in the notation of [7], reads

$$\ddot{U}_j = l^2(U_{j+1} - 2U_j + U_{j-1}) - U_j - \phi(2\alpha\nu + 2\alpha\dot{U}_j)$$
$$+ \gamma_c(\dot{U}_{j+1} - 2\dot{U}_j + \dot{U}_{j-1}), \qquad (24)$$

where U_j is the dimensionless displacement of the jth block, ν is the dimensionless pulling velocity, $l^2 = k_c/k_p$ the ratio of the slipping time to the loading time and α is the rate of velocity-weakening in the scaled frictional force ϕ. (See Fig. 14b). The solid line has the form used in [7] which we refer as Coulomb form, for convenience. The dashed curve uses a resistive creep branch ending at v_0 ($\sim 10^{-7}$ here) beyond which the velocity weakening law operates.) γ_c is the scaled dissipation coefficient. The continuum limit of (24) exists for the creep branch (even in the absence of $R(t)$) which ensures a length scale below which all perturbations are damped [7]. Such a length scale is absent for the Coulomb case even as the continuum limit exists due to the additional dissipative term. (We have retained the same symbol for the dimensionless time.)

Equation (24) without the last term has been extensively studied [6, 7, 8]. Different types of stick-slip events ranging from one-block event to those extending over the entire fault (occurring roughly once in a loading period $\tau_L \sim 2/\nu$) are seen in the steady state. These earthquake-like events mimic the empirical Gutenberg – Richter law.

We have solved (24) using a fourth-order Runge-Kutta method with open boundary condition for both types of frictional laws shown in Fig. 14b. Random initial conditions are imposed. After discarding the initial transients, long data sets are recorded when the system has reached a stationary state. The parameters used here are $l = 10, \alpha = 2.5, N = 100, 200$ for $\nu = 0.01$ and 0.001 and a range of values of γ_c. The modified BK model produces the same statistics of slip events as that without the last term in (24) as long as the value of γ_c is small, typically $\gamma_c < 0.5$. The results presented here are for $N = 100$ and $\gamma_c = 0.02$.

We first note that the rate of energy dissipated [47] due to local accelerating blocks is given by $dE_{ae}/dt = -2R(t)$. As stated earlier, we interpret $R(t)$ as the acoustic energy dissipated. The calculated $R(t)$ exhibits bursts similar to the AE signals observed in experiments. A plot of the mean kinetic energy and $R(t)$ are shown respectively in Fig. 15a,b when the frictional law with a creep branch is used. For the case when the Coulomb law is used, $R(t)$ is considerably more noisy.

We first analyse the statistics of the energy bursts $R(t)$. Denoting A to be the amplitude of $R(t)$ (i.e., from a maximum to the next minimum), we find that the distribution of the magnitudes $D(A)$, shows a power law $D(A) \sim A^{-m}$. Instead of a single power law anticipated, we find that the distribution shows two regions, one for relatively smaller amplitudes and another for large values shown by the two distinct plots in the same plot Fig. 16a ($\circ$ for large and $\diamond$ for small amplitudes). (Similar results are obtained when the coulomb frictional law has a creep branch.) The value of m for the small amplitudes region (typically $<10^{-4}$) is $\sim 1.78 \pm 0.01$, significantly smaller than that for large amplitudes ($>10^{-4}$) which is $\sim 2.09 \pm 0.02$, consistent with the recent experimental result [73]. The increase in the exponent value mimics a similar observation for large magnitude earthquakes (>7.0 on the Richter scale [74]).

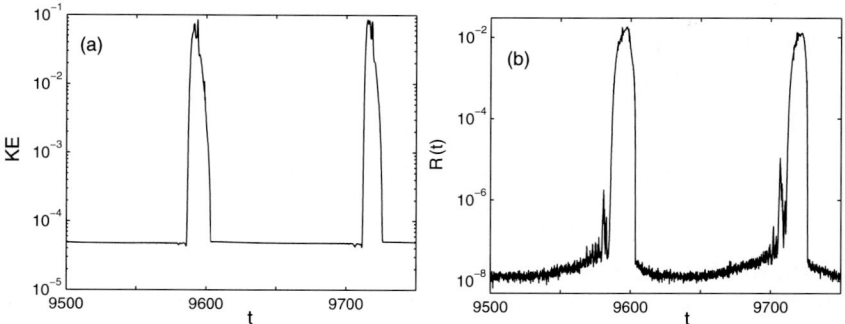

Fig. 15. (a) Mean kinetic energy as a function of time. (b) Dissipation $R(t)$ as a function of time t [63]

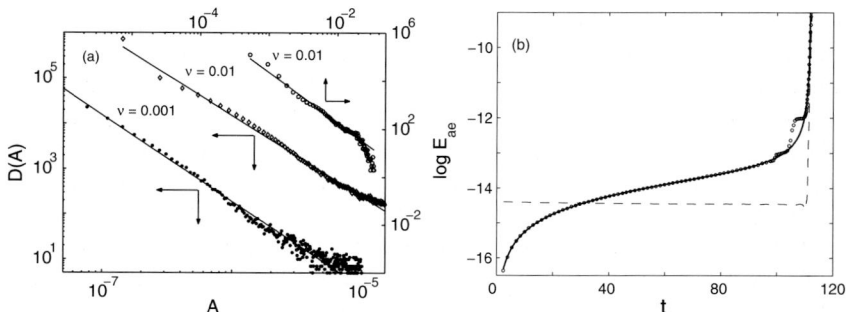

Fig. 16. (a) Distribution of amplitudes of $R(A)$, $D(A)$ versus A for small amplitudes ($\diamond$ for $\nu = 0.01$ and $\bullet$ for $\nu = 0.001$) and large amplitudes (o for $\nu = 0.01$ shifted up for clarity).(b) $log E_{ae}$ versus t (o) along with the fit (*solid line*). *Dashed line* corresponds to the mean kinetic energy. (See [63])

In contrast, we note that one single exponent fits the power law for the seismic moments in the BK model.

Recent experiments by Yabe et al. [75] report that the exponent value corresponding to relatively small amplitude regime increases with decreasing deformation rate while that for the large amplitude regime is found to be insensitive. To check this, we have calculated $D(A)$ for $\nu = 0.001$ shown in Fig. 16a curve ($\bullet$) which shows that the exponent for small amplitude increases to 1.91 ± 0.02. However, we find that the exponent for larger amplitude regime is insensitive (not shown) to the changes in ν. This result can be physically explained by analyzing the influence of the pulling velocity on slip events of varying sizes. We first note that $R(t)$ depends on the difference in the velocities of neighboring blocks. The velocity of "microscopic" events (small number of blocks) has been shown to proportional to ν [7]. For single block events, as the neighboring blocks are at rest, the number of such events are fewer in proportion to the pulling speed, both of which are evident from Fig. 16a. For the two block events, the contribution comes mostly from the edges as the difference in the velocities of the two blocks are of similar magnitude. In a similar way, it can be argued that for slip events of finite size, the extent of the contribution to $R(t)$ is decided by the ruggedness of the velocity profile within the slipping region; the magnitude of $R(t)$ is lower if the velocity is smoother. The ruggedness of the velocity profile, however, is itself decided by how much time the system gets to "relax". At lower values of ν, there is sufficient time for the blocks to attain nearly the same velocities as the neighboring blocks compared to that at higher ν values. Thus, the larger slip events contribute lesser to $R(t)$ for smaller ν values and hence the slope of $log D(A)$ increases for smaller value of ν.

Now we consider the possibility of a precursor effect. A plot of $R(t)$ is shown in Fig. 15b for the frictional law with a creep branch. *A gradual increase in the activity of the energy dissipated can be seen which accelerates just prior to the occurrence of a "major" slip event.* The rapid increase in $R(t)$ coincides with the abrupt increase in the mean kinetic energy (KE) shown by the dashed line. Here, we have used KE as a measure of event size as it is a good indicator of the magnitude of the slip events. We find similar increase in $R(t)$ for all "major" slip events. (In our simulations, the KE of observable events ranges from 10^{-4} to 0.2. We refer to all such events as "major" events.) This suggests that $R(t)$ can be used as a precursor for the onset of a major slip event. As $R(t)$ is noisy, a better quantity for analysis is the cumulative energy dissipated $E_{ae}(t)$ ($\propto \int_0^t R(t') dt'$). $E_{ae}(t)$ grows in a stepped manner with their magnitudes increasing as we approach a major event. A plot of $log E_{ae}$ is shown in Fig. 16b along with a fit (continuous curve) having the functional form,

$$log E_{ae}(t) = -a_1 t^{-\alpha_1}[1 - a_2|(t-t_c)/t_c|^{-\alpha_2}]. \qquad (25)$$

Here, t is the time measured from some initial point after a major event. The constants a_1, a_2, α_1 and α_2 and t_c are adjustable. The crucial parameter t_c is the time of occurrence of a major slip event (often referred to as the "failure" point). It is clear that the fit is striking. Given a reasonable stretch of the data, the initial increasing trend in $log E_{ae}$ is easily fitted to a "stretched exponential," i.e., $-a_1 t^{-\alpha_1}$. The additional term is introduced to account for the observed rapid increase in the activity as we approach the major event. In contrast, the mean kinetic energy abruptly increases as a major event is reached (dashed line in Fig. 16b). It is clear that the estimated t_c agrees quite well with that of the mean KE of the event.

Now, we address the question of predictability of major slip events using $E_{ae}(t)$. This is equivalent to determining the correct t_c ahead of the event. Given $E_{ae}(t)$ over a reasonable initial stretch of time, say till t_1 (the first arrow in Fig. 17a), we find that the four constants a_1, a_2, α_1 and α_2 are already well determined (within a small error bar). These change very little with time. (Only t_c changes.) A fit to (25) also gives $t_c^{(1)}$ at t_1 (which can only

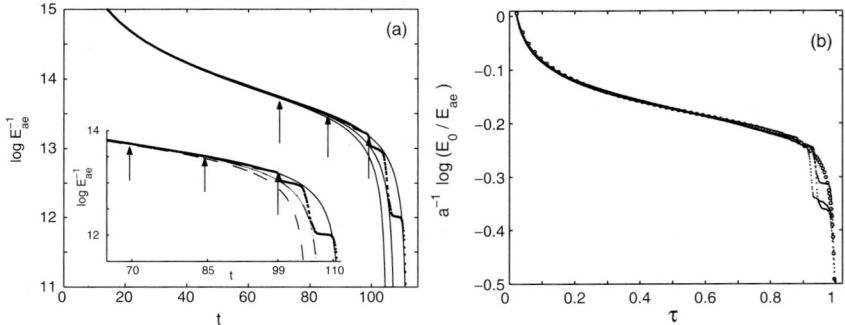

Fig. 17. (a) A plot of $logE_{ae}^{-1}$ versus t. Inset shows the enlarged section at time t_1 $(--)$, t_2 $(-\cdot)$ and t_3 (*solid line*). Data shown $(\cdot)$ is indistinguishable from the fit for t_3. (b) Collapsed data using $a_1^{-1}log(E_{ae}(0)/E_{ae}(\tau))$ vs. τ for three different events along with the fit shown by ($\circ$). (See [63])

be considered as an estimate based on the data till t_1). One such curve is shown by the dashed line with the first arrow at t_1. Note that it is more useful to use $logE_{ae}^{-1}$ as plotted in Fig. 17a. However, as time progresses, the data accumulated later usually deviates from the predicted curve if t_c is inaccurate as is the case for the fits till t_1 and t_2 for instance (Fig. 17a). If on the other hand, the deviation of the predicted curve from the accumulated data decreases with passage of time within the error bar (as is the case for the region just before t_3), then, the value of t_c is likely to be accurate. Indeed, the extrapolated continuous curve corresponding to the data fit till $t = t_3$ (the third arrow in Fig. 17a) with the predicted $t_c^{(3)}$ is seen to follow the data very well. (Usually, the data deviates from the predicted curve with a sudden decrease in E_{ae}^{-1} which is again an indication of a coherent slipping of several blocks before the onset of a fully delocalized event. But the general trend soon follows the extrapolated curve.) Then, t_3 can be taken as the warning time for the onset of the major event. The actual t_c read off from the kinetic energy plots is 110.4 where as the predicted t_c is 111.6 giving a percentage error in the prediction $\sim 1\%$.

If the approach to all slip events is described by the scale invariant form, then one should expect to find a data collapse for different events. Indeed, in terms of a scaled time $\tau = t/t_c$, we find that the data corresponding to different events collapses into a single curve given by

$$a_1^{-1}\log\left(\frac{E_{ae}(0)}{E_{ae}(\tau)}\right) = \tau^{-\alpha_1}[1 - a_2|(\tau-1)|^{-\alpha_2}] + a_1^{-1}logE_{ae}(0). \quad (26)$$

A plot corresponding to three different events is shown in Fig. 17b along with the fit. The results are similar when the Coulomb frictional law is used except that $R(t)$ is more noisy and hence prone to slightly larger errors in predicted t_c.

In summary, the model mimics the bursts of acoustic energy. The model also shows that the exponent values in the power law state for the energy dissipated $R(t)$ for small and large amplitude regimes are different, with the former being more sensitive to the pulling speed. The dependence of the exponent on the pulling speed has been traced to the form of $R(t)$, namely, the gradient of the local velocity. More significantly, the analysis shows that it is possible to predict a major event fairly accurately. At the first sight, the predictability aspect appears to be surprising considering the fact that the statistics of the seismic (slip) events exhibits a power law. However, the data collapse for different events clearly suggests that the dynamics of approach to major events is universal. Further, we note that a SOC state demands that all observable quantities should follow a scale invariant form which is clearly respected by (26) representing the approach to all events. We note the underlying physical mechanism used in modeling the precursor feature is quite generic to any system where the front moves at high velocities once threshold friction is crossed. As there is no correspondence between seismic moments or the mean KE of the events with $R(t)$ (which depends on the difference between the velocities of neighboring elements), our analysis cannot predict the magnitudes of the slip events. (In some cases we find $R(t)$ is larger for a smaller slip event.) We stress that this precursor effect is absent in the total kinetic energy or seismic moments. We point out here that the gradual increase in the energy dissipated as we approach a major event is different from that reported in the context of the BK model [78]. Our work also differs from the approach of Huang et al. [59] in the sense that in their analysis, the hierarchical structure and long range interaction are necessary ingredients for the power law approach to failure with log periodic corrections. Thus, although these results are obtained in the context of the BK model, we expect this to be applicable to other situations of failure of materials and structures. As far as we know, this is first model which explains several unexplained experimental results on AE on rock samples.

7 Conclusions

In summary, we have considered three different physical situations, frictional sliding, the Portevin–Le Chatelier effect and peeling of an adhesive tape, that can be classified as stick-slip systems to examine the universal features. The idea of considering these three different physical systems is to demonstrate different features of stick-slip that can be illustrated.

As we have seen, while the physical mechanisms in each of these cases are different, they do exhibit several common features. The first feature that is subsumed in the name itself is the separation of time scales and is reflected in the fact that the stuck state lasts considerably longer time than the slipping state. Indeed such systems are traditionally called relaxation oscillations. In all such systems one has a fast time scale (the slip mode) and one slow time scale corresponding to the loading cycle. Another feature that is also related to this is the bi-stable nature of the system with the slip occurring between two stable states corresponding to the resistive branches. This also means that there is an unstable branch which is not available to the system. It is this property that gives rise to "N" shaped curve for the force – drive rate relation. The unstable branch itself is usually not measurable except in the case of kinetic friction. It is worth emphasizing again here that as this branch is unstable, measurement of this branch is nontrivial in the sense that the pull rate has to match exact stationary state value. This mean that the rate at which the system is being pulled must be fully imparted to the system which is done by using a stiff spring to couple the measuring device.

The discussion on the PLC effect shows that it is possible to give a dynamical interpretation for the negative strain rate sensitivity of the flow stress. Indeed, the two stable branches have been shown to correspond to the two pieces of the slow manifold S_1 and S_2. The S_2 part of the stable manifold where most dislocations are in the pinned configuration corresponds to low strain rate stable branch. In the model, the pinning of mobile dislocations occurs due to solute atoms. In contrast, once these dislocations are unpinned, the mechanism of slowing down of the mobile dislocations is gathering of solute atoms around the mobile dislocations. It is this process that corresponds the S_1 part of the slow manifold. The process of unpinning corresponds to the unstable branch of the negative SRS. As this is only example where such a connection between the physical processes and negative SRS has been established, it would nice if similar modeling could be carried out in other situations.

In general, the dynamics of stick-slip systems are carried out using the negative flow rate characteristic feature as an input as in the case of the peeling of adhesive tape and Burridge-Knopoff model for earth quakes. In both these cases, one does not have a deeper understanding of the origin of the branches as in the case of the PLC effect. Even so, the peeling problem showed that just the inclusion of NFRC is not adequate and one needs to consider other time scales for understanding the complex dynamics.

Lastly, we have also shown that predictability of individual events in a power law state is not forbidden as it has been believed so far. The approach to the point of slip itself obeys a power law and hence does not violate the scale invariance property. This was demonstrated within the frame work of an extension of Burridge-Knopoff model for earthquakes.

There are other gains form the analysis of stick-slip dynamics. One that is interesting is the modeling acoustic emission process during peeling of an adhesive tape as well as from rock samples. A strain rate dependent dissipative functional introduced appears to capture the essential features of acoustic emission during the peeling and deformation of rock samples. The second gain is the possibility of regularizing singular situations. From a mathematical point of view, the additional time scale that was introduced to lift the singular nature of algebraic – differential equations in the peeling problem may suggest a possible means of regularizing similar situations.

Finally, very often, the signature of stick-slip comes from measuring a scalar signal. However, this gives no clue about the possible physical mechanisms that lead to these relaxational oscillations. Thus, it is necessary to understood whether these serrations are of stochastic or dynamical origin. This can only be answered when a dynamical analysis of the signal time series is pursued along the lines carried out for the PLC effect [22, 23, 24]. This is perhaps a prerequisite for developing an appropriate model for the situation under study. However, it must be stated that noise is generally superposed on the signal even when they are of deterministic origin and a proper dynamical analysis of time series should help in modeling the physical situation.

Finally, we shall very briefly mention a few potential stick-slip problems that need further investigation. As stated earlier, stick-slip is a generic phenomenon appearing in variety of systems ranging all the way from geophysical scales to laboratory scales. Very often however, the underlying physics is not fully understood and thus some of them have not even been recognized as stick-slip situations. Here we consider very briefly a few examples that are good candidates for stick-slip systems.

The first example we mention is the charge density waves (CDW), a problem that has been extensively studied in condensed matter physics [79]. Even so, certain aspects of CDW such as the transport properties have remained ill understood. For example, the transition from Ohmic to non-Ohmic regime is not well understood. Specifically, under the action of an electric field, one finds a transition from the Ohmic regime ($E < E_t$) to the non-Ohmic one ($E > E_t$) and above the threshold, the current – voltage characteristic is nonlinear 18 (a). In addition, one also finds extremely low frequency voltage pulses (~ 1 Hz) in certain compounds such as blue bronzes $K_{0.30}MoO_3$. (Voltage fluctuations are seen within the Ohmic regime as well [80]. See Fig. 18a).

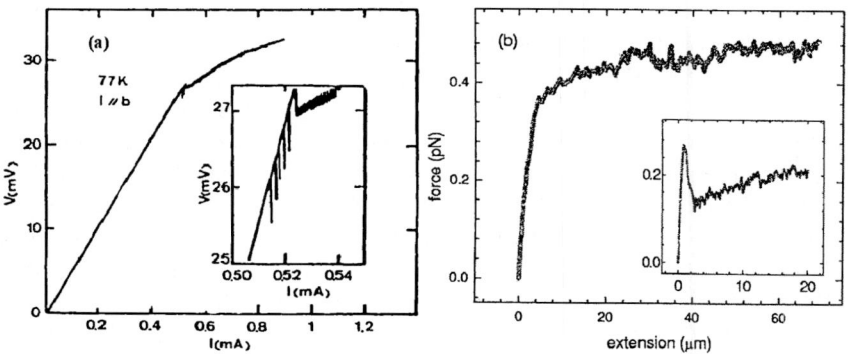

Fig. 18. (a) A plot of dc voltage-current that shows a sharp threshold for the onset of non-Ohmic regime (after [80]). (b) Force – extension curve for membrane tubulation. The inset shows another plot which is very similar to a sharp yield point follwed by serrations (after [87])

At low value of the electric fields, it is clear that the CDW as a whole cannot move, yet one does find current in the sample. In order to explain this, in analogy with plastic deformation where generation of dislocations and movement under applied stress contributes the flow, Lee and Rice [81] suggested that phase dislocations (PD) of the CDW could carry current at electric fields too low to depin the CDW as a whole. The threshold E_t can be viewed as the onset of plastic flow of the charged phase dislocations in the CDW lattice. Below the threshold, the PD's are unable to move, being pinned by impurities of crystal defects while above the threshold the motion of the PD's are cooperative and nonlinear. The very low frequency voltage pulses (~ 1 Hz) observed in many systems are quite similar to the serrations observed in the Portevin-Le Chatelier effect. Thus, they can be attributed to successive and cooperative pinning and unpinning of phase dislocations in the CDW lattice. The analogy between a tensile test at a constant stress rate and voltage measurement at constant current rate sets up a correspondence between stress σ_a and voltage V and strain ϵ and the current I. The voltage fluctuations at different current rates look similar to the stress-strain curves for various applied strain rates in Fig. 4. Further, one can associate the plastic strain rate $\dot{\epsilon}_p$ which is proportional to the number N_m of mobile dislocations and their velocity V_m with the nonlinear current J_{CDW}, the latter being is proportional to the number N_d and velocity V_d of moving PD's. Indeed, several mechanisms of dislocations carry over to phase dislocations. Some of the striking similarities between the CDW and the jerky flow has been drawn [82, 83]. A brief mention of the similarities between the PLC effect and CDW systems has been given in [84]. In this sense, the above mapping of the voltage fluctuations with the stress drop fluctuations in the PLC effect does recognize the phenomenon to be stick-slip dynamics.

As a second example we consider a problem of bilogical interest, namely the deformation of multilamellar vesicles. Membrane vesicles and tubules play an important role in cell biology. With the availability experimental tools that enable studies at nano-length scales, their properties are being studied. Tubules are expected to play an important role in intercellular transport [85]. Recent studies have demonstrated that localized picoNewton forces are exerted by molecular motors which in turn can induce tubule formation [86]. Recently, some efforts have been focused on the formation of tubules from multi-lamellar vesicles [87, 88]. Roopa and Shivashankar [87] have carried out deformation of multi-lamellar vesicles into tubules by using optical tweezer technique. In fact, they have studied in detail both the deformation dynamics as well as the frequency response during the process of deformation [88].

A typical force versus displacement curve during nanotubulation is shown in Fig. 18b. The serrations in force versus displacement is evidently similar to the stress-strain curve in the PLC effect.The typical force is of the order picoNewtons. Experiments performed in a range of pull rates show these serrations. The existing theories on the formation of tabulation are basically "steady state" calculations [86, 89, 90, 91, 92, 93, 94]. However, serrations can only arise under competing time scales of internal relaxation and imposed time scale of deformation. Thus, it is clear that the serrated nature cannot be explained by steady state theories. An appropriate framework is lacking and such an efforts are in progress.

Acknowledgements

GA would like to acknowledge the award of Raja Ramanna Fellowship.

References

1. F.P. Bowden and D. Tabor, Friction and Lubrication of Solids, (Clarendon, Oxford, 1950).
2. E. Rabinowicz, Friction and Wear of Materials (Wiley, New York, 1995).
3. F. Heslot, T. Baumberger, B. Perrin, B. Caroli, and C. Caroli, Phys. Rev., E **49**, 4973 (1994).
4. C.H. Scholz, The Mechanics of Earthqukes and Faulting, Cambridge University, (2002) p. 250.
5. T. Baumberger, Solid State Commun., **102**, 175 (1997).
6. R. Burridge and L. Knopoff, Bull. Seissmol. Soc. Am. **57**, 341 (1967).
7. J.M. Carlson and J.S. Langer, Phys. Rev. Lett. **62**, 2632 (1989); Phys. Rev. A **40** 6470 (1989).
8. J. Schmittbuhl, J.P. Vilotte and S. Roux, J. Geophys. Res. **101**, B12, 27 714 (1996).
9. L.P. Kubin, C. Fressengeas, and G. Ananthakrishna, in *Dislocations in Solids*, Edited by F.R.N. Nabarro and J.P. Hirth, (Elsevier Science, Amsterdam, 2002), Vol. 11.
10. F. Le Chatelier, Rev. de Métal **6**, 914 (1909). A. Portevin and F. Le Chatelier, C.R. Acad. Sci. Paris **176**, 507 (1923).
11. P. Penning, Acta. Metall. **20**, 1169 (1972).
12. D. Maugis and M. Barquins, in *Adhesion 12*, edited by K.W. Allen, (Elsevier, London, 1988), p. 205.
13. D.C. Hong and S. Yue, Phys. Rev. Lett. **74**, 254 (1995).
14. G. Ananthakrishna and M.C. Valsakumar, J. Phys. D, **15**, L171 (1982).
15. S. Rajesh and G. Ananthakrishna, Phys. Rev. E **61**, 3664 (2000).
16. A.H. Cottrell, *Dislocations and Plastic Flow in Crystals*. (Clarendon Press, Oxford, 1953).
17. A. Van den Beukel, Phys. Stat. Solidi A**30**, 197 (1975).
18. L.P. Kubin and Y. Estrin, Acta. Metall. **33**, 397 (1985).
19. K. Chihab, Y. Estrin, L.P. Kubin and J. Vergnol, Scripta. Metall. **21**, 203 (1987).
20. A. Rosen and S.R. Bodner, J. Mech. Phys. Solids, **15**, 47 (1967); S.R. Bodner and A. Rosen, J. Mech. Phys. Solids, **15**, 63 (1967).
21. L.P. Kubin and Y. Estrin, J. de Physique **46**, 497 (1986).
22. G. Ananthakrishna et al., Scripta. Metall. **32**, 1731 (1995).
23. S.J. Noronha et al., Int. Jl. of Bifurcation and Chaos **7**, 2577 (1997).
24. G. Ananthakrishna et al., Phys. Rev. E **60**, 5455 (1999).

25. S. Rajesh and G. Ananthakrishna, Physica D **140**, 193 (2000).
26. M.S. Bharathi et al., Phys. Rev. Lett. **87**, 165508 (2001).
27. G. Ananthakrishna and M.S. Bharathi, Phys. Rev. E **70**, 026111 (2004).
28. M.S. Bharathi, S. Rajesh and G. Ananthakrishna, Scripta Mater. **48**, 1355 (2003).
29. J. Bikermann, J. Appl. Phys. **28**, 1484 (1957).
30. D.H. Kaeble, J. Colloid Sci. **19**, 413 (1963).
31. D.W. Aubrey and M. Sheriif, J. Polym. Sci. Part A-1, **18**, 2597 (1980).
32. M. Ciccotti, B. Giorgini and M. Barquins, Int. J. of Adhes. and Adhes. **18**, 35 (1998).
33. C. Gay and L. Leibler, Phys. Today **52**, Issue 11, 48 (1999).
34. M. Barquins and M. Ciccotti, Int. J of Adhes. and Adhes. **17**, 65 (1997).
35. M.C. Gandur, M.U. Kleinke and F. Galembeck, J. Adhes. Sci. Technol. **11**, 11 (1997).
36. D.C. Hong (private communication).
37. R. De, A. Maybhate and G. Ananthakrishna, Phys. Rev. E **70**, 046223 (2004).
38. R. De and G. Ananthakrishna, Phys. Rev. E **71**, 055201(R) (2005).
39. E. Hairer, C. Lubich and M. Roche, *Numerical Solutions of Differential-algebraic Systems by Runge-Kutta Methods* (Springer-Verlag, Berlin, 1989).
40. R.B. Bird, R.C. Armstrong and O. Hassager, Dynamics of Polymeric Liquids (Wiley, New York, 1987) Vol. 1.
41. E. Lemaire, P. Levitz, G. Daccord and H. Van Damme, Phys. Rev. Lett. **67**, 2009 (1991).
42. L.P. Kubin, K. Chihab and Y. Estrin, Acta. Metall. **36**, 2707 (1988).
43. M.A. Lebyodkin, Y. Brechet, Y. Estrin and L.P. Kubin, Phys. Rev. Lett. **74**, 4758 (1995).
44. E.C.G. Sudarshan and N. Mukunda, Classical Dynamics: A Modern Perspective (John Wiley and Sons, New York, 1974).
45. M. Diener, The Mathematical Intelligence, **6**, 38 (1984).
46. H.D.I. Abarbanel, *Analysis of Observed Chaotic Data* (Springer-Verlag, New York, 1996).
47. L.D. Landau and E.M. Lifschitz, *Theory of Elasticity* (Pergamon, Oxford, 1986).
48. P. Bak, C. Tang and K. Wiesenfeld, Phys. Rev. Lett. **59**, 381 (1987).
49. H.J. Jensen, *Self-Organised Criticality* (Cambridge University Press, Cambridge, 1998).
50. I. Main, Rev. Geophys. **34**, 433 (1996).
51. Nature debate, http://helix.nature.com/debates/earthquake (1999).
52. C.H. Scholz, J. Geophys. Res. **73**, 1417 (1968).
53. A. Lockner, J. Acous. Emission **14**, S88 (1996).
54. P.R. Sammonds, P.G. Meredith and I.G. Main, Nature **359**, 282 (1992).
55. A. Gaurino et al. Eur. Phys. B **26**, 141 (2002).
56. A. Johansen and D. Sornette, Eur. Phys. B **18** 163 (2000).
57. I.G. Main and P.G. Meredith, Tectonophysics **167**, 273 (1989).
58. C.H. Scholz, The Mechanics of Earthquakes and Faulting (Cambridge University Press, 1990).
59. Y. Huang, H. Saleur, C. Sammis and D. Sornette, Europhys. Lett. **41**, 43 (1998).
60. J. Rosendahl, M. Vekic and J.E. Rutledge, Phys. Rev. Lett. **73**, 537 (1994).
61. S.L. Pepke and J.M. Carlson, Phys. Rev. E **50**, 236 (1994).
62. M. Acharyya and B.K. Chakrabarti, Phys. Rev. E **53**, 140 (1996).

63. R. De and G. Ananthakrishna, Europhys. Lett. **66**, 715 (2004).
64. A. Lockner et al., Nature **350**, 39 (1991).
65. K. Mogi, Bull. Earthquake Res. Inst. **40**, 125 (1962).
66. C.H. Scholz, Bull. Seismol. Soc. Am. **58**, 399 (1968).
67. B. Gutenberg and C.F. Richter, Ann. Geofis. **9**, 1 (1596).
68. S. Hainzl, G. Zoller and J. Kurths, Geophys. Res. Lett. **27**, 597 (2000).
69. P. Diodati, F. Marchesoni and S. Piazza, Phys. Rev. Lett. **67**, 2239 (1991).
70. A. Petri et al., Phys. Rev. Lett. **73**, 3423 (1994).
71. A. Garcimartin et al., Phys. Rev. Lett. **79**, 3202 (1997).
72. M.M. Carmen et al., Nature **410**, 667 (2001).
73. Y. Yabe et al., Pure Appl. Geophys **160**, 1163 (2003).
74. J.F. Pacheco, C.H. Scholz and L.R. Sykes, Nature **355**, 71 (1992).
75. Y. Yabe, Geophys. Res. Lett. **29**, 10.1029/2001GL014369 (2002).
76. R. Ahluwalia and G. Ananthakrishna, Phys. Rev. Lett. **86**, 4076 (2001); S. Sreekala and G. Ananthakrishna, Phys. Rev. Lett. **90**, 135501 (2003).
77. J. Rice, J. Geophys. Res. **98**, 9885 (1993); J. Rice and Ben-Zion, Proc. Natl. Acad. Sci. USA **93**, 3811 (1996).
78. B.E. Shaw, J.M. Carlson and J.S. Langer, J. Geophys. Res. **97**, 479 (1992).
79. S. Brazovski and T. Nattermann, Adv. Phys. **53**, 117 (2004).
80. J. Dumas et al., Phys. Rev. Lett. **50**, 757 (1983).
81. P.A. Lee and T.M. Rice, Phys. Rev. B **19**, 3970 (1979).
82. J. Dumas and D. Feinberg, Europhys. Lett. **2**, 555 (1986).
83. J. Dumas, C. Schlenker, J. Marcus and R. Buder, Phys. Rev. Lett. **50**, 757 (1983): J. Dumas and C. Schlenker, Lecture Notes in Physics, **217**, 439 (1985).
84. G. Ananthakrishna, J. Indian Inst. **78**, 165 (1998).
85. K. Hirschberg et al., J. Cell. Biol. **143**, 1485 (1998).
86. A. Roux et al., PNAS, **99**, 5394 (2002).
87. T. Roopa and G. Shivashankar, Appl. Phys. Lett. **82**, 1631 (2003).
88. T. Roopa et al., Biophys. J. **87**, 974 (2004).
89. O. Rossier et al., Langmuir, **19**, 575 (2003).
90. I Derenyi, F. Julicher and J. Prost, Phys. Rev. Lett. **66**, 238101 (2002).
91. B. Bozic et al., Biophys. J. **61** 963 (1992).
92. V. Heinrich et al., Biophys. J. **76** 2056 (1999).
93. T.R. Powers, G. Huber and R.E. Golstein, Phys. Rev. E **65**, 041901 (2002).
94. R.E. Waugh and R.M. Hochmuth, Biophys. J. **52** 391 (1987).

Search for Precursors in Some Models of Catastrophic Failures

S. Pradhan[1] and B.K. Chakrabarti[2]

[1] Norwegian University of Science and Technology, Trondheim, Norway
   pradhan.srutarshi@ntnu.no
[2] Theoretical Condensed Matter Physics Division and Centre for Applied Mathematics and Computational Science, Saha Institute of Nuclear Physics, Kolkata, India
   bikask.chakrabarti@saha.ac.in

1 Introduction

In this chapter we review the precursors of catastrophic avalanches (global failures) in several failure models, namely (a) Fibre Bundle Model (FBM), (b) Random Fuse Model (RFM), (c) Sandpile Models and (d) Fractal Overlap Model. The precursor parameters identified here essentially reflect the growing correlations within such systems as they approach their respective failure points. As we show, often they help us to predict the global failure points in advance.

Needless to mention that the existence of any such precursors and detailed knowledge about their behavior for major catastrophic failures [1] like earthquakes, landslides, mine/bridge collapses, would be of supreme value for our civilization. So far, we do not have any established set of models for these major failure phenomena. However, as discussed in various chapters of this book, several reasonable models of failure have already been developed in various contexts. We review here some of the precursors of global failures in these models.

2 Precursors in Failure Models

2.1 Composite Material Under Stress: Fibre Bundle Model

Fibre bundle model represents various aspects of fracture-failure process in composite materials. A bundle of parallel fibres, clamped at both ends (Fig. 1 (left)) represents the model where fibres are assumed to obey Hookean elasticity up to the breaking point (Fig. 1 (right)). The model needs three basic ingredients: (a) a discrete set of N elements (b) a probability distribution of

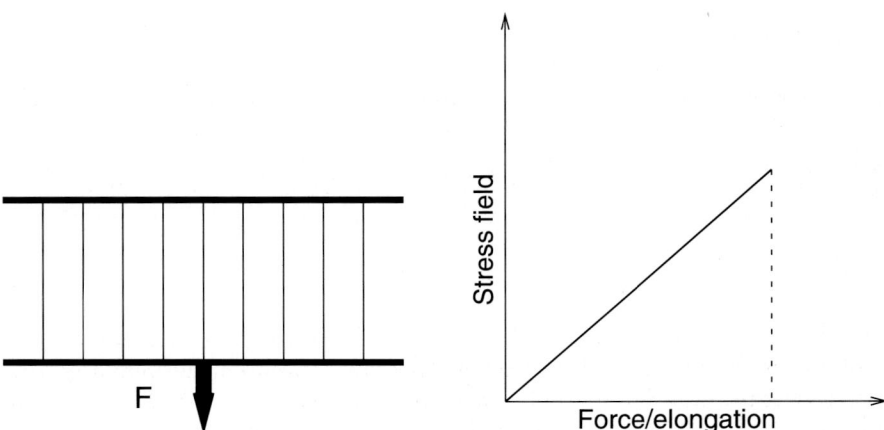

Fig. 1. Fibre bundle model (*left*) and the force-response of a single fibre (*right*)

the strength of fibres (c) a load-transfer rule. Peirce [2] initiated the model study in the context of testing the strength of cotton yarns. Since then this model has been studied from different views. Fibre bundles are of two classes with respect to the time dependence of fibre strength: The "static" bundles [2, 3, 4, 5, 6, 7, 8, 9, 10] contain fibres whose strengths are independent of time, where as the "dynamic" bundles [11, 12, 13] are assumed to have time dependent elements to capture the creep rupture and fatigue behaviours. According to the load sharing rule, fibre bundles are being classified into two groups: equal-load-sharing (ELS) bundles and local-load-sharing (LLS). In ELS bundles, intact fibres bear the applied load equally and in LLS bundles, the terminal load of the failed fibre is given equally to all the intact neighbours. With steadily increasing load, a fibre bundle approaches the failure point obeying a dynamics determined by the load sharing rule. The phase transition [7] and dynamic critical behaviour of the fracture process in such democratic bundles has been established through recursive formulation [8, 9] of the failure dynamics. The exact solutions [9, 10] of the recursion relations suggest universal values of the exponents involved. Attempts have also been made [11] to study the ELS and LLS bundles from a common framework introducing a single parameter which determines the load transfer rule.

We discuss here, failure of static fibre bundles under global load sharing (ELS) for two different loading conditions: (a) load increment by equal amount and (b) quasi-static loading–which follows weakest link failure at each step of loading. We show analytically the variation of precursor parameters with the applied stress, which help to estimate the failure point accurately.

Recursive Dynamics in ELS Bundle: Precursors

ELS model assumes that the intact fibres share the applied load equally. The strength threshold of a fibre is determined by the stress value it can bear, and beyond which it fails. Usually, thresholds are taken from a randomly distributed normalised density $p(x)$ within the interval 0 and 1 such that

$$\int_0^1 p(x)dx = 1 . \tag{1}$$

The breaking dynamics starts when an initial stress σ (load per fibre) is applied on the bundle. The fibres having strength less than σ fail instantly. Due to this rupture, total number of intact fibres decreases and effective stress increases and this compels some more fibres to break. Such stress redistribution and further breaking of fibres continue till an equilibrium is reached, where either the surviving fibres are strong enough to bear the applied load on the bundle or all fibres fail.

This self organised breaking dynamics can be represented by recursion relations [8, 9] in discrete time steps. Let U_t be the fraction of fibres in the initial bundle that survive after time step t, where time step indicates the number of occurrence of stress redistribution. Then the redistributed stress after t time step becomes

$$\sigma_t = \frac{\sigma}{U_t} ; \tag{2}$$

and after $t+1$ time steps the surviving fraction of fibre is

$$U_{t+1} = 1 - P(\sigma_t) ; \tag{3}$$

where $P(\sigma_t)$ is the cumulative probability of corresponding density distribution $p(x)$: $P(\sigma_t) = \int_0^{\sigma_t} p(x)dx$. Now using (2) and (3) we can write the recursion relations which show how σ_t and U_t evolve in discrete time:

$$\sigma_{t+1} = \frac{\sigma}{1 - P(\sigma_t)} ; \quad \sigma_0 = \sigma \tag{4}$$

and

$$U_{t+1} = 1 - P(\sigma/U_t); \quad U_0 = 1 . \tag{5}$$

At the equilibrium or steady state $U_{t+1} = U_t \equiv U^*$ and $\sigma_{t+1} = \sigma_t \equiv \sigma^*$. This is a fixed point of the recursive dynamics. Equations (4) and (5) can be solved at the fixed point for some particular distribution. At the fixed-point, the solutions [8, 9, 10] assume universal form

$$\sigma^*(\sigma) = C - (\sigma_c - \sigma)^{1/2} ; \tag{6}$$

$$U^*(\sigma) = C + (\sigma_c - \sigma)^{1/2} ; \tag{7}$$

where σ_c is the critical stress and C is a constant. From the recursions and their solutions we can derive the following response quantities:

Susceptibility: Which is defined as the amount of change in U when the external stress changes by a infinitesimal amount and can be derived as

$$\chi = \left|\frac{dU^*(\sigma)}{d\sigma}\right| = \frac{1}{2}(\sigma_c - \sigma)^{-\beta}\,; \quad \beta = \frac{1}{2}\,. \tag{8}$$

Relaxation Time: This is the number of step the bundle needs to come to a stable state after the application of an external stress. From the solution of recursive dynamics we get [8, 9] near σ_c

$$\tau \propto (\sigma_c - \sigma)^{-\theta}\,; \quad \theta = \frac{1}{2}\,. \tag{9}$$

Inclusive Avalanche: This is the amount of avalanche per step of stress redistribution

$$\frac{dU}{dt} = U_t - U_{t+1} \tag{10}$$

At $\sigma = \sigma_c$, the dynamics becomes "critically slow" as

$$U_t \sim t^{-\gamma}\,; \quad \gamma = 1\,; \tag{11}$$

which suggests that at σ_c inclusive avalanches follow a power law (exponent -2) decay with time.

Prediction of Global Failure Point

(A) Using χ and τ

We found that susceptibility (χ) and relaxation time (τ) follow power laws with external stress and both of them diverge at the critical stress. Therefore if we plot χ^{-2} and τ^{-2} with external stress, we expect a linear fit near critical point and the straight lines should touch X axis at the critical stress. We indeed found similar behaviour (Fig. 2) in simulation experiments.

For application, it is always important that such prediction can be done in a single sample. We have performed the simulation taking a single bundle having very large number of fibres and we observe similar response of χ and τ. The prediction of failure point is also quite satisfactory (Fig. 3).

(B) Using Inclusive Avalanches

When we put a big load on a material body, sometimes it becomes important to know whether that body can support the load or not. The similar question can be asked in FBM. We found that if we record the inclusive avalanche, i.e, the amount of failure in each load redistribution – then the pattern of inclusive burst clearly shows whether the bundle is going to fail or not. For any stress below the critical state, inclusive avalanche follow exponential decay (Fig. 4 (right)) with time step and for stress values above critical stress it is a power

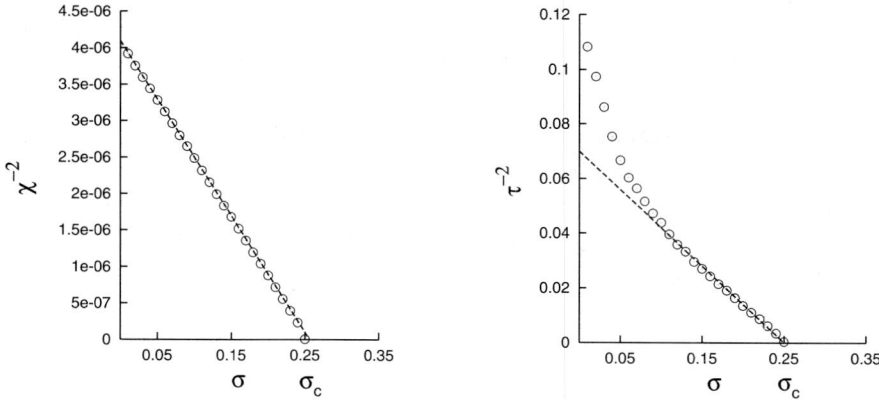

Fig. 2. Variation of χ^{-2} and τ^{-2} with applied stress for a bundle having $N = 50000$ fibres and averaging over 1000 samples. We consider uniform distribution of fibre threshold

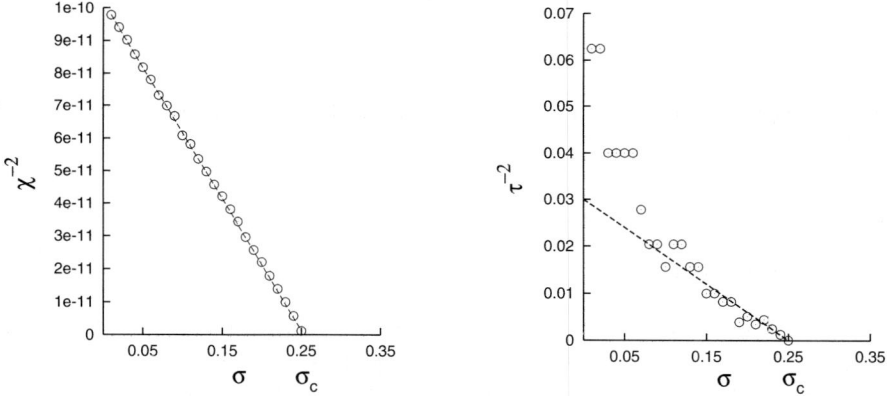

Fig. 3. Variation of χ^{-2} and τ^{-2} with applied stress for a single bundle having $N = 10000000$ fibres with uniform distribution of fibre threshold

law followed by a gradual rise (Fig. 4 (left)). Clearly at critical stress it follows a robust power law with exponent -2 that we already get analytically.

(C) Using Avalanche Distribution

An "Avalanche" is the amount of failure occurs as the system moves from one stable state to the next stable state when the load is increased quasi-statically. We can simply measure it by counting the failure elements between two consecutive load increment. Such a series of avalanches has been shown in Fig. 5 up to the final failure point. If we record all the avalanches till final failure, the avalanche distribution follows a universal power law [4]. This

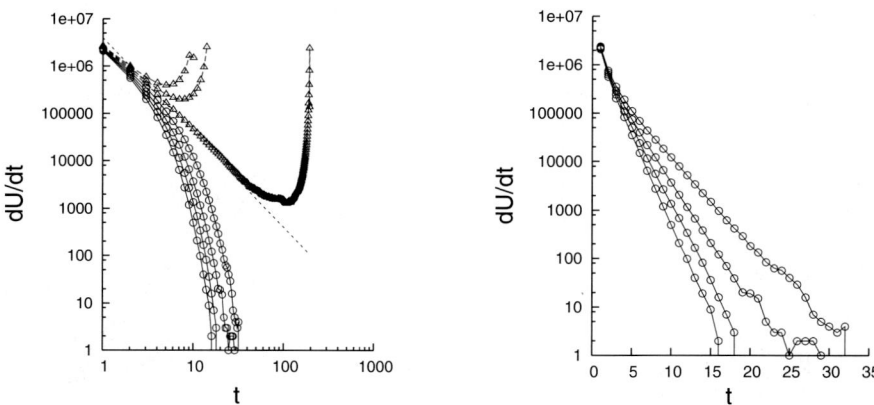

Fig. 4. Log-log plot of inclusive avalanche with step of load redistribution for 7 different stress values [*left*] and log-normal plot of the same for 4 different stress values below the critical stress [*right*]. The simulation has been performed for a single bundle with $N = 10000000$ fibres having uniform distribution of fibre threshold

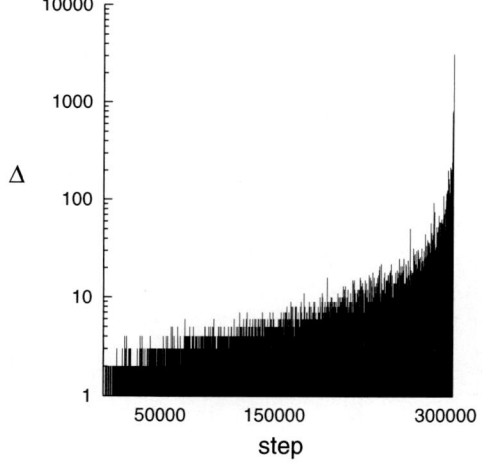

Fig. 5. Magnitude of avalanches with step of load increment in FBM

avalanches can be recorded experimentally measuring the acoustic emissions during fracture-failure of materials. We want to study whether the avalanche distribution changes if we start gathering the avalanches from some intermediate states of the breaking process (Fig. 6 (left)). In simulations we see that the exponent of avalanche distribution shows a crossover between two values ($-5/2$ and $-3/2$) and the crossover point (length) depends on the starting position (x_0) of our measurement (Fig. 6 (right)). Now we are going to explain the above observation analytically for quasi-static loading. For a bundle

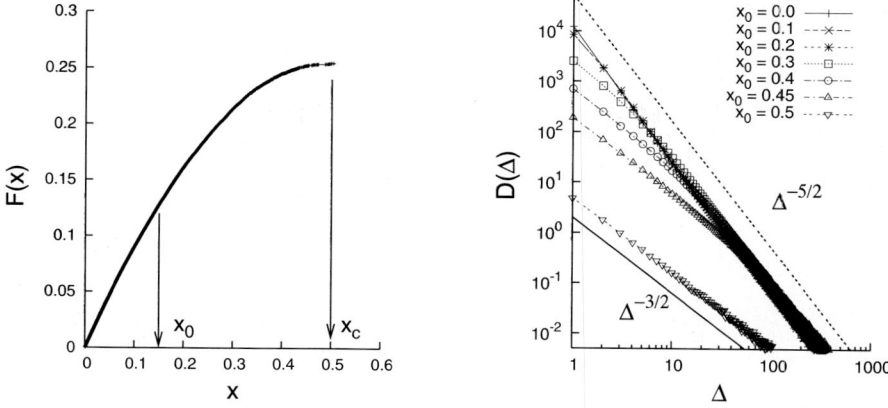

Fig. 6. Average force versus elongation curve (*left*) and simulation results of avalanche distributions for different starting position (*right*) in FBM

of many fibres the average number of bursts of magnitude Δ is given by [4]

$$\frac{D(\Delta)}{N} = \frac{\Delta^{\Delta-1}}{\Delta!} \int_0^{x_c} p(x) \left[1 - xp(x)/Q(x)\right] \left[xp(x)/Q(x)\right]^{\Delta-1} \quad (12)$$
$$\times \exp\left[-\Delta xp(x)/Q(x)\right] dx, \quad (13)$$

where $Q(x) = \int_x^\infty p(y)\,dy$ is the fraction of total fibres with strength exceeding x and x_c is the critical value beyond which the bundle fails instantly. For uniform distribution (having upper bound x_m), if we start recording avalanches from an intermediate point x_0, the avalanche distribution becomes

$$\frac{D(\Delta)}{N} = \frac{\Delta^{\Delta-1}}{\Delta!(x_m - x_0)} \times \int_{x_0}^{x_c} \frac{x_m - 2x}{x} \left[\frac{x}{x_m - x} e^{-x/(x_m - x)}\right]^\Delta dx. \quad (14)$$

Introducing the parameter $\epsilon = \frac{x_c - x_0}{x_m}$ and a new integration variable $z = \frac{x_m - 2x}{\epsilon(x_m - x)}$, we obtain

$$\frac{D(\Delta)}{N} = \frac{2\Delta^{\Delta-1} e^{-\Delta} \epsilon^2}{\Delta!(1 + 2\epsilon)} \times \int_0^{4/(1+2\epsilon)} \frac{z}{(1 - \epsilon z)(2 - \epsilon z)^2} e^{\Delta[\epsilon z + \ln(1 - \epsilon z)]} dz. \quad (15)$$

For small ϵ, i.e., close to the critical threshold distribution, we can expand

$$\epsilon z + \ln(1 - \epsilon z) = -\frac{1}{2}\epsilon^2 z^2 - \frac{1}{3}\epsilon^3 z^3 + \ldots, \quad (16)$$

with the result

$$\frac{D(\Delta)}{N} \simeq \frac{\Delta^{\Delta-1} e^{-\Delta} \epsilon^2}{2\Delta!} \times \int_0^4 e^{-\Delta \epsilon^2 z^2/2} z\,dz = \frac{\Delta^{\Delta-2} e^{-\Delta}}{2\Delta!} \left(1 - e^{-8\epsilon^2 \Delta}\right). \quad (17)$$

Using Stirling approximation $\Delta! \simeq \Delta^\Delta e^{-\Delta}\sqrt{2\pi\Delta}$, this becomes

$$\frac{D(\Delta)}{N} \simeq (8\pi)^{-1/2}\Delta^{-5/2}\left(1 - e^{-\Delta/\Delta_c}\right), \tag{18}$$

with

$$\Delta_c = \frac{1}{8\epsilon^2} = \frac{x_m^2}{8(x_c - x_0)^2}. \tag{19}$$

Clearly, there is a crossover [5] at a burst length around Δ_c, so that

$$\frac{D(\Delta)}{N} \simeq \begin{cases} (8/\pi)^{1/2}\epsilon^2\,\Delta^{-3/2} & \text{for} \quad \Delta \ll \Delta_c \\ (8\pi)^{-1/2}\Delta^{-5/2} & \text{for} \quad \Delta \gg \Delta_c \end{cases} \tag{20}$$

For $x_0 = 0.40x_m$, we have $\Delta_c = 12.5$ uniform distribution. We found a clear crossover near $\Delta = \Delta_c = 12.5$ in the simulation experiment (Fig. 7).

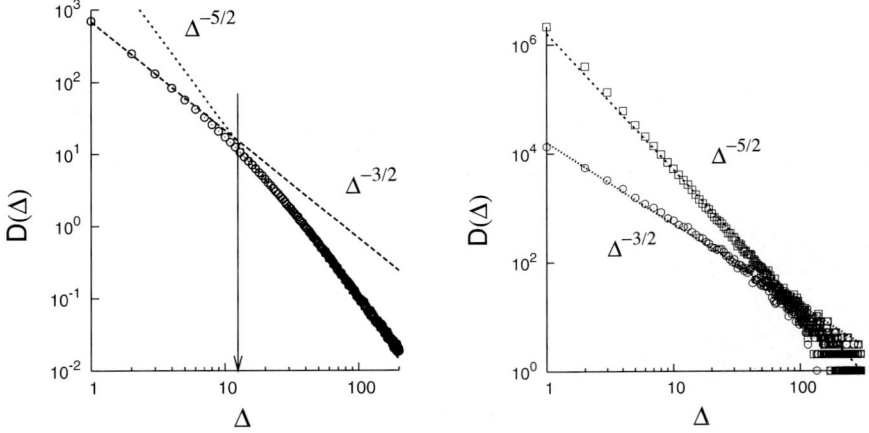

Fig. 7. Crossover in avalanche distributions: bundle with 10^6 fibre and $x_0 = 0.8x_c$ (averaging over 20000 samples) [*left*] and a single bundle with 10^7 fibre; $x_0 = 0$ (*squares*) and $x_0 = 0.9x_c$ (*circles*) [*right*]. Dotted straight lines are the best fits to the power laws

The simulation results shown in the figures are based on *averaging* over a large number of samples. For applications it is important that crossover signals are seen also in a single sample. We show (Fig. 7) that equally clear power laws are seen in a *single* fibre bundle when N is large.

This crossover phenomenon is not limited to the uniform threshold distribution. The $\xi = 3/2$ power law [5] in the burst size distribution dominates over the $\xi = 5/2$ power law whenever a threshold distribution is non-critical, but close to criticality. Therefore, the magnitude of the crossover length correctly inform us how far the system is from the global failure point.

2.2 Electrical Networks within a Voltage Difference: Random Fuse Model

Random fuse model [1] describes the breakdown phenomena in electrical networks. It consists of a lattice in which each bond is a fuse, i.e., an Ohmic resistor as long as the electric current it carries is below a threshold value. If the threshold is passed, the fuse burns out irreversibly (Fig. 8 (right)). The threshold (th) of each bond is drawn from an uncorrelated distribution $p(th)$. All fuses have the same resistance. The lattice is a two-dimensional square one placed at 45° with regards to the bus bars (Fig. 8 (left)) and an increasing current is passed through it. Numerically, the Kirchhoff equations are solved at each point of the system. For the entire failure process avalanches of different sizes occur (Fig. 9 (left)) and the system finally comes to a point where no

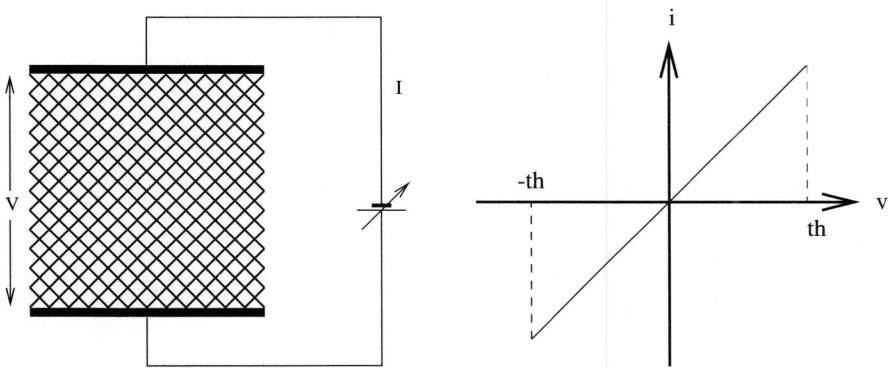

Fig. 8. Random fuse model (*left*) and the current-response of a single Ohmic resistor (*right*)

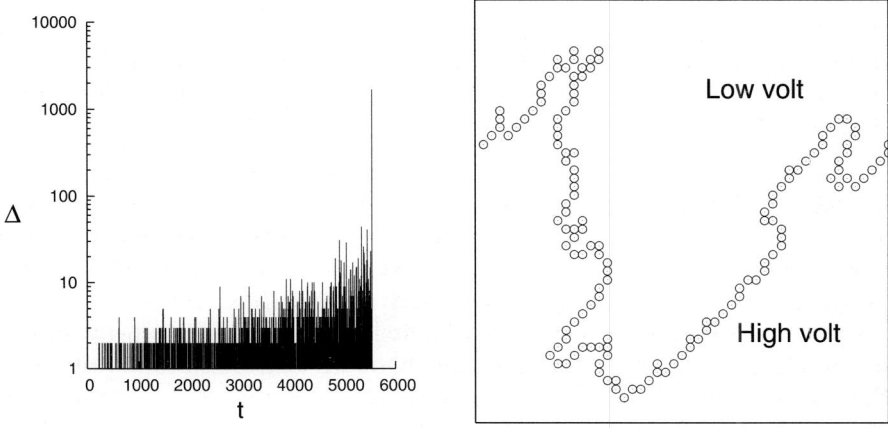

Fig. 9. Avalanches in fuse model (*left*) and the ultimate fractured state (*right*)

current passes through the system (Fig. 9 (right)) – we call it the global failure point. The average avalanche size has been measured [7] for such system and it follows a power law with increasing current or voltage:

$$m \sim (I_c - I)^{-\gamma} \text{ or } m \sim (V_c - V)^{-\gamma}; \gamma = 1/2$$

Therefore if we plot m^{-2} with I or V we can expect a linear fit which touches X axis at the critical current value I_c or critical voltage value V_c (Fig. 10).

If we record all the avalanches, the avalanche distribution follows a universal power law with $\xi \simeq 3$. But what will happen if we start recording the burst at some intermediate state of the failure process? We observe [5] that for a system of size 100×100, 2097 fuses blow on the average before catastrophic failure sets in. When measuring the burst distribution only after the first 2090 fuses have blown, a different power law is found, this time with $\xi = 2$. After 1000 blown fuses, on the other hand, ξ remains the same as for the histogram recording the entire failure process (Fig. 10).

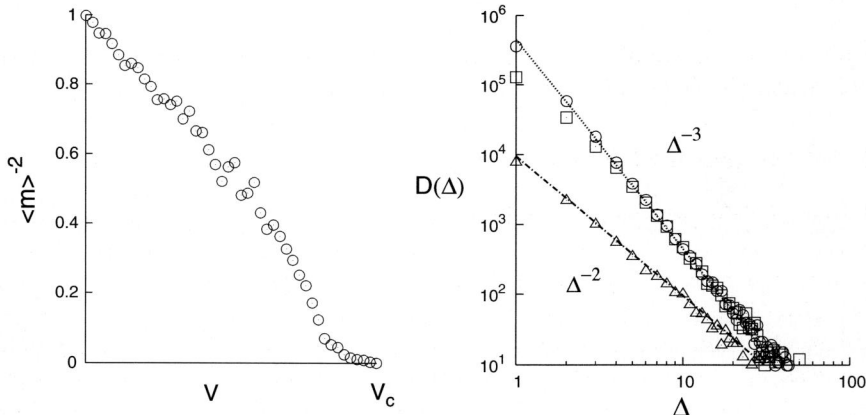

Fig. 10. Plot of m^{-2} with V (*left*) and the crossover in avalanche distributions in a fuse model (*right*)

2.3 SOC Models of Sandpile

Growth of a natural sandpile is a nice example of self-organised criticality [15, 16, 17, 18, 19]. If sand grains are added continuously on a small pile, the system gradually approaches towards a state at the boundary between stable and unstable states where the system shows long-range spatio-temporal fluctuations similar to those observed in equilibrium critical phenomena. This special state has been identified as the critical state of the pile where the response to addition of sand grains becomes unpredictable: Avalanche of any

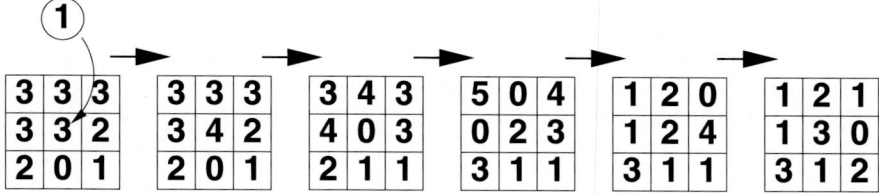

Fig. 11. A real sandpile and the sandpile model [16] on a square lattice

size is equally probable at this state. Therefore, we can expect system spanning avalanche (global failure) only at this critical state.

BTW Model and Manna Model

Bak et al. [15] proposed a sandpile model on square lattice which captures correctly the properties of a natural sandpile. At each lattice site (i,j), there is an integer variable $h_{i,j}$ which represents the height of the sand column at that site. A unit of height (one sand grain) is added at a randomly chosen site at each time step and the system evolves in discrete time. The dynamics starts as soon as any site (i,j) has got a height equal to the threshold value $(h_{th} = 4)$: that site topples, i.e., $h_{i,j}$ becomes zero there, and the heights of the four neighbouring sites increase by one unit

$$h_{i,j} \to h_{i,j} - 4, h_{i\pm 1,j} \to h_{i\pm 1,j} + 1, h_{i,j\pm 1} \to h_{i,j\pm 1} + 1 \ . \qquad (21)$$

If, due to this toppling at site (i,j), any neighbouring site become unstable (its height reaches the threshold value), the same dynamics follows. Thus the process continues till all sites become stable ($h_{i,j} < h_{th}$ for all (i,j)). When toppling occurs at the boundary of the lattice, extra heights get off the lattice and are removed from the system. With continuous addition of unit height (sand grain) at random sites of the lattice, the avalanches (toppling) get correlated over longer and longer ranges and the average height (h_{av}) of the system grows with time. Gradually the correlation length (ξ) becomes of the order the system size L as the system attains the critical average height $h_c(L)$. On average, the additional height units start leaving the system and the average height remains stable there (see Fig. 12a). The distributions of

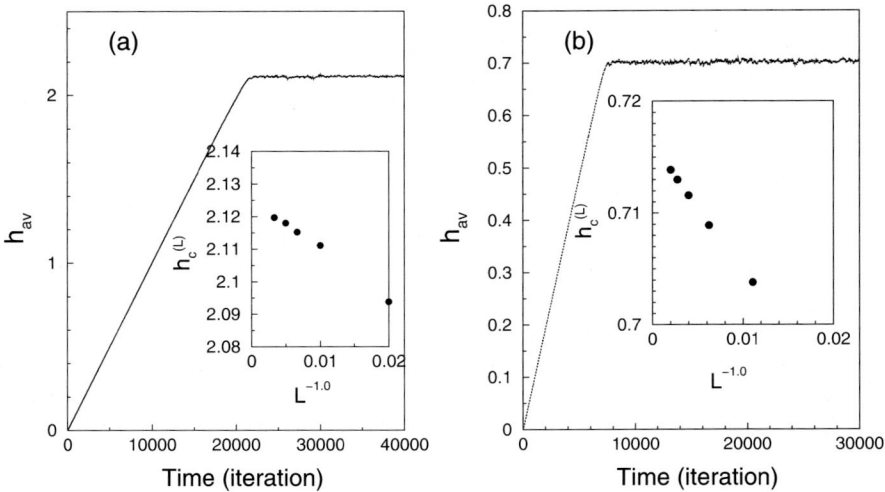

Fig. 12. Growth of BTW (a) and Manna (b) sandpiles. Inset shows the system size dependence of the critical height

the avalanche sizes and the corresponding life times follow robust power laws [17], hence the system becomes critical here.

We can perform a finite size scaling fit $h_c(L) = h_c(\infty) + CL^{-1/\nu}$ (by setting $\xi \sim |\ h_c(L) - h_c(\infty)\ |^{-\nu} = L$), where C is a constant, with $\nu \simeq 1.0$ gives $h_c \equiv h_c(\infty) \simeq 2.124$ (see inset of Fig. 12a). Similar finite size scaling fit with $\nu = 1.0$ gave $h_c(\infty) \simeq 2.124$ in earlier large scale simulations [16].

Manna proposed the stochastic sand-pile model [16] by introducing randomness in the dynamics of sand-pile growth in two-dimensions. Here, the critical height is 2. Therefore at each toppling, two rejected grains choose their host among the four available neighbours randomly with equal probability. After constant adding of sand grains, the system ultimately settles at a critical state having height h_c. A similar finite size scaling fit $h_c(L) = h_c(\infty) + CL^{-1/\nu}$ gives $\nu \simeq 1.0$ and $h_c \equiv h_c(\infty) \simeq 0.716$ (see inset of Fig. 12b). This is close to an earlier estimate $h_c \simeq 0.71695$ [19], made in a somewhat different version of the model. The avalanche size distribution has got power laws similar to the BTW model, however the exponent seems to be different [16, 17], compared to that of BTW model.

Sub-Critical Response: Precursors

We are going to investigate the behaviour of BTW and Manna sandpiles when thay are away from the critical state ($h < h_c$). At an average height h_{av}, when all sites of the system have become stable (dynamics have stopped), a fixed number of height units h_p (pulse of sand grains) is added at any central point of the system [8, 18]. Just after this addition, the local dynamics starts and it takes a finite time or iterations to return back to the stable state ($h_{i,j} < h_{th}$ for all (i,j)) after several toppling events. We measure the response parameters: $\Delta \rightarrow$ number of toppling $\tau \rightarrow$ number of iteration and $\xi \rightarrow$ correlation length which is the distance of the furthest toppled site from the site where h_p has been dropped.

(A) In BTW Model

We choose $h_p = 4$ for BTW model to ensure toppling at the target site. We found that all the response parameters follow power law as h_c is approached: $\Delta \propto (h_c - h_{av})^{-\lambda}$, $\tau \propto (h_c - h_{av})^{-\mu}$, $\xi \propto (h_c - h_{av})^{-\nu}$; $\lambda \cong 2.0$, $\mu \cong 1.2$ and $\nu \cong 1.0$. Now if we plot $\Delta^{-1/\lambda}$, $\tau^{-1/\mu}$ and $\xi^{-1/\nu}$ with h_{av} all the curve follow straight line and they should touch the x axis at $h_{av} = h_c$. Therefore by a proper extrapolation we can estimate the value of h_c and we found $h_c = 2.13 \pm .01$ (Fig. 13) which agree well with direct estimates of the same.

(B) In Manna Model

Obviously we have to choose $h_p = 2$ for Manna model to ensure toppling at the target site. Then we measured all the response parameters and they seem to follow power law as h_c is approached: $\Delta \propto (h_c - h_{av})^{-\lambda}$, $\tau \propto (h_c - h_{av})^{-\mu}$, $\xi \propto (h_c - h_{av})^{-\nu}$; $\lambda \cong 2.0$, $\mu \cong 1.2$ and $\nu \cong 1.0$. As in BTW model, all the curve follow straight line if we plot $\Delta^{-1/\lambda}$, $\tau^{-1/\mu}$ and $\xi^{-1/\nu}$ with h_{av} and proper extrapolations estimate the value of $h_c = 0.72 \pm .01$ which is again a good estimate (Fig. 14).

Our simulation results suggest that although BTW and Manna models belong to different universality class with respect to their properties at the critical state, both the models show similar sub-critical response or precursors. A proper extrapolation method can estimate the respective critical heights of the models quite accurately.

2.4 Fractal Overlap Model of Earthquake

It has been claimed recently that since the fractured surfaces have got well-characterized self-affine properties, the distribution of the elastic energies released during the slips between two fractal surfaces (earthquake events) may follow the overlap distribution of two self-similar fractal surfaces [20, 21, 22, 23]. To support this idea, Chakrabarti and Stinchcombe [24] have analytically shown using renormalization group technique that for regular fractal overlap (Cantor sets and carpets) the contact area distribution follows power law. This claim has also been verified by extensive numerical simulations [25]. If one cantor set moves uniformly over other, the overlap between the two fractals (Fig. 15) change quasi-randomly with time. In this section we analyse the time series data of such overlaps to find the prediction possibility of a next large overlap.

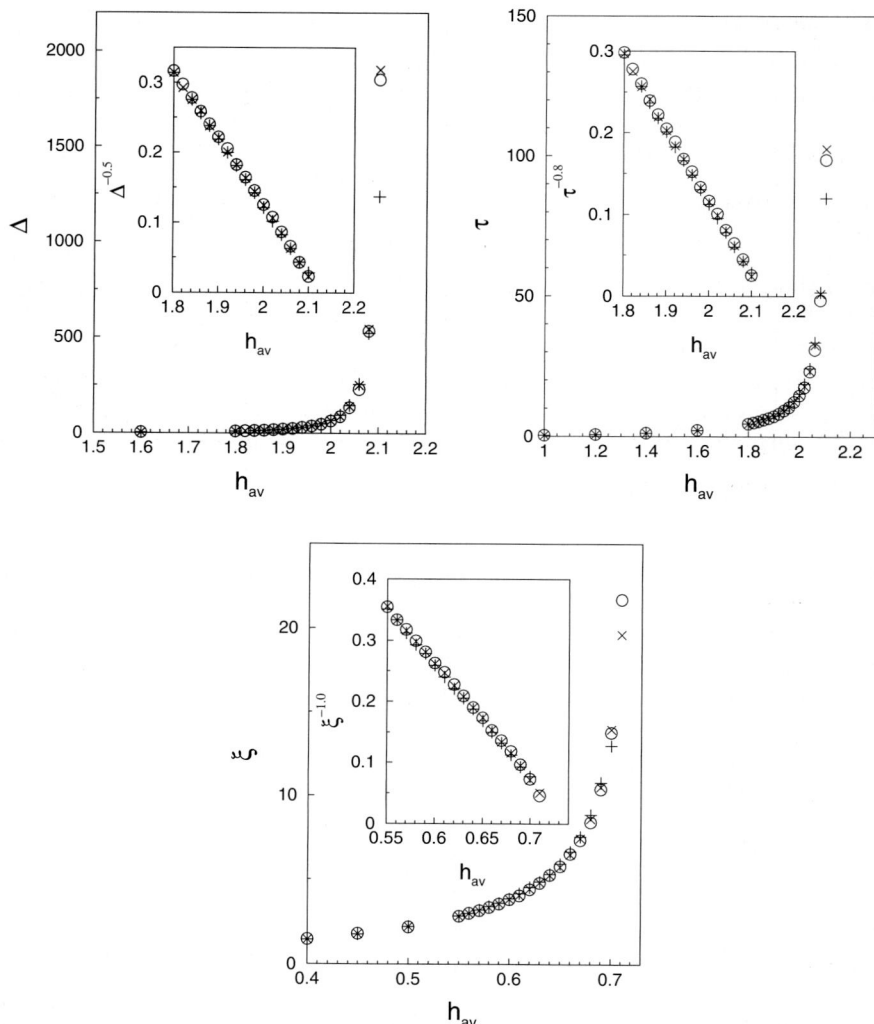

Fig. 13. Precursor parameters and prediction of critical point in BTW model

The Time Series Data Analysis

We consider now the time series obtained by counting the overlaps $m(t)$ as a function of time as one Cantor set moves over the other (periodic boundary condition is assumed) with uniform velocity. The time series are shown in Fig. 16., for finite generations of Cantor sets of dimensions $\ln 2/\ln 3$ and $\ln 4/\ln 5$ respectively.

We calculate the cumulative overlap size $Q(t) = \int_o^t m dt$. We see that "on average" $Q(t)$ is seen to grow linearly with time t for regular Cantor sets.

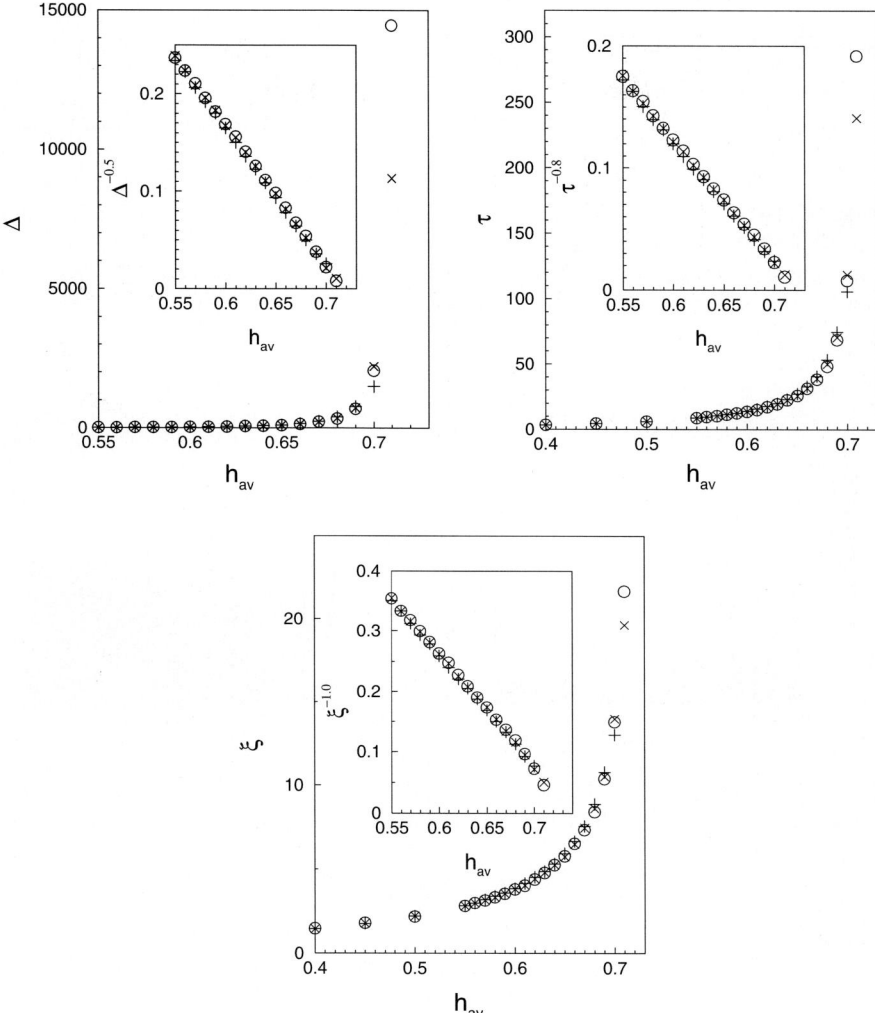

Fig. 14. Precursor parameters and prediction of critical point in Manna model

This gives a clue that instead of looking at the individual overlaps $m(t)$ series one may look for the cumulative quantity. In fact, for the regular Cantor set of dimension $\ln 2/\ln 3$, the overlap m is always 2^k, where k is an integer. However the cumulative $Q(t) = \sum_{i=0}^{t} 2^{k_i}$ can not be easily expressed as any simple function of t. Still, we observe $Q(t) \simeq ct$, where c is dependent on M.

We first identify the "large events" occurring at time t_i in the $m(t)$ series, where $m(t_i) \geq M$, a pre-assigned number. Then, we look for the cumulative overlap size $Q(t) = \int_{t_i}^{t_{i+1}} m dt$, where the successive large events occur at times t_i and t_{i+1}. The behaviour of Q_i with time is shown in Fig. 17 for regular

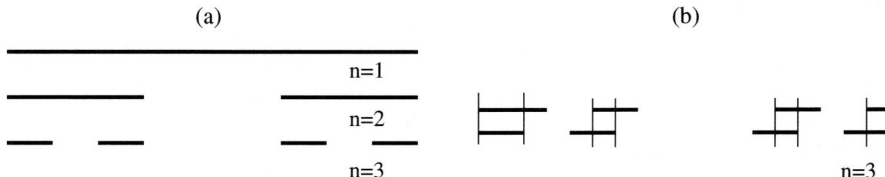

Fig. 15. (a) A regular Cantor set of dimension $\ln 2/\ln 3$; only three finite generations are shown. (b) The overlap of two identical (regular) Cantor sets, at $n = 3$, when one slips over other; the overlap sets are indicated within the *vertical lines*, where periodic boundary condition has been used

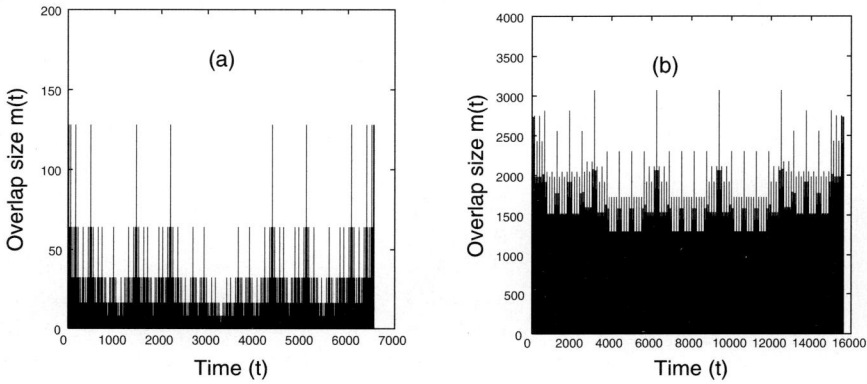

Fig. 16. The time (t) series data of overlap size (m) for regular Cantor sets: (a) of dimension $\ln 2/\ln 3$, at 8th generation: (b) of dimension $\ln 4/\ln 5$, at 6th generation

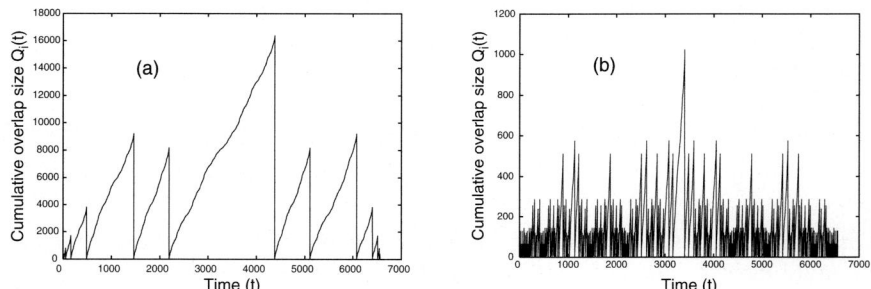

Fig. 17. The cumulative overlap size variation with time (for regular Cantor sets of dimension $\ln 2/\ln 3$, at 8th generation), where the cumulative overlap has been reset to 0 value after every big event (of overlap size $\geq M$ where $M = 128$ and 32 respectively)

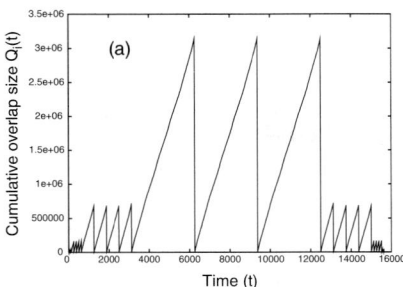 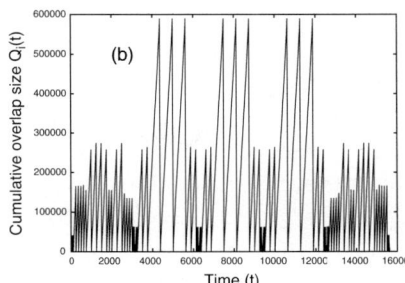

Fig. 18. The cumulative overlap size variation with time (for regular Cantor sets of dimension $\ln 4/\ln 5$, at 6th generation), where the cumulative overlap has been reset to 0 value after every big event (of overlap size $\geq M$ where $M = 2400$ and 2048 respectively)

cantor sets with $d_f = \ln 2/\ln 3$ at generation $n = 8$. Similar results are also given for Cantor sets with $d_f = \ln 4/\ln 5$ at generation $n = 6$ in Fig. 18. It appears that there are discrete (quantum) values up to which Q_i grows with time t and then drops to zero value.

Finally we found that if one fixes a magnitude M of the overlap sizes m, so that overlaps with $m \geq M$ are called "events" (or earthquake), then the cumulative overlap Q_i grows linearly with time up to some discrete quanta $Q_i \cong lQ_0$, where Q_0 is the minimal overlap quantum, dependent on M and l is an integer.

3 Conclusions

Knowledge of precursors sometimes help to estimate precisely the location of the global failure or critical point through a proper extrapolation procedure. Therefore precursors which are available long before the global failure, can be used to resist an imminent global failure.

In all the dynamical systems studied here, we find that long before the occurrence of global failure, the growing correlations in the dynamics of constituent elements manifest themselves as various precursors. In Fibre Bundle Model, the breakdown susceptibility χ and the relaxation time τ, both diverge as the external load or stress approaches the global failure point or critical point. The distribution of avalanches exhibits a crossover in power law exponent values when the system comes closer to the failure point. Also the pattern of inclusive avalanches can tell us whether a bundle can support the applied load or not. In Fuse model, divergence of susceptibility and crossover in avalanche power law exponent are the signature of imminent breakdown of the system. Whereas in both BTW and Manna sandpile models the number of toppling Δ, relaxation time τ and the correlation length ξ grow and diverge following power laws as the systems approach their respective critical points h_c from the sub-critical states. Therefore these parameters directly help to predict the critical point in advance. However in fractal overlap model the time series data analysis suggest that the cumulative overlap sizes can assume some quantized values which indirectly helps to speculate whether a large overlap (event) is imminent or not.

Acknowledgments

S.P. thanks the Research Council of Norway (NFR) for financial support through Grant No. 166720/V30.

References

1. B.K. Chakrabarti and L.G. Benguigui, *Statistical Physics of Fracture and Breakdown in Disorder Systems*, Oxford Univ. Press, Oxford (1997); H.J. Herrmann and S. Roux (Eds), *Statistical Models for the Fracture of Disordered Media*, North Holland, Amsterdam (1990); P. Bak, *How Nature Works*, Oxford Univ. Press, Oxford (1997).
2. F.T. Peirce, J. Textile Inst. **17**, T355–368 (1926).
3. H.E. Daniels, Proc. R. Soc. London A **183**, 405 (1945).
4. P.C. Hemmer and A. Hansen, J. Appl. Mech. **59**, 909 (1992); A. Hansen and P.C. Hemmer, Phys. Lett. A **184**, 394 (1994); M. Kloster, A. Hansen and P.C. Hemmer, Phys. Rev. E **56**, 2615 (1997).
5. S. Pradhan, A. Hansen and P.C. Hemmer, Phys. Rev. Lett. **95**, 125501 (2005); S. Pradhan and A. Hansen, Phys. Rev. E **72**, 026111 (2005).
6. D. Sornette, J. Phys. A **22**, L243 (1989); D. Sornette, J. Phys. I (France) **2**, 2089 (1992); A.T. Bernardes and J.G. Moreira, Phys. Rev. B **49**, 15035 (1994).
7. S. Zapperi, P. Ray, H.E. Stanley and A. Vespignani, Phys. Rev. Lett. **78**, 1408 (1997); Y. Moreno, J.B. Gomez and A.F. Pacheco, Phys. Rev. Lett. **85**, 2865 (2000).
8. S. Pradhan and B.K. Chakrabarti, Phys. Rev. E **65**, 016113 (2001).
9. S. Pradhan, P. Bhattacharyya and B.K. Chakrabarti, Phys. Rev. E **66**, 016116 (2002);
10. P. Bhattacharyya, S. Pradhan and B.K. Chakrabarti, Phys. Rev. E **67**, 046122 (2003);
11. A. Hansen and P.C. Hemmer, Trends Stat. Phys. **1**, 213 (1994). R.C. Hidalgo, F. Kun and H.J. Herrmann, Phys. Rev. E **64**, 066122 (2001); S. Pradhan, B.K. Chakrabarti and A. Hansen, Phys. Rev. E **71**, 036149 (2005).
12. B.D. Coleman, J. Appl. Phys. **27**, 862 (1956); B.D. Coleman, Trans. Soc. Rheol. **1**, 153 (1957); B.D. Coleman, Trans. Soc. Rheol. **2**, 195 (1958).
13. S.L. Phoenix, SIAM (Soc. Ind. Appl. Math.) J. Appl. Math. **34**, 227 (1978); S. Roux, Phys. Rev. E **62**, 6164 (2000); R. Scorretti, S. Cilibreto and A. Guarino, Europhys. Lett. **55**, 626 (2001).
14. S. Pradhan and B.K. Chakrabarti, Phys. Rev. E **67**, 046124 (2003).

15. P. Bak, C. Tang and K. Wiesenfeld, Phys. Rev. Lett. **59**, 381 (1987); P. Bak, C. Tang and K. Wiesenfeld, Phys. Rev. A. **38**, 364 (1988).
16. S.S. Manna, J. Stat. Phys. **59**, 509 (1990); P. Grassberger and S.S. Manna, J. Phys. France **51**, 1077 (1990); S.S. Manna, J. Phys. A: Math. Gen. **24**, L363 (1991).
17. D. Dhar, Physica A **270**, 69 (1999); D. Dhar, Physica A **186**, 82 (1992); D. Dhar, Physica A **263**, 4 (1999).
18. M. Acharyya and B.K. Chakrabarti, Physica A **224**, 254 (1996); Phys. Rev. E **53**, 140 (1996).
19. A. Vespignani, R. Dickman, M.A. Munoz and S. Zapperi, Phys. Rev. E **62**, 4564 (2000); A. Chessa, E. Marinari and A. Vespignani, Phys. Rev. Lett. **80**, 4217 (1998).
20. Gutenberg, B. and Richter, C.F., *Seismicity of the Earth and Associated phenomena*, Princeton University Press, Princeton, N.J. (1954).
21. R. Burridge, L. Knopoff, Bull. Seis. Soc. Am. **57**, 341 (1967).
22. J.M. Carlson, J.S. Langer, Phys. Rev. Lett. **62**, 2632 (1989)
23. V. De Rubeis et al, Phys. Lett. **76**, 2599 (1996).
24. B.K. Chakrabarti, R.B.Stinchcombe, Physica A **270**, 27 (1999).
25. S. Pradhan, B.K. Chakrabarti, P. Ray and M.K. Dey, Phys. Scr. T **106**, 77 (2003); S. Pradhan, P. Chaudhuri and B.K. Chakrabarti, in Continuum Models and Discrete Systems, Ed. D. Bergman, E. Inan, Nato Sc, Series, Kluwer Academic Publishers (Dordrecht) 245 (2004).

Part IV

Miscellaneous Short Notes

Nonlinear Analysis of Radon Time Series Related to Earthquake

N.K. Das[1], H. Chauduri[2], R.K. Bhandari[1], D. Ghose[2], P. Sen[2], B. Sinha[1,2]

[1] Variable Energy Cyclotron Center, 1/AF, Bidhannagar, Kolkata-700064
   nkdas@veccal.ernet.in
[2] Saha Institute of Nuclear Physics, 1/AF, Bidhannagar, Kolkata-700064
   prasanta.sen@saha.ac.in

1 Introduction

Plethora of seemingly random phenomena on the earth are governed by some complex mechanism and find themselves non-random in statistical sense. It has been established that many apparently random phenomena in nature, such as occurrence of earthquakes, fault system, function of heart and brain besides several other events have ingrained pattern that can be well construed following nonlinear analysis [1, 2, 3]. Radon emanating from seismo-active thermal spring at Bakreswar, West Bengal, India, is continuously monitored in situ so as to look into the variational behaviour of radon concentration and to decipher the physical implications contained therein with regard to seismic events. Radon issues as a component of two phase fluid flow through network of fractures and fissures encrusted within the solid earth constituting a rather complex geodynamic system. The variational disposition of concentration with time happens to be the realization of certain ongoing dynamical process within the earth and the interactions experienced by up-streaming fluids while migrating to the surface. The macroscopic interactions of the fluids with the physical parameters of the adjoining crust bring about microscopic alterations in the compositions of the emerging gases. Estimation of fractal dimension, d_f [5], along with other invariants such as correlation dimension and lyapunov exponent are proven means for describing the temporal variations and help to understand the governing mechanism accountable for gaseous emission. The trajectory traced by the process of gas release such as radon from a deep seated thermal spring over the time invariably arise from its source parameters located far away from the test point together with other intervening factors involving interactions during migration with space and time. The nonlinear analysis of real time series enables us to conceptualizing the evolutionary mechanism pertinent to the anomalous fluctuation of radon concentration that are consistent with an onset of seismic tremor [6].

The presence of noise in the time series under study could mask the actual dynamical information of the system. The finite length of the experimental data coupled with the complexity of data result in a relatively short scaling region. Consequently, , the precision of the determination of nonlinear quantifiers is affected. To do away with the noise present in the signals Singular Value Decomposition (SVD) technique is used. Since the small singular values primarily represent noise, a filtered signal, after removal of small SVD's, with less noise is derived.

The method of surrogate data is employed to test whether the time series is in the state exhibiting nonlinear characteristics or consistent with the linear correlated noise. Surrogate data provides fairly reasonable test for characterizing nonlinearity in real time data set. Statistical quantifiers are computed for the radon time series and that are compared with the corresponding values evaluated for surrogates. If the test statistics obtained for the experimental data is significantly different from that estimated for the surrogates, the null hypothesis is liable to be rejected and the experimental data set is regarded nonlinear [7]. In the present case, ensemble of stochastic surrogates are generated by phase randomization keeping the mean as well as the variance and power spectrum same as that of the original experimental time series.

In an attempt to explore the underlying structure and the governing mechanism of manifestly aberrant appearance of the radon time series, nonlinear statistical approach has been employed on the experimental radon data. This paper describes the nonlinear structure of radon time series with special reference to earthquake. The data presented corresponds to an earthquake (M = 7.2) occurred in Japan on August 16, 2005. There are two data sets which we have conveniently named as RnO, which includes the anomalous components ($>2\sigma$) [8] and RnT after removing the anomalous data points from the same sequence. Total data points in RnO is 13245 and in RnT it is 12648 respectively, having a sampling interval of 10 mins.

2 Methods of Analysis

2.1 Phase Space Plot

Any time series generated by a nonlinear process can be considered as the projection on the real axis of a higher dimensional geometrical object that describes the behaviour of the system under study. A series of scalar measurements, $x(t)$, can be used to define the orbits describing the evolution of the states of the system in a **m** dimensional Eucleadian space [9, 10]. The orbits will then consist of points X with coordinates,

$$X^m(t) = x(t), x(t+\tau), \ldots, x(t+(m-1)\tau) \tag{1}$$

where τ is referred as the delay time for a digitized time series while **m** is termed as the embedding dimension.

We choose the method of false nearest neighbour (FNN) to determine the optimum embedding dimension needed to reconstruct the phase-space trajectory of the time series. The first minimum of the percentage FNN gives an estimate of the embedding dimension. As the minimum value of **m** is reached, the fraction of FNN should converge and this minimum dimension provides an upper bound estimate of the topological dimension of the phase-space attractor. In the present study, calculated embedding dimensions are 7 and 6 emerges to be the acceptable values for RnO and RnT series respectively. The concept of constant time delay (τ) is introduced because in an experiment the sampling interval is, in general, chosen without an accurate knowledge of characteristic time scales involved in the evolutionary process. The delay time for attractor construction was computed by average mutual information (AMI) method [11]. The first minimum of the AMI is considered to be the preferred delay time.

The dimension of an attractor provides important information about the nature of the system giving off temporal signal and the effective degrees of freedom that are present in the physical system under consideration. Two different approaches for determining the attractor dimension are pursued; one based on geometric properties, the capacity dimension, D_b and the other resulting from probabilistic approach, the correlation dimension D_2. The method of is capable of representing a very important diagnostic procedure for distinguishing between determinism and stochasticity. As opposed to the stochastic process, D_2 tends to be a constant as the embedding dimension increases in the case of deterministic process. Depending on the size of the correlation dimension relative to the embedding dimension used to calculate it, we can determine whether the signal behaves like a random or chaotic process. The capacity dimension is calculated using the so-called "box counting" algorithm. In this procedure the properly embedded phase space is successively divided into equal hypercubes and the logarithm of the fraction of hypercubes that are occupied with data points is plotted versus the logarithm of the normalized linear dimension of the hypercubes. On the other hand, the correlation dimension is computed by the sphere-counting algorithm by Grassberger-Procaccia [12]. Briefly, it is obtained by counting the data points inside hyper-spheres of various radii centered on each data point in a reconstructed phase space with appropriate embedding dimension. Its value is close to the capacity dimension, but is generally a more accurate measure. As a consequence of the Kaplan-Yorke conjecture this value must be greater than 2.0 so as to identify deterministic traits of the attractor.

2.2 Fractal Dimension

The rescaled range analysis also referred as range over standard deviation (R/S) anaysis is employed to determine the fractal dimension (d_f) for characterization of the time series data [5]. The R/S analysis is performed on the discrete time series data (x_i) of dimension N by calculating the mean $\bar{x}(N)$, standard deviation, $S(N)$ and the deviation from the mean, which is termed as the accumulated departure, $X(n, N)$, according to the following formulas;

$$S(N) = \left[\frac{1}{N}\sum_{i=1}^{N}(x_i - \bar{x}(N))^2\right]^{1/2} \quad (2)$$

and

$$R(N) = max\{X(n, N)\} - min\{X(n, N)\} \quad (3)$$

where

$$X(n, N) = \sum_{i=1}^{n}(x_i - \bar{x}(N)), \quad 0 < n \leq N$$

The procedure adopted is to calculate R/S_1 for the whole data set of number N. Subsequently the data set is divided into half and R/S is calculated for each part and then averaged to get R/S_2. This is repeated for N/4, N/8 ... and so on. Thus a series of R/S_i are obtained from the time series data for the respective range of divisions and are then plotted as a function of number of data points/window interval on a log-log scale. The hurst coefficient, H, of the data is thus derived from the slope of the straight line fitted to these points in least square sense. H for the time sequence is related to its fractal dimension d_f by $d_f = 2 - H$. R/S analysis is a simple yet robust method for fast fractal estimation. Time series yielding $H > 0.5$ carries the signature of persistence behaviour [4].

2.3 Correlation Dimension

Correlation dimension (D_2) is one of the most widely used measure of attractor dimension. The idea is to construct a function $C(r)$ which is the probability of finding those pairs within a circle of radius r, the separation of two arbitrary points x_i and x_j of the reconstructed phase space. This is done by calculating the separation between every pair of N data points and sorting them into bins of width **dr** proportional to **r** [12]. A correlation dimension can be evaluated using the distances between each pair of points in the set of N number of points,

$$s(i, j) = |x_i - x_j| \quad (4)$$

A correlation function, $C(r)$ is then calculated as

$$C(r) = \frac{1}{N^2} \times [\text{number of pairs } (i, j) \text{ such that } s(i, j) < r] \quad (5)$$

Alternatively, it may be represented by

$$C(r) = \frac{1}{N^2} \times \sum_{i=1}^{N}\sum_{j=i+1}^{N}\Theta[r - s(i, j)] \quad (6)$$

Where, $\Theta(f)$ is the Heavyside function. In the limit of an infinite number of data and for vanishingly small r, the correlation sum should scale like a power law, $C(r) \propto r^D$ and the correlation dimension is then represented as,

$$D_2 = \lim_{r \to 0} \frac{\log C(r)}{\log(r)} \tag{7}$$

The D_2 for each time series was computed using lag reconstructed signals with the pre-determined embedding dimension by the FNN method. We plotted log $C(r)$ as a function of Log (r) and computed D_2 from the slope of a linear fit.

2.4 Largest Lyapunov Exponent (λ)

Lyapunov exponents are recognised to be the most reliable indicators to distinguish a chaotic from a non-chaotic process. It quantifies chaos by means of estimating the sensitivity to perturbations in initial conditions. The exponent (λ) also represents the nature of time evolution of close trajectories in the phase space and is recognized as the hallmark for chaotic behaviour. If the initial separation of two close trajectories grow exponentially, the value of λ becomes positive and implies that the dynamics underlying the measured signals capture the signature of chaos. Because of the sensitive dependence on the initial condition, even infinitesimal measurement error, noise or perturbation from external effects to initial conditions will expand exponentially over time. The average rate of exponential divergence describes instability of the system and the rapid loss of information as time evolves. We have estimated the largest Lyapunov exponent to probe the nature of temporal evolution of the observed time series. To evaluate the Lyapunov exponents it is necessary to monitor the behaviour of two closely neighbouring points in phase space as a function of time. The initial separation of two points in the Euclidian sense is determined and then the separation is examined over several orbits. Had the two near by trajectories started off with an initial separation of $d_0(t=0)$ then the trajectories diverges with time satisfying the relation, $d_t = d_o e^{\lambda t}$, where d_t is the separation at time t. As a matter of fact, λ is used to calculate the logarithm of time, $\lambda = \frac{\log d_0}{\log d_t}$. For periodic motion λ_{max} is equal to zero, indicating that the limit cycle conserves its information in time. The algorithm used in the present study to calculate the largest Lyapunov exponent is due to Wolf et al. [13]. This algorithm involves embedding the time series in m-dimensional space and monitoring the divergence of trajectories initially close to each other in that phase space. The largest Lyapunov exponent can then be estimated from

$$\lambda_{max} = \frac{1}{\Delta t} \sum_{k=1}^{M} \log_2 \frac{d(t_k)}{d_0(t_{k-1})} \tag{8}$$

where $d_0(t_{k-1})$ is the separation distance of two initially close points in the trajectory. After a fixed evolution time Δt, the final distance between the two points evolves to $d(t_k)$ and M is the replacement s-Prsen.eps.

3 Results and Discussions

The profiles for radon concentrations apparently varies in irregular fashion with spike like fluctuations. This may be ascribed to several factors such as nonuniform volatile emission from the underground reservoir subjected to hypogene forces, earth's vibration and physico-chemical interactions while spouting through the fractures and fissures of the earth's crust.

The phase space plots, as shown in Figs. 1 and 2 corresponding to radon concentration indicate that the process of radon emission is clearly non-random with certain degree of correlation. Figures 3, 5 and Figs. 4, 6 demonstrate the fractional dimension of the experimental data and its somewhat persistent nature coupled with broadband power spectra with its exponential decay towards the higher frequency. These features are highly suggestive of a underlying nonlinear chaotic process. Moreover, the nonlinear invariants, such as D_2, D_b and λ, computed from the experimental time series and those obtained for the surrogate data are found to be considerably different from each other implying nonlinear structure of the radon time series. Correlation dimension and capacity dimension in conjunction with the Lyapunov exponent are estimated to be higher in the seismically quiescent period rather than in seismically disturbed phase, Table 1. The analysis exhibits a discernible deterministic nonlinearity in radon time series and leads to the inference that the effect may contain or is caused by a deterministic mechanism. These observations obviously imply the influence of seismic phenomena that leads to trimming down of the strength of mechanism yielding deterministic chaos. Table 1 displays the computed statistical measures of the nonlinear analysis of the time sequence during the seismically normal state along with the period when anomalous fluctuations were recorded.

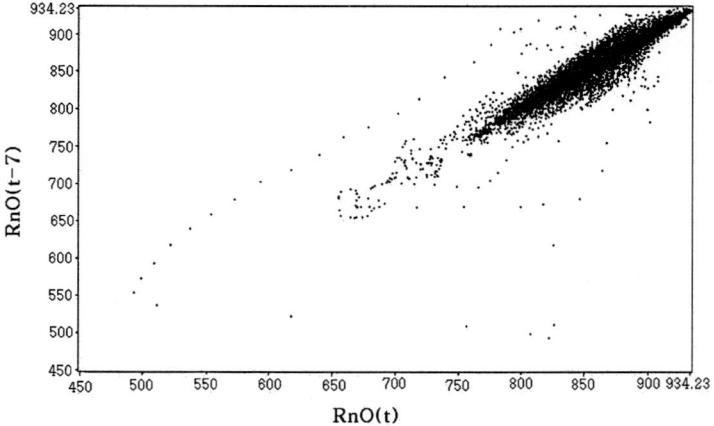

Fig. 1. State space plot of RnO series

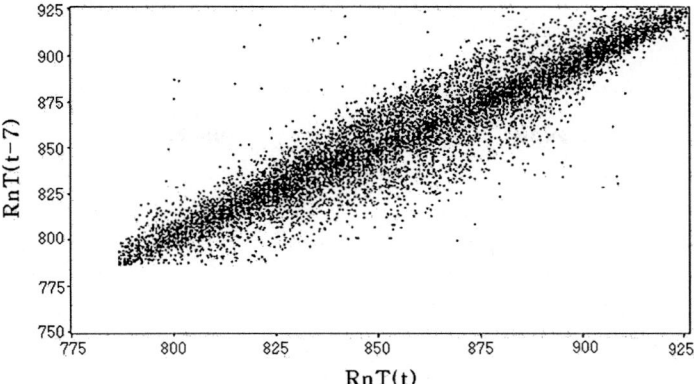

Fig. 2. State space plot of RnT series

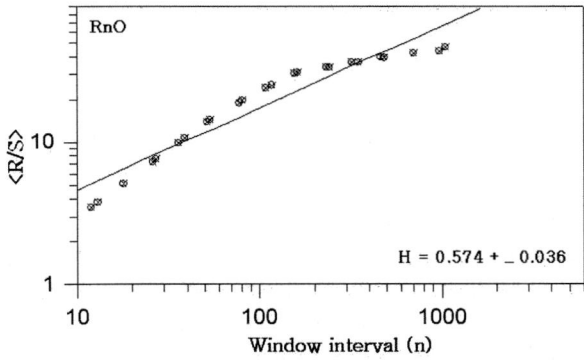

Fig. 3. Plot of Log(R/S) vs. Log (n) to evaluate Hurst coefficient for RnO series

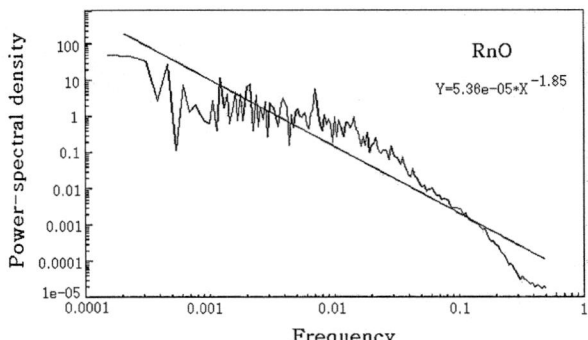

Fig. 4. Power spectrum of RnO series

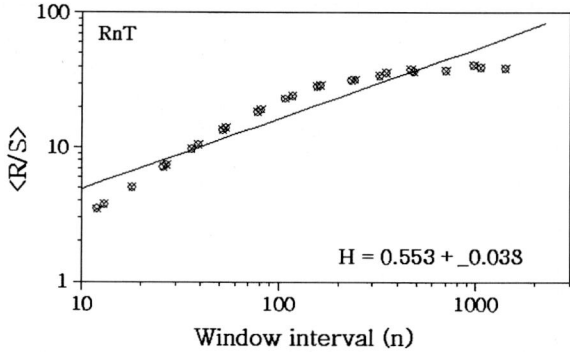

Fig. 5. Plot of $\text{Log}(R/S)$ vs. $\text{Log}(n)$ to evaluate Hurst coefficient for RnT series

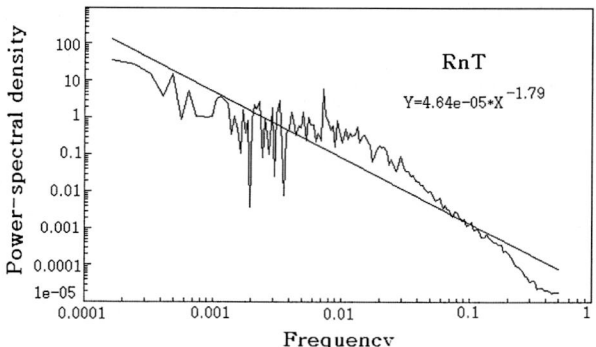

Fig. 6. Power spectrum of RnT series

Table 1. Nonlinear statistical measures for RnO and RnT

Time Series	Hurst Coefficient	Correlation Dimension	Capacity Dimension	Lyapunov Exponent
RnO	0.57±0.036	2.54±0.312	1.72 ±0.107	0.23±0.016
SRnO	0.45±0.032	3.37±0.449	1.23 ±0.078	0.29±0.012
% Variation	26.66	32.67	39.83	2.53
RnT	0.55±0.038	2.89 ± 0.378	1.96 ±0.122	0.33 ±0.013
SRnT	0.47±0.022	3.51±0.262	1.43 ±0.089	0.27±0.012
% Variation	17.02	21.45	37.06	22.22

It is found that the measures are conspicuously lower when the system is subjected to some anomalous effects as compared to that obtained for seismically normal period. The results reveal that the seismic tremor induces a perceptible reduction of the strength of correlation dimension, capacity dimension and Lyapunov exponent. This, in turn, expresses the of the time series data under the influence of seismic events. Observed discrepancy between for original data set (RnO) and truncated data set (RnT) probably stem from the

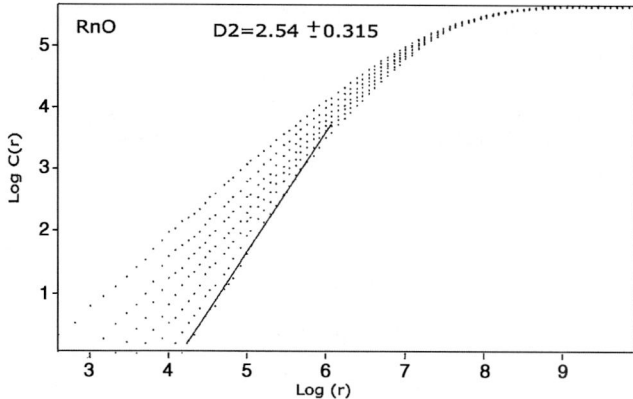

Fig. 7. Plot of Log $C(r)$ vs. Log (r) to estimate correlation dimension for RnO series

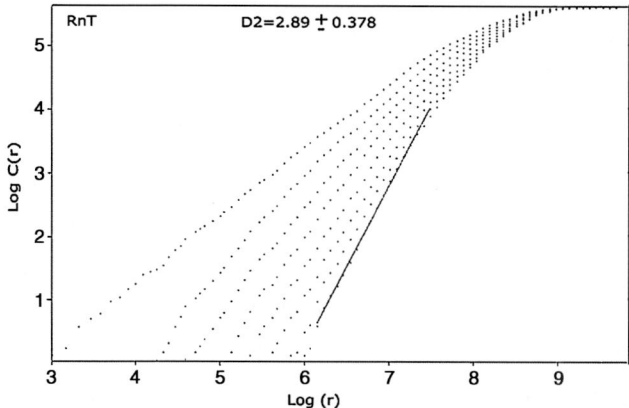

Fig. 8. Plot of Log $C(r)$ vs. Log (r) to estimate correlation dimension for RnT series

close interplay between the seismic wave and the ongoing dynamical process within the fluid reservoir.

4 Conclusion and Outlook

Estimation of nonlinear invariants unravel the inherent traits of radon signals affirming the nonlinear and chaotic attributes of the process causing radon emission. The governing mechanism underlying the dynamics of radon fluctuation stands to be nonlinear and shows sensitive dependence on the initial conditions, capturing the impression of a nonlinear . External perturbations upon the under-earth fluid reservoir due to seismic events account for the observed disparity in the nonlinear measures of two time sequences derived under two different geo-physical conditions of the same system. Even though the elevated radon concentration prior to major seismic event is not adequately substantiated through mathematical modelling, yet the small changes in initial conditions due to micro-seismicity and associated geological alterations, in all likelyhood, result in the observed increase of radon concentration even at large distances substantially [14].

Acknowledgements

We sincerely thank to the Department of Science & Technology (IS-STAC) and the Department of Atomic Energy, Government of India for sponsoring the project.

References

1. D.L. Turcotte *Fractals and Chaos in Geology and Geophysics*, Cambridge Unversity Press, New York (1992).
2. A. Das, P. Das and A.B. Roy, Complexity Int., **09**, 1–8, (2002).
3. J. Planinic, B. Vukovic and V. Radolic, J. Environ. Radioactv., **75**, 35–45, (2004).
4. P. Barat, N.K. Das, D. Ghose et al., Physica A, **262**, 9–15, (1999).
5. H.E. Hurst, Trans. Am. Soc. Civ. Eng., **116**, 770, (1951).
6. N.K. Das, R.K. Bhandari, D. Ghose et al., Appl. Rad. Isot., **64**, 144–148, (2006).
7. J. Theiler, S. Eubank, A. Longtin et al., Physica D, **58**, 77 (1992).
8. C.H. Scholz, L.R. Sykes and Y.P. Aggawal, Science, **181**, 803–810, (1973).
9. F. Takens *Lecture Notes in maths.*, Springer, New York, **898**, (1981).
10. T. Saucer, J. Yorke and M. Casdagli J. Stat. Phys. **65**, 579–616, (1994).
11. H. Kantz and T. Schreibre, *Nonlinear Time Series Analysis*, Cambridge Univesity Press, United Kingdom, 131–133 (2002).
12. P. Grassberger and I. Procaccia, Phys. Rev.Lett. **50**, 346–349, (1983).
13. A. Wolf, J.B. Swift, H.L. Swanney et al.: Physica D, **16**, 285–317, (1985).
14. N.K. Das, R.K. Bhandari, D. Ghose et al.: Curr. Sci., **89** (8), 1399–1404, (2005).

A Thermomechanical Model of Earthquakes

B. Bal and K. Ghosh

Saha Institute of Nuclear Physics, 1/AF Bidhannagar, Kolkata-700064, India
bijaybhushan.bal@saha.ac.in
kuntal.ghosh@saha.ac.in

1 Introduction

Do earthquakes have anything to do with oscillations? The question naturally arises since the phenomenon of earthquake appears to be a random, chaotic and unpredictable one, while the term *oscillation* seems to indicate more towards a periodic and hence predictable event. But we shall presently see that in spite of this apparent contradiction, at least some aspects of earthquakes can indeed be understood better in terms of a particular type of oscillation discussed later. Understanding the dynamics of any system is the primary condition for making any prediction about the system. The efforts to explain tectonic plate movement in terms of oscillations is not absolutely new [1] as we shall soon see, but the present approach is a new effort in this direction with the help of some simple table-top experiments based on apparatus like pressure-cooker and pop-pop boat.

We all know that the term *earthquake* covers different types of vibration of the earth's surface due to natural causes. Presently, earthquakes are classified with respect to their origin into three different types: tectonic, volcanic and colluese or denudation. Of these, the last mentioned type represents the earthquakes occurring due to the collapse of considerable amount of mass like a large landslide occurring in mountain region. The earthquakes associated with volcanic activity are also understandable from a similar approach, but a clear picture about the first type i.e. the dynamics of tectonic movement which lead to earthquakes and associated devastations, is yet to come out. The most prevalent hypothesis on the primary cause of tectonic movement is *the convection cell hypothesis*. According to this, unequal heat distribution in the mantle may produce convection cells driven by heat from radioactive decay in the earth's interior below the lithosphere. Hot material rises, spreads laterally, cools and sinks deeper into the mantle to be reheated. While the Shallow Convection Cell Model claims that the convection cells are restricted to the asthenosphere, the Deep Convection Cell Model is based on the assumption that the entire mantle is involved in convection. This movement in the mantle is of the order of a few centimeters per year. However, these models are not adequate to explain some observed phenomena like: (a) Development of stress with time on tectonic plate and subsequent breakdown. (b) A major earthquake is associated with pre-shaking and post-shaking . (c) The possibility of major earthquakes is to recur with time.

Wesson [2] has also pointed out as many as 74 objections to gradualist plate tectonics. He has shown that there is no convection in mantle at all. Beloussov [3], Meyerhoff and Meyerhoff [4] have also pointed out similar inconsistencies in the theory of tectonic plate movement based on convection cell hypothesis. The tussle between the oscillatory model and the gradualist model has got a long and chequered history in tectonic movement understanding. For example, *the elevation hypothesis* advanced in the eighteenth century by Lomonosov and Hutton, or *the contraction hypothesis* put forward by Elie de Beuamont in the middle of the nineteenth century based on the cosmogenic hypothesis of Kant-Laplace were two very strong gradualist models [1]. The challenge to these gradualist ideas came from *the pulsation hypothesis* which was introduced by Butcher in 1920. It is interesting to note that although this hypothesis falls in the oscillatory line, it in fact utilized the basic principles of *the contraction hypothesis* [1]. As a matter of fact the modern *convection cell hypothesis*, in its turn also tries to take into cognizance alternate regions of stretching and contraction due to alternately oriented convection currents [1]. The model proposed here, is a new oscillatory model through which we shall try to answer the inadequacies of *the convection cell hypothesis*, mentioned above. The same model, we shall also see, helps us to get an insight into the dynamics of volcanic activity too.

2 Earthquakes and Relaxation Oscillation

Whenever we speak of an oscillation, what we most frequently come across in literature, is the sinusoidal oscillation shown in Fig. 1a which represents a phenomenon continually varying with time. However, it is also well-known to us that there are other type of oscillations where unlike the sine or cosine wave, the functional form changes its state in an abrupt manner. Two such examples, one of a square wave and the other a saw tooth wave, have been shown in Figs. 1b and 1c. Such waveforms may represent a system, which when energized by a time-independent source, produces an output that is not only time-dependent and periodic, but also one that changes its state in an abrupt manner as shown in these examples. Such a system is termed as a relaxation oscillation based system [5]. We shall now take the help of two specific examples of relaxation oscillation for a better understanding of the phenomena of earthquake and tectonic movement in the next two subsections.

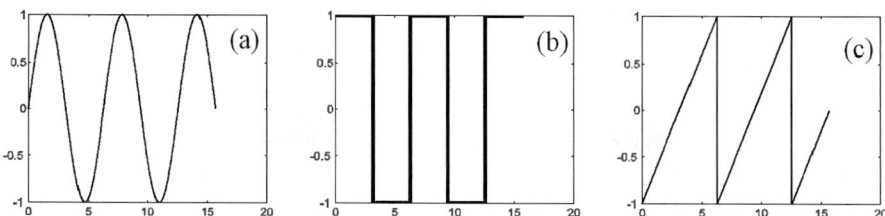

Fig. 1. (a) A sine wave (b) A square wave (c) A sawtooth wave

2.1 Relaxation Oscillation in Pressure Cooker

In Fig. 2, we see the schematic diagram of a pressure cooker, as used in our everyday life. The maximum pressure at which the jet starts operating may be termed as the barrier height of the cooker. This height, is determined by the weight placed at the top. On the other hand, the frequency of steam ejection through the jet will be a function of both the weight as well as the rate of heating by the heating system. In other words if we change the rate of heating, the frequency of ejection will only change and not the maximum amplitude in the pressure meter. But if the weight is adjusted then both frequency as well as the maximum amplitude reached, will undergo changes. This is a typical example of relaxation oscillation comparable with many natural events, relevant to our discussion [5]. In Fig. 3a and 3b we show the experimental results of the variation of pressure and temperature respectively, as measured in the pressure meter and thermometer shown in Fig. 2. The meter reading shown is shown here after steam pressure has already developed to almost 5 lb/in^2 and

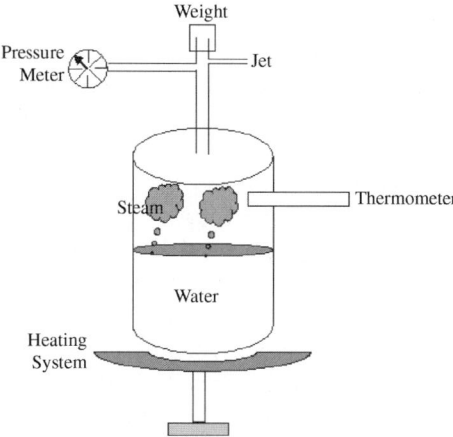

Fig. 2. A schematic diagram of the pressure cooker used for conducting the experiment

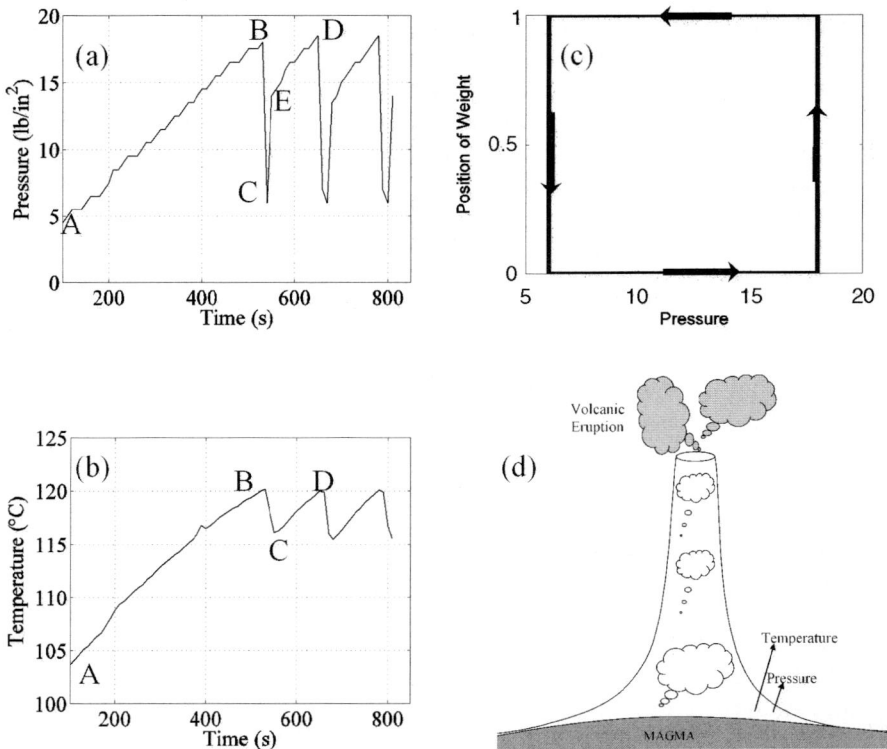

Fig. 3. (a) The Pressure-time profile (b) The Temperature-time profile (c) The Weight position-Pressure hysteresis loop (d) A typical volcanic eruption

temperature to about 105^0. Below we give a description of the findings from Figs. 3a and 3b: AB – Represents the first phase of the development of temperature and pressure. BC – Represents the start of jet flow and immediate fall of pressure and temperature. CD – As the jet flow stops pressure suddenly jumps to E and then with time reaches D, although temperature rise in that region is not abrupt. The pressure falls down immediately as the jet starts operating, but the weight is retained at its uplifted position as long as the jet flow continues. Then as soon as the jet flow stops, the pressure immediately jumps to E to begin its journey to E. This means that the system is associated with a hysteresis loop as is evident from Fig. 3c. The position of the weight is binary, i.e. either 0 (rest position)or 1 (ejection position).

The sudden fall in both temperature and pressure, in these graphs, is only natural because of the abrupt drop in the number of gaseous particles inside the pressure cooker with each jet emission. The situation may be compared to the phenomenon of volcanic eruption (Fig. 3d). The change of pressure and temperature inside the volcano is expected to follow a similar pattern of relaxation oscillation. Moreover, the amplitude and frequency of occurrence of volcanic activity will depend upon the barrier height that is to be surpassed. So the future activity of the volcano may be predicted by studying the nature of barrier height, which in turn depends on a study of the sedimentation at the mouth resulting from the previous eruption (Fig. 4) and the history of earlier eruption in the same place.

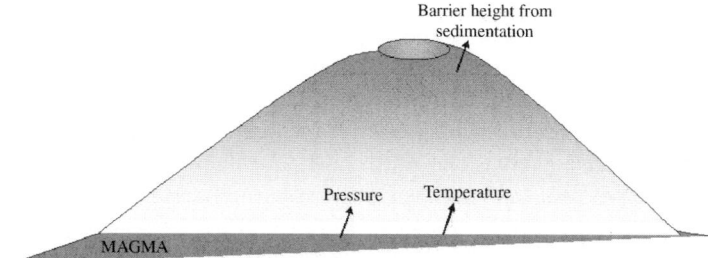

Fig. 4. Volcanic barrier height resulting from sedimentation at the mouth

2.2 Relaxation Oscillation in Pop-pop Boat

We shall now try to understand tectonic plate movement that is responsible for earthquakes, in the light of relaxation oscillation as exhibited by a pop-pop boat. In Fig. 5, we see the schematic diagram of a pop-pop boat, a childhood toy for many. As the name indicates the boat does not actually move continually, but does so in discrete steps with a pop-pop sound. Water enters into the reactor through one of the nozzles (marked with the same shade as

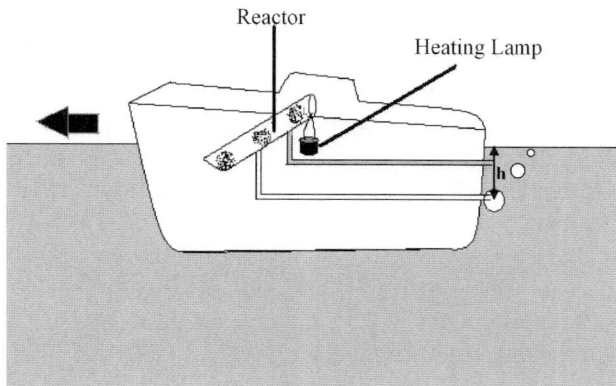

Fig. 5. The schematic diagram of a pop-pop boat

water), gets heated and the resultant steam forms a bubble at the outlet of the other nozzle, also under water. Once the pressure developed within the bubble exceeds the pressure exerted by water at that height (hdg – where h is the depth of the outlet nozzle under water and d the specific gravity of water), the bubble is released [5] and the boat moves forward under its impact. Water enters through the previous nozzle and the cycle continues. The barrier height (hdg) ensures that this phenomenon is another instance of relaxation oscillation.

We shall now see that tectonic plate movement can be explained through this approach as is evident from Fig. 6. Here the liquid is the fluid magma on which the plates are floating. Since a part of the tectonic plate is submerged in magma, magma vapour ejected continuously from magma, gets trapped in the intermediate space between the plate and the magma level. With time, this pressure increases and at some critical pressure (barrier height), the system finds some path (the nozzle shown in Figure) to release the pressure and the plate feels a jerk and moves in some direction. Such sudden tectonic movements may lead to earthquakes. Moreover, due to such sudden intermittent movement, as opposed to a continuous one, it can even place itself over another plate (plate 2 in Fig. 6). In course of time this will develop a stress on plate 2 that may result in plate subduction, as happened with the Indian and Burma plate when the devastating Tsunami was generated on December 26, 2004.

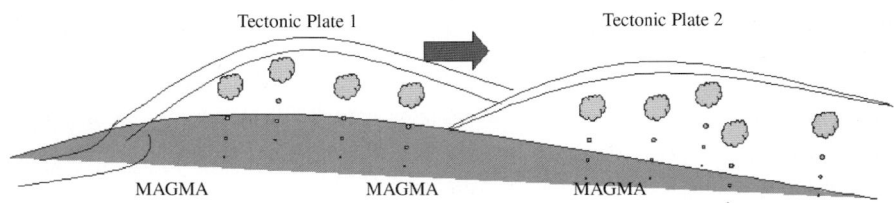

Fig. 6. Movement of tectonic plate in the light of pop-pop model

3 Earthquake Prediction

We shall first discuss the phenomenon of pre-shaking and post-shaking associated with earthquake occurrences and explain their possible causes in the light of the present model. From such an approach, we shall also try to predict the nature of earthquake in a region.

Let us recapitulate our pressure cooker model again and have a look at Fig. 7a. We find that instead of one there are two weights one of which is double the mass of the other. When such a system is heated which of the two

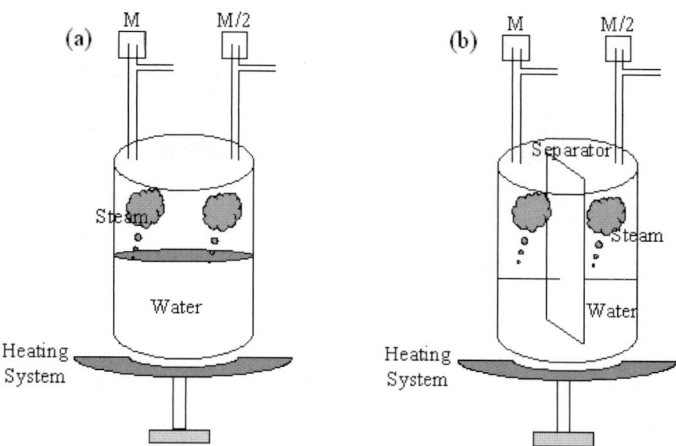

Fig. 7. (a) The pressure cooker with two different weights (b) The same pressure cooker with a partition in between

weights will be lifted? Clearly, the lower weight will always act as the barrier height and hence the jet connected to the more massive one will in fact remain silent throughout. If the mass of the lower weight is further reduced its jet will operate more frequently. The analogy in case of tectonic plates however, may not be always that simple. This is evident in Fig. 7b, where the same pressure cooker is separated from top to deep inside the water level by a partition. In such a case prediction about which weight will come into picture and when, becomes more complicated and as a matter of fact, both these weights may get lifted and consequently both the jets may be operative from time to time depending upon the rate of heating and the nature of fluctuation of pressure near jet. Here two relaxation oscillators are weakly coupled through water.

If we now come to the phenomenon of pre-shaking associated with earthquakes with the analogy of a lid placed on a bowl of water that is placed on a heating system (Fig. 8a), we find that the lid is found to vibrate frequently due to the leakage of steam through non-uniform contact points of the lid and the vessel, prior to the final forceful ejection of the lid. Such small amplitude movements are comparable to the small oscillations observed in many cases prior to major earthquakes. They are also observable in pressure cookers as short intermittent hissing sounds just before the final ejection of steam through the jet. Now, in a real situation, once a severe earthquake occurs, the plate (lid in our analogy) is likely to rapture at many places due to the impact. If the lid in our experiment also breaks falling on the ground due to its forceful ejection and if we place such a fractured lid atop the bowl with the help of a net (Fig. 8b), then the new amplitude and frequency of oscillation (which in relaxation oscillation are functions of barrier height) of such a fractured lid will be governed by the path of minimum impedance,

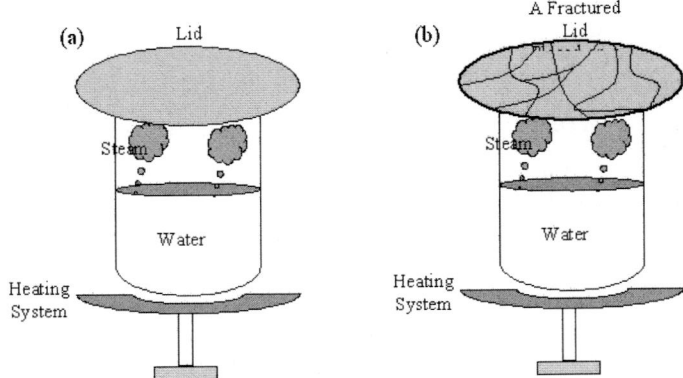

Fig. 8. (a) A vessel of water covered by a lid (b) The vessel is covered by fractures of the same lid placed carefully over a net

i.e. the smallest fractured portion of the lid. This is how we can explain the lower amplitude but more frequent aftershocks that follow a high magnitude earthquake.

To conclude, a major earthquake will probably be associated with a number of new fractures in the plate, the number being of course dependent upon the composition as well as the morphology of the plate at that position. If a large number of such fractures are created, then the possibility of another major earthquake in that region diminishes, although small magnitude earthquakes will occur more frequently. The exact amplitude and frequency of such occurrences will be governed by how the trapping arrangement of fluid magma changes within the new tectonic pieces after the formation of the fractures. In course of time the resultant movements of the tectonic pieces due to such new arrangement may in turn lead to the accumulation of newer stresses that may again usher in the possibility of another major earthquake there.

References

1. A.F. Yakushova, Factors Conditioning the Tectonic Movements and Development of the Earth's Crust. Tectonic Hypotheses. In: *Geology with the elements of Geomorphology*, translated by G.C. Egorov (Mir Publishers, Moscow 1986) pp 370–385
2. P.S. Wesson, J. Geology **80**, pp. 185–197 (1972)
3. V.V. Beloussov, EOS, American Geophysical Union **60**, pp. 207–210 (1979)
4. A.A. Meyerhoff and H.A. Meyerhoff, American Association of Petroleum Geologists (AAPG) Bulletin **56**, pp. 269–336 (1972)
5. M. Katti and B. Bal: Physica D **112**, pp. 451–458 (1998)

Fractal Dimension of the 2001 El Salvador Earthquake Time Series

Md. Nurujjaman, R. Narayanan, and A.N.S. Iyengar

Saha Institute of Nuclear Physics, 1/AF, Bidhannagar, Kolkata-700 064, India

1 Introduction

Earthquakes occur on the earth's surface as a result of rearrangement of terrestrial cortex or higher part of the mantle. The energy released in this process propagates over long distances in the form of elastic seismic waves [1]. In order to predict earthquakes many models have been proposed [2, 3]. Dynamics of an earthquake is so complicated that it is quite difficult to predict using available models. Seismicity is a classic example of a complex phenomenon that can be quantified using fractal concepts [4].

In this paper we have estimated the fractal dimension, maximum, as well as minimum of the singularities, and the half-width of the multifractal spectrum of the El Salvador Earthquake signal at different stations. The data has been taken from the California catalogue (http://nsmp.wr.usgs.gov/nsmn_ eq-data.html). The paper has been arranged as follows: In Sect. 2 the basic theory of multifractality has been discussed, and the results have been presented in Sect. 3.

2 Multifractal Analysis

The Hölder exponent of a time series $f(t)$ at the point t_0 is given by the largest exponent such that there exists a polynomial $P_n(t - t_0)$ of the order of n satisfying [5–7]

$$|f(t) - p_n(t - t_0)| \leq C|t - t_0|^{\alpha(t_0)} \tag{1}$$

The polynomial $P_n(t - t_0)$ corresponds to the Taylor series of $f(t)$ around $t = t_0$, up to n. The exponent α measures the irregularities of the function f. Higher positive value of $\alpha(t_0)$ indicates regularity in the function f. Negative α indicates spike in the signal. If $n < \alpha < n + 1$ it can be proved that the function f is n times differentiable but not $n + 1$ times at the point t_0 [8].

All the Hölder exponents present in the time series are given by the singularity spectrum D(α). This can be determined from the Wavelet Transform Modulus Maxima(WTMM). Before proceeding to find out the exponents α using wavelet analysis, we discuss about the wavelet transform.

2.1 Wavelet Analysis

In order to understand wavelet analysis, we have to first understand '*What is a wavelet?*'. A wavelet is a waveform of effectively limited duration that has an average value of zero, shown in the Fig. 1 (bottom). The difference of wavelets to sine waves, which are the basis of Fourier analysis, is that sinusoids do not have limited duration, but they extend from minus to plus infinity. And where sinusoids are smooth and predictable, wavelets tend to be irregular and asymmetric. For example, 'gaus4' wavelet (Fig. 1(bottom)) is defined as $\psi(t) = \frac{d^N}{dt^N} e^{-t^2/2}$, where $N = 4$. Fourier analysis breaks up a signal into sine waves of various frequencies. Similarly, wavelet analysis breaks up a signal into shifted and scaled versions of the original (or mother) wavelet. It can be intuitively understood that signals with sharp changes might be better analyzed with an irregular wavelet than with a smooth sinusoid. Local features can be described better with wavelets that have local extent.

Wavelet transform can be defined as

$$Wf(s,b) = \frac{1}{s} \int_{-\infty}^{+\infty} f(t)\psi\left(\frac{x-b}{s}\right) dt \qquad (2)$$

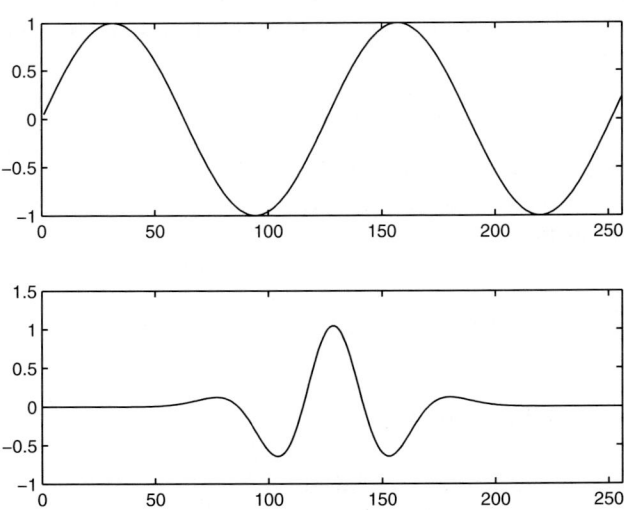

Fig. 1. A sinusoidal signal (*up*), and the gaussian Wavelet having four vanishing moments (*bottom*). Sinusoid has no limitation of duration where as wavelet has a limited duration

where s, and b are the scale and time respectively. In order to detect singularities we will further require ψ to be orthogonal to some low-order polynomials [8]:

$$\int_{-\infty}^{+\infty} t^m \psi(t) dt = 0, \quad \forall m, \quad 0 \leq m < N \tag{3}$$

for example, the wavelet in Fig. 1 has four vanishing moments, i.e. $N = 4$.

2.2 Singularity Detection

Since the wavelet has N vanishing moments, so $\int_{-\infty}^{+\infty} P_n(t - t_0)\psi(t) = 0$, (if $n < N$), and therefore, the wavelet coefficient only detects the singular part of the signal.

$$Wf(s, b) \sim s^{\alpha(t_0)}, \quad a \to 0^+, \tag{4}$$

So, as long as, $N > \alpha(t_0)$ the Hölder exponents can be extracted from log-log plot of the (4).

2.3 Wavelet Transform Modulus Maxima

Let $[u_p(s)]_{p \in Z}$ be the position of all maxima of $|Wf(b, s)|$ at a fixed scale s. Then the partition function Z is defined as [11]

$$Z(q, s) = \sum_p |Wf(b, s)|^q \tag{5}$$

Z will be calculated from the WTMM. Drawing an analogy from thermodynamics, one can define the exponent $\tau(q)$ from the power law behavior of the partition function [9, 10, 11] as

$$Z(q, s) \sim a^{\tau(q)}, \quad a \to 0^+, \tag{6}$$

The log-log plot of (6) will give the $\tau(q)$ of the signal.

Now the multifractal spectrum $D(\alpha(q))$ vs $\alpha(q)$ can be computed from the Legendre transform

$$D(\alpha) = min_q(q\alpha - \tau(q)) \tag{7}$$

where, the Hölder exponent $\alpha = \frac{\partial \tau(q)}{\partial q}$.

3 Results and Discussion

In the present paper we have analyzed the El Salvador earthquake data recorded at different stations as shown in the Table 1. In this table we have arranged the stations according to their distances from the epicenter.

Wavelet analysis of the data recorded at different stations shows that the major events of the earthquake have taken place at short time scales. For eg.

Table 1. In this table the fractal dimension (3-rd column), minimum and maximum (4-th and 5-th column) values of the singularities have been shown for different stations according to their distance from the epicentral distance of the earthquake

Earthquake Recording Station	Epicentral Distance(Km)	Fractal Dim	Singularity(α)	
			α_{max}	α_{min}
Santiago de Maria	52.50648	0.81	2.23	1.46
Presa 15 De Septiembre Dam	63.85000	0.83	2.85	1.14
San Miguel	69.95400	0.88	2.85	1.68
Sensuntepeque	90.50100	0.84	2.53	1.40
Observatorio	91.02397	0.82	2.76	1.52
Cutuco	96.63410	0.84	2.60	1.48
Santa Tecia	97.99589	0.91	2.58	1.69
Acajutia Cepa	139.41800	0.89	3.05	1.83
Santa Ana	142.01300	0.86	3.32	1.49
Ahuachapan	157.35800	0.75	2.48	1.68
Cessa Metapan	165.78460	0.93	3.44	1.58

Fig. 2 (top) shows a burt of activity in a short duration and the corresponding Continuous Wavelet Transform (CWT) in Fig. 2 (bottom) for the time series recorded at Santa Tacia station. In this figure (Fig. 2 [bottom]) the maximum correlation is shown by white color (which indicates maximum correlation), which occurs between 15 to 25 seconds approximately shown in Fig. 2. CWT of the recorded data also shows that pseudo frequencies of the major events are less than 2 Hz. For Santa Tacia data it is few hundred mHz to 2 Hz. From

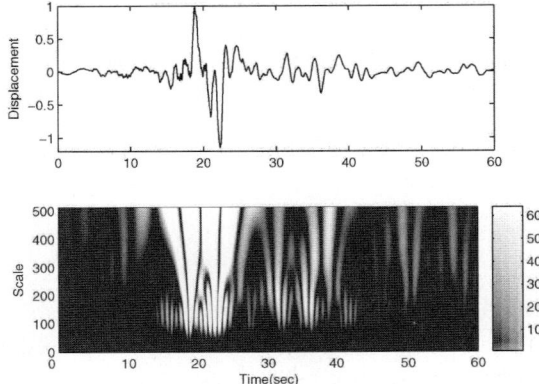

Fig. 2. Typical raw data recorded at station Santa Tecia (*Top*) and its Continuous wavelet transform (cwt) (*bottom*). From cwt it is clear that the major earthquake events occurs within few seconds (in between 15–20 sec)

the same figure (Fig. 2 [bottom]) it is also clear that the high frequencies i.e. 1–2 Hz come in very short range (1–4 seconds), and mHz frequencies comes with relatively longer durations (about 10 seconds). Multifractal analysis of the earthquake data recorded at different stations of increasing distances from the El Salvador earthquake epicenter of 2001 has been carried out. In the table 1 the first column represents the station according to their distance from the earthquake epicenter (distances shown in the second column is in km). In order to get the multifractal spectrum we first calculated the WTMM tree shown in the Fig. 3 as described in Subsect. 2.3. Using Legendre transform method we have obtained the multifractal spectrum shown in the Fig. 4. From multifractal analysis it is clear that the fractal dimension of the singularity support is around one. Lower bound and upper bound of the singularity increases with the distances of the station from the earthquake epicenter shown in Table 1 and in Fig. 4. It indicates the signal becomes smoother with distance, but the half width of the singularity support has random variation with distances.

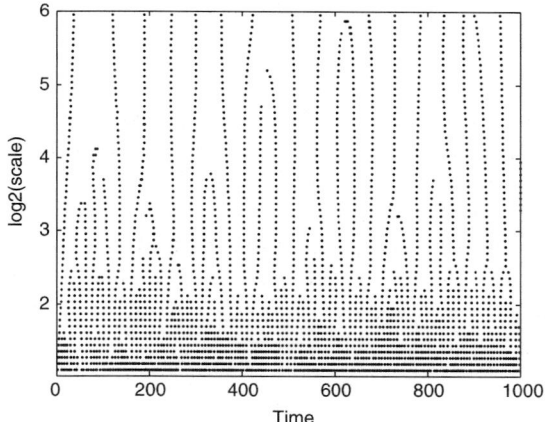

Fig. 3. WTMM skeleton of the data taken at Santa Tacia station (Raw data Fig. 2)

In conclusion, the data shows a multifractal behavior, and the major event takes place in a short duration.

Acknowledgments

Some of the MATLAB function of Wavelab has been used in this analysis (address: http://www-stat.stanford.edu/~wavelab).

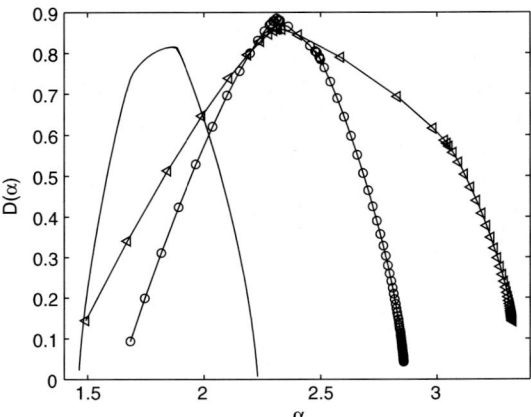

Fig. 4. Multifractal analysis analysis of El Salvador earthquake. In the above figure $-$, $-0-$, and $-\triangleleft-$, are the singularity spectrums of the data recorded at the stations Santiago de Maria, Santa Tacia, and Santa Ana respectively

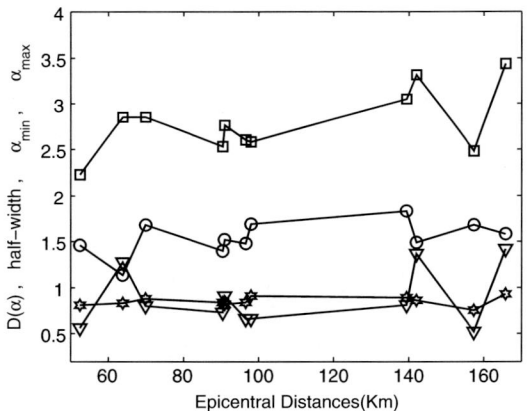

Fig. 5. From this figure it is clear that the fractal dimension of singularity support is around 1 ($-*-$), lower bound and upper bound of singularity increases with the epicentral distances ($-0-$, $-\square-$ and respectively), and half-width has not such incremental behavior ($-\nabla-$)

References

1. R. Yulmetyev, F. Gafarov, P. Hanggi, R. Nigmatullin, and S. Kayumov, Phys. Rev. E **64**, 066132 (2001).
2. S. Pradhan and B.K. Chakrabarti, Int. J. Mod. Phys. B, **17**, 5565–5581 (2003).
3. R. Burridge and L. Knopoff, Bull. Seis. Soc. Am. **57**, 341 (1967).
4. K. Nanjo, H. Nagahama , Chaos, Solitons and Fractals **19**, 387397 (2004).
5. J.F. Muzy, E. Bacry, and A. Arneodo, Phys. Rev. Lett. **67**, 3515 (1991).

6. E. Bacry, J.F. Muzy, and A. Arneodo, J. Statist. Phys **70**, 635 (1993).
7. J.F. Muzy, E. Bacry, and A. Arneodo, Phys. Rev. E **47**, 875 (1993).
8. A. Arneodo, E. Bacry, and J.F. Muzy , Physica A **213**, 232 (1995).
9. Z.R. Struzik, Physica A **296**, 307 (2001)
10. A. Arneodo, Y. d'Aubenton-Carafa, E. Bacry, P.V. Garves, J.F. Muzy, and C. Thermes, Physica D **96**, 219–320 (1996).
11. Stéphane Mallat: *a Wavelet tour of Signal processing*, 2nd edn., Academic Press, pp 163–219 (2001).

Approach in Time to Breakdown in the RRTN Model

A.K. Sen[1] and S. Mozumdar[2]

[1] TCMP Division, Saha Institute of Nuclear Physics, 1/AF Bidhannagar, Kolkata 700064, India
[2] South Point High School, 82/7A Ballygunge Place, Kolkata 700019, India
asokk.sen@saha.ac.in

1 Introduction

The phenomenon of breakdown is seen in widely different varieties of *driven systems*, starting from failures of mechanical systems (like fractures of materials, avalanches, (earth)-quakes) to biological systems (like denatured proteins etc.) [1]. Electrical breakdown itself can be of two different types. One is the fuse type breakdown due to the Joule heating through the ohmic conductors, and hence it is an irreversible phenomenon. The other is dielectric breakdown. If an insulating material (made of microscopic disordered metallic and dielectric, i.e., insulating, phases) is placed between two electrodes and a voltage V is applied across them, such that the electric field $E = V/L$ (L being the length of the sample in the direction of the field) has a low value, no current flows through the solid. The islands of conducting phase without the external force cannot provide for a continuous path for a current to flow through the macroscopic sample. However, if the field is higher than some sample-dependent critical value $E_c = V_c/L$, then some dielectric regions may break under their local field (electrical stress) thereby making extra pathways for current to flow through, and the solid becomes a conductor. If E is brought below E_c from higher values, the solid becomes insulating again. Hence this type of breakdown is reversible. At a low enough temperature and in the presence of disorder (or, other scattering mechanisms, not considered here), quantum mechanical tunneling (or hopping) between the sites or bonds, may become important, and thus contribute to a breakdown (dielectric) of the system.

This report relates to our early attempt to understand the dynamics of a reversible breakdown or fracture using a bond percolation model. We believe that some of our results may have some relevance for many generalized breakdown/fracture processes, and in particular, for an (earth)-quake, where mechanical fracture is involved. If a field above E_c is applied to a macroscopically insulating sample, which is a composite made of microscopic insulating and conducting domains, the breakdown of some dielectric/insulating regions into conducting zones in such a system, propagates in time through the sample until the whole sample acquires a geometrical connectivity of conducting regions, and starts conducting. In other words, the dynamics of the system leading to such a phase transition, is extremely important to understand.

1.1 The Model

We work with an electrical network based on a semi-classical percolation model [2]. This is a Random Resistor Network (RRN) on an underlying lattice where the conducting bonds (o-bonds) are thrown randomly with a predetermined volume fraction p. Additionally, one can imagine the possibility of tunneling to occur wherever two o-bonds are geometrically separated by an insulating bond having the length of a lattice constant. These specially positioned insulating bonds in a completely deterministic (i.e., perfectly correlated) manner within a given random configuration of the RRN, are called tunneling bonds (t-bonds). For simplicity, tunneling is considered negligible beyond one lattice constant. Further, one assumes for simplicity again, a t-bond to be active, with a non-zero conductance g_t only when the voltage difference across it is greater than some threshold voltage v_g. Clearly g_t is explicitly non-linear, or piecewise linear, and there is a cusp singularity at v_g. This network of random o-bonds and correlated t-bonds is called a Random Resistor cum Tunneling-bond Network (RRTN).

2 Some Features of the Model

2.1 Percolation Threshold

The geometrical bond-percolation threshold is defined to be the minimum concentration $p = p_c$ of the o-bonds (i.e., a RRN) at which a conducting pathway is created from one side of the system to its opposite side. For 2d-square lattice, $p_c = 0.5$. The presence of the t-bonds adds some more parallel paths of transport to the already existing ones (if there is any). As one can envisage, not a single t-bond would "fire" (i.e., become active), if the external potential $V < v_g$. Now, if the RRN is insulating (i.e., non-percolating), then it may require some minimum number of t-bonds to fire, to make the corresponding RRTN conducting (or, to make the breakdown occur). Clearly, this gives rise to a macroscopic threshold voltage (V_g), or equivalently, a breakdown voltage (V_B) for that particular RRTN configuration. Further, one notes that it does not take all the possible t-bonds to fire, for the breakdown to take place. A particular RRTN configuration in which all the possssible t-bonds are active, is called a *maximal RRTN*. Thus, we may think of the maximal RRTN issue as a very specific *correlated bond percolation* problem. Once the positions of the o-bonds for a particular RRN configuration are known, the positions of all the possible t-bonds are automatically determined for the corresponding maximal RRTN configuration. This new tunneling-enhanced bond percolation threshold for the maximal RRTN's has been found to be $p_{ct} = 0.181$ [3]. We did also find from the calculation of the critical exponents that the problem belongs to the same universality class as that of the pure (uncorrelated) bond or site percolation.

2.2 Non-linear Response

As the externally applied potential is increased beyond the macroscopic threshold (V_g or V_B) of a specific RRTN sample, some of the t-bonds overcome their microscopic thresholds (a fixed v_g) and increase the overall conductance of the system since the process leads to newer parallel connectivities for the whole macroscopic composite. We find [2, 4] that for $p > p_c$, the $I - V$ characteristic is linear upto a certain macroscopic voltage (V_g) for the given sample, beyond which the non-linearity shows up. For $p < p_c$ (but larger than p_{ct}), there is no current below a threshold/breakdown voltage (V_B) and non-linear conduction appears for $V > V_B$ [4]. Non-linearity is always there in the $I - V$ response for any value of p in the interval $p_{ct} < p < 1$. However, for $p_{ct} < p < p_c$, there is no system spanning path with the o-bonds only (in an average sense), i.e., through the RRN, and the response (average current) is zero for small voltages ($V < V_B$). For $p > p_c$, one thus finds a lower ohmic region for $V < V_B$ and an upper ohmic region for $V \to \infty$. Obviously, this V_B (or, $E_B = V_B/L$), as a driver of criticality [5], is both p as well as the configuration dependent.

2.3 Dielectric Breakdown as a Paradigm

As stated earlier, the response of the system to an externally applied voltage is non-linear and the RRTN model, as a *driven system*, can show reversible breakdown if the underlying RRN is an insulator and if the external voltage $V > V_B$ [5]. As one approaches p_c from below, one obiserves a criticality and a power law as

$$E_B \sim (p_c - p)^{t_B} , \qquad (1)$$

where, t_B is called the breakdown exponent. For the RRTN, we had found that $t_B = 1.42$ [5], which is not very different from the value of $t_B = 1.33$ for the RRN.

3 Approach to Breakdown in Time

Here we present the results of our preliminary study on the dynamics of breakdown in the RRTN. As discussed in the [5], we work with a concentration $p = 0.3$ which satisfies $p_{ct} < p < p_c$. We continue to work with a square lattice and choose a $E = 2.0$, well above E_B, the breakdown field for the chosen samples. For example, with $p = 0.4$, $L = 20$, E_B is around 0.068 on average [5].

For ease of numerical calculations, the conductance of the insulating bonds (as well as that of inactive t-bonds) is chosen to be 10^{-50} mho (instead of exact zero) for $v < v_g$ and to be the fixed non-zero value $g_t = 10^{-2}$ mho, for $v > v_g$. The initial condition is chosen to be that the voltages at all the nodes is zero except the ones on the side of the active brass-bar, where the voltage is V. To be able to observe the progress of the fracture/breakdown phenomenon here, the voltage at each node is updated using a time evolution algorithm (basically the continuity equation alongwith Kirchoff's Law), as done since our original work [2] which we call as *lattice Kirchoff's dynamics*:

$$v(j,k,t+1) = v(j,k,t) + \frac{\sum I_{mn}(t)}{\sum g_{nn}}, \qquad (2)$$

where g_{nn} are the various microscopic conductances of the nn bonds around the node (j,k). The time taken to update the voltages, of all the nodes once, is taken as unity. This *scaled* unit of time varies with size of the sample, as in one of our recent studies [6] on the non-Debye relaxation in the RRTN. After each unit time, the current in the layer opposite to the brass bar is computed.

A typical response curve is shown in the Fig. 1. There is no response detected for some time, initially. However, disconnected, brittle fractures start to build up inside the sample in the form of local clusters (see Figs. 2a–2d). This sequence may be called pre-response, which is like the fore-shocks in an earthquake. The current suddenly starts building up from a certain time (see Fig. 1). At this time, one of the clusters, that has been formed with the o- and the t-bonds, spans the whole system (see Fig. 2e). This corresponds to the breakdown of the whole system (analogous to the main shock in an

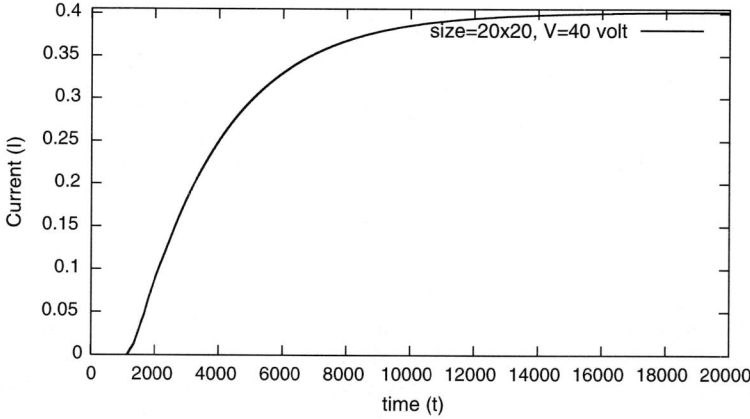

Fig. 1. The growth of current (hence, the growth of dielectric breakdown) with time in a 20×20 size square lattice RRTN at a fixed driving voiltage

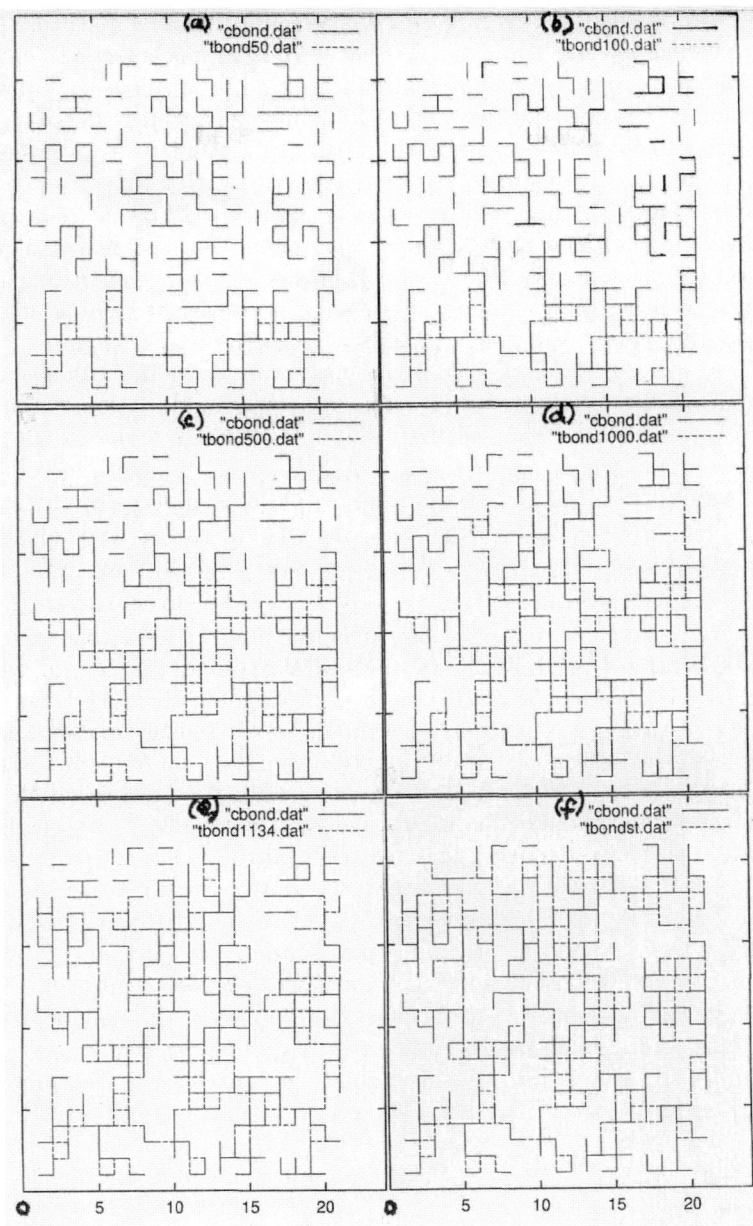

Fig. 2. A sequence of snapshots of an electrically "fracturing" 20×20 square lattice at times (**a**) $t = 50$, (**b**) $t = 100$, (**c**) $t = 500$, (**d**) $t = 1000$, (**e**) $t = 1134$, and (**f**) the time to reach the steady state with a given accuracy. Here the time is already scaled in units of one updating scan through all the nodes of the lattice. The dielectric breakdown or the macroscopic, system-spanning fracture appears for the first time at $t = 1134$ units. So, $\tau_B = 1134$ in this case

earthquake). We denote this time by τ_B, and call it the breakdown time. With further passage of time, some of the other local clusters continue to grow and either join the main cluster that spans the system or form separate paths for percolation (viz., after-shock events). A few clusters may still remain local and cannot grow to contribute to the path of percolation under the given field (see Fig. 2f). Through all this time, the system relaxes with an inverse power-law in time [6]. Once all the possible paths of percolation at that voltage have formed (no more t-bond breaks down), the system approaches a steady state condition following exponential (Debye) relaxation. To determine the steady state, we monitor the fractional difference of currents through the first and last layer and check out when it tends to zero (here, less than 10^{-8}) as a function of time. In other words, if the current entering the system is equal to the current leaving the system within a pre-decided tolerance factor, the system is considered to be in a steady state. The time to arrive at the steady state is denoted by τ_{st}.

The values of τ_B and τ_{st} vary from sample to sample with the same p. Both of them however increase with the size of the system. The values of τ_B and τ_{st} were found for a number of samples with different sizes.

4 Analysis of Data and Conclusion

The ratio of the two times scales, i.e., $\kappa = \tau_{st}/\tau_B$ was plotted with the number of samples in the form a histogram. The ratio κ was found to show a distribution. The Fig. 3 shows such a histogram for $L = 20$, in an unnormalised (raw) form. The distribution indicates a large enough tail as well. The spread of the distribution seems to decrease, and the peak position shifts towards a lower (non-zero) value of κ with an increase in size of the system. For example, for $L = 20, 30, 40, 60$, and 80, κ equals $50, 45, 40, 37$, and 34 respectively.

The two times τ_B and τ_{st} are indicative of two important time-scales in the breakdown of the RRTN model. Since the ratio seems to tend to a constant value, only one (independent) time-scale exists. However, this can be predicted more strongly only by taking a sufficiently large number of samples and by taking finite size effects into consideration. The finite size scaling analysis is currently under progress.

Acknowledgements

The authors thank Prof. B.K. Chakrabarti and the CAMCS (SINP) for giving the opportunity of presenting our early thoughts on the dynamics of fracture or breakdown processes in nature. SM expresses his gratefulness to the SINP for letting him use some of its facilities (e.g., Library, computer section, etc.) required for this work.

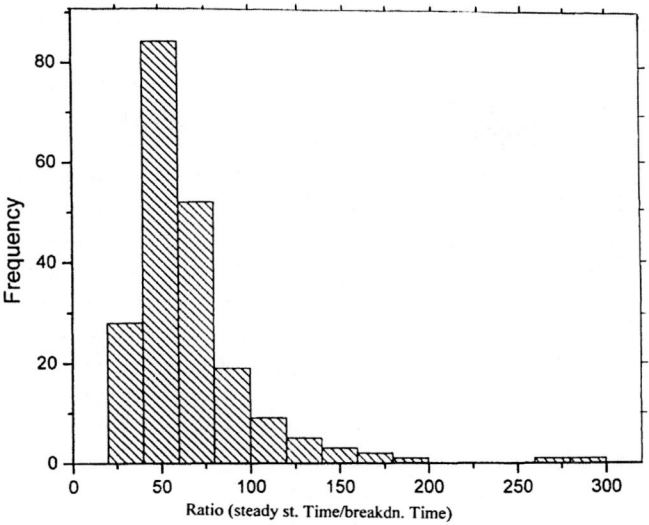

Fig. 3. The histogram for the ratio of two time scales, $\kappa = \tau_{st}/\tau_B$, using in excess of 200 configurations for the same 20×20 square lattice as in the Fig. 2. This histogram has a prominent peak, but a large enough tail as well. For larger size systems, the tails seems to get shorter (not shown here)

References

1. D. Bhattacharya, R. Basu, A. Ghosh, A. Manna, A. Nandy and P. Nandy, Biophys. J. **64**, 550 (1993)
2. A.K. Sen and A. Kar Gupta, in K.K. Bardhan, B.K. Chakrabarti and A. Hansen, Eds., *Non-linearity and breakdown in soft condensed matter*, Lecture Notes in Physics, vol. **437**, p. 271, Springer, Heidelberg (1994)
3. A. Kar Gupta and A.K. Sen, Physica **A 215**, 1 (1995)
4. A. Kar Gupta and A.K. Sen, Phys Rev **B 57**, 3375 (1998)
5. A. Kar Gupta and A.K. Sen, Physica **A 247**, 30 (1997)
6. S. Bhattacharya and A.K. Sen, Europhys. Lett. **71**, 797 (2005)

Critical Behaviour of Mixed Fibres with Uniform Distribution

U. Divakaran[1] and A. Dutta[2]

[1] Department of Physics, Indian Institute of Technology, Kanpur 208016, India
   udiva@iitk.ac.in
[2] Department of Physics, Indian Institute of Technology, Kanpur 208016, India
   dutta@iitk.ac.in

The breakdown phenomena of materials is a very complex process which involves a study of formation of microscopic cracks, their evolution to the other regions of the materials via some interaction and the ultimate collapse of the structure [1]. Random Fibre Bundle Model (RFBM) [2, 3] is a paradigmatic model, simple yet subtle, which captures the essential physics of breakdown processes. It consists of N parallel, vertical fibres whose ends are connected to two horizontal rods. The upper rod is fixed and a load F is applied to the lower one so that a stress $\sigma(= F/N)$ is generated on each fibre. The fibres are assigned a threshold strength chosen randomly from a cumulative distribution function $P(\sigma)$ with a corresponding density function $\rho(\sigma)$:

$$P(\sigma) = \int_0^\sigma \rho(x)dx \ .$$

If the generated stress σ is greater than the threshold strength of a fibre, the fibre breaks. The additional load due to breaking of the fibres is distributed to the remaining intact fibres with some load sharing rule. This model can be compared with the *Ising Model* of magnetic transitions where the load sharing rule represents the interaction amongst the spins. The two extreme cases of load sharing rules are: Global Load Sharing (GLS) [2] where the broken fibre redistributes its stress equally to the remaining intact fibres, and Local Load Sharing (LLS) [4] where the stress is redistributed only to the nearest intact neighbours of the broken fibre. GLS is equivalent to a mean field Ising model with infinite-range interactions among the elements of the system. As the external load is increased, more and more fibres break and their stress is redistributed to the other intact fibres following the prescribed load sharing rule. Eventually, the complete failure of the RFBM takes place at a critical load or stress. The study of this phase transition phenomena from a state of partial failure to that of complete failure has been a topic of interest since last 80 years.

The dynamical phase transition of random fibre bundles with uniformly distributed (UD) threshold strength ($0 < \sigma_{th} < 1$) and global load sharing rule has been studied extensively in recent years [3, 5, 6]. Models of this class undergo a phase transition at the critical stress σ_c (=0.25). The critical stress and the exponents associated with the dynamical phase transition can be derived exactly using recursive equations for the number of unbroken fibres and the redistributed stress. It has been shown using different distributions of threshold strengths (e.g., linearly increasing, linearly decreasing and Weibull) that the failure of fibre bundle with GLS defines a universality class with same set of critical exponents [6, 7].

However, whether there is a critical behavior in RFBM following the LLS rule is still a question [6]. In one dimension, the critical stress σ_c is found to vanish in the thermodynamic limit as $\sigma_c \sim (1/\ln L)$ for both Weibull distribution (WD) and UD of threshold strength [4, 8]. In the case of hierarchical load distribution, the critical stress vanishes as $(1/\ln \ln L)$[9]. Whether there is a well-defined phase transition in a RFBM with LLS on a higher dimensional lattice is still not clear.

Recently, we have studied the critical behaviour of mixed fibre bundle with uniform distribution of threshold strength and GLS [8]. The bundle is called *mixed* as a fraction of weak (class A) fibres are mixed with the strong fibres (class B) in the bundle. The resulting density of threshold distribution $\rho(\sigma_{th})$ of any fibre has a finite discontinuity. Let's assume that a fraction x of the total fibres belong to class A with threshold strength uniformly distributed between 0 and σ_1. The remaining $(1-x)$ fraction are uniformly distributed between σ_2 and 1 such that $\sigma_1 < \sigma_2$. The uniformity of the distribution is ensured and it leads to a relation between σ_1, σ_2 and x so that only two among the three variables are independent. This model shows dynamical a phase transition to a completely failed state at a critical stress $\sigma_c = 1/4(1 - \sigma_2 + \sigma_1)$ with mean field (GLS) exponents. Interestingly, σ_c can be tuned by changing the discontinuity in the threshold distribution i.e., $(\sigma_2 - \sigma_1)$. It can be argued that for the analytical method to hold good, one must identify the appropriate region in the parameter space of x, σ_1 and σ_2. In the Weibull version (where $P(\sigma) = 1 - \exp[-(\sigma)^\rho]$) of the mixed RFBM, a fraction x belongs to class A ($\rho = 2$) and the remaining to class B ($\rho = 3$). The critical exponents show that mixed RFBM with weibull distribution of threshold strength also belongs to the universality class of GLS. The most interesting result in this case is the linear variation of the critical stress as a function of x.

To probe the universality class of RFBM further, we have studied the critical behaviour of random fibres with UD of threshold strength placed on a random graph with co-ordination number $z = 3$. Consider the fibres placed on a circle with periodic boundary conditions such that the fibre at site i is connected to site $i+1$. Apart from being connected to its neighbours, a fibre is also connected randomly to one more unique fibre thus making the co-ordination number of each fibre equal to 3. This model has a non-zero critical stress σ_c at the thermodynamic limit. Using the finite size scaling approach [10], we have obtained the critical stress and the exponents of the transition. On a random graph, even with LLS rule, the exponents associated with the transition are Mean field or GLS. It can be shown [11] that a random graph with the above mentioned connections is equivalent to a Bethe lattice with the corresponding co-ordination number in the thermodynamic limit ($N \to \infty$). These results clearly resemble the magnetic transition in the nearest neighbour Ising model on a Bethe lattice which has the same critical exponents as the infinite-range model.

In this article, we report the critical behaviour of the most general version of mixed random fibres with UD. The normalised probability distribution is sketched below:

$$\begin{aligned} \rho(\sigma_{th}) &= \frac{1}{\sigma_4 - \sigma_3 + \sigma_2 - \sigma_1} & \sigma_1 < \sigma_{th} \leq \sigma_2 \\ &= 0 & \sigma_2 < \sigma_{th} < \sigma_3 \\ &= \frac{1}{\sigma_4 - \sigma_3 + \sigma_2 - \sigma_1} & \sigma_3 \leq \sigma_{th} \leq \sigma_4 \, . \end{aligned} \qquad (1)$$

As described before, the model has weak ($\sigma_1 \leq \sigma_{th} \leq \sigma_2$, class A) and strong ($\sigma_3 \leq \sigma_{th} \leq \sigma_4$, class B) fibres (see Fig. 1) with no fibres having threshold between σ_2 and σ_3. To ensure the uniformity of the model, we must have

$$\frac{1}{\sigma_4 - \sigma_3 + \sigma_2 - \sigma_1} \int_{\sigma_1}^{\sigma_2} d\sigma = x$$

$$\text{or,} \quad \frac{\sigma_2 - \sigma_1}{\sigma_4 - \sigma_3 + \sigma_2 - \sigma_1} = x \, . \qquad (2)$$

The above (2) provides a relationship between the five variables namely x, $\sigma_1, \sigma_2, \sigma_3,$ and σ_4 out of which four are independent. At the same time, for the distribution to be meaningful, we must have $\sigma_1 \leq \sigma_2 \leq \sigma_3 < \sigma_4$. Let us now consider the dynamics of the model. The application of an external force F generates a stress $\sigma = F/N$, where N is the total number of fibres. A fraction

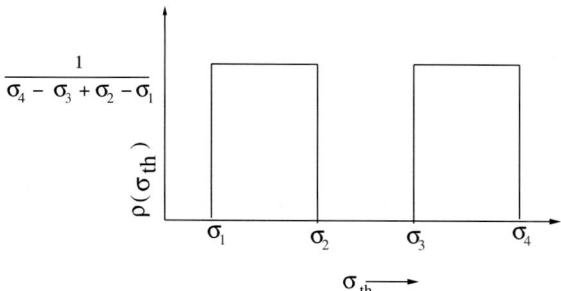

Fig. 1. Mixed Uniform Distribution

$P(\sigma)$ of the total fibres break due to this stress σ. The additional load as a result of breaking of fibres is distributed globally to the remaining $N(1-P(\sigma))$ intact fibres thus causing further failures. Let us define $U_t = N_t/N$ as the fraction of unbroken fibres after a time step t. Then [3]

$$\sigma_t = \frac{\sigma}{U_t}. \tag{3}$$

The recursive relation between U_t, U_{t+1} and between σ_t, σ_{t+1} (for a given applied stress σ) is given by

$$U_{t+1} = 1 - P(\sigma_t) = 1 - P\left(\frac{\sigma}{U_t}\right)$$

$$\sigma_{t+1} = \frac{\sigma}{U_{t+1}} = \frac{\sigma}{(1-P(\sigma_t))}. \tag{4}$$

More failures take place as the broken fibres redistribute their stress until a fixed point is reached at which no further breaking of fibres occur (i.e., $U_t = U_{t+1} = U^*$). To look at the complete failure of the bundle, the redistributed stress must cross σ_3 so that some fibres from the strong class also break. For convenience, lets denote $a = \sigma_4 - \sigma_3 + \sigma_2 - \sigma_1$. With this notation, the recursive relation at time t when the redistributed stress has already crossed σ_3 takes the form

$$U_{t+1} = 1 - \left[\frac{\sigma_2 - \sigma_1}{a} + \frac{1}{a}\int_{\sigma_3}^{\sigma_t} d\sigma\right]$$

$$= 1 - \left[\frac{\sigma_2 - \sigma_1}{a} + \frac{\sigma_t - \sigma_3}{a}\right] \tag{5}$$

Similarly,

$$\sigma_{t+1} = \frac{\sigma}{1 - \left[\frac{\sigma_2 - \sigma_1}{a} + \frac{\sigma_t - \sigma_3}{a}\right]}. \tag{6}$$

At fixed point, $U_{t+1} = U_t = U^*$ and $\sigma_{t+1} = \sigma_t = \sigma^*$, and we obtain

$$U^{*2} - \frac{\sigma_4 U^*}{a} + \frac{\sigma}{a} = 0$$

and $\sigma^{*2} - \sigma_4 \sigma^* + a\sigma = 0$ \hfill (7)

The stable solutions to the above two equations are:

$$U^* = \frac{\sigma_4}{2a} + \frac{1}{\sqrt{a}}\left(\frac{\sigma_4^2}{4a} - \sigma\right)^{1/2} \tag{8}$$

and $\sigma^* = \frac{\sigma_4}{2} - \sqrt{a}\left(\frac{\sigma_4^2}{4a} - \sigma\right)^{1/2}$ \hfill (9)

with $\sigma_c = \frac{\sigma_4^2}{4a}$ \hfill (10)

Clearly, if $\sigma > \sigma_c$, U^* (or σ^*) is imaginary. But U^* is the fraction of unbroken fibres and hence imaginary U^* is unphysical. This implies that σ_c is the critical stress beyond which if the external stress is applied, no fixed point exists and complete failure of the bundle takes place. Also, σ_c must be greater than σ_1. The redistributed stress at the fixed point of the critical stress is $\sigma_4/2$ (see (9)). Since we have assumed that the redistributed stress crosses σ_3 before the complete failure, σ_3 must be less than $\sigma_4/2$ for the calculations to be valid. Thus all the distributions which satisfy the following criteria should have a critical stress as given by (10).

$$\sigma_1 \leq \sigma_2 \leq \sigma_3 < \sigma_4$$

$$\sigma_3 < \frac{\sigma_4}{2}$$

It can be checked in each of the above distributions that

$$U^*(\sigma_c) + P(\sigma^*(\sigma_c)) = 1$$

At this point, one should probe the range of x for which the above analytical approach holds good. From (2), we find that

$$\sigma_4 - \sigma_3 = \frac{(1-x)}{x}(\sigma_2 - \sigma_1).$$

We have also found that for the redistributed stress to reach class B fibres at or before the critical point, σ_3 must be smaller than $\sigma_4/2$. These two conditions taken together lead to a restriction given as

$$\sigma_2 > \sigma_1 + \frac{x}{(1-x)}\frac{\sigma_4}{2}. \tag{11}$$

For a discontinuity to be present in the threshold distribution, as assumed already, we must have $\sigma_3 > \sigma_2$. Equation (11) and the condition $\sigma_3 > \sigma_2$ can not be simultaneously satisfied for any value of $x \geq 1/2$. This rules out the possibility of having half or more than half fibres in class A. For a given x ($< 1/2$), one can however choose appropriately the values of σ_1, σ_2 and σ_3 so that all the above conditions are satisfied.

A few examples of the allowed distribution of the threshold strength are:

x	σ_1	σ_2	σ_3	σ_4	σ_c
0.3	0.1	0.3	0.4	0.87	0.28
0.4	0.05	0.3	0.35	0.725	0.21

The order parameter for this model is appropriately defined as [5] $U^* - U^*(\sigma_c)$ where $U^*(\sigma_c) = \sigma_4/2a$. As the critical point is approached, the order parameter goes to zero continuously following a power law $(\sigma_c - \sigma)^\beta$, with the critical exponent $\beta = 1/2$ as seen from (8). Susceptibility χ can be defined as the increment in the number of broken fibres due to an infinitesimal increase in the external load [6] and diverges as $(\sigma_c - \sigma)^{-\gamma}$ as the critical point is approached. The exponents obtained are same as that of mean field (GLS) with $\beta = 1/2$ and $\gamma = 1/2$.

It is interesting to note that the critical stress of the model can be tuned by varying the parameters of the distribution. At different limits, the model reduces to the already obtained results [6]. For example, (i) If $\sigma_1 = 0$ and $\sigma_4 = 1$, the model is same as studied in [8] with $\sigma_c = 1/4(1 - \sigma_3 + \sigma_2)$. (ii) If $(\sigma_2 - \sigma_1) = 0$, or equivalently $x = 0$, we have only strong class B fibres in the bundle. The elastic to plastic deformation and the critical phenomena of this model is discussed in [5]. The critical stress (10) reduces to $\sigma_4^2/4(\sigma_4 - \sigma_3)$. (iii) When $\sigma_1 = 0, (\sigma_3 - \sigma_2) = 0$ and $\sigma_4 = 1$, the model reduces to simple uniform distribution with threshold between 0 and 1, the value of critical stress being 0.25. It is to be noted that if the class B fibres are absent (or $\sigma_4 - \sigma_3 = 0$), the expression for the critical stress has to be modified. This is due to the fact that we have assumed in our calculations that the redistributed stress reaches the class B fibres. Thus the presence of class B fibres play a crucial role. The interesting burst avalanche distribution of the generalised model is the subject of our present research.

Acknowledgments

The authors thank S.M. Bhattacharjee, P. Bhattacharyya and B.K. Chakrabarti for their suggestions and useful comments.

References

1. B.K. Chakrabarti and L.G. Benguigui, *Statistical Physics of Fracture and Breakdown in Disordered Systems*, Oxford Univ. Press, Oxford (1997).
2. F.T. Peirce, J. Text. Inst. **17**, 355 (1926); H.E. Daniels, Proc. R. Soc. London A **183**, 404 (1945); B.D. Coleman, J. Appl. Phys. **29**, 968 (1958); R. da Silveria, Am. J. Phys. **67**, 1177 (1999).
3. S. Pradhan, P. Bhattacharyya and B.K. Chakrabarti, Phys. Rev. E**66**, 016116 (2002)
4. J.B. Gomez, D. Iniguez and A.F. Pacheco, Phys. Rev. Lett. **71**, 380 (1993)
5. P. Bhattacharyya, S. Pradhan and B.K. Chakrabarti, Phys. Rev. E**67**, 046122 (2003).
6. S. Pradhan and B.K. Chakrabarti, Int. J. Mod. Phys. B **17**, 5565 (2003).
7. Y. Moreno, J.B. Gomez and A.F. Pacheco, Phys. Rev. Lett. **85**, 2865 (2000)
8. Uma Divakaran and Amit Dutta, cond-mat/0512205 (2005).
9. W.I. Newman and A.M. Gabrielov, Int. J. Fract. **50**, 1 (1991)
10. B.J. Kim, Europhys. Lett. **66**, 819 (2004)
11. D. Dhar, P. Shukla and J.P. Sethna, J. Phys. A: Math. Gen. **30**, 5259 (1997)

Index

actual contact area 171, 173, 174, 177, 178
actual contact point 170–178
aeolian transport 363–365, 367, 369–371, 373–375, 377, 379, 381, 383, 385, 386
aerodynamics 365, 366, 371, 372, 374, 385
   entrainment 373
   direct 374
aftershock 259, 260, 264, 270, 275–278
aftershocks 114, 125–128, 282, 285
Amontons-Coulomb law 171, 173
Ananthakrishna model 428, 431
angle of repose 379, 384
anomalous scaling of rough surfaces 106
avalanche 29, 33, 37, 47, 70, 71, 387, 389, 390, 392, 394–396, 398–400, 402, 403, 405, 409–413, 416, 463, 467
avalanche distribution 29, 35, 36
average force 42
average wavelet coefficient method 95, 108

Bak-Tang-Wiesenfeld model 169, 469
Barabasi-Albert network 62
barrier height 493, 495–497
beams 82
bending deformation 82
Benioff strain 283, 286
binomial distribution 164
box-counting dimension 100

breakdown 61, 507–513
breakdown process 27
breaking criterion 83
breaking modes 82
Brownian motion 93
Brownian particle 93
Burridge-Knopoff model 169, 181, 182, 184, 186, 187, 225–233, 235–237, 239, 244–246, 248–256, 428, 444–446, 452
burst 62, 66, 88
burst distribution 47, 52

Cantor set 19–23, 25, 156, 157, 472
catastrophic failure 33, 52
Chakrabarti–Stinchcombe model 19, 25, 157, 352, 471
Chapman-Kolmogorov equation 287
characteristic earthquake 183–185
cluster 75
clustering 193
clustering structure, secondary 202
complex-system perspective 194
constitutive behavior 61, 67, 84, 86
contact age 174, 178
continuous damage 65, 66
continuous phase transitions 62, 80
convection cell 491
correlation 211
   combined space-time 267, 276, 278
   dimension 262, 266, 268, 270, 273, 274, 276, 277, 483
coupled-map lattice 387, 404
crack

edge   134, 143, 144, 146
  propagation   153
creep   77
  rupture   77
critical phenomena   62
critical point   45, 462, 468
crossover   39, 49, 62, 464, 468
cumulative distribution   98
cumulative overlap   473

data collapse   195
decimation   207
density of seismic rates   197
dependence of the recurrence time with history   214
detection tool   40
deterministic process   489
detrended fluctuation analysis   95
diffusion equation   94
dilatancy   394, 402
dimensionless recurrence time   198, 201
distribution of mean seismic rates   219
distribution of minimum seismic rates   219
domino effect   47
Doppler effect   138, 146
double-couple source   113, 114, 116, 117, 307
dune   363, 367, 369–371, 373–375, 378–387, 413, 420
  Barchan   363, 364, 367, 368, 381, 383, 386

earthquake   39, 97, 113–115, 121, 127, 140, 153, 155, 169, 175, 178, 182–187, 191–194, 196, 199–208, 212, 214–216, 221, 223–246, 250, 252, 254–256, 259–262, 264, 267, 269, 272, 277, 278, 281–286, 288, 289, 291–299, 303–323, 325–345, 347–359, 459, 475, 481, 482, 499, 501–504, 510, 512, 523
  focal mechanism   303, 305–308, 310, 317, 332–334, 336, 342, 344, 345, 349, 355, 356
  forecasting   338, 340, 347
  size   303–306, 310, 317–321, 332, 334, 340

  time   303, 310
effective breaking threshold   86
elastic response   49
embedding dimension   100
equal load sharing   5, 13, 16, 17, 24, 28, 60, 460
event size distribution   390, 393, 397–400, 402–404
expected residual recurrence time   204
extended self-similarity   281, 286, 289, 298
extreme statistics   4, 11, 12, 18, 19, 23, 24, 98

f-α formalism   105
fatigue   61
fibre bundle model   4, 5, 12, 27, 29–31, 33, 35, 37, 39, 41, 43, 45, 47, 49–53, 55, 57–63, 65, 67, 69, 71, 73, 75, 77, 79, 81, 83, 85, 87, 89, 91, 459, 460, 516
  earthquake   58
  materials failure   58
fibre reinforced composites   63
first order phase transitions   62
first return method   95
fixed-point   29, 43, 44, 461
  equation   209
Fokker-Planck equation   95
Fourier power spectrum method   95
fractal   7, 19–23, 25, 100, 471
  dimension   3, 10, 22, 100, 481, 484, 499, 501–505
  overlap   7, 19
fractal-overlap model   157
fractional noise   95
fracture   3–5, 7–11, 19, 20, 23, 303–305, 312, 313, 316, 332, 345, 350–352, 355, 357, 358, 507, 510–512
  brittle   7, 9, 97
  statistics   11
  stress   4, 11, 18
  surfaces   97
friction   169–179, 182, 183, 186
  adhesion theory   171
  constitutive law   186
  dynamic   114, 115, 118, 139, 140, 151
  solid   424, 428
  static   114, 115, 123, 139, 140, 149

fuse model   39, 467

gamma distribution   198
Gaussian distribution   94
Gaussian hill   369
geological fault   155
global load sharing   60
glued interfaces   82
gradual cracking   67
grains   388–392, 399, 400, 402–405, 407, 409–417
  mobile   402, 405, 410, 414
  stuck   411, 412
granular material   363
Griffith   7
  formula   9, 23
  law   11
  stress   9
growth models   103
Gumbel
  distribution   12
  statistics   4, 11, 23
Gutenberg–Richter law   6, 7, 20, 25, 114, 121, 155, 169, 170, 178, 182, 186, 187, 192, 193, 195, 196, 207, 208, 210, 217, 223, 224, 228, 229, 232, 233, 250, 255, 285
  instantaneous fulfillment of   196

Hölder exponent   499–501
hazard rate   203
heterogeneous materials   61
Hurst
  coefficient   487, 488
  exponent   94
    effective   102
    global   109
    local   109

imminent failure   39
impact threshold   373
inclusive avalanche   462
independence of magnitude with history   215
infinite hierarchy of exponents   105
interface   82
interfacial failure   82
intermediate load sharing   48

Kelvin-Voigt element   78

Landers earthquake   201
Langevin equation   95
law of corresponding states   196
lifetime   79
Lifshitz
  argument   19
  scale   12, 24
load transfer function   72
local load sharing   18, 24, 28, 60, 460
localization   52
Lyapunov exponent   485, 486, 488
Lévy
  distribution   99
  flight   98

mainshock   40, 223, 225, 229, 239–255
Manna model   470
Markov
  process   281, 286–288, 292
  time scale   281, 286, 287, 292, 298
matrix   63
mean-field   30
micromechanical model   68
modified Omori law   192
moment   76
multiaffine surface   105
multiaffinity   105
multifractal   103

Navier Stokes equation   369, 375
  turbulent   366
negative aging   205
Newtonian fluid   364
noise, power-law distributed   98
nonhomogeneous Poisson process   196, 217
nonlinear complexity   488
nonstationary Poisson process   196

Omori law   126–128, 192, 224, 228, 241, 247, 249, 250, 254
order parameter   44
order statistics   96
overhang   102

percolation   11, 24, 54
  statistics   12
  threshold   4, 10, 12
plastic flow   428, 454
plasticity   9, 10, 16, 17, 24, 65, 68

plateau 68
Poisson distribution 32
Portevin-Le Chatelier effect 423, 424, 428, 454
post-shaking 492, 496
power law 36, 48, 52, 62, 71, 72, 81, 88, 463
  distribution 192, 197, 198
pre-shaking 492, 496, 497
precursor 224, 250, 254, 255, 459, 471
prediction 462
probability distribution 28
pulses 146

quaternions 303, 304, 310, 334, 347

Radon gas emission 481, 486
random walk 37, 41
recurrence time 194
  probability density 194
recursive dynamics 29, 43, 461
relaxation 45, 79, 462
  oscillation 491, 497
renewal process 210
renormalization-group transformation 207, 209
Reynolds number 364, 371, 375
ripple 387, 413–419
rotating cylinder 387, 388, 390–392, 404, 405
roughness exponent 94
RRTN model 509, 512
rupture 27, 43

saddle point approximation 104
sand storm 373
sandpile 387–405, 407, 409–412, 414, 420, 468
  rotated 390, 396
scale invariance 93, 261, 262, 270, 303–305, 309, 310, 313–315, 328, 332, 334, 341, 349
scaling function 195
scaling law 195
seismic
  catalog 226, 229
  correlations 239, 242, 246
  rate 192
    mean 192

seismicity 303, 304, 306, 307, 309–319, 321, 322, 325, 328, 332, 336–338, 340–342, 345–349, 351–358
  spatial heterogeneity 219
self affinity 93
self organized criticality 62, 169, 181, 186
self-affine 471
  function 94
  scaling 7, 19, 20
separation bubble 369, 370, 383
shear 82, 113, 115, 116, 128, 130, 136–140, 150, 153
  failure 82
Shields parameter 373
shock 125–128
simulation technique 69
SOC 468
solid on solid approximation 103
splash process 374
spring and block model (see also Burridge-Knopoff model) 114
state variable 174, 175
stationary seismicity 198
stationary state 79
statistical moment 98
statistical stationarity 94
stick-slip dynamics 223–225, 232, 423–429, 431, 433–437, 439, 441, 443–445, 447, 449, 451–455, 457
stochastic methods 95
stochastic point process 303, 338, 339, 341
stochastic process 287
strain hardening 66
strength threshold 28
stress 28
  concentration 8, 13
stress-transfer function 73
sub-critical 470
superlubricity 171, 172
susceptibility 45, 462

tectonic
  movement 496
  motion 7, 491
  plate 5–7, 19, 21, 23, 25
  shell 5
thermal noise 62

thermal spring   481
thermally activated cracking   62
thermodynamic limit   74
thinning   207
threshold distribution   34, 42
tilt   388, 391, 411, 416
time irreversibility of seismicity   214
time-series   159, 472
transition   73, 78

unified scaling law   217
uniform distribution   28, 39, 43, 47, 465
universal   45
universality   200
   class   16, 25, 46, 48, 54, 89, 471
   exponent   5, 7, 20, 24

viscoelastic   78
viscoelastic fibre bundle model   79
volcanic activity   491, 492, 495
von Kármán's constant   365, 368

waiting time   80, 194
   paradox   205
wavelet   95, 107
   Daubechies basis   109
   transform modulus maxima   500, 501
Weibull
   distribution   4, 11, 12, 28, 39, 60, 68, 73, 74, 79, 80, 87, 516
   modulus   11
   statics   23
white noise   97

Lecture Notes in Physics

For information about earlier volumes
please contact your bookseller or Springer
LNP Online archive: springerlink.com

Vol.659: E. Bick, F. D. Steffen (Eds.), Topology and Geometry in Physics

Vol.660: A. N. Gorban, I. V. Karlin, Invariant Manifolds for Physical and Chemical Kinetics

Vol.661: N. Akhmediev, A. Ankiewicz (Eds.) Dissipative Solitons

Vol.662: U. Carow-Watamura, Y. Maeda, S. Watamura (Eds.), Quantum Field Theory and Noncommutative Geometry

Vol.663: A. Kalloniatis, D. Leinweber, A. Williams (Eds.), Lattice Hadron Physics

Vol.664: R. Wielebinski, R. Beck (Eds.), Cosmic Magnetic Fields

Vol.665: V. Martinez (Ed.), Data Analysis in Cosmology

Vol.666: D. Britz, Digital Simulation in Electrochemistry

Vol.667: W. D. Heiss (Ed.), Quantum Dots: a Doorway to Nanoscale Physics

Vol.668: H. Ocampo, S. Paycha, A. Vargas (Eds.), Geometric and Topological Methods for Quantum Field Theory

Vol.669: G. Amelino-Camelia, J. Kowalski-Glikman (Eds.), Planck Scale Effects in Astrophysics and Cosmology

Vol.670: A. Dinklage, G. Marx, T. Klinger, L. Schweikhard (Eds.), Plasma Physics

Vol.671: J.-R. Chazottes, B. Fernandez (Eds.), Dynamics of Coupled Map Lattices and of Related Spatially Extended Systems

Vol.672: R. Kh. Zeytounian, Topics in Hyposonic Flow Theory

Vol.673: C. Bona, C. Palenzula-Luque, Elements of Numerical Relativity

Vol.674: A. G. Hunt, Percolation Theory for Flow in Porous Media

Vol.675: M. Kröger, Models for Polymeric and Anisotropic Liquids

Vol.676: I. Galanakis, P. H. Dederichs (Eds.), Half-metallic Alloys

Vol.677: A. Loiseau, P. Launois, P. Petit, S. Roche, J.-P. Salvetat (Eds.), Understanding Carbon Nanotubes

Vol.678: M. Donath, W. Nolting (Eds.), Local-Moment Ferromagnets

Vol.679: A. Das, B. K. Chakrabarti (Eds.), Quantum Annealing and Related Optimization Methods

Vol.680: G. Cuniberti, G. Fagas, K. Richter (Eds.), Introducing Molecular Electronics

Vol.681: A. Llor, Statistical Hydrodynamic Models for Developed Mixing Instability Flows

Vol.682: J. Souchay (Ed.), Dynamics of Extended Celestial Bodies and Rings

Vol.683: R. Dvorak, F. Freistetter, J. Kurths (Eds.), Chaos and Stability in Planetary Systems

Vol.684: J. Dolinšek, M. Vilfan, S. Žumer (Eds.), Novel NMR and EPR Techniques

Vol.685: C. Klein, O. Richter, Ernst Equation and Riemann Surfaces

Vol.686: A. D. Yaghjian, Relativistic Dynamics of a Charged Sphere

Vol.687: J. W. LaBelle, R. A. Treumann (Eds.), Geospace Electromagnetic Waves and Radiation

Vol.688: M. C. Miguel, J. M. Rubi (Eds.), Jamming, Yielding, and Irreversible Deformation in Condensed Matter

Vol.689: W. Pötz, J. Fabian, U. Hohenester (Eds.), Quantum Coherence

Vol.690: J. Asch, A. Joye (Eds.), Mathematical Physics of Quantum Mechanics

Vol.691: S. S. Abdullaev, Construction of Mappings for Hamiltonian Systems and Their Applications

Vol.692: J. Frauendiener, D. J. W. Giulini, V. Perlick (Eds.), Analytical and Numerical Approaches to Mathematical Relativity

Vol.693: D. Alloin, R. Johnson, P. Lira (Eds.), Physics of Active Galactic Nuclei at all Scales

Vol.694: H. Schwoerer, J. Magill, B. Beleites (Eds.), Lasers and Nuclei

Vol.695: J. Dereziński, H. Siedentop (Eds.), Large Coulomb Systems

Vol.696: K.-S. Choi, J. E. Kim, Quarks and Leptons From Orbifolded Superstring

Vol.697: E. Beaurepaire, H. Bulou, F. Scheurer, J.-P. Kappler (Eds.), Magnetism: A Synchrotron Radiation Approach

Vol.698: S. Bellucci (Ed.), Supersymmetric Mechanics – Vol. 1

Vol.699: J.-P. Rozelot (Ed.), Solar and Heliospheric Origins of Space Weather Phenomena

Vol.700: J. Al-Khalili, E. Roeckl (Eds.), The Euroschool Lectures on Physics with Exotic Beams, Vol. II

Vol.701: S. Bellucci, S. Ferrara, A. Marrani, Supersymmetric Mechanics – Vol. 2

Vol.702: J. Ehlers, C. Lämmerzahl, Special Relativity

Vol.703: M. Ferrario, G. Ciccotti, K. Binder (Eds.), Computer Simulations in Condensed Matter Systems Volume 1

Vol.704: M. Ferrario, G. Ciccotti, K. Binder (Eds.), Computer Simulations in Condensed Matter Systems Volume 2

Vol.705: P. Bhattacharyya, B.K. Chakrabarti (Eds.), Modelling Critical and Catastrophic Phenomena in Geoscience

Printing: Krips bv, Meppel
Binding: Stürtz, Würzburg